Multiscale Soft Tissue Mechanics and Mechanobiology

Gerhard A. Holzapfel • Raymond W. Ogden
Editors

Multiscale Soft Tissue Mechanics and Mechanobiology

State-of-the-Art Modeling

Previously published in *Journal of Elasticity* Volume 129, Issues 1–2, 2017

Springer

Editors
Gerhard A. Holzapfel
Center of Biomedical Engineering
Graz University of Technology Institute of Biomechanics
Graz, Austria

Raymond W. Ogden
School of Mathematics and Statistics
University of Glasgow
Glasgow, UK

ISBN 978-94-024-1218-5
Library of Congress Control Number: 2017960425

Printed on acid-free paper

This Springer imprint is published by Springer Nature
The registered company is Springer Science+Business Media B.V.
The registered company address is: Van Godewijckstraat 30, 3311 GX Dordrecht, The Netherlands

Contents

J Elast (2017) 129:1
DOI 10.1007/s10659-017-9645-2

Foreword

Roger Fosdick[1]

The *Journal of Elasticity: The Physical and Mathematical Science of Solids* is dedicated to the timely publication of advancements in special areas of the physical and mathematical science of solids that stimulate mechanics research at the fundamental level and promote interaction across disciplines in engineering, science and mathematics. The collection of 12 papers contained in this special volume on soft tissue mechanics is dedicated to multi-scale modelling and mechanobiology issues which relate to arteries, the eye, the brain and the heart. Interactions between the mechanical environment of soft tissues within the body and their cellular biological properties is explored and discussed in these works. The contributions of this volume expose fundamental scaling problems in mechanobiology that could benefit from the attention of the mathematical science and computer science communities, and many modeling issues clearly involve nonlinear continuum mechanics and constitutive theory from the physical sciences.

The guest editors of this special volume, Gerhard Holzapfel and Ray Ogden, have put together an excellent collection of 12 state-of-the-art articles from leading experts in the field of soft tissue mechanics. The breadth and depth of aspects covered in these pages is testament to the rapidly emerging cutting edge of discovery in the field, and to the importance of maintaining interaction with the wider mechanics community of researchers in the physical and mathematical science of solids.

✉ R. Fosdick
fosdick@aem.umn.edu

[1] Department of Aerospace Engineering and Mechanics, University of Minnesota, Minneapolis, MN 55455-0153, USA

J Elast (2017) 129:3–5
DOI 10.1007/s10659-017-9646-1

Preface

Gerhard A. Holzapfel[1,2] · Ray W. Ogden[3]

In the year 2000 a series of papers on 'Cardiovascular Soft Tissue Mechanics' was published in a special issue of the Journal of Elasticity, with guest editors Stephen C. Cowin and Jay D. Humphrey. This compilation of seven papers has been very successful and very influential in the further development of the mechanics of fibrous tissues (over 2,000 citations as in May 2017, according to the Web of Science). Since 2000 there has been much progress in the modeling and simulation of soft tissue mechanics and it is now considered an appropriate time to produce a further special issue, but now focusing on broader aspects of soft tissue mechanics. We have therefore brought together a collection of 12 papers from selected leading experts in the field to provide particular reference to multiscale modeling and mechanobiology for a range of different soft tissues, such as arteries, the eye, the brain, and the heart.

Of particular importance these days is the development of multiscale models that are able to link two or more length, time or energy scales to inform the macroscopic response of tissues. This needs advances not only from the constitutive modeling point of view but also from the computational perspective because the transition from the nano- to the micro- to the macro-level is a key aspect which needs further development. Mechanobiology is concerned with studies of the cellular biological responses due to changes in the mechanical environment and has gained significant importance in the last 15 years, in particular in relation to soft biological tissues. Indeed, the word 'mechanobiology' was not mentioned in any of the papers in the 2000 special issue. Hence, modeling in mechanobiology has much promise for the future understanding of mechanical and biological interactions.

We now briefly describe the contents of each of the papers in this issue. The paper by Lanir [1] provides an extensive review of the characteristic features of soft tissue structures

✉ G.A. Holzapfel
holzapfel@tugraz.at

1 Graz University of Technology, Graz, Austria

2 Norwegian University of Science and Technology, Trondheim, Norway

3 University of Glasgow, Glasgow, UK

and their constitutive modeling. It includes a brief survey of the multiscale and multiphasic soft tissue structure and the mechanical characteristics of both soft tissues and their constituents. Using structure-function relationships, the means by which tissue structure remodels during growth in order to adapt to physiological function is discussed. As a particular example of multiscale modeling the paper by Holzapfel and Ogden [2] reviews fiber dispersion models that are used for capturing the biomechanical response of fibrous tissues with particular reference to excluding compressed fibers within the models. The analysis suggests that the predictive powers of the general structure tensor and the angular integration models are very similar, contrary to some claims in the literature.

The next few papers deal with different aspects of growth for various tissues. The article by Nims and Ateshian [3] provides several examples to illustrate an approach to modeling the solid matrix of biological tissues using reactive constrained mixtures including chemical kinetics. Exchange of mass between various constituents of a mixture is used to describe tissue growth and remodeling. The next two papers by Ambrosi *et al.* [4] and Rausch and Humphrey [5] describe multiscale models of biological, chemical, and mechanical contributions to growth. The first is an application to solid tumors treated as poroelastic solids, while the second one concerns an application to the growth and remodeling of a thrombus. The following two papers by Pinsky and Cheng [6] and Grytz and El Hamdaoui [7] focus on computational models of the eye, lens-induced myopia and the cornea. Pinsky and Cheng [6] develop a constitutive model which accounts for the swelling pressure in hydrated connective tissue with particular reference to swelling in the human cornea, and the model predictions were found to be in good agreement with experimental data. Grytz and El Hamdaoui [7] develop a mechanical model for the sclera in which growth and remodeling were considered to be independently controlled mechanisms during eye development. This model was also found to be in agreement with experimental data.

The next three papers are related to modeling brain bulging, buckling of layers of grey and white matter and axonal buckling in the brain. In particular, the paper by Weickenmeier *et al.* [8] models the swelling-induced deformation in bulging brains. The proposed approach has the potential to inform surgeons of possible ways to reduce brain damage and to improve the outcomes of the procedure of decompressive craniectomy. The study by Ben Amar and Bordner [9] proposes a simple model of growth for a bilayer of grey and white matter using a neo-Hookean model in the short and large wavelength limits. By a post-buckling analysis the gyrification index is predicted and compared with experimental data. Lang *et al.* [10] test the hypothesis that localized stretching is a possible explanation for the regional buckling of axons modeled as Euler beams on an elastic foundation, where the elastic foundation models the surrounding brain tissue. The solutions of the model are consistent with experimental observations of the undulations induced in stretched axons.

The last two papers of this special issue are related to growth and remodeling of cardiac tissue. The study by Witzenburg and Holmes [11] examines eight constitutive laws and compares their relative predictions on the basis of a simple one-dimensional test-bed. Wang *et al.* [12] summarize recent state-of-the-art modeling studies of cardiac growth and remodeling in health and disease, and use a finite element approach to simulate growth in an embryonic linear heart tube with the aim of better understanding the role of structural remodeling in heart failure.

In multiscale modeling and mechanobiology much progress has clearly been made in the last few years, but these subjects are still in their infancy and require much more development. In particular, much more detailed experimental data are needed at the nano- and macro-levels in order to inform the modeling and to better understand the underlying mechanisms. Moreover, improvements are needed in the mathematical and computational methods for upscaling from the nano- through to the macro-scale. We hope that this special issue provides some stimulus to the scientific community to help the further development of these exciting subjects.

References

1. Lanir, Y.: Multi-scale structural modeling of soft tissues: mechanics and mechanobiology. J. Elast. (2016). doi:10.1007/s10659-016-9607-0
2. Holzapfel, G.A., Ogden, R.W.: On fiber dispersion models: exclusion of compressed fibers and spurious model comparisons. J. Elast. (2016). doi:10.1007/s10659-016-9605-2
3. Nims, R.J., Ateshian, G.A.: Reactive constrained mixtures for modeling the solid matrix of biological tissues. J. Elast. (2017). doi:10.1007/s10659-017-9630-9
4. Ambrosi, D., Pezzuto, S., Riccobelli, D., Stylianopoulos, T., Ciarletta, P.: Solid tumors are poroelastic solids with a chemo-mechanical feedback on growth. J. Elast. (2017). doi:10.1007/s10659-016-9619-9
5. Rausch, M.K., Humphrey, J.D.: A computational model of the biochemomechanics of an evolving occlusive thrombus. J. Elast. (2017). doi:10.1007/s10659-017-9626-5
6. Pinsky, P.M., Cheng, X.: A constitutive model for swelling pressure and volumetric behavior of highly-hydrated connective tissue. J. Elast. (2017). doi:10.1007/s10659-016-9616-z
7. Grytz, R., El Hamdaoui, M.: Multi-scale modeling of vision-guided remodeling and age-dependent growth of the tree shrew sclera during eye development and lens-induced myopia. J. Elast. (2016). doi:10.1007/s10659-016-9603-4
8. Weickenmeier, J., Saez, P., Butler, C.A.M., Young, P.G., Goriely, A., Kuhl, E.: Bulging brains. J. Elast. (2016). doi:10.1007/s10659-016-9606-1
9. Ben Amar, M., Bordner, A.: Mimicking cortex convolutions through the wrinkling of growing soft bilayers. J. Elast. (2017). doi:10.1007/s10659-017-9622-9
10. Lang, G., Waters, S., Vella, D., Goriely, A.: Axonal buckling following stretch injury. J. Elast. (2017). doi:10.1007/s10659-016-9604-3
11. Witzenburg, C.M., Holmes, J.W.: A comparison of phenomenologic growth laws for myocardial hypertrophy. J. Elast. (2017). doi:10.1007/s10659-017-9631-8
12. Wang, V.Y., Hussan, J.R., Yousefi, H., Bradley, C.P., Hunter, P.J., Nash, M.P.: Modelling cardiac tissue growth and remodelling. J. Elast. (2017). doi:10.1007/s10659-017-9640-7

J Elast (2017) 129:7–48
DOI 10.1007/s10659-016-9607-0

Multi-scale Structural Modeling of Soft Tissues Mechanics and Mechanobiology

Yoram Lanir[1]

Received: 6 March 2016 / Published online: 15 November 2016

Abstract Soft tissues account for a major fraction of the body volume and mass. They are present in all non-skeletal organs, being responsible for protecting the body, maintaining internal homeostasis, and allowing for mobility. Their function in different organs is highly diverse, as are their properties which are optimally suited for their specific tasks. From a mechanical perspective, specificity of structure and properties is acquired via evolutionary adaptation of the tissue composition and multi-scale structure. In modeling tissue mechanics and mechano-biology, it is thus natural to seek the structural determinants of tissues and their evolution (the "structural approach"). Earlier models were exclusively phenomenological, based either on the general principles of non-linear continuum mechanics or alternatively, on empirical mathematical expressions that fit specific response patterns. In the late 1970's, structural models were introduced to tissue mechanics (Lanir in J. Biomechanics 12(6): 423–436, 1979; Lanir in J. Biomechanics 16(1): 1–12, 1983). Ever since, a gradually increasing number of structural models have been developed for different types of tissues, and today, it is the method of choice (Cowin and Humphrey in J. Elasticity 61: ix–xii, 2000). The structural approach was recently extended to incorporate a mechanistic formulation of mechano-biological pathways by which tissue structures remodel during growth (Lanir in Biomech Model Mechanobiol, 14(2): 245–266, 2015). Here, the characteristic features of soft tissue structures and their constitutive modeling are reviewed. The presentation starts with a brief survey of the multi-scale and multi-phasic soft tissues structure. The global mechanical characteristics of soft tissues and of their constituents are then briefly reviewed. These two aspects form the basis for structural constitutive formulation via the multi-scale structure-function link. Based on established criteria for model validity, predictions of the formulated theory are contrasted against measured response characteristics. Using this structure-function relationship, the evolutionary pathway by which tissue structure and mechanics remodel during growth to adapt to their physiological function, is laid down. The review concludes with an account of the state of the art, the big picture, and future research challenges in tissue mechanobiological modeling.

✉ Y. Lanir
yoramlanir@yahoo.com

[1] Faculty of Biomedical Engineering, Technion – IIT, Haifa 32000, Israel

Keywords Soft biological tissues · Finite deformation · Non-linear mechanics · Growth and remodeling · Multi-scale structural modeling · Mechanobiology

Mathematics Subject Classification 2.010 · 2.080

Abbreviations

CSK Cytoskeleton
CVS Cardiovascular system
ECM Extra-cellular matrix
G&R Growth and remodeling
GAG Glycosaminoglycans
LM Light microscopy
LVE Linear viscoelasticity
LV Left ventricle
OA Opening angle
PC Preconditioning
PG Proteoglycans
QLV Quasilinear viscoelasticity
REV Representative elementary volume
RPC Recruitment preconditioning
RS Residual stress
RVE Recruitment viscoelasticity
SEM Scanning electron microscopy
VE Viscoelasticity
1D, 2D, 3D One-, two-, and three-dimensional

1 Introduction

The foundations of modern soft tissues mechanics were established by Y.C. Fung, starting in the late 1960s [1]. Over the years, Fung introduced the concepts of pseudo-elasticity, exponential stress-strain models, quasilinear viscoelasticity, preconditioning and residual stress, which are cornerstones of the field till today. Fung's models are based on non-linear continuum mechanics, and are phenomenological in nature. As such, their material parameters bear no physical meaning. In view of the large inter-specimen biological variability, where each tissue sample can have its own mechanical properties, lack of physical meaning implies that there is no common denominator which can serve to unify and project between different specimen and tissues. In the late 1970s structure-based multi-scale approach was introduced to tissue modeling wherein parameters have clear physical or structural meaning and structure serves as the common determinant of properties.

This review is dedicated to structural modeling of soft tissues mechanics and mechanobiology. Structural models are important not only in their potential to unify apparently unrelated observations into a coherent framework but also in facilitating insight into the underlying processes and their interactions, thereby providing the groundwork for hypotheses testing of the role and significance of the system processes. Space limitations did not allow for including a review of various semi-structural and structure-motivated model which are partially phenomenological. Two recent reports provide historical background of the structural approach to tissue characterization. One is a personal account of the conception of the structural approach [2]. The other [3] is an insightful review of the history of structural constitutive models.

1.1 Soft Tissue Multi-scale Structure

Soft tissues exhibit a variety of structures, compositions and properties which render them highly suitable for their physiological functions. The structure of each tissue is a multi-scale hierarchy of elements from the molecules through cells, micro-fibrils and fibers, to the tissue level. Mechanically, the predominant elements are the fibers, the fluid-like matrix, and the excitable muscle cells in active tissues. The protein fibers are primarily non-excitable collagen and elastin fibers, and fibronectins and laminins micro-fibers which are mechanically less significant. Collagen is the structural building block of tendons, skin, blood vessels, cornea, sclera, cartilage, intervertebral discs, bones, fascia and uterus cervix. Elastin fibers are found in vertebrate skin, walls of arteries and veins, the lung parenchyma, and in the ligamentum nuchae of horses and cows. Each fiber is itself a multi-scale structure. Collagen is made of distinct fibrils, each of which is a multi-scale system ranging from the fibril down to the sub-fibrils, the highly structured micro-fibrils, and ending in the triple-helical tropo-collagen molecule composed of three amino acid alpha chains [4]. Elastin fibers consist of two proteins, an amorphous insoluble elastin and an elastic micro-fibril composed primarily of the glycoprotein fibrillin [5]. The structure of the fiber network is highly variable. In some tissues the fibers are organized in parallel orientation (collagen in tendons and ligaments, elastin in ligament nuchae), while in others the networks are planar (the skin, heart valve leaflets, body membranes), or are in association with the tissue cells in a more complex there-dimensional manner (e.g., elastin and collagen in blood vessels, collagen and myocytes in the heart [6, 7]).

The matrix of soft tissues is comprised of a water solvent containing solutes of ions, protein molecules and very large proteoglycans (PGs) which are encaged within the tissue fiber networks. PGs are complex macromolecules which are made of a protein core with covalently bound negatively charged glycosaminoglycan (GAG) chains. The fixed charges of the GAG chains endow the matrix with osmotic pressure which attracts water and swells the tissue up to equilibrium where the matrix pressure is mechanically counter-balanced by tension in the tissue fibers [8]. Hence soft tissues are well hydrated thereby benefiting from a plasticizing effect [9].

The cells in soft tissues are of different types and functions. They can have distinct disposition within the tissue system. Whereas some smooth muscle cells (depending on the tissue type), and other non-contractile cells are often found dispersed within the fiber network, arterial smooth muscle cells (SMCs) are encaged within the media inter-lamellar elastin network with approximately circumferential disposition. In the bladder the SMCs are organized in three distinct, closely packed sub-layers within the detrusor layer of the wall. In the heart wall, the myocytes are circumferentially organized parallel to each other in an orientation which changes gradually between the inner endocardium to the outer epicardium surfaces. In skeletal muscle, the cells are long, and span the entire length of the muscle.

Each cell is by itself a multi-scale system, ranging from the molecular level, via the cytoskeleton (CSK), to the whole-cell level. In non-muscle cells, the dynamic CSK consists of actin and microtubule filaments and of dispersed acto-myosin motor units, which enable active contraction of the cells. Muscle cells contain highly organized contracting sarcomere units made of inter-digitating actin and myosin filaments, and of non-contracting titin protein filaments which protect the sarcomeres from over-stretch.

1.2 Soft Tissues Mechanical Properties

Soft tissues undergo finite reversible deformations, at times stretching by 100 % or more (e.g., skin, mesentery membrane). Their stress-strain relationship is convexly non-linear, where the stiffness increases with tensile strain. Their response is in general anisotropic as well as time-dependent, subject to reversible (with time) viscoelasticity, and to irreversible preconditioning adaptation [10–13]. Viscoelasticity is manifested as creep under constant stress, stress relaxation under constant strain and onset of a hysteresis loop between loading and unloading responses. Tissue preconditioning is seen under repeated cycles of stretch, where the stress decays asymptotically with cycle number until reaching a stable response. Under *in vitro* conditions the effects of preconditioning are only partially reversible after a long rest period. Importantly, the tissue response depends on its preconditioning protocol, implying that any change in the testing protocol may require a new round of preconditioning [14, 15].

Residual Stress and Strain (*RS*): In their unloaded state, tissues are not stress-free but are subjected to internal residual stress and to its associated residual strain. In arteries, it was found [16, 17] that when an artery ring is cut longitudinally, it springs open with an opening angle (OA) whose magnitude was taken to be a measure of the RS level. A similar phenomenon was found in left ventricle (LV) myocardial rings [18, 19].

Importantly, RS must be incorporated in tissue stress analysis since the true fibers' strain (and the resulting true tissue stress) is the one referenced to the stress-free state at which all fibers are un-stretched and stress-free [20, 21]. The true strain and associate true stress determine the actual mechanical environment of the tissue cells, to which they can respond by remodeling their extra-cellular matrix (ECM). In addition, Fung and others showed that RS significantly affects the tissue's stress and strain distributions, and carries substantial functional benefits by reducing stress concentration near the internal surface of arteries and LV and by increasing the compliance of the latter during the diastolic feeling phase, thereby improving its blood pumping performance [16, 19, 22, 23].

1.3 Mechanics of Tissues' Constituents

Tissue fibers are thin and long. When contracted, they buckle under very low load and assume an undulated (wavy) disposition. This is often seen in collagen fibers in unloaded tissues. The level of fiber undulation is non-uniformly distributed amongst the fibers. Upon stretch, the fibers become gradually straight and load-bearing [24–26], thereby increasing the tissue stiffness. The straight collagen fiber is stiff, with an axial modulus ranging (depending on the species and tissue) between 100 MPa and 1200 MPa [27]. It can be reversibly stretched up to 5–8 %. Collagen fibers are viscoelastic and undergo preconditioning adaptation under repeated stretch cycles. Elastin is close to being purely elastic, and is highly extensible being able to reversibly stretch by 100 % and more. Its stress-strain response is linear up to approximately 60 % stretch and then curves up to a convex response. In the linear range, the elastin stiffness is approximately (depending on the tissue) two orders of magnitude lower than that of the collagen, i.e., approximately 1 MPa, as measured in bovine ligamentum nuchae [28] and as estimated for elastin in the skin [14].

Tissue cells are of two classes, muscle and non-muscle cells. In non-muscle cells such as fibroblasts, the CSK is highly dynamic, and constantly subjected to actin filaments polarization/depolarization, and to spontaneous activity of its acto-myosin motor units. The cells need a homeostatic mechanical environment in order survive and proliferate [29]. They achieve this by adapting their extra-cellular fiber matrix (ECM) and by exerting traction on the ECM fibers, at a level which is cell- and tissue-specific [30, 31]. Traction is produced either passively by the intra-cellular actin cytoskeleton [32], or actively by the motor units [33]. In muscle cells, all three types (skeletal, myocytes and smooth muscle) share the actin-myosin contractile unit. Yet, there are important mechanical differences between them. In response to a single short stimulus (twitch), a skeletal muscle contracts and subsequently relaxes rapidly (it has a short refractory period). Successive twitches add up to produce stronger response (by wave summation) which increases with frequency up to a tetanized state at which the force can be maintained for a long period [34]. Cardiac myocytes on the other hand, cannot tetanize. They contract rapidly but have a longer refractory period during which they cannot be re-excited, a pre-requisite feature for the rhythmic contraction and blood pumping function of the heart. While some SMCs properties are similar to those of skeleton and cardiac muscle [35], they contract slowly. Some contract rhythmically in organs producing peristaltic motion (e.g., ureter and intestine) while others (e.g., in the arterial wall) maintain constant contraction force for very long periods, thereby maintaining a controlled vessel diameter and blood supply as required by the organ. Smooth muscle lining the walls of some hollow organs possess a malleable contractile apparatus—their cytoskeleton and sarcomeres can modify both their structure and number in series to adapt to large length changes. This length adaptation does not change their passive tension nor their active isomeric force [36–40]. An extreme example are smooth muscle cells in the bladder wall, which increase their length by a factor of up to 2.0–2.2 under low urine pressure as the bladder is gradually filled up to 7–10 times its empty volume. When activated, these cells contract and reduce their reference length back to its initial level within seconds as the bladder is emptied.

2 Multi-scale Structural Modeling of Tissue Mechanics

Updated reviews of soft tissues structure-motivated and structural models have recently been published [3, 41–53]. Here, the focus will be on fully structural modeling.

The determinants of soft tissue response are its constituents' structure and mechanics, the manner by which structure changes with deformation, the kinematic response of each constituent to the deformation, and the way by which the contributions of all constituents are integrated to yield the global tissue response.

Experimental observations revealed that in the stretched tendon [26, 54] and skin [24], there is a gradual straightening (recruitment) with stretch of initially undulated collagen fibers, and this is accompanied by an increase in the tissue stiffness. A similar gradual straightening was observed in bi-axially stretched mesentery [25]. In addition to their undulation, observations have revealed that in flat tissues the fibers' orientation is dispersed [25], and that in response to tissue deformation the fibers rotate towards the direction of highest stretch [24, 55]. The quantitative correlates of these observations which are needed for structural constitutive formulation, are based on the following assumptions [56, 57] which are derived from direct observations and measured data.

2.1 Basic Assumptions

Assumption B1 Soft tissues consist of fibers and cells embedded in a matrix of fluid ground substance.

Assumption B2 There are no voids in the tissue space.

Assumption B3 All constituents are incompressible, and so is the tissue as a whole under short loading protocols during which there is insufficient time for fluid to filtrate in or out of the tissue space.

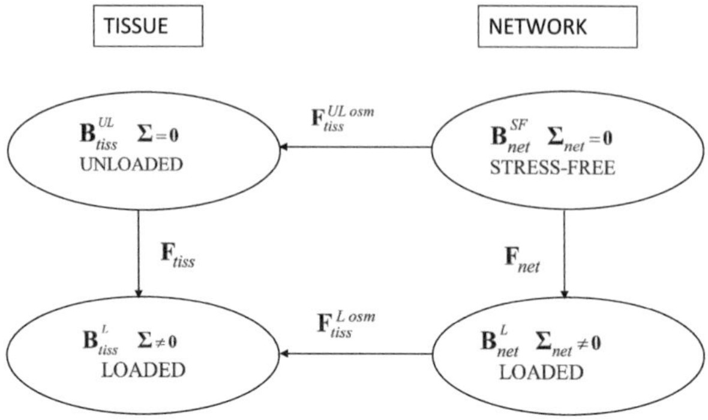

Fig. 1 Mapping scheme between the stress-free, unloaded and loaded fiber network and tissue. The $\boldsymbol{B}$'s designate the tissue and network configurations F's are the deformation gradients, and $\boldsymbol{\Sigma}$'s are the Cauchy stresses. Superscript *SF*—stress free, UL—Unloaded

Assumption B4 Tension in the fibers and muscle cells are functions of their axial deformation. Being thin and long, the fibers buckle under contraction thereby losing their stiffness.

Assumption B5 The strain in each constituent is equal to the global tissue strain (affine deformation) [52, 58].

Assumption B6 The global tissue response to loading equals the sum contributions of its solid constituents and of the matrix hydrostatic pressure (rule of mixtures).

2.2 Tissue Kinematics

The formulation considers a tissue representative elementary volume (REV) whose dimensions are small enough compared to those of the whole tissue/organ so that its strain can be regarded as locally homogeneous, yet large enough (compared with the tissue cells and fibers) to be statistically representative of the local tissue structure, kinematics and properties. Since the tissue is initially swollen by the matrix pressure from a stress-free to an unloaded state, two reference configurations and two loaded ones are considered (Fig. 1). The mapping between each two REV configurations is prescribed by the associated deformation gradient $\mathbf{F} \equiv \partial \mathbf{x}/\partial \mathbf{X}$ where $\mathbf{X}$ and $\mathbf{x}$ are respectively the REV coordinates prior to and following the deformation. Associated with each deformation gradient $\mathbf{F}$ are the right Cauchy-Green deformation tensor $\mathbf{C} = \mathbf{F}^T\mathbf{F}$ and the Lagrangian Green strain $\mathbf{E} = (\mathbf{C} - \mathbf{I})/2$. In view of the scheme in Fig. 1 and to simplify the notation, unless otherwise specified, these deformation metrics relate to the fiber network.

2.3 Mechanics of Single Fibers

In most soft tissues the collagen fibers are buckled and undulated in the stress-free state. A single fiber is characterized by its straightening stretch λ_s under which it becomes straight and tension bearing. When stretched in the loaded state by stretch ratio λ, its true stretch λ_f is

$$\lambda_f = \begin{Bmatrix} \lambda/\lambda_s & \forall\lambda > \lambda_s \text{ stretched,} \\ 1 & \forall\lambda \le \lambda_s \text{ buckled, unstretched.} \end{Bmatrix} \tag{2.1}$$

The strains corresponding to each of the stretches $\lambda_s, \lambda, \lambda_f$ are related to them by $e = (\lambda^2 - 1)/2$. Substitution into Eq. (2.1) yields the true fiber strain in terms of its global (e) and straightening strains (e_s)

$$e_f = \begin{cases} (e - e_s)/(1 + 2e_s) & \forall e > e_s \text{ stretched,} \\ 0 & \forall e \le e_s \text{ buckled, unstretched.} \end{cases} \tag{2.2}$$

If the fiber is hyper-elastic then its strain energy function under adiabatic or isothermal conditions depends solely on its true strain, i.e., $w_f = w_f(e_f)$. Its second Piola-Kirchoff stress is $s_f(e_f) = \partial w_f / \partial e_f$. In view of its incompressibility (Assumption B3) the fiber's first Piola-Kirchhoff and Cauchy's stress are related to its second Piola-Kirchhoff stress by $t_f = \lambda_f \cdot s_f$ and $\sigma_f = \lambda_f^2 \cdot s_f$, respectively.

2.4 Mechanics of Uniaxial Fiber Bundles

A fiber bundle consists of the network fibers sharing a similar orientation in the initial reference configuration. By affine deformation (Assumption B5), the bundle and its fibers rotate and stretch during deformation in mutually similar magnitudes and thus, share similar deformed orientation. Hence, for all bundle fibers $\lambda = \Lambda_{bun}$, $e = E_{bun}$. In addition, a bundle (and its fibers) having an initial orientation $\mathbf{N}$ (a unit vector with a solid angle Ω with respect to a given Cartesian coordinate system) will assume an orientation $\hat{\mathbf{n}}$ (a unit vector) in the deformed configuration, where

$$\hat{\mathbf{n}} = \mathbf{n}/|\mathbf{n}|, \quad \mathbf{n} = \mathbf{F} \cdot \mathbf{N}. \tag{2.3}$$

Based again on affine deformation, the bundle and its fibers' stretch and strain in the loaded state are:

$$\Lambda_{bun}(\mathbf{N}) = [\mathbf{N} \cdot \mathbf{C}\mathbf{N}]^{1/2}, \qquad E_{bun} = \big((\Lambda_{bun})^2 - 1\big)/2. \tag{2.4}$$

The straightening strains are non-uniformly distributed over the bundle fibers by a density distribution function $D(q, \mathbf{N})$ which is defined as follows: in the bundle of stress-free reference orientation $\mathbf{N}$, the fraction of fiber volume becoming straight between strain levels $e = q$ and $e = q + dq$ is equal to $D(q, \mathbf{N}) \cdot dq$. The term $\mathbf{N}$ in $D(q, \mathbf{N})$ denotes that the recruitment distribution may be orientation-dependent. Normalization requires that $\int_0^{+\infty} D(q, \mathbf{N}) \cdot dq = 1$ for every $\mathbf{N}$.

Since $\lambda = \Lambda_{bun}$, and Λ_{bun} is related to the global strain by affine deformation (Eq. (2.4)) then so is also the fiber global stretch λ. Yet, the fiber true deformation (Eq. (2.1)) is not affine since it is related to the global deformation by the dispersed fiber straightening stretch λ_s when stretched, and is buckled and collapsed when contracted.

In the hyper-elastic case, the bundle strain energy per unit stress-free tissue volume is the sum of strain energies of all its straight and stretched fibers:

$$W_{bun}(E_{bun}, \mathbf{N}) = \phi_f^0 \int_0^{E_{bun}} w_f(e_f) \cdot D(q, \mathbf{N}) \cdot dq \quad \forall \mathbf{N}, \tag{2.5}$$

where ϕ_f^0 is the total fibers volume fraction in the tissue which converts the fibers' strain energy per fibers volume to its level per tissue volume. The bundle second Piola-Kirchhoff stress is derived from its strain energy by chain differentiation:

$$S_{bun}(E_{bun}, \mathbf{N}) = \frac{dW_{bun}}{dE_{bun}}(E_{bun}, \mathbf{N}) = \phi_f^0 \int_0^{E_{bun}} D(q, \mathbf{N}) \cdot \left(\frac{dw_f}{de_f} \cdot \frac{de_f}{dE_{bun}} \right) \cdot dq, \tag{2.6}$$

which in view of Eq. (2.2) and $e = E_{bun}$ yields:

$$S_{bun}(E_{bun}, \mathbf{N}) = \phi_f^0 \cdot \int_0^{E_{bun}} \frac{D(q, \mathbf{N})}{(1+2q)} \cdot s_f(e_f) \cdot dq. \tag{2.7}$$

2.5 Mechanics of Three-Dimensional Networks and Tissues

In 3D tissues made of a single type of fibers, each REV has a 3D network of fibers. Each fiber, in addition to its level of undulation (or alternatively, its straightening strain e_s) in the initial reference configuration (before deformation), is characterized by its orientation **N** which features a solid angle Ω. This orientation angle is distributed over the un-deformed fiber population with a characteristic orientation density $\hat{\mathbb{R}}(\Omega)$, such that the volume fraction of the network fibers oriented between the solid angles Ω and $\Omega + d\Omega$ is $\hat{\mathbb{R}}(\mathbf{N}) \cdot d\Omega$. Normalization requires that $\int_\Omega \hat{\mathbb{R}}(\mathbf{N}) \cdot d\Omega = 1$.

In a spherical polar coordinate system (R, Φ, Θ) the fiber's reference orientation **N** is expressed as

$$\mathbf{N} = \sin\Phi \cdot \cos\Theta \cdot \mathbf{1}_1 + \sin\Phi \cdot \sin\Theta \cdot \mathbf{1}_2 + \cos\Phi \cdot \mathbf{1}_3, \tag{2.8}$$

where $\mathbf{1}_i$ $(i = 1, 2, 3)$ are unit vectors along the three Cartesian axes. The density normalization condition is expressed as[1,2]

$$\int_0^\pi \int_0^\pi \mathbb{R}(\Phi, \Theta) \cdot d\Phi \cdot d\Theta = 1, \qquad \mathbb{R}(\Phi, \Theta) = \hat{\mathbb{R}}\big(\mathbf{N}(\Phi, \Theta)\big) \cdot \sin\Phi, \tag{2.9}$$

where the $\sin\Phi$ term in $d\Omega = \sin\Phi \cdot d\Theta \cdot d\Phi$ was incorporated into $\mathbb{R}(\Phi, \Theta)$, the modified orientation distribution. The axial strain of the bundle and its fibers is related to the global REV strain **E** by affine deformation (Assumption B5), namely

$$\begin{aligned} E_{bun} = \mathbf{N} \cdot \mathbf{E}\mathbf{N} = {} & E_{11} \sin^2\Phi \cdot \cos^2\Theta + E_{22} \sin^2\Phi \cdot \sin^2\Theta + E_{33} \cos^2\Phi \\ & + 2E_{12} \sin^2\Phi \cdot \sin\Theta \cdot \cos\Theta + 2E_{13} \sin\Phi \cdot \cos\Phi \cdot \cos\Theta \\ & + 2E_{23} \sin\Phi \cdot \cos\Phi \cdot \sin\Theta, \end{aligned} \tag{2.10}$$

where E_{ij} $(i, j = 1, 2, 3)$ are the components of the Green-Lagrange strain **E** in a Cartesian system.

The total strain energy in the network fibers equals the sum contributions of all the fibers (Assumption B6), or alternatively, all the fiber bundles. Expressed per un-deformed tissue volume, this strain energy is given by:

$$W_{net}(\mathbf{E}) = \phi_f^0 \sum_{\mathbf{N}} W_{bun}(E_{bun}, \mathbf{N}) = \phi_f^0 \int_\Omega \mathbb{R}(\mathbf{N}) \cdot W_{bun}(E_{bun}, \mathbf{N}) \cdot d\Omega. \tag{2.11}$$

The second Piola-Kirchhoff network stress is derived from the strain energy by chain differentiation as follows:

$$\mathbf{S}_{net}(\mathbf{E}) = \frac{\partial W_{net}}{\partial \mathbf{E}} = \int_\Omega \mathbb{R}(\mathbf{N}) \cdot \frac{\partial W_{bun}}{\partial E_{bun}} \cdot \frac{\partial E_{bun}}{\partial \mathbf{E}} \cdot d\Omega. \tag{2.12}$$

[1] The integration limits must account for the orientation symmetry since each fiber has two opposite orientations. One possibility for limits is $0 \le \Phi \le \pi$, $0 \le \Theta \le \pi$, another one is $0 \le \Phi \le \pi/2$, $0 \le \Theta \le 2\pi$.

[2] For example, in an isotropic network $\mathbb{R}(\Phi, \Theta) = \sin\Phi/2\pi$.

By using Eqs. (2.6), (2.7) and (2.10) it is seen that

$$\mathbf{S}_{net}(\mathbf{E}) = \phi_f^0 \int_{\Omega} \mathbb{R}(\mathbf{N}) \cdot \left[\int_0^{E_{bun}(\mathbf{N})} \frac{D(q,\mathbf{N})}{(1+2q)} \cdot s_f(e_f) \cdot dq \right] \cdot \mathbf{N} \otimes \mathbf{N} \cdot d\Omega, \tag{2.13}$$

where $\mathbf{N} \otimes \mathbf{N}$ is the dyadic product of $\mathbf{N}$ with itself. In polar coordinates Eq. (2.13) is expressed as

$$\mathbf{S}_{net}(\mathbf{E}) = \phi_f^0 \int_0^{\pi} \int_0^{\pi} \mathbb{R}(\Phi,\Theta) \cdot \left[\int_0^{E_{bun}(\Phi,\Theta)} \frac{D(q,\mathbf{N}(\Phi,\Theta))}{(1+2q)} \cdot s_f(e_f) \cdot dq \right] \cdot \mathbf{N} \otimes \mathbf{N} \cdot d\Theta \cdot d\Phi. \tag{2.14}$$

The total tissue second Piola-Kirchhoff stress is the sum contributions of its fiber network and matrix pressure P_{mat}. Under short loading protocols the tissue is incompressible (Assumption B3), so that

$$\mathbf{S}_{tiss}(\mathbf{E}) = \mathbf{S}_{net}(\mathbf{E}) - P_{mat}\mathbf{C}^{-1}. \tag{2.15}$$

The network first Piola-Kirchhoff stress expressed in terms of the corresponding fibers stress is given by

$$\begin{aligned} \mathbf{T}_{net}(\mathbf{E}) &= \mathbf{FS} \\ &= \phi_f^0 \mathbf{F} \left\{ \int_{\Omega} \mathbb{R}(\mathbf{N}) \cdot \Lambda_{bun}^{-1}(\mathbf{N}) \cdot \left[\int_0^{E_{bun}(\mathbf{N})} \frac{D(q,\mathbf{N})}{(1+2q)^{1/2}} \cdot t_f(e_f) \cdot dq \right] \cdot \mathbf{N} \otimes \mathbf{N} \cdot d\Omega \right\}, \end{aligned} \tag{2.16}$$

where the inner integrand is obtained by inserting $t_f = \lambda_f \cdot s_f$ and $\lambda_f = \Lambda_{bun}/\lambda_s$ (Eq. (2.1)). The global tissue first Piola-Kirchhoff stress is

$$\mathbf{T}_{tiss}(\mathbf{E}) = \mathbf{T}_{net}(\mathbf{E}) - P_{mat}\mathbf{F}^{-T}. \tag{2.17}$$

Finally, the network and tissue Cauchy stresses are, respectively,

$$\begin{aligned} \Sigma_{net}(\mathbf{E}) &= \mathbf{F} \cdot \mathbf{S} \cdot \mathbf{F}^T \\ &= \phi_f^0 \cdot \mathbf{F} \cdot \left\{ \int_{\Omega} \mathbb{R}(\mathbf{N}) \cdot \Lambda_{bun}^{-2}(\mathbf{N}) \cdot \left[\int_0^{E_{bun}(\Omega)} D(q,\mathbf{N}) \cdot \sigma_f(e_f) \cdot dq \right] \cdot \mathbf{NN} \cdot d\Omega \cdot \mathbf{F}^T \right\}, \end{aligned} \tag{2.18}$$

$$\Sigma_{tiss}(\mathbf{E}) = \Sigma_{net}(\mathbf{E}) - P_{mat}\mathbf{I}, \tag{2.19}$$

where $\Lambda_{bun} = (1+2E_{bun})^{1/2}$ is derived from the REV strain $\mathbf{E}$ by using Eq. (2.10).

2.6 Mechanics of Tissues with Multiple Types of Fibers

In tissues with multiple types of fibers, in general each fiber type i has its own volume fraction $\phi_{f,i}^0$ and mechanical properties $s_f^i(e_f)$. Its network has its own recruitment $D^i(q,\mathbf{N})$ and orientation $\mathbb{R}^i(\mathbf{N})$ distributions. In addition, each network may have its own stress-free reference configuration. The tissue stress-free configuration corresponds to that of the fibers network with the lowest stress-free dimensions (at this configuration no fiber of any type is stretched). For the purpose of constitutive formulation, all networks recruitment and orientation densities should be referenced to that stress-free configuration. The stress derivation is carried out by summing the contributions of all networks and of the matrix pressure. For example, the second Piola-Kirchhoff stress in the i type network is derived from its fibers' mechanics and densities by:

$$\mathbf{S}_{net}^{i}(\mathbf{E}) = \phi_{f,i}^{0} \int_{\Omega} \mathbb{R}^{i}(\mathbf{N}) \cdot \left[\int_{0}^{E_{bun}(\mathbf{N})} \frac{D^{i}(q,\mathbf{N})}{(1+2q)} \cdot s_{f}^{i}(e_{f}) \cdot dq \right] \cdot \mathbf{N} \otimes \mathbf{N} \cdot d\Omega. \tag{2.20}$$

The total tissue second Piola-Kirchhoff stress is the sum of all networks stresses and of the matrix pressure,

$$\mathbf{S}_{tiss}(\mathbf{E}) = \sum_{i} \left[\mathbf{S}_{net}^{i}(\mathbf{E}) \right] - J P_{mat} \mathbf{C}^{-1}. \tag{2.21}$$

The first Piola-Kirchhoff and Cauchy stresses are derived in a parallel manner to Eqs. (2.16)–(2.19).

2.7 Validity of Structural Hyper-Elastic Models

Criteria for validity of the constitutive formulation are of two types: general ones which all models must satisfy, and specific ones relating to the model fit to the response of a specific material under a general deformation scheme. Two important general criteria are model convexity (which guarantees physical plausibility) and objectivity (coordinate frame indifference). In contrast to some other hyper-elastic tissue models [59, 60], structural hyper-elastic models were shown [61] to always be convex. The tensor nature of all terms in these models guarantees their objectivity.

Regarding the model fit to response data, experimental tissue characterization should take into consideration the non-linearity of tissue responses and the significant inter-sample biological variability. Non-linear responses can be characterized by a number of models which are indistinguishable in their fit to a given data set (Albert Green, private communication). The challenge is to determine which model is the "true" one. Two measures should be taken to meet this challenge. The first is testing under as many different protocols as possible (e.g., 1D, 2D, and 3D tests under several deformation schemes and strain levels)—only the model which fits well to the response data under all protocols is acceptable. To cope with inter-sample variability, all tests should be carried out on the same sample. Failure to adopt these measures is common in the published literature, thereby casting doubt on the reliability of the models proposed.

The study of Hollander and co-workers on the coronary artery media [62] could serve as an example of multi-scale, multi-phasic structural tissue modeling. The passive media was considered as a 3D structure consisting of helical lamellar collagen and elastin fibers interconnected by inter-lamellar elastin (Fig. 2). Structural and mechanical parameters were estimated from 3D data of inflation, extension, and twist tests under various combinations of axial force, internal pressure and torsion [63, 64]. Sensitivity analysis revealed that four parameters are sufficient to reliably represent the response under all 3D protocols (the descriptive performance—Fig. 3), and to reliably predict the tissue response under protocols not used for parameter estimation (the predictive performance—Fig. 4). These parameters are the helical angle of the lamellar fibers, two parameters of these fibers non-linear stress-strain law, and the stiffness of the inter-lamellar elastin. In a comparative study [65], this structural model out-performed earlier phenomenological and semi-structural (structurally motivated) models in both its descriptive and predictive capabilities (Fig. 5). While both the phenomenological and semi-structural models provide reasonable fit to data at certain deformation regions, they deviate from the data in other regions in both pattern and proximity to

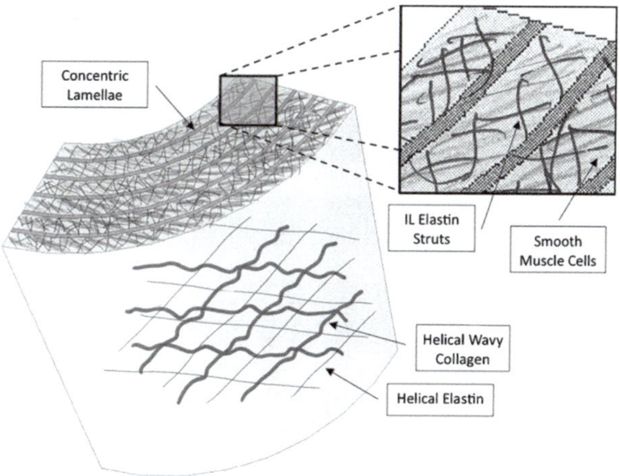

Fig. 2 A scheme of the vessel wall media structure including the lamellae helical elastin-collagen fibers, the inter-lamellar (IL) struts, and the smooth muscle cell. From [62] with permission

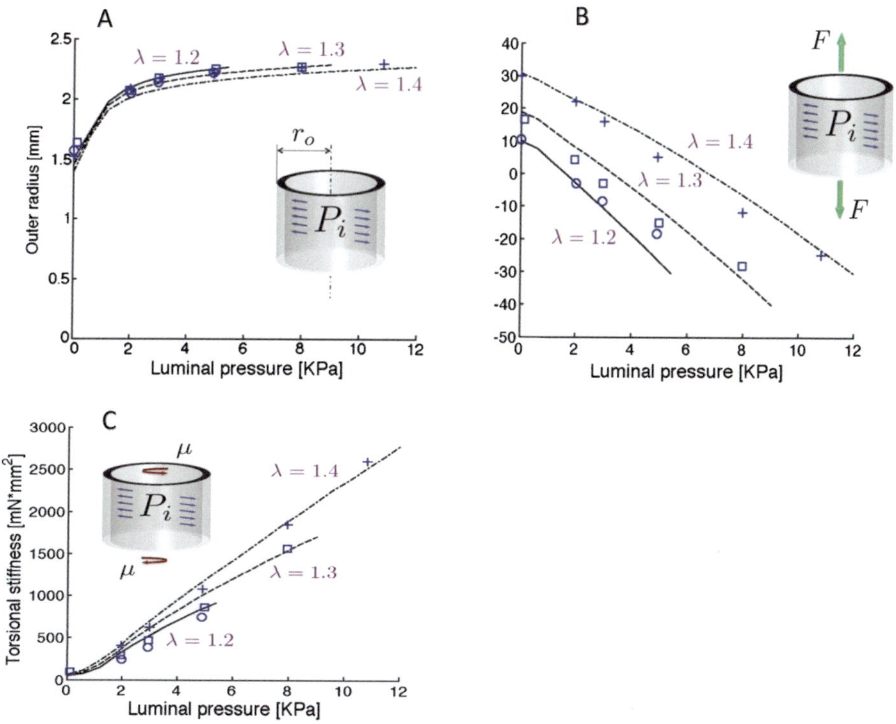

Fig. 3 The structural model descriptive power: predictions (*lines*) compared with coronary media experimental data (*symbols*) of (**A**) outer radius r_o, (**B**) axial force F, and (**C**) torsion stiffness μ, all as functions of the inner luminal pressure P_i, under three axial stretch ratios $\lambda = 1.2$ (*circle*), $\lambda = 1.3$ (*square*), $\lambda = 1.4(+)$. From [62] with permission

the data. The fully structural model provided good fit under all conditions tested. Moreover, in their predictive performance, only the structural model was able to reliably predict the tissue response under protocols not used for parameter estimation.

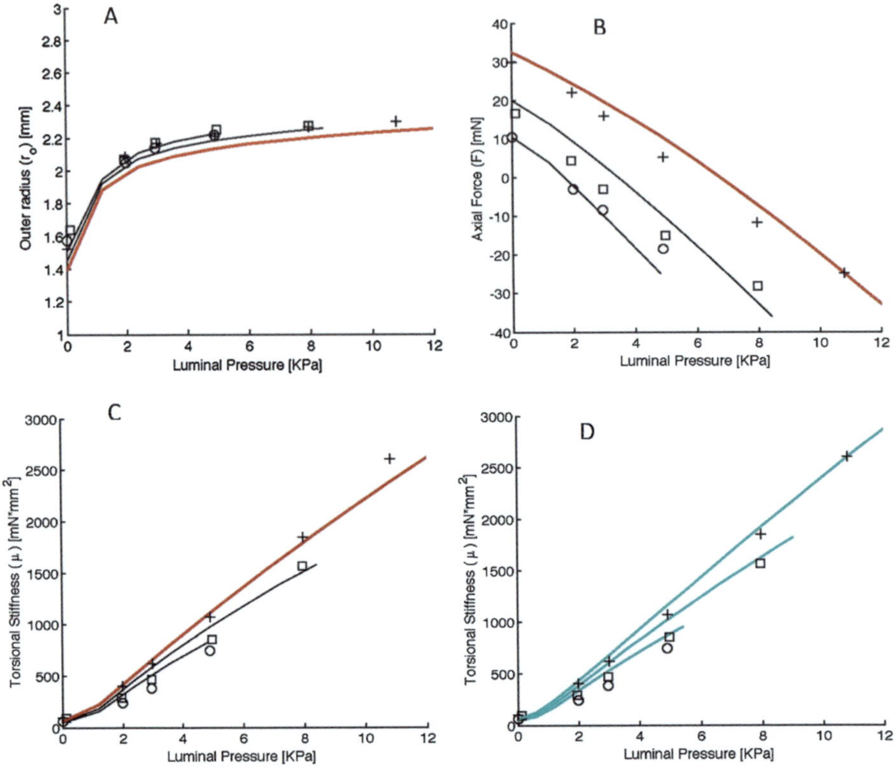

Fig. 4 The structural model predictive power: (**A**, **B**, and **C**)—comparison of coronary media data under three stretch ratios (*symbols* as in Fig. 3) with predictions using parameters estimated from partial data under stretch ratios 1.2 and 1.4 (*black lines*), and comparison with data of stretch ratio 1.3 (*red line*) not used for parameter estimation. (**D**) comparison of torsion stiffness vs. internal pressure data under three stretch ratios (*symbols* as in Fig. 3) with predictions (*blue lines*) of torsional stiffness with parameters estimated from inflation-extension data. From [62] with permission

2.8 The Matrix Osmotic Pressure

The tissue matrix consists of mobile molecules of water, anions and cations and negatively charged immobile PG macromolecules. The matrix osmotic properties arise from the thermodynamic conditions that under equilibrium (no flow of ions and water), the chemical potentials of the water in the matrix and in the external fluid must be equal. In thermodynamics the chemical potential of the water is the sum of its driving forces (the forces that can move the water) given by [66, 67]

$$\mu^w = \mu_0^w + RT \cdot \ln(a^w) + P\bar{V}^w, \tag{2.22}$$

where μ_0^w is the water reference chemical potential, which is a universal constant, a^w is the water chemical activity, P and $\bar{V}^w$ are respectively the water hydrostatic pressure and molar volume, and R and T are respectively the universal gas constant and absolute temperature. Equality of the internal and external chemical potentials of the water implies that $\mu_{mat}^w = \tilde{\mu}^w$, where the upper tilde designates external fluid. Let $\tilde{\mu}^w, \tilde{a}^w$, and $\tilde{P}$ be the corresponding terms of Eq. (2.22) in the external fluid, and μ_{mat}^w, a_{mat}^w, and P_{mat} in the tissue matrix. Then

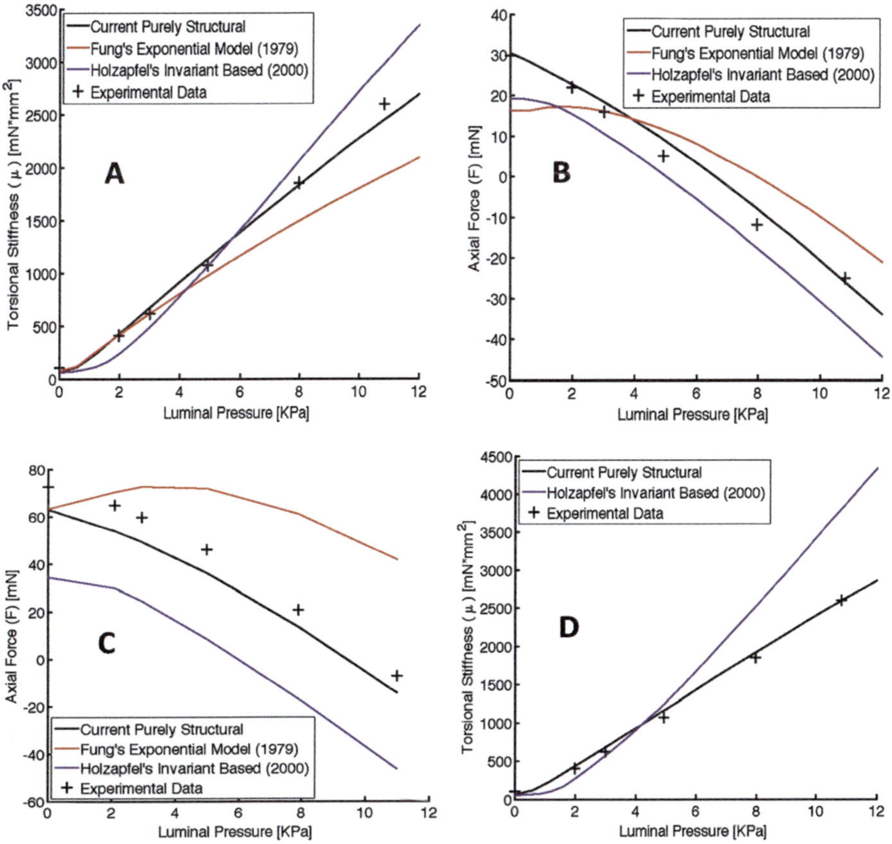

Fig. 5 Comparison of the structural model (*black line*) to Fung [144] exponential phenomenological model (*brown line*) and to Holzapfel et al. semi-structural (invariant-based) [59] model (*purple line*) by contrasting the models' predictions against 3D inflation/extension/twist response data of coronary artery media layer. (**A**, **B**) The models descriptive performance (fit to data used for parameter estimation) in predicting axial force and torsional stiffness responses, respectively, under axial stretch of 1.4 and increasing luminal pressure. (**C**) The models predictive performances (fit to data not used for estimation) in predicting the axial force under axial stretch of 1.4 and increasing luminal pressure based on parameter estimation from corresponding data under axial stretches of 1.2 and 1.3. (**D**) The models predictions of the torsional stiffness under axial stretch of 1.4 and increasing luminal pressure based on parameters estimated from 2D inflation/extension data under the same axial stretch. Fung's model torsion predictions could not be compared here since its torsional stiffness is a separate parameter that can be estimated only from twist data

by Eq. (2.22), the equality of chemical potentials implies that

$$P_{mat} - \tilde{P} \equiv \Delta P = -RT \cdot \left[\ln\left(a^w\right) - \ln\left(\tilde{a}^w\right)\right] / \bar{V}^w. \tag{2.23}$$

The water activity $-RT \cdot \ln(a^w)/\bar{V}^w$ is termed "osmotic pressure", designated by π. Equation (2.23) implies that under equilibrium, $\Delta P \equiv P_{mat} - \tilde{P} = \Delta\pi \equiv \pi_{mat} - \tilde{\pi}$, i.e., the interior/exterior hydrostatic and osmotic pressures differences are mutually equal.

The water activity is related to its concentration by $a^w = y^w c^w$, where y^w is the water activity coefficient, which is taken to be equal in the external and internal fluids, and c^w is the water molar concentration, equal to the number of moles of water in the solution divided by the solution volume. Insertion into Eq. (2.23) yields $\Delta\pi = -RT \cdot [\ln(c^w) - \ln(\tilde{c}^w)]/\bar{V}^w$. The presence of ions implies that the water concentration is less than one ($c^w, \tilde{c}^w < 1$), so that the osmotic pressure terms are always positive. In addition, the interior ion concentration is higher than the exterior one, since in addition to the negative free ion charges in both internal/external fluids, an excess of positively charged cations exists in the matrix fluid to counter-balance the fixed negative charges of the PGs. Higher ion concentration means a lower water concentration, so that the interior water concentration is smaller than the exterior one. This implies that both hydrostatic and osmotic pressure differences are positive, ($\Delta\pi = \Delta P > 0$).

The matrix pressure required to balance the tissue osmotic pressure difference builds up as water filtrates into the tissue, thereby swelling it and stretching the tissue fibers. In an unloaded tissue at equilibrium, two conditions are satisfied: thermodynamic equilibrium between matrix hydrostatic and osmotic pressures ($\Delta\pi = \Delta P$), and mechanical equilibrium between the fibers' stress produced by their tension, and the matrix hydrostatic pressure. Following the mixture theory [68], the total tissue stress is the weighted sum of the fibers stress and of the fluid matrix pressure (the rule of mixtures—Assumption B6). Hence, in terms of Cauchy stresses, under equilibrium, the tissue stress is the sum of all networks stresses and the matrix pressure, namely

$$\boldsymbol{\Sigma}_{tiss}(\mathbf{E}) = \sum_i \left[\boldsymbol{\Sigma}^i_{net}(\mathbf{E})\right] - P_{mat}\mathbf{I}, \qquad P_{mat} = \tilde{P} + \Delta\pi. \tag{2.24}$$

2.9 The Matrix Hydrostatic Pressure

The term P_{mat} representing the hydrostatic pressure in the fluid matrix, has a clear physical meaning. Equation (2.24) has a similar form to expressions proposed in a number of previous modeling studies where the term P (instead of P_{mat} in Eq. (2.24)), is a Lagrange multiplier which arises from the constraint of tissue incompressibility. Yet soft tissues are, in general, compressible even though their constituents are incompressible. In the transition between equilibrium configurations under two different loadings, water may filtrate in or out of the tissue thereby changing its volume, until a new steady state is reached, which satisfies both mechanical and thermodynamic ($\Delta\pi = \Delta P$) equilibriums. Yet, Eq. (2.24), which results from the rule of mixtures, still holds regardless of possible volume changes. Under short loading protocols, there is insufficient time for fluid filtration and the volume changes may be negligible. The tissue can then be regarded as approximately incompressible. Under each short loading the tissue can be under mechanical but not under thermodynamic equilibrium. The pressure P can then be considered a Lagrange multiplier. Importantly however, its value is determined by the loading boundary conditions and is, in general, different from its level under both mechanical and thermodynamic equilibrium (second term in Eq. (2.24)).

2.10 Tissue Swelling and Residual Stress

Tissue swelling by the matrix pressure has a number of important consequences. First, tissues are kept well hydrated, a requisite for the cells' survival and a distinct mechanical advantage associated with the water plasticizing effect. In addition, swelling increases tissue rigidity and stiffness [69]. Un-swollen tissues (under vanishing matrix pressure) are stress-free since the fibers become un-stretched. Since they have no contraction stiffness (Assumption B4), the fibers buckle, assuming a variety of undulated configurations, and the tissue becomes floppy, showing no resistance to loading. It has no fixed stress-free nor unloaded configurations and cannot develop residual stress [20]. Under non-vanishing matrix pressure, the tissue swells, its fibers become stretched and the tissue acquires compressive as well as bending stiffness. Swelling and its resulting tissue stiffness lead to the onset of residual stress (RS), which carries distinct functional advantages in a number of tissues and organs.

An examination of the mechanisms of tissue RS [20] revealed that there is a hierarchy of RS producing mechanisms. At the micro-level (tissue interstitium), RS is induced by local interactions between the tissue constituents (fibers, cells, ground substance matrix). It can stem either from direct contact stresses between the tissue solid constituents, which result from their incompatible growth and remodeling [70, 71], or from the tissue swelling by the osmotic matrix. Analysis of published observations and measured data [20] suggested that under *in vivo* conditions, the osmotic driven swelling is the predominant micro RS-producing mechanism. At the meso-level, additional RS, results from internal interactions between non-homogeneities in either the tissue properties [72] or in its micro-structure and composition (e.g., between the media and adventitia in the arterial wall [73], or between sub-layers of the wall [74, 75]). The macro-level (organ) RS arises from organ-level kinematic constraints on the tissue structure which produce additional internal loading (e.g., forces and bending moments required to close the open artery from its unloaded open configuration). Based on these considerations, it was proposed that the onset of macro (OA-related) and meso (heterogeneities- related) RS depends on the presence of a micro RS. Without micro-RS, the tissue would be floppy and incapable of developing contact stresses between in-homogeneities nor bending resistance to closure of the opening angle. The crucial role of tissue swelling has been experimentally supported by the observed significant effect of osmolarity on tissue swelling and its OA level (a metric of RS magnitude) in the left ventricle, aorta and other arteries in rats [18, 73, 76], and by the finding that a true stress-free state in a carotid artery can only be reached by enzymatic destruction of the swelling-resisting elastin fibers in the media layer of the vessel wall [74]. Such destruction removes the fiber network in which the PG are encaged, thereby allowing them to exudate from the tissue space, thus abolishing the associated micro-RS and with it, the other RS components.

3 Tissue Viscoelasticity and Preconditioning

Soft tissues are not hyper-elastic but rather viscoelastic (VE). They also undergo preconditioning (PC) adaptation to loading [34]. These time-dependent response aspects are inherent properties of tissues in the sense that they significantly affect their function. Yet they are often ignored in tissue modeling. There is one important exceptional case in which tissues can be considered close to hyper-elastic: cardiovascular (CVS) tissues function *in vivo* under long-term constant cyclic loading imposed by the dynamic blood pressure. Under these circumstances, CVS tissues are fully preconditioned to their loading and, in addition, their VE is nearly exhausted, thereby demonstrating close to elastic response. In general, however, tissue VE and PC must be integrated into their constitutive formulation and failure to do so results not only in inability of the model to account for the observed time-dependent response but also in protocol-dependence of the models and their estimated parameters, thus lacking generality and reliability under general deformation schemes and losing their usefulness in analysis and design.

Previous non-elastic tissues models have addressed solely the VE aspect. Linear viscoelastic (LVE) and quasi-linear ones (QLV) models assumed that the viscoelastic stress relaxation function is strain-independent. While LVE considers a linear immediate ("elastic") stress-strain relationship (the tissue response under infinitely fast step deformation), the widely used QLV theory [11, 34] admits non-linear immediate response in line with experimental observations. For uniaxial tissue bundles, such as in tendons and ligaments, the second Piola-Kirchhoff QLV stress is expressed as:

$$S_{bun}^{QLV}(e,t) = \int_0^t G(t-\tau) \cdot \frac{dS_{bun}^e(e)}{d\tau} \cdot d\tau = \int_0^t G(t-\tau) \cdot \frac{dS_{bun}^e}{de} \cdot \frac{de}{d\tau} \cdot d\tau, \qquad (3.1)$$

where $G(t)$ is the reduced relaxation function, and $S_{bun}^e(e)$ is the immediate bundle response.

QLV validity has been questioned and is not generally accepted. While some studies demonstrated good predictive performance of QLV, others did not. In particular, ligament creep and stress relaxation responses were found to be mutually incompatible [77]. Fung speculated [34] that the tissues' creep response may be more non-linear than its relaxation response. We proposed [78] that these disagreements stem from inadequacies in the experimental procedures used and from failure to consider the tissues dynamic structure in modeling the tissue response. For example, in some studies, samples were not fully preconditioned thereby introducing bias due to uncontrolled combination of PC and VE effects on the tissue response. This occurs since both PC and VE exhibit stress decay under repeated deformation cycles. VE and PC differ, however, in that *in vitro* VE decay is reversible after rest period, while PC effects are at least partially irreversible.

To shed light on the true nature of time-dependent tissue responses, an experimental procedure aimed to separate between PC and VE effects, was developed. The results in sheep digital tendons show that following PC of stretch cycles repeated up to a stable response, a pure viscoelastic response was subsequently obtained provided the highest strain used was 1 % lower than the strain used for preconditioning [78].

From the modeling aspect, it is natural to apply structural models for studying VE and PC, since these non-elastic features are properties of the tissue fibers. As such, they can be readily integrated into the recruiting fibers modeling in the structural approach. This can be readily seen in Eq. (2.20). The only mechanical term in this equation is the fibers' hyperelastic stress-strain law $s_f^i(e_f)$. Hence, in cases of time-dependence, the sole modification needed is to replace this term with the corresponding time-dependent term $s_f[e_f(q,t),t]$, as elaborated below.

3.1 Theory of Recruitment Viscoelasticity (RVE)

The recruitment viscoelasticity (RVE) theory was developed [78] in an attempt to reconcile conflicting observations and paradigms regarding tissue VE. It incorporates the fiber' VE in the framework of a recruiting tissue structure. Viidik [54] found that when a tendon is stretched, its non-uniformly undulated collagen fibers at reference gradually become straight with stretch while the tendon stiffness increases (the "toe" region), until all fibers are stretched and the force-stretch response becomes linear (the linear region). Unloading produces a reverse response, but is subjected to hysteresis. This suggests that the tissue fibers are viscoelastic. Assuming that their reduced relaxation function G depends solely on time and is stretch-independent, the fiber's second Piola-Kirchhoff stress is

$$s_f\big[e_f(q,t),t\big] = \left\{ \begin{array}{lll} \displaystyle\int_0^t G(t-\tau) \cdot \frac{ds_f^e(e_f)}{d\tau} \cdot d\tau & \forall e_f > 0 & \text{(stretched fiber)}, \\ 0 & \forall e_f \le 0 & \text{(buckled fiber)}, \end{array} \right\} \qquad (3.2)$$

where $s_f^e(e_f)$ is the fiber's immediate response [11]. Replacing $s_f(e_t)$ in the recruiting bundle stress formula (2.7) by this expression, while ignoring, for now, the orientation dependence of $D(q)$, gives

$$S_{bun}^{RVE}\left[E_{bun}(t)\right] = \begin{cases} \phi_f^0 \int_0^{E_{bun}} \frac{D(q)}{(1+2q)} \cdot dq \cdot \int_0^t G(t-\tau) \cdot \frac{ds_f^e(e_f(q,\tau))}{d\tau} \cdot d\tau & \forall e_f > 0, \\ 0 & \forall e_f \le 0. \end{cases} \tag{3.3}$$

The elastic response s_f^e depends on time solely via the fiber strain $e_f(q,t)$. Hence by chain differentiation, Eq. (3.3) assumes the form

$$S_{bun}^{RVE}\left[E_{bun}(t)\right] = \begin{cases} \phi_f^0 \int_0^{E_{bun}} \bar{D}(q) \cdot dq \cdot \int_0^t G(t-\tau) \cdot \frac{ds_f^e(e_f)}{de_f} \cdot \frac{\partial e_f(q,\tau)}{\partial \tau} \cdot d\tau & \forall e_f > 0, \\ 0 & \forall e_f \le 0, \end{cases} \tag{3.4}$$

where $\bar{D}(q) = D(q)/(1+2q)$ is a modified fibers recruitment distribution. Equation (3.4) is the 1D version of the RVE theory.

3.2 Validity of the Recruitment Viscoelasticity Theory

Validity testing of RVE alone was carried out [78] on each specimen in order to avoid bias arising from inter-sample biological variability. The fibers' relaxation function was taken to be a three-term exponential Prony series expressed by[3]

$$G(t) = k_0 + k_1 \,\mathrm{Exp}\left(\frac{-t}{\tau_1}\right) + k_2 \,\mathrm{Exp}\left(\frac{-t}{\tau_2}\right) + k_3 \,\mathrm{Exp}\left(\frac{-t}{\tau_3}\right), \tag{3.5}$$

where the three terms represent respectively short, medium, and long relaxation processes. The fibers' recruitment density was assumed to be a beta function[4]. The model parameters were estimated from data of sheep digital extensor tendons, subjected to a sequence of short and long stress relaxation protocols to decreasing stretch levels of 6 %, 5 % and 4 %. RVE validity was first tested for its descriptive performance (fit to data used for parameter estimation). The results show (Fig. 6) that while in the first 6 % domain, where the tissue was not fully preconditioned, RVE fit was poor, while in the subsequent 5 % and 4 % series, during which the tissue was fully PC and the response was therefore purely VE, RVE predictions fit the data very well. Another validity test addressed RVE predictive performance (fit to data not used for parameter estimation). With parameters estimated from stress relaxation tests, the predicted creep response fit the data very well provided the stretch level remained slightly lower (by 1 %) than the PC stretch, a strain range under which there are no PC effects and the response is purely VE. Hence, the apparent previously reported incompatibility [77] of stress relaxation versus creep responses, is resolved by the structural theory. Further support of the RVE theory comes from combined VE and PC studies presented below.

---

[3] The Prony series in Eq. (3.5) is not related to the underlying mechanisms which give rise to the fiber's viscoelasticity. These mechanisms are insufficiently known. In common with other phenomenological VE models, this Prony representation is not unique.

[4] The beta function was preferred over the normal one since it is physically more realistic and more general: it is bounded, can be symmetric and non-symmetric, and can assume different shapes.

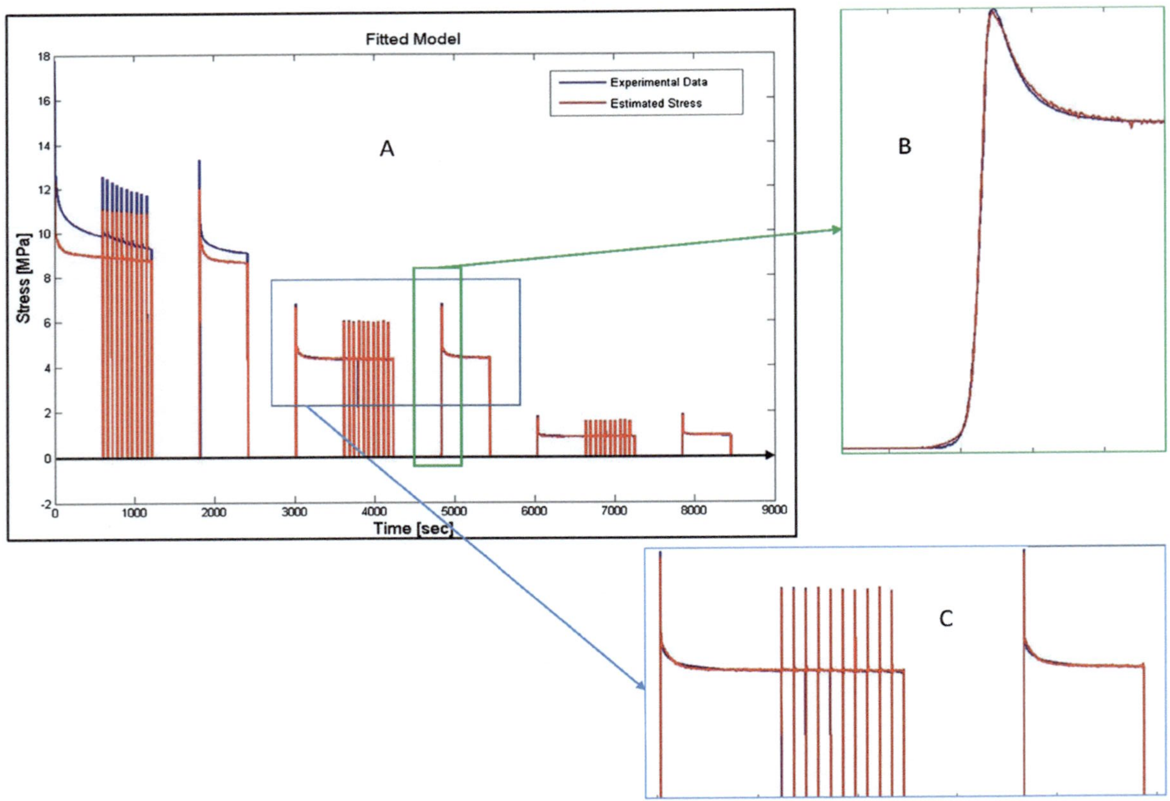

Fig. 6 (**A**) RVE descriptive performance: comparison of the tendon data (*blue line*) under a sequence of short and long stress relaxation test to stretch levels of 6 %, 5 %, and 4 % to the model predictions (*red line*). The fit is excellent under 5 % stretch where the tissue is fully preconditioned (*red* and *blue lines* in *insets* **B** and **C** are indistinguishable), but not good at 6 % stretch where the tissue is not fully preconditioned (**A**). Modified from [78] with permission

3.3 Recruitment Versus Quasilinear Viscoelasticity

Examination of the RVE relative to the QLV theory can shed light on the differences between them and elucidate the reasons for QLV success under some circumstances and failure under others. Analytically, in the RVE stress expression (Eq. (3.3)), if and only if the integrand $\frac{ds_f^e(e_f(q,\tau))}{d\tau}$ is a single-single valued function, then the order of integration can be changed so that

$$S_{bun}^{RVE}(e,t) = \int_0^t G(t-\tau) \cdot \int_0^e \frac{D(q)}{1+2q} \cdot \frac{ds_f^e(e_f(q,\tau))}{d\tau} \cdot dq \cdot d\tau. \tag{3.6}$$

By Eq. (2.7), the inner integral is equal to the time derivative of the bundle elastic response S_{bun}^e. Hence,

$$S_{bun}^{RVE}(e,t) = \int_0^t G(t-\tau) \cdot \frac{dS_{bun}^e}{dt} \cdot d\tau. \tag{3.7}$$

Equation (3.7) is identical to the QLV equation (Eq. (3.1)). However, the condition of single-single valued integrand is only valid when all fibers in the bundle are stretched. If however, some fibers contract and buckle during the stretch protocol, then their elastic stress vanishes along the entire range of negative fiber stretch, and the condition of single-single value is no longer maintained.

It is thus seen that the QLV theory is equivalent to RVE theory and is valid only if no fibers buckle during the deformation protocol. This condition is met by both creep and stress relaxation tests. Yet, under physiological conditions, the strain is likely to be dynamic and fibers are likely to contract and buckle during activity. In such cases the QLV theory is invalid. RVE, on the other hand, incorporates fiber buckling and is thus valid under general protocols. The case of quick release tests demonstrates this point. Experimentally, soft tissues which function under tension cannot support compressive load (Fig. 7). Model realization showed, however, that QLV predicts negative stress under quick release, while RVE correctly predicts zero stress (Fig. 8).

RVE can be readily generalized to three-dimensional networks as was done for the 1D case. When the fiber VE model (Eq. (3.2)) replaces the fiber hyper-elastic model in the three dimensional network stress equation (Eq. (2.14), then the following network VE stress is obtained

$$\mathbf{S}_{net}^{RVE}(\mathbf{E},t) = \phi_f^0 \int_\Omega \mathbb{R}(\mathbf{N}) \cdot \left[\int_0^{E_{bun}(\Omega)} [D(q,\mathbf{N})/(1+2q)] \cdot \int_0^t G(t-\tau) \cdot \frac{ds_{fq}^e(e_f)}{d\tau} \cdot d\tau \cdot dq \right] \cdot \mathbf{N} \otimes \mathbf{N} \cdot d\Omega. \tag{3.8}$$

3.4 Recruitment Preconditioning (RPC)

Experimental observations showed that tendon PC under repeated stretch cycles is associated with increase in their unloaded reference length. Although the effects of PC *in vivo* are fully reversible, under *in vitro* conditions they are only partially reversible after a long rest. Experience has shown that in all soft tissues, PC effects are strain-dependent and have two components, one is plastic (time-independent), the other is viscous (time-dependent). It is logical to assume that since the tendon properties derive from its parallel fibers, PC in a single collagen fiber is due to increased fiber reference length, and that the PC effects have plastic and viscous components. Hence, in the structural model of sheep digital extensor

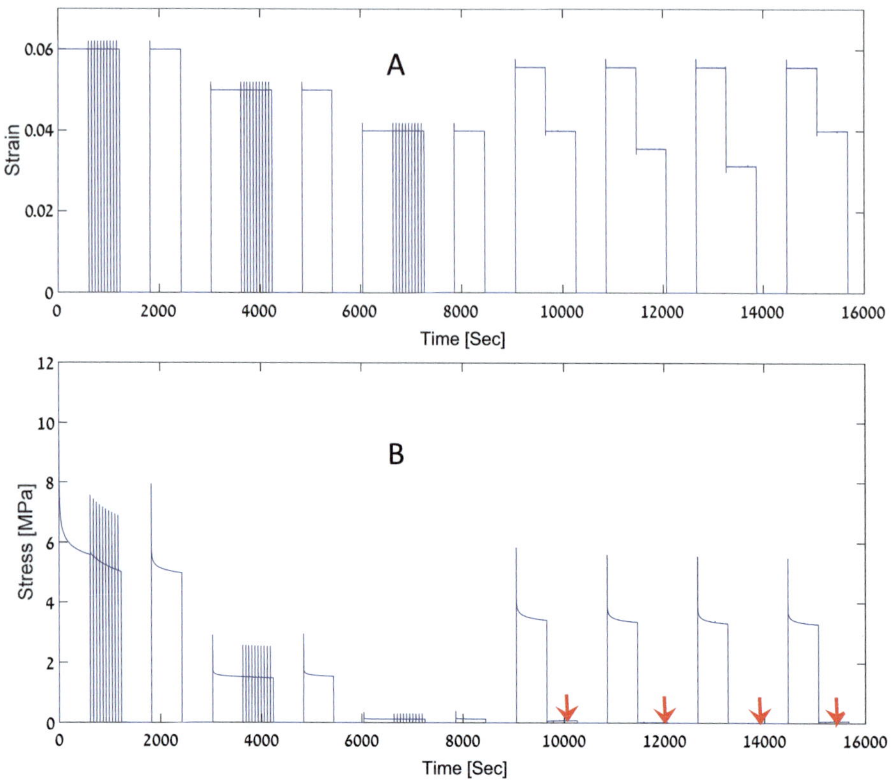

Fig. 7 Response of the fully preconditioned tendon to quick release protocols under decreasing levels of stretch (**A**). No negative stress was demonstrated in any of the tests (*red arrows* in **B**) demonstrating absence of compressive rigidity in the tissue. From [78] with permission

Fig. 8 Simulated RVE and QLV responses (**B**) to quick release to a strain e3 at which some fiber buckle (**A**). X0 is the recruitment strain of longest fiber which buckle under strain lower than x0. While QLV predicts onset of negative stress, no such stress is predicted by RVE, in line with experimental evidence (**B**). From [78] with permission

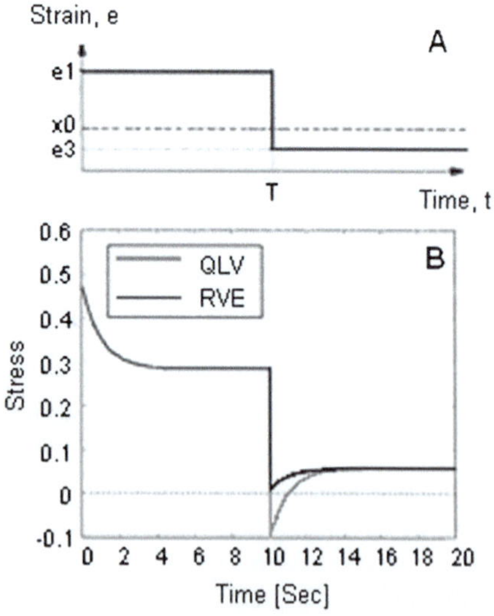

tendons, PC in collagen fibers was modeled [15] by a combination of linear viscous and linear plastic components expressed respectively by:

$$\frac{de_s}{dt} = \begin{Bmatrix} p_{1c}(e_f - p_{2c}), & e_f > p_{2c}, \\ 0 & \text{otherwise}, \end{Bmatrix} \tag{3.9}$$

$$\frac{de_s}{dt} = \begin{cases} \frac{de_f(t)}{dt}, & e_f > p_{3c} > 0, \frac{de_f(t)}{dt} > 0, \\ 0, & \text{otherwise}, \end{cases} \tag{3.10}$$

where e_s is the fiber straightening strain, p_{2c} and p_{3c} are strain thresholds below which no viscous or plastic lengthening occurs, and p_{1c} is a viscous constant. These PC laws are applied to all collagen fibers, whether they are in a uniaxial bundle or in a multi-dimensional network.

In contrast to the tendon, uniaxial skin tests showed that while PC affects the stress response at low strain levels, it has an insignificant effect on the reference length [79]. Since elastin fibers dominate the skin low strain response, it is reasonable to assume that PC affects the elastin fiber stiffness but does not increase its length. In view of its observed initial linearity (up to approximately 60 % stretch) which is followed by convex non-linearity at high strain levels, the elastin stress-strain response was taken to be a combined linear and non-linear functions of strain as in $s_e(e) = A_e \cdot e + B_e \cdot e^{Ne}$. In most studies the strain does not reach the non-linear elastic range. Hence it was assumed that PC affected the linear constant A_e, while the non-linear constants B_e and N_e, remained unchanged. The reduction of elastin stiffness from its initial (A_e^0) to its fully preconditioned (A_e^{eq}) value was assumed to depend on the fiber's highest PC strain $e_{\max}$ (the Mullins effect [80] known in rubbers and polymers), following the equation

$$A_e^{eq} = A_e^0[1 + L \cdot e_{\max} + F \cdot e_{\max}^M], \quad L, F < 0, \tag{3.11}$$

where L, F, and M are material constants which were estimated from the data.

3.5 Validity of the Combined Preconditioning-Viscoelastic Recruitment Formulation

As discussed above in connection with hyper-elastic models, two important general validity criteria of constitutive models (model convexity and objectivity) are satisfied by hyper-elastic structural models. Assigning VE and PC properties to the fibers does not violate this validity. At each point in time, the fibers' forces are integrated to produce the tissue REV stress tensor in the same way as in the hyper-elastic case. The structural constitutive laws are thus both objective and convex also with time-dependence of the fibers properties.

Two studies have demonstrated the essential role of VE and PC in tissue constitutive formulations. One study was performed with sheep digital extensor tendons [15] which consist of a bundle of non-uniformly undulated (gradually recruiting) collagen fibers. The recruitment dispersion was assumed to follow a beta distribution. The fiber VE was modeled by RVE and their PC, by Eqs. (3.9), (3.10). The combined recruitment, PC and VE aspects were used to account for the tendon response under a sequence of short and long stress relaxation protocols at increasing (4 %, 6 %, 8 %) and then decreasing (7 %, 6 %) stretch levels (Fig. 9). Sensitivity analysis and visual inspection revealed that a remarkably good fit and very good predictive capabilities of the response under this complex protocol are

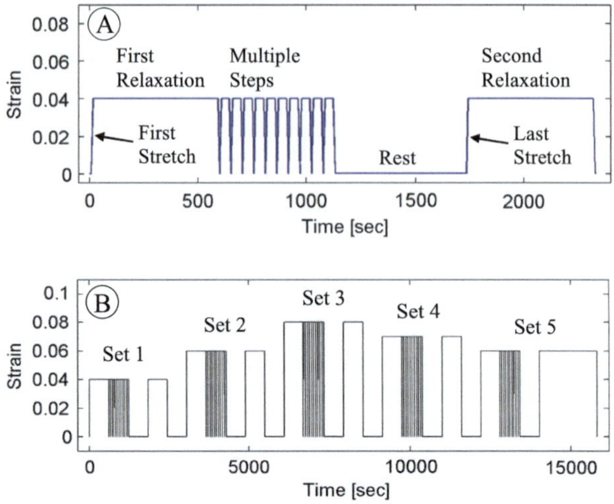

Fig. 9 The tendon composite test protocol consisting of a sequence of short and long-term relaxation tests at a constant stretch (4 % in **A**) which was then repeated at increasing (6 %, 8 %) then decreasing (7 %, 6 %) stretch levels (**B**). From [15] with permission

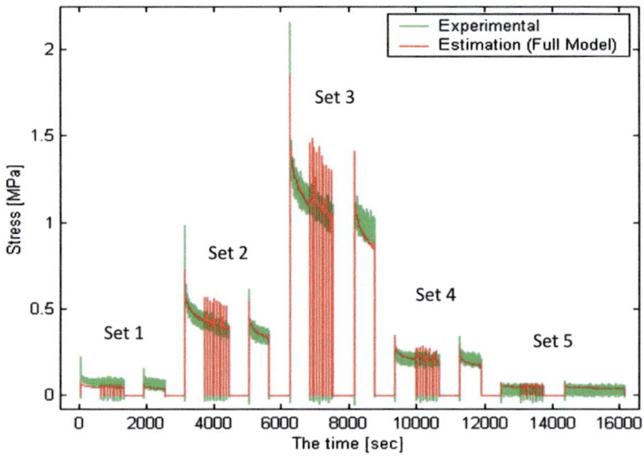

Fig. 10 Descriptive performance of the combined RVE and RPC uniaxial tendon model [15]. Comparison of data to model predictions under the protocol in Fig. 9. The model (*red line*) successfully predicts the data (*green line*), including the significant PC induced difference between the first (*Set 2*) and second (*Set 5*) 6 % response data. From [15] with permission

obtained when both collagen PC and VE features are included (Fig. 10). In another study [14], the rabbit skin non-linear and time-dependent response data [13] under various combinations of biaxial stretch protocols (Fig. 11), were modeled by a multi-fiber (collagen and elastin) structural model incorporating both the fibers orientation and recruitment dispersions. The collagen PC and VE were modeled as in Eqs. (3.5), (3.9), (3.10). The elastin PC was modeled by Eq. (3.11) and its VE by Eq. (3.5). Sensitivity analysis revealed that both elastin non-linearity (the $B_e \cdot e^{Ne}$ term) and its VE have little effect on the skin biaxial response. Comparison with measured data showed a good fit to data under all biaxial protocols (Fig. 12), and good predictive performance (Fig. 7 and Tables 4 and 5 in reference [14]) when the collagen and elastin PC and collagen VE were both included in the fibers constitutive models.

It should be stressed that the indispensable role of VE and PC in the tissue constitutive formulation is not unique to structural modeling. Tong and Fung [81] attempted to apply a non-linear hyper-elastic phenomenological model to the same skin biaxial data [13] used in the aforementioned structural study [14] and showed that the model fit well to the data of each individual protocol. However, when the constants estimated in one experiment under a specific preconditioned protocol, were used to predict the stress-strain response of the same

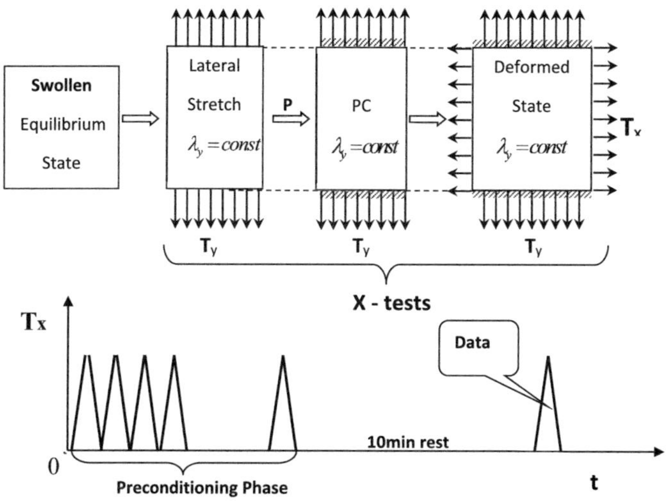

Fig. 11 The biaxial stretch protocol of a flat skin rectangular specimen [13]. In the X-tests (*upper panel*) the rectangular sample was first stretched in the lateral (Y) direction to a constant stretch, then preconditioned by repeated stretch at constant rate of stretch in the main direction (X in this test) to a set stretch level. The sample was then allowed to rest for 10 min and was then stretched to same level while data was recorded (*lower panel*). A similar protocol was applied in the Y-tests. From [14] with permission

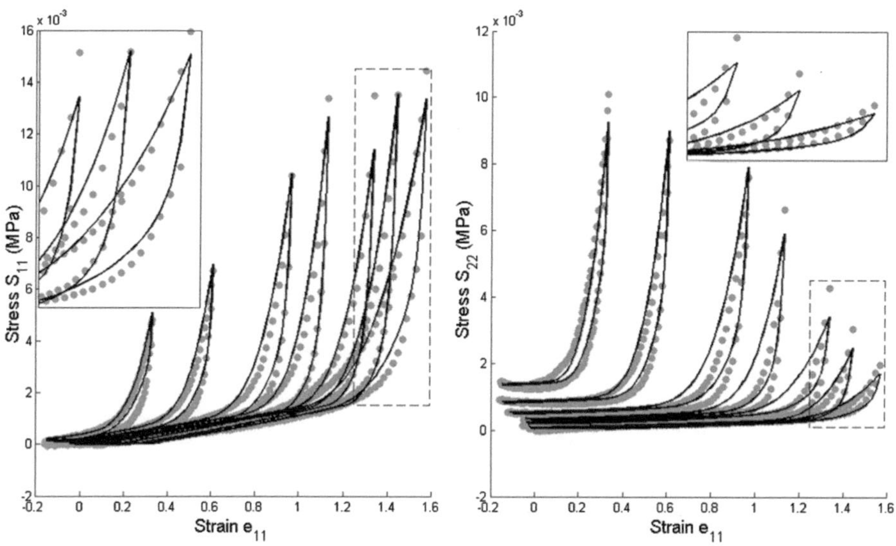

Fig. 12 Descriptive performance of the combined VE & PC biaxial structural model [14]. Comparison of skin 2D stress data under X-stretch at various fixed Y-stretch levels (*grey dots*) to model predictions (*black lines*). *Left panel—main* (X) stress, *right panel—lateral* (Y) stress responses. Predictions fit the data well for all combinations of biaxial stretches. From [14] with permission

specimen under another preconditioned protocol, success was not uniform. They concluded that since the estimated model parameters changed from one experiment to another, the tissue is not elastic. Similar inter-protocol parameter variations were observed in a variety of soft tissues (e.g., [82–86]) again demonstrating that they are not elastic.

4 Structural Modeling of Active Tissues

Multi-scale structural modeling can readily incorporate active contraction of muscular fibers and cells. The active muscle properties are subject of most extensive research published in many journals and a large number of dedicated books. In spite of substantial progress over several decades, there are still no generally accepted models for all muscle types. In general, the active muscle force depends on its length, contraction velocity, time of activation, and muscle contractility (inotropic state). But the specifics vary greatly between both the three muscle types and locations in the body. A review of this vast topic is outside the scope of the present report. Instead, the approach here is to use the framework of the structural theory, assume that the active force model of a given single muscle cell is known, and indicate the generic manner by which it is integrated and summed up with the contributions of the passive constituents in the muscle to produce the total muscle stress.

The myocardium serves here as an example. Reviews of myocardial modeling in general, and on application of constitutive models in three dimensional stress analysis of the heart, have been presented in Yin [87] and more recently by Glass et al. [88] and Schmid and Hunter [89].

Activated muscle fibers (such as in skeletal muscle) apply contraction force in their axis direction. Muscle cells (myocytes and smooth muscle cells) contract along the direction of their sarcomere which usually coincides with their long axis.

4.1 Myocardium Structure

The myocardium is a fibrous tissue in which the myocytes occupy about 70 % of the volume while the remaining ∼30 % is occupied by the fluid matrix (∼20 %), other types of cells (e.g., fibroblasts) and other interstitial components [90]. Collagen fibers occupy only about 5 % of the weight [91] but are an important determinant of the tissue passive properties [92]. Light microscopy (LM) images showed that the left ventricle (LV) wall structure is both locally anisotropic and heterogeneous across the wall thickness. The myocytes are elongated cells arranged in a layered disposition and parallel to each other with a circumferential angle (Γ in Fig. 13) which varies gradually from the inner endocardium to the outer epicardium wall surfaces [93]. Longitudinally, myocytes are interconnected via dense intercalated discs which provide the cells with mechanical, electrical and chemical inter connections, thereby

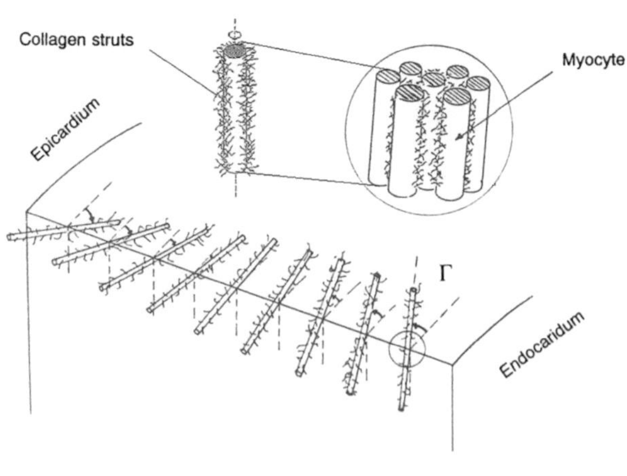

Fig. 13 Scheme of the parallel myocytes orientation distribution across the myocardial wall and of the transverse collagen network interconnecting neighboring myocytes (*inset*)

forming fiber-like structures and facilitating functional integrity as a syncytium. Scanning electron microscopy (SEM) images demonstrated that the collagen component forms both crisscross weaves which surround individual or groups of myocytes, and a dense array of transverse collagen struts which interconnect contiguous myocytes [94–97], thereby maintaining structural integrity of the myocardium.

4.2 Mechanics of the Passive Myocardium

Based on LM and SEM observations [93, 94, 96–98], a hyper-elastic structural model was developed for the passive myocardium [99, 100]. It contains an ensemble of REVs, each representing a local structural unit consisting of a single myocyte, symmetrically interconnected with its neighbors by a 3D network of initially undulated transverse collagen fibers (Fig. 13). In addition, each REV incorporated a system of passive fibers running parallel to the direction of the myocytes, representing the longitudinal contribution of the crisscross fibrous weaves which surround the myocytes, combined with that of the passive inter-myocytes titin filaments which run along the sarcomeres and the myocytes long axes [101]. Structural characterization of the myocardial wall entails specification of the orientation distribution of the transmural myocytes [93], the orientation distribution of the transverse collagen fibers relative to their REV myocyte to which they are connected, and the recruitment distribution of the collagen fibers. With these structural attributes specified, the strain field in each REV across the wall and the strain in each collagen fiber in the REV system were evaluated [99, 100] from the global myocardial strain field, based on affine deformation (Assumption B5). The passive stress in each fiber was then calculated from its strain. By the rule of mixtures, the sum of passive contributions of all transverse and parallel fibers was calculated and assigned as the overall myocardial passive fibers stress.

4.3 Mechanics of the Active Myocardium

The active stress of the myocardium results from and is equal to the sum contributions of all its myocytes. In the framework of structural modeling, the law governing the development of the active myocyte tension is assumed to be known and is a given input to the model. The active tension in each myocyte acts along its length. Stress summation across all myocytes is performed after each myocyte stress is mapped to a common global coordinate system. The resultant global active stress is then added to the global passive stress and the matrix pressure to yield the total myocardium stress.

Myocyte activation is a complex event affected by a number of determinants. The generated active force depends on external determinants (length, velocity and electric excitation), as well as on internal factors such as the kinetics of both the actin-myosin cross-bridges and of troponin and Ca^{+2} ions, which control cross-bridge attachment, and on cellular control mechanisms. A number of models have been proposed, each highlighting different aspects of the activation event. However, regardless of the model chosen, the generality of the structural approach allows any model to be integrated into the stress analysis. As a simple example, an active stress law of the form

$$\sigma_{mus}^{act}(\tau, t) = \sigma_{\max} \cdot T(\tau) \cdot L(\lambda_{mus}) \cdot V(v) \tag{4.1}$$

was used [100] in LV stress analysis along the entire cardiac cycle (diastole and systole), where $\sigma_{\max}$ is the highest possible active stress under optimal muscle stretch, $T(\tau)$ is an activation function of the time τ passed since the myocytes were activated, $L(\lambda_{mus})$ is a function of muscle stretch, representing the dependence of the stress on the sarcomeres actin-myosin overlap, and $V(v)$ is a velocity function which reflects the hyperbolic relationship between active stress and the shortening velocity [102]. Equation (4.1) was applied to each sarcomere, then mapped to the global myocardium coordinate system and summed up across all REVs. Details on the realization and comparison of the model predictions to the LV pressure-volume data along the entire cardiac cycle are given in reference [100].

5 Multi-scale Modeling of Tissue Growth and Remodeling (G&R)

Living tissues are unique in their ability to adapt to altered mechanical environments by changing their size, shape, structure, and mechanical properties. This adaptation, termed Growth and Remodeling (G&R), occurs not only during growth, but also in adulthood in response to physical activity, pregnancy, dietary changes and pathologies, such as hypertension. There is strong experimental support to the notion that G&R results from the tissue cells' strive to maintain a homeostatic mechanical environment required for survival and proliferation (e.g., [29]). The cells do so by adapting their extra-cellular fibrous matrix (ECM) via turnover—degradation of some fibers and production and deposition of new ones.

G&R models are useful not only in providing a coherent framework and facilitating insight into the underlying processes occurring in health and under pathological conditions, but also due to their key role in tissue engineering in design and fabrication of biological constructs. For example, mechanical conditioning was found not only to stimulate matrix production but also to be a major determinant of the construct evolution towards target structures and properties [103–105].

Previous G&R models can be classified into three categories: early phenomenological models focused on the consequences of G&R, based on classical continuum mechanics. In arteries for example, G&R is manifested globally by changes in diameter and wall thickness in an attempt to maintain homeostatic wall hoop strain and luminal wall shear stress [106–109]. The constraint mixture approach was later developed [110] to facilitate separate monitoring of the evolution and contribution to the global response of each of the tissue constituents. Structure-motivated models considered the G&R-induced evolution of specific structural features such as the helical angle of collagen fibers in arterial media. Other tissue constituents (such as the media elastin and matrix) were phenomenologically modeled. A common characteristic of previous G&R models is their ad-hoc assumptions as to the phenomenology of G&R-associated geometrical or structural alterations. Alternatively, since tissue properties derive from their constituents, it is advantageous to analyze G&R in terms of the specific turnover processes experienced by each constituent in the tissue space, based on the constituent's loading history and its specific mechano-biological nature (the mechanistic approach [111]). The structural multi-scale framework presented here is highly suited for this goal, since it incorporates and monitors the mechanical environment and resulting mechano-biological turnover of each constituent, everywhere in the tissue space, and then integrates the effects of all these local events to predict the evolution of the tissue structure and associated properties. To demonstrate this approach, the case of tissues with one type of uni-directional bundle of fibers (e.g., collagen in tendons and ligaments) will be formulated in detail [112]. This will be followed by generalization to tissues with multiple types of fibers.

5.1 Basic Assumptions of the Structural G&R Theory

The multi-scale tissue G&R theory is based on similar assumptions that underlie the multi-scale theory of inanimate tissue mechanics, supplemented with the following mechano-biological assumptions regarding constituent turnover:

Assumption GR1 The characteristic time of tissue loading (seconds to hours) is several orders of magnitude shorter than the characteristic time of G&R-related biological processes (weeks to months [111]). Hence at any G&R time, the tissue response to short loading can be analyzed independent of ongoing G&R processes.

Assumption GR2 Fibers turnover occurs through degradation of some fibers while others are synthesized and deposited.

Assumption GR3 The biological processes of fiber degradation and synthesis are first-order reactions.

Assumption GR4 The reaction rates are specific to each constituent and have both loading-independent (basal) and loading-dependent components.

Assumption GR5 Synthesized fibers are deposited on extant fibers, under a stretch within the homeostatic range.

Assumption GR6 During G&R the fibers and matrix volume fractions are maintained at constant levels [70, 113].

The detailed theory formulation is laid down in reference [112]. A brief review of the main features is presented here.

5.2 The Tissue and Fiber Kinematics

The growth-induced evolution of the tissue structure, combined with tissue swelling under the matrix pressure, introduce multiple reference and loaded configurations, which are interrelated by corresponding mappings (Fig. 15). During G&R, the uniaxial tissue (Fig. 14) is progressively stretched from its original unloaded reference length $L_{tiss}^{UL}(0)$ to its growth induced length $L_{tiss}^{GR}(t)$ by a growth protocol $\Lambda_{tiss}^{GR}(t)$ which is an input to the model. The fiber bundle is simultaneously stretched from its original stress-free reference length $L_{bun}^{SF}(0)$, by an associated bundle stretch $\Lambda_{bun}^{GR}(t)$. Due to fiber turnover, growth induces changes in both the unloaded tissue and stress-free bundle reference configurations ($L_{tiss}^{GR}(t)$ and $L_{bun}^{SF}(t)$, respectively). The ratio between the tissue growth-induced length $L_{tiss}^{GR}(t)$ and its remodeling-induced unloaded reference length $L_{tiss}^{UL}(t)$ is the pre-stretch $\Lambda_{tiss}^{PS}(t)$. Finally, at each configuration during growth the tissue and bundle lengths are interrelated by the corresponding osmotic induced tissue swelling Λ_{tiss}^{osm}.

Importantly, of the several mappings between the various tissue and bundle configurations, only two are given inputs to the model, namely the tissue growth protocol $\Lambda_{tiss}^{GR}(t)$ and its short testing protocol $\Lambda_{tiss}^{test}(\tau)$. The initial tissue swelling $\Lambda_{tiss}^{osm}(0)$ is determined from the assumed initial bundle configuration and the matrix pressure. The other mappings ($\Lambda_{tiss}^{GR\ osm}(t)$, $\Lambda_{tiss}^{ref\ osm}(t)$, $\Lambda_{tiss}^{PS}(t)$, $\Lambda_{bun}^{PS}(t)$) can be evaluated only after the G&R-remodeled

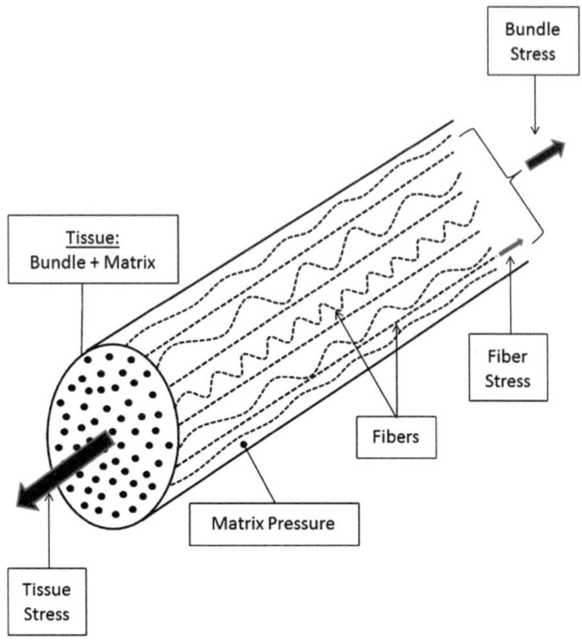

Fig. 14 A scheme of the uniaxial tissue structure with its fluid-like matrix and non-uniformly undulated fibers. Only stretched fibers are stressed. From [112] with permission

bundle stress-free configuration $L_{bun}^{SF}(t)$ has been determined from the turnover-induced evolution of the bundle structure.

The various tissue and bundle mappings are interrelated by multiplicative decomposition only when the tissue configuration (structure and composition) is fixed. For example, the bundle and tissue test stretches are inter-related by (Fig. 15):

$$\Lambda_{bun}^{test}(t^*, \tau) = \Lambda_{tiss}^{test}(t^*, \tau) \cdot \Lambda_{tiss}^{ref\ osm}(t^*), \tag{5.1}$$

where $\Lambda_{tiss}^{ref\ osm}(t^*)$ is the tissue swelling under the matrix pressure in the remodeled configuration at the test time t^*. In contrast, tissue configurational changes which include growth cannot be prescribed by multiplicative decomposition of the deformation gradients since growth itself, which involves mass changes, cannot be represented by a corresponding deformation gradient as $\mathbf{F}^{\mathbf{G}} = \partial \mathbf{x}^{\mathbf{G}} / \partial \mathbf{X}^{\mathbf{G}}$ [114]. Such representation implies that the coordinates of each material point prior to ($\mathbf{X}^{\mathbf{G}}$) and following ($\mathbf{x}^{\mathbf{G}}$) growth are inter-related by bijective motion (one-to-one and onto). Bijective motion is not valid for G&R during which some fibers are degraded and new ones are produced since such fibers do not have both $\mathbf{X}^{\mathbf{G}}$ and $\mathbf{x}^{\mathbf{G}}$ coordinates.

The bundle fibers are variably stretched with the tissue depending on each fiber "birth" length l_f^b, or its nominal birth stretch $\lambda_f^b = l_f^b / \Lambda_{bun}^{SF}(0)$. The birth stretch determines the true fiber stretch (which in turn determines its tension and its contribution to the tissue stress). The fiber true stretch is given by

$$\lambda_f(\lambda_f^b, t) = \Lambda_{bun}^{nom}(t) / \lambda_f^b. \tag{5.2}$$

Hence, although the bundle and its fibers global deformations are affine, the fibers true deformations are not.

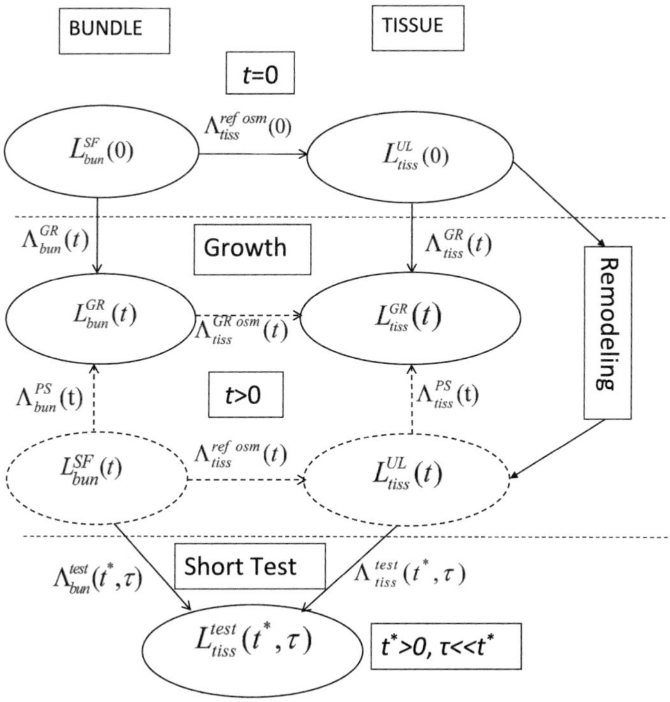

Fig. 15 A scheme of the evolving bundle and tissue stress-free (SF), unloaded (UL), growth-induced ($G\&R$), and test-loaded (test) configurations and of the mappings between them. Λ_{tiss}^{*osm} designate the tissue swelling under the matrix osmotic driven pressure. L_{bun}^{GR}, L_{tiss}^{GR} stand for the bundle and tissue growth induced dimensional changes, and $L_{bun}^{SF}(t)$, $L_{tiss}^{UL}(t)$ are the remodeled bundle stress-free and tissue unloaded configurations, respectively. Λ_{bun}^{PS}, Λ_{tiss}^{PS} are the bundle and tissue growth induced pre stretches. Λ_{bun}^{test}, Λ_{bun}^{test} are short test protocols aimed to monitor the evolution of the mechanical properties

5.3 Volume and Mass

The total fiber and matrix volumes are respectively $V_f(t)$ and $V_{mat}(t)$. The corresponding volume fractions are $\phi_f(t) = V_f(t)/V(t)$ and $\phi_{mat}(t) = V_{mat}(t)/V(t)$, where $V(t)$ is the total tissue volume. Since there are no voids (Assumption B2), then $V(t) = V_f(t) + V_{mat}(t)$, and $\phi_f(t) + \phi_{mat}(t) = 1$. Since the volume fractions remain constant during G&R (Assumption GR6 [70, 113]), it follows that $\phi_f(t) = \phi_f(0) \equiv \phi_f^0$, and likewise for the matrix.

The evolving distribution of fiber mass over its birth stretch is $m(\lambda_f^b, t)$ such that at time t the mass of fibers with birth stretches between λ_f^b and $\lambda_f^b + d\lambda_f^b$ is $m(\lambda_f^b, t) \cdot d\lambda_f^b$. The total fiber mass is therefore $M_f(t) = \int_0^{+\infty} m(\lambda_f^b, t) \cdot d\lambda_f^b$. The distributions of fibers mass and volume are interrelated by $m(\lambda_f^b, t) = \rho_f \cdot v(\lambda_f^b, t)$ where ρ_f is the fiber mass density.

5.4 Turnover Kinetics

Both fiber degradation and production are first-order reactions where each has deformation-dependent and deformation-independent components (Assumptions GR3 and GR4). The corresponding equations for the degradation are

$$\frac{\partial^{dd} m_f}{\partial t}\left(\lambda_f^b, t\right) = -B^{dd}(\lambda_f)\cdot m_f\left(\lambda_f^b, t\right), \qquad \frac{\partial^{db} m_f}{\partial t}\left(\lambda_f^b, t\right) = -B^{db}\cdot m_f\left(\lambda_f^b, t\right), \quad (5.3)$$

where $\partial^{dd}/\partial t$ and $\partial^{db}/\partial t$ are respectively, the deformation-dependent and deformation-independent degradation time derivatives. $B^{dd}(\lambda_f)$ and B^{db} are the corresponding reaction rate constants. Similar equations hold for fiber production with corresponding rate constants $B^{pd}(\lambda_f)$ and B^{pb} where the indices *pd* and *pb* designate respectively, the deformation-dependent and basal production rate constants.

5.5 Fiber and Bundle Mechanics

Cauchy stress in a single hyper-elastic fiber is $\sigma_f(\lambda_f)$, where the fiber true stretch at time t $\lambda_f(\lambda_f^b, t)$ depends on its birth stretch (Eq. (5.2)). In the model realization, the fiber stress-strain relationship is taken to be linear, i.e., $\sigma_f = K(\lambda_f - 1)$, where K is the stiffness. The bundle Cauchy stress Σ_{bun} is the weighted sum of all the fiber contributions

$$\Sigma_{bun}(t) = \left(\Phi_f(t)/M_f(t)\right)\cdot \int_0^{+\infty} m_f\left(\lambda_f^b, t\right)\cdot \sigma_f(\lambda_f)\cdot d\lambda_f^b, \quad (5.4)$$

where $\Phi_f(t)(=\Phi_f^0)$ transforms the stress from its level per bundle unit area to its level per tissue unit area. By the rule of mixtures, the total tissue Cauchy stress is

$$\Sigma_{tiss}(t) = \Sigma_{bun}(t) - P_{mat}. \quad (5.5)$$

Equation (5.4) cannot be used as it stands, since at a given growth time t the fibers' birth stretches and their distribution are unknown. The experimentally measurable quantity at growth time t is the distribution of the fiber recruitment stretches $D(x,t)$, which plays a key role in the multi-scale modeling of inanimate tissue mechanics presented above[5]. It can be readily shown [112] that the distributions $m(\lambda_f^b, t)$ and $D(q,t)$ are interrelated by

$$D(q,t) = \Lambda_{tiss}^{ref}\cdot m_f\left(\lambda_f^b, t\right)/M_f(t). \quad (5.6)$$

5.6 Feasibility of the Multi-scale Structural G&R Theory

To demonstrate the feasibility and utility of the multi-scale structural G&R theory, a specific case [112] of a uniaxial tissue whose linear elastic fibers were initially straight and unstretched, was analyzed. The tissue was subjected to a growth protocol of a constant stretch rate of 0.5 per year for a period of two years. Kinetic and mechanical parameters values were adopted from published data. Figure 16 depicts the predicted evolution of the tissue structure in terms of fiber recruitment stretch distribution. The initial uniform distribution evolves rapidly to a bell shaped form, with no trace of the initial uniform distribution after approximately 5 % of the total growth time. The bell shaped recruitment stretch distribution is a commonly observed feature of soft tissues [53, 115–118]. The predicted tissue stress-strain response evolves from an initial linear relationship of the uniformly stretched fibers to a convex non-linear response, in which the tissue stiffens with strain (Fig. 17). Again, this feature is common to all soft tissues. A third observed characteristic is the onset of pre-strain and associated pre-stress which evolve with growth (Fig. 18).

[5] $D(q,t)$ has the same physical meaning at a given growth time t as the function $D(q,\mathbf{N})$ in Eq. (2.5) in a uniaxial fiber network (no dependence on $\mathbf{N}$).

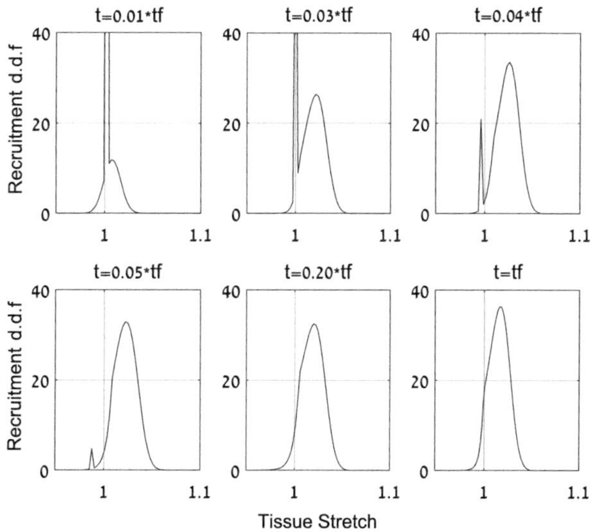

Fig. 16 Evolution with growth time of the fibers recruitment stretch density distribution. Fibers having recruitment stretch lower than unity are already stretched in the unloaded length thereby balancing the matrix pressure. The symbol "*tf*" stands for the total growth period duration. From [112] with permission

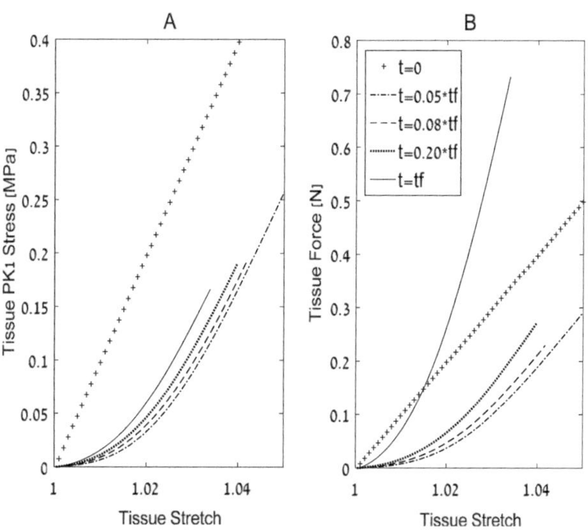

Fig. 17 Evolution of the tissue (fibers and osmotic active matrix) stress-strain relationship at selected growth times of $t = 0(+)$, $t = 0.05 * t_f$, $t = 0.08 * t_f$, $t = 0.20 * t_f$, $t = t_f$ (t_f being the duration of the growth period). (**A**) The tissue first Piola-Kirchhoff stress as function of its stretch. (**B**) The corresponding tissue total force. The initial linear response (at $t = 0$) of the uniformly straight fibers, evolves with growth time to convexly non-linear ones. From [112] with permission

5.7 Evolutionary Determinants of Tissues Structure and Mechanics

Scientifically, it is both intriguing and of potential practical importance to determine if the evolutionary attributes of tissue structure and mechanics which are depicted in Figs. 16–18 are genetically determined. The results presented here show that these characteristics could well evolve as a result of the mechano-biology of the tissue constituents. It is shown that based solely on the turnover processes of the constituents, these three key characteristics of structure and mechanics evolve with growth towards their observed patterns.

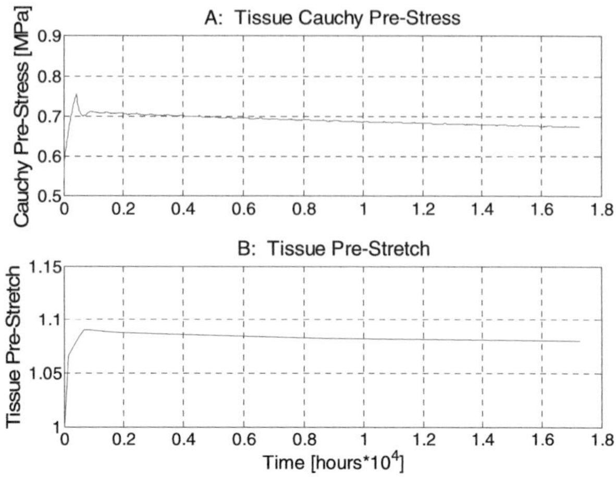

Fig. 18 Evolution with growth time of the in situ tissue pre-stress (**A**) and pre-stretch (**B**). From [112] with permission

6 State of the Art, the Big Picture, Future Challenges

To summarize, an account will be given of the main merits and limitations of the multi-scale structural tissue modeling. This will be followed by a personal outlook regarding tissues characterization, along with future research challenges and prospects in the field.

6.1 Merits of the Multi-scale Structural Approach

The structural theory is of a physical nature, in the sense that it simulates real processes occurring in realistic tissue structures. The theory parameters have clear physical meaning and can often be measured independently using methodologies unelated to tissue mechanical testing (e.g., testing of individual fibers, histological staining and ultrastructural imaging). The structural approach can readily incorporate contributions of tissue constituents which are very different in their nature (e.g., fluid matrix, passive and active solid fibers) and integrate all into a clear and precisely orchestrated global tissue response. Due to its physical nature, the theory is free of the problem of indistinguishability between candidate non-linear models which is often encountered in non-linear phenomenological characterization. Moreover, and probably for the same reason, experience has shown that the inter-specimen variance between parameter estimates of structural models is significantly lower than in phenomenological models. Another useful feature of structural models is their aforementioned guaranteed physical plausibility. In terms of reliability and fit to data, detailed comparison [65] revealed that structural models are superior to non-structural models both as descriptors and predictors of 3D response characteristics of coronary arteries media. Most importantly, structural models have the unique ability to expand by smoothly incorporating additional tissue constituents with different properties, or more complex properties of constituents, such as fiber non-linearity, time-dependence, active contraction and mechano-biological turnover. It was shown above that perplexing dilemmas in tissue characterization, such as the apparent incompatibility between viscoelastic stress relaxation and creep responses [77, 78], the ranges of validity and interplay between preconditioning and viscoelasticity [78], and the kinematic limitations and physically unrealistic response predictions occasionally encountered with nonstructural models [60, 119], can be readily resolved using structural modeling.

6.2 Limitations of the Multi-scale Structural Approach

It has been claimed that implementation of the structural approach in finite element stress analysis, presents an excessively heavy computational load due to the need to integrate the contributions of fibers over their entire recruitment stretch distribution and 3D orientation dispersion. Various attempts have been made to circumvent this problem but, so far, none have proven to reliably represent tissue response characteristics under any general loading (see discussion below in "Reduced Structure Representation"). On the other hand, in practice, with today's powerful fast mainframes and parallel coding interfaced with computer clusters, the computational load is tractable, as shown for the complex case of 3D finite element analysis of the LV with passive and active myocardium in a real geometry model [120].

Another concern raised is that by assuming deformation affinity, interactions between tissue fibers are not considered in structural models [3]. Two recent studies directly addressed this issue by experimentally testing deformation affinity [52, 58]. The results show that in the flat tissues tested (pericardium, aortic valve leaflet and mitral valve anterior leaflet), the fiber kinematics was affine. As the authors correctly conclude, affinity suggests that the fibers are largely non-interacting and deform with the bulk tissue.

6.3 The Big Picture

The research in tissue mechanics has been dominated in the past almost exclusively by phenomenological models within the powerful frame of classical continuum mechanics of bulk materials, using concepts and paradigms adopted from this discipline. Biological aspects were often adapted to fit continuum framework. Phenomenological models served as a starting point and contributed significantly to progress in tissue mechanics. However, in recent years, it is becoming clear that soft tissue characterization requires a frame of thinking that transcends the scope of continuum mechanics. The paradigm should be to model tissues based on the real structures and biological processes occurring in the tissue system, and to link the resulting formulation to continuum mechanics.

A number of reasons underlie this need for paradigm change. First, tissues are multi-scale, multi-phasic structures, exhibiting complex time-dependent responses. In addition, tissues are thermodynamically open, rather than closed systems. In the living tissue, mass, energy and information are constantly exchanged with the environment, and with the whole body. Muscle tissues are excitable and develop active contractions. All *in vivo* tissues change via biological processes, to adapt to their external environment. Hence, tissues do not fit the standard requirements of continuum mechanics. For instance, classical conservation laws are only valid in living tissues for short time periods and under passive states, but are generally violated. For one or more of these reasons, phenomenological models have proven to be limited in their ability to reliably represent the complex properties of live tissues under general conditions and deformation schemes.

It is thus submitted that future prospects in tissue biomechanics research requires a new paradigm. For tissues *in vitro* much progress has been achieved by moving from phenomenology to structure-function and casual connections. In live tissues, there is a need to expand the outlook to bodies which remodel via biological processes. These changes will require massive experimental effort to elucidate hitherto unknown features of tissue mechano-biology and to develop testing protocols suitable for reliable model validation. While preliminary steps have been taken, much remains to be done.

From the modeling aspect, the multi-scale structural approach presented here is a platform suitable for realizing this new paradigm. On the one hand (as shown above), it can incorporate biological effects and resulting changes in mass, structure and properties. On the other hand, it bridges between the mechano-biological processes and continuum mechanics in a coherent manner. It thus affords clearer and richer insights and opens new possibilities in a significantly wider horizon compared to continuum models of inanimate materials.

6.4 Future Challenges

The realism and versatility of multi-scale tissue characterization pave the way for future research challenges, both theoretical and experimental, several of which are presented here.

Experimental Structural Characterization: Tissue structure is the cornerstone of the structural approach. In the absence of pertinent information, previous structural characterizations were for the most part indirect. Based on qualitative histological observations, general forms of orientation and waviness distributions (such as the beta function) were assigned to both features, and parameters of these distributions were estimated to best fit given mechanical response data. Attempts to directly determine the distribution functions based on morphological observations, have been impeded by difficulties in reliable extraction of the tissue 3D structural features, and by the inherent experimental uncertainties associated with the need to process the specimen, or by the use of differential digestion, thereby distorting the tissue's micro-structure. Recent developments in non-linear optics (e.g., multi-photon microscopy and coherent anti-Stokes Raman scattering microscopy), coupled with computer-interfaced sequential optical sectioning, hold great promise for 3D structural characterization. The advantage of these methodologies is that tissue samples are maintained in their native state (i.e., unstrained and unprocessed). In addition, the application of two types of nonlinear laser/tissue interaction—two photon excited fluorescence (TPEF) and second-harmonic generation (SHG),—allow for retrieval of distinct structural data for the elastin and collagen type I fibers [121, 122], respectively.

Reduced Structure Representation: Implementation of the structural approach in finite element stress analysis presents a heavy computational load due to the associated need to integrate the contributions of fibers over their entire 3D orientation distribution function (also termed "integration on a sphere"). Although modern fast computing facilities render this task tractable (e.g., in the case of the left ventricle [120]), attempts are being made to develop a computationally more efficient representation. Although promising first steps have been made, a methodology which is general enough to incorporate any tissue structure and any fiber material law is still lacking. The quest for such representation is a challenging theoretical/numerical undertaking. One current approach relies on application of generalized structural tensor to represents the tissue's 3D structure [123, 124], instead of the fiber orientation distribution. Higher computational efficiency was achieved with the structure tensor since in application, it was multiplied by a weighted average of the fiber stretches, thus ignoring dependence of the latter on the fiber orientation. Comparison with the exact solution revealed that this methodology is only valid when all fibers are in tension and when the fiber orientation span is small [125, 126]. Recently, another drawback of the structural tensor was revealed. It turns out that except for the case of uniform or close to uniform orientation dispersion, the distributions recovered from the structure tensor fit poorly to the data. This stems from the fact that the second rank generalized structure tensor is associated with just the lowest ranked term in a Fourier series expansion of the orientation distribution based on spherical functions, and it contains the least amount of information on the distribution compared to higher order terms in the expansion [51].

A second class of methodologies used to reduce the structural representation, attempts to simulate the tissue properties by a discrete set of fiber-bundles. One method was developed for amorphous polymers, and represents the tissue properties by an eight-chain, three-dimensional rectangle unit cell, whose shape determines the tissue orthotropy [127]. While this unit cell is a structure, it is unrelated to the real tissue structure. The two other methods represent the tissue's continuous 3D orientation distribution by a discrete set of fibers. In the first, the fibers in the set have fixed orientations, and weight is assigned to each set to optimize the fit to the tissue response [128]. This approach was shown, however, to introduce undesirable anisotropy to an isotropic material [129, 130]. The second variant applies the spherical t-design [131, 132], which is a set of N points on a sphere (or their equivalent fiber orientation), such that the average value of any polynomial f of degree t or less on this set equals the integral of f over the sphere. Hence, integration over the 3D orientation distribution is replaced by a sum of discrete values of that function. The design degree t depends on the orientation distribution function and on the fiber material law [133]. Hence, the polynomial degree t must be evaluated for each case [134] and may change as parameter estimates evolve during iterative search for optimal parameters during the process of tissue material characterization.

To summarize, at present, there seem to be no reliable reduced representation of tissue structure which could serve as an alternative to the orientation distribution function. This challenge is yet to be met.

Clinical Applications of the G&R Model: A number of pathologies are closely related to a remodeling transition from stable healthy state into an unstable diseased state. Two examples are cardiac hypertrophy (CH) and abdominal aortic aneurism (AAA). Untreated maladaptive CH leads to heart failure which is a leading cause of morbidity and has a 5-year mortality rate of 50 %. CH is a response to stress overload or to a disease. There are two types of CH. Eccentric hypertrophy is associated with increased chamber volume without change in the wall thickness, and can result from myocardial infarct (MI) damage to the myocardium or from mitral valve regurgitation. Structurally, increased chamber volume results from remodeling of the collagen network of the heart (which is predominant in its passive diastolic function), and an associated increase in the number and length of sarcomeres in series, thus leading to myocytes lengthening. The second type is concentric hypertrophy, which is associated with increased wall thickness and may arise from increased systolic ventricle pressure under hypertension or from aortic valve stenosis. Structurally, the wall thickens due to over-stress-induced increased in the number of parallel sarcomeres in the myocytes. Clinically, it is well known that mild to moderate left ventricle hypertrophy such as observed in training athletes, need not be pathological and may actually enhance cardiac performance (adaptive hypertrophy). The challenges are to identify the mechanical and biological culprits which lead to a pathological transition, and to establish the clinical threshold of transition from adaptive to maladaptive hypertrophy. Multi-scale structural characterization provides a predictive modeling platform for myocardial G&R that is realistically based on the physical tissue constituents and incorporates their measured mechano-biology into predictable and measurable function, thereby facilitating reliable clinical decisions.

AAA is an enlargement of the aorta, which result from deterioration of the elastin network in the vessel wall. The elastin function in resisting the dynamic luminal blood pressure can be compensated by increased production of the other wall constituents (SMCs and collagen). The important clinical goal is to establish the thresholds of transition from a stable into an unstable state, which may lead to dissection and life-threatening massive internal bleeding. Here again, multi-scale modeling provides a straightforward platform for incorporation of measurable mechano-biological features of tissue constituents and their integration into global clinical measures.

Cellular Mechano-Biology: Cells are constantly exposed to mechanical forces in the body. In fact cells require a homeostatic mechanical environment in order to survive and proliferate [29]. Forces affect and control cell shape and biological functions, including gene expression, metabolism, secretion of biological factors, apoptosis (programed cell death), and modification of their ECM structure and properties, via turnover of fibers. Cells are exquisite sensors of their mechanical environment. Depending on the species, tissue and cell type, cells are sensitive to the nature of applied forces (tension/compression, pressure, shear), and to their magnitude, waveform and duration. For example, proteoglycan synthesis in chondrocytes (cells of the articular cartage) was found to be suppressed by static compression of up to 50 % strain, but cyclic compression has varied effects, depending on the loading regimen [135–138]: dynamic loading of 15 % at 0.3 Hz of chondrocytes seeded in agarose, decreased GAG synthesis, while under 1 Hz frequency increased it [139]. Many functions of endothelial cells (cells lining the blood vessel inner surface) are regulated by the blood shear stress [140]. Cyclic stretch was found to enhance the proliferation of human myocytes seeded in a gelatin matrix [141], and to increase elastin production by SMCs seeded in polymeric scaffold [142].

Cells mechano-biology is of fundamental importance in tissue engineering (TE) which aims to replace diseased native tissues with *in vitro* generated functional biological constructs that possess suitable structure and mechanical properties. Yet, a major obstacle to production of engineered tissues is the limited understanding of the mechanics-biology interrelationships between forces and cells' functions. The specific forces and regimes (magnitude, frequency, duration and waveform) which induce specific biological response are insufficiently understood. Comprehensive understanding can be achieved through well controlled experiments and realistic computational analysis of the mechanical forces experienced by the cells.

Design and Fabrication of Engineered Tissue: The goal of tissue engineering is to develop biological substitutes that restore, maintain, or improve function of native tissues, or of a whole organ [143]. Generation of an engineered tissue featuring specific structure and mechanical properties is an endeavor complementary to cell mechano-biology research. Knowledge gained from the research can be integrated into G&R-related modeling and design of the requisite materials and bioreactor protocols, which should lead to evolution of *in vitro* manufactured constructs to target engineered tissues. While the concepts and frameworks of this emerging sub-discipline are just beginning to be formulated, it is reasonable to expect that the multi-scale G&R structural theory presented above, will play a central role due to its versatility and realism in simulating the structure and biological remodeling that occur in the tissue substitute. With this achieved, the art of tissue engineering will be established on principles parallel to those of classical engineering design.

6.5 Concluding Comments

The research field of tissue mechanics is in its formative period. It is characterized by revolutionary transformation from the realm of classical continuum mechanics, typical of the early years of the field just four decades ago, towards its own new identity. New concepts and tools are being developed to cope with the special challenges presented by the complex behavioral features of tissues, concepts which were hitherto unknown in classical mechanics. The present period is reminiscent of the formative developments in 1950s for the Theory of Plasticity, in the 1960s for Fracture Mechanics, and in the 1970–80s for Composite Materials. The field is experiencing an ongoing merge of biology, anatomy, physiology, bio-chemistry and physical chemistry into concepts and tools of non-linear continuum and structural mechanics. This results not only in improved realism of tissue characterization, but also in expansion of scope and creation of new fronts for classical mechanics towards more general response features such as activation, remodeling, and changing body mass.

Research in tissue mechanics is both important and rewarding. It holds the promise for most significant potential contributions to community health, by deciphering underlying mechanisms of physio-pathological processes, thereby paving the way towards development of protective and therapeutic measures, anti-aging strategies, and tissue engineered substitutes. The research work itself is highly rewarding, offering new, challenging and rich experience, due to its inter- and multi-disciplinary nature. It incorporates and integrates non-linear mechanics, mathematics, physiology, biology and chemistry. It is not surprising to find that as a result of these new horizons, interesting challenges in mechanics and material characterization present themselves. The topics presented here are just a few examples.

Acknowledgement The comments and advice of Prof. David Durban are gratefully acknowledged. This research was supported by the National Institutes of Health (grant NIH R01 HL117990).

References

1. Fung, Y.C.: Elasticity of soft tissues in simple elongation. Physiol. Entomol. **213**(6), 1532–1544 (1967)
2. Lanir, Y.: Autobiographical postscript. In: Kassab, G.S., Sacks, M.S. (eds.) Structure-Based Mechanics of Tissues and Organs. Springer, New York (2016). pp. v–xxi
3. Humphrey, J.D.: From stress-strain relations to growth and remodeling theories: a historical reflection on microstructurally motivated constitutive relations. In: Kassab, G.S., Sacks, M.S. (eds.) Structure-Based Mechanics of Tissues and Organs, pp. 123–133. Springer, New York (2016)
4. Fratzl, P. (ed.) Collagen Structure and Mechanics, p. 508. Springer, New York (2008)
5. Ross, R., Fialkov, P.J., Altman, L.K.: The Moorphogenesis of Elasic Fibers. In: Sandberg, L.B., Gray, W.R., Franzblau, C. (eds.) Elastin and Elastic Tissue, pp. 7–17. Plenum Press, New York (1977)
6. Borg, T.K., Caulfield, J.B.: Collagen in the heart. Tex. Rep. Biol. Med. **39**, 321–333 (1979)
7. Borg, T.K., Caulfield, J.B.: Association of collagen struts with cardiac myocytes. J. Cell Biol. **87**(2), A126–A126 (1980)
8. Maroudas, A.: Balance between swelling pressure and collagen tension in normal and degenerate cartilage. Nature **260**(5554), 808–809 (1976)
9. Eckert, C.E., Fan, R., Mikulis, B., Barron, M., Carruthers, C.A., Friebe, V.M., Vyavahare, N.R., Sacks, M.S.: On the biomechanical role of glycosaminoglycans in the aortic heart valve leaflet. Acta Biomater. **9**(1), 4653–4660 (2013)
10. Emery, J.L., Omens, J.H., McCulloch, A.D.: Strain softening in rat left ventricular myocardium. J. Biomech. Eng. **119**(1), 6–12 (1997)
11. Fung, Y.C.: Stree-strain-history relations of soft tissues in simple elongation. In: Fung, Y.C., Perrone, N., Anliker, M. (eds.) Biomechanic—Its Foundations and Objectives, pp. 181–208. Prentice-Hall, Englewood Cliffs (1972)
12. Gregersen, H., Emery, J.L., McCulloch, A.D.: History-dependent mechanical behavior of Guinea-pig small intestine. Ann. Biomed. Eng. **26**(5), 850–858 (1998)
13. Lanir, Y., Fung, Y.C.: 2-dimensional mechanical-properties of rabbit skin. 2. Experimental results. J. Biomech. **7**(2), 171 (1974)
14. Lokshin, O., Lanir, Y.: Micro and macro rheology of planar tissues. Biomaterials **30**(17), 3118–3127 (2009)
15. Sverdlik, A., Lanir, Y.: Time-dependent mechanical behavior of sheep digital tendons, including the effects of preconditioning. J. Biomech. Eng. **124**(1), 78–84 (2002)

16. Chuong, C.J., Fung, Y.C.: On residual stresses in arteries. J. Biomech. Eng. **108**(2), 189–192 (1986)
17. Vaishnav, R.N., Vossoughi, J.: Estimation of residual strains in aortic segments. In: Hall, C.W. (ed.) Biomedical Engineering II: Recent Developments, pp. 330–333. Pergamon, New York (1983)
18. Lanir, Y., Hayam, G., Abovsky, M., Zlotnick, A.Y., Uretzky, G., Nevo, E., BenHaim, S.A.: Effect of myocardial swelling on residual strain in the left ventricle of the rat. Am. J. Physiol., Heart Circ. Physiol. **270**(5), H1736–H1743 (1996)
19. Omens, J.H., Fung, Y.C.: Residual strain in rat left ventricle. Circ. Res. **66**(1), 37–45 (1990)
20. Lanir, Y.: Mechanisms of residual stress in soft tissues. J. Biomech. Eng. **131**(4) (2009)
21. Lanir, Y.: Osmotic swelling and residual stress in cardiovascular tissues. J. Biomech. **45**(5), 780–789 (2012)
22. Fung, Y.C.: What are the residual stresses doing in our blood vessels? Ann. Biomed. Eng. **19**(3), 237–249 (1991)
23. Fung, Y.C., Liu, S.Q.: Strain distribution in small blood vessels with zero-stress state taken into consideration. Physiol. Entomol. **262** (2 Pt 2), H544–H552 (1992)
24. Brown, I.A.: A scanning electron microscope study of the effects of uniaxial tension on human skin. Br. J. Dermatol. **89**(4), 383–393 (1973)
25. Chu, B.M., Frasher, W.G., Wayland, H.: Hysteretic behavior of soft living animal tissue. Ann. Biomed. Eng. **1**(2), 182–203 (1972)
26. Viidik, A., Ekholm, R.: Light and electron microscopic studies of collagen fibers under strain. Z. Anat. Entwickl. Gesch. **127**, 154–164 (1968)
27. Butler, D.L., Grood, E.S., Noyes, F.R., Zernicke, R.F., Brackett, K.: Effects of structure and strain measurement technique on the material properties of young human tendons and fascia. J. Biomech. **17**(8), 579–596 (1984)
28. Aaron, B.B., Gosline, J.M.: Elastin as a random-network elastomer—a mechanical and optical analysis of single elastin fibers. Biopolymers **20**(6), 1247–1260 (1981)
29. Grinnell, F.: Fibroblasts, myofibroblasts, and wound contraction. J. Cell Biol. **124**(4), 401–404 (1994)
30. Eastwood, M., Porter, R., Khan, U., McGrouther, G., Brown, R.: Quantitative analysis of collagen gel contractile forces generated by dermal fibroblasts and the relationship to cell morphology. J. Cell. Physiol. **166**(1), 33–42 (1996)
31. Harris, A.K., Stopak, D., Wild, P.: Fibroblast traction as a mechanism for collagen morphogenesis. Nature **290**(5803), 249–251 (1981)
32. Pourati, J., Maniotis, A., Spiegel, D., Schaffer, J.L., Butler, J.P., Fredberg, J.J., Ingber, D.E., Stamenovic, D., Wang, N.: Is cytoskeletal tension a major determinant of cell deformability in adherent endothelial cells? Physiol. Entomol. **274**(5 Pt 1), C1283–C1289 (1998)
33. Kumar, S., Maxwell, I.Z., Heisterkamp, A., Polte, T.R., Lele, T.P., Salanga, M., Mazur, E., Ingber, D.E.: Viscoelastic retraction of single living stress fibers and its impact on cell shape, cytoskeletal organization, and extracellular matrix mechanics. Biophys. J. **90**(10), 3762–3773 (2006)
34. Fung, Y.C.: Biomechanics: Mechanical Properties of Living Tissues, 2nd edn. Springer, New York (1993). xviii, 568 p.
35. Murphy, R.A.: Mechanics of smooth muscle. In: Bohr, D.F., Somlyo, A.P., Sparks, A.V.J. (eds.) Handbook of Physiology, Sect. 2: The Cardiovascular System, pp. 325–351. American Physiological Society Bathesda (1980)
36. Bosse, Y., Solomon, D., Chin, L.Y., Lian, K., Pare, P.D., Seow, C.Y.: Increase in passive stiffness at reduced airway smooth muscle length: potential impact on airway responsiveness. Am. J. Physiol., Lung Cell. Mol. Physiol. **298**(3), L277–L287 (2010)
37. Herrera, A.M., McParland, B.E., Bienkowska, A., Tait, R., Pare, P.D., Seow, C.Y.: 'Sarcomeres' of smooth muscle: functional characteristics and ultrastructural evidence. J. Cell Sci. **118**(Pt 113), 2381–2392 (2005)
38. Seow, C.Y., Pratusevich, V.R., Ford, L.E.: Series-to-parallel transition in the filament lattice of airway smooth muscle. J. Appl. Physiol. **89**(3), 869–876 (1985). 2000
39. Speich, J.E., Almasri, A.M., Bhatia, H., Klausner, A.P., Ratz, P.H.: Adaptation of the length-active tension relationship in rabbit detrusor. Am. J. Physiol., Ren. Fluid Electrolyte Physiol. **297**(4), F1119–F1128 (2009)
40. Wang, L., Pare, P.D., Seow, C.Y.: Selected contribution: effect of chronic passive length change on airway smooth muscle length-tension relationship. J. Appl. Physiol. **90**(2), 734–740 (1985). 2001
41. Sacks, M.D.: In: Kassab, G.S., Sacks, M.S. (eds.) Finite Element Implementation of Structural Constitutive Models in Structure-Based Mechanics of Tissues and Organs, pp. 347–363. Springer, New York (2016)
42. Nguyen, T.D.: Biomechanics of the cornea and sclera. In: Kassab, G.S., Sacks, M.S. (eds.) Structure-Based Mechanics of Tissues and Organs, pp. 285–315. Springer, New York (2016)

43. Lee, L.C., Wenk, J., Klepach, D., Kassab, G.S., Guccione, J.M.: Structure-based models of ventricular myocardium. In: Kassab, G.S., Sacks, M.S. (eds.) Structure-Based Mechanics of Tissues and Organs, pp. 249–263. Springer, New York (2016)
44. Krishnamurthy, A., Coppola, B., Tangney, J., Kerckhoffs, R.C.P., McCulloch, A.D.: A microstructurally based multi-scale constitutive model of active myocardial mechanics. In: Kassab, G.S., Sacks, M.S. (eds.) Structure-Based Mechanics of Tissues and Organs, pp. 439–460. Springer, New York (2016)
45. Jor, J.W.Y., Babarenda Gamage, T.P., Nielsen, P.M.F., Nash, M.P., Hunter, P.J.: Relationship between structure and mechanics for membranous tissues. In: Kassab, G.S., Sacks, M.S. (eds.) Structure-Based Mechanics of Tissues and Organs, pp. 135–173. Springer, New York (2016)
46. Grytz, R., Meschke, G., Jonas, J.B., Downs, C.: Glaucoma and structure-based mechanics of the lamina cribrosa at multiple scales. In: Kassab, G.S., Sacks, M.S. (eds.) Structure-Based Mechanics of Tissues and Organs, pp. 93–122. Springer, New York (2016)
47. Gasser, T.C.: Histomechanical modeling of the wall of abdominal aortic aneurysm. In: Kassab, G.S., Sacks, M.S. (eds.) Structure-Based Mechanics of Tissues and Organs, pp. 57–78. Springer, New York (2016)
48. Cortes, D.H., Elliott, D.M.: Modeling of collageneous tissues using distributed fiber orientation. In: Kassab, G.S., Sacks, M.S. (eds.) Structure-Based Mechanics of Tissues and Organs, pp. 15–39. Springer, New York (2016)
49. Chen, H., Zhao, X., Lu, X., Kassab, G.S.: Microstructure-based constitutive models for coronary artery adventitia. In: Kassab, G.S., Sacks, M.S. (eds.) Structure-Based Mechanics of Tissues and Organs, pp. 225–248. Springer, New York (2016)
50. Bilston, L.E.: The influence of microstructure on neural tissue mechanics. In: Kassab, G.S., Sacks, M.S. (eds.) Structure-Based Mechanics of Tissues and Organs, pp. 1–14. Springer, New York (2016)
51. Lanir, Y., Namani, R.: Reliability of structure tensors in representing soft tissues structure. J. Mech. Behav. Biomed. Mater. **46**, 222–228 (2015)
52. Fan, R., Sacks, M.S.: Simulation of planar soft tissues using a structural constitutive model: finite element implementation and validation. J. Biomech. **47**(9), 2043–2054 (2014)
53. Rezakhaniha, R., Agianniotis, A., Schrauwen, J.T., Griffa, A., Sage, D., Bouten, C.V., van de Vosse, F.N., Unser, M., Stergiopulos, N.: Experimental investigation of collagen waviness and orientation in the arterial adventitia using confocal laser scanning microscopy. Biomech. Model. Mechanobiol. **11**(3–4), 461–473 (2012)
54. Viidik, A.: Simultaneous mechanical and light microscopic studies of collagen fibers. Z. Anat. Entwicklungsgesch. **136**(2), 204–212 (1972)
55. Evans, J.H., Barbenel, J.C., Steel, T.R., Ashby, A.M.: Structure and mechanics of tendon. Symp. Soc. Exp. Biol. **34**, 465–469 (1980)
56. Lanir, Y.: A structural theory for the homogeneous biaxial stress-strain relationships in flat collagenous tissues. J. Biomech. **12**(6), 423–436 (1979)
57. Lanir, Y.: Constitutive equations for fibrous connective tissues. J. Biomech. **16**(1), 1–12 (1983)
58. Lee, C.H., Zhang, W., Liao, J., Carruthers, C.A., Sacks, J.I., Sacks, M.S.: On the presence of affine fibril and fiber kinematics in the mitral valve anterior leaflet. Biophys. J. **108**(8), 2074–2087 (2015)
59. Holzapfel, G.A., Gasser, T.C., Ogden, R.W.: A new constitutive framework for arterial wall mechanics and a comparative study of material models. J. Elasticity **61**(1–3), 1–48 (2000)
60. Humphrey, J.D.: An evaluation of pseudoelastic descriptors used in arterial mechanics. J. Biomech. Eng. **121**(2), 259–262 (1999)
61. Lanir, Y.: Plausibility of structural constitutive equations for swelling tissues—implications of the C-N and S-E conditions. J. Biomech. Eng. **118**(1), 10–16 (1996)
62. Hollander, Y., Durban, D., Lu, X.A., Kassab, G.S., Lanir, Y.: Experimentally validated microstructural 3D constitutive model of coronary arterial media. J. Biomech. Eng. **133**(3) (2011)
63. Lu, X., Yang, J., Zhao, J.B., Gregersen, H., Kassab, G.S.: Shear modulus of porcine coronary artery: contributions of media and adventitia. Am. J. Physiol., Heart Circ. Physiol. **285**(5), H1966–H1975 (2003)
64. Wang, C., Garcia, M., Lu, X., Lanir, Y., Kassab, G.S.: Three-dimensional mechanical properties of porcine coronary arteries: a validated two-layer model. Am. J. Physiol., Heart Circ. Physiol. **291**(3), H1200–H1209 (2006)
65. Hollander, Y., Durban, D., Lu, X., Kassab, G.S., Lanir, Y.: Constitutive modeling of coronary arterial media-comparison of three model classes. J. Biomech. Eng. **133**(6) (2011)
66. Katchalsky, A.C., Peter, F.: Nonequilibrium Thermodynamics in Biophysics. Cambridge University Press, Cambridge (1967)
67. Richards, E.G.: An Introduction to the Physical Properties of Large Molecules in Solution. Cambridge University Press, Cambridge, U.K. (1980)

68. Truesdell, C.: Mechanical basis of diffusion. J. Chem. Phys. **37**, 2336–2344 (1962)
69. Lanir, Y.: Biorheology and fluid flux in swelling tissues. 1. Bicomponent theory for small deformations, including concentration effects. Biorheology **24**(2), 173–187 (1987)
70. Rodriguez, E.K., Hoger, A., McCulloch, A.D.: Stress-dependent finite growth in soft elastic tissues. J. Biomech. **27**(4), 455–467 (1994)
71. Skalak, R., Zargaryan, S., Jain, R.K., Netti, P.A., Hoger, A.: Compatibility and the genesis of residual stress by volumetric growth. J. Math. Biol. **34**(8), 889–914 (1996)
72. Taber, L.A., Humphrey, J.D.: Stress-modulated growth, residual stress, and vascular heterogeneity. J. Biomech. Eng. **123**(6), 528–535 (2001)
73. Azeloglu, E.U., Albro, M.B., Thimmappa, V.A., Ateshian, G.A., Costa, K.D.: Heterogeneous transmural proteoglycan distribution provides a mechanism for regulating residual stresses in the aorta. Am. J. Physiol., Heart Circ. Physiol. **294**(3), H1197–H1205 (2008)
74. Greenwald, S.E., Moore, J.E. Jr., Rachev, A., Kane, T.P., Meister, J.J.: Experimental investigation of the distribution of residual strains in the artery wall. J. Biomech. Eng. **119**(4), 438–444 (1997)
75. Stergiopulos, N., Vulliemoz, S., Rachev, A., Meister, J.J., Greenwald, S.E.: Assessing the homogeneity of the elastic properties and composition of the pig aortic media. J. Vasc. Res. **38**(3), 237–246 (2001)
76. Guo, X.M., Lanir, Y., Kassab, G.S.: Effect of osmolarity on the zero-stress state and mechanical properties of aorta. Am. J. Physiol., Heart Circ. Physiol. **293**(4), H2328–H2334 (2007)
77. Thornton, G.M., Oliynyk, A., Frank, C.B., Shrive, N.G.: Ligament creep cannot be predicted from stress relaxation at low stress: a biomechanical study of the rabbit medial collateral ligament. J. Orthop. Res. **15**(5), 652–656 (1997)
78. Raz, E., Lanir, Y.: Recruitment viscoelasticity of the tendon. J. Biomech. Eng. **131**(11), 111008 (2009)
79. Eshel, H., Lanir, Y.: Effects of strain level and proteoglycan depletion on preconditioning and viscoelastic responses of rat dorsal skin. Ann. Biomed. Eng. **29**(2), 164–172 (2001)
80. Mullins, L.: Softening of rubber by deformation. Rubber Chem. Technol. **42**, 339–362 (1969)
81. Tong, P., Fung, Y.C.: The stress-strain relationship for the skin. J. Biomech. **9**(10), 649–657 (1976)
82. Chuong, C.J., Fung, Y.C.: Compressibility and constitutive equation of arterial wall in radial compression experiments. J. Biomech. **17**(1), 35–40 (1984)
83. Fung, Y.C., Fronek, K., Patitucci, P.: Pseudoelasticity of arteries and the choice of its mathematical expression. Physiol. Entomol. **237**(5), H620–H631 (1979)
84. Lanir, Y.: Biaxial stress-strain relationship in the skin. Isr. J. Technol. **17**(2), 78–85 (1979)
85. Vawter, D.L., Fung, Y.C., West, J.B.: Constitutive equation of lung-tissue elasticity. J. Biomech. Eng. **101**(1), 38–45 (1979)
86. Humphrey, J.D., Vawter, D.L., Vito, R.P.: Pseudoelasticity of excised visceral pleura. J. Biomech. Eng. **109**(2), 115–120 (1987)
87. Yin, F.C.: Ventricular wall stress. Circ. Res. **49**(4), 829–842 (1981)
88. Glass, L., Hunter, P.J., McCulloch, A.D. (eds.): Theory of Heart. Springer, New York (1991)
89. Schnid, H., Hunter, P.J.: Multi-scale modeling of the heart. In: Holzapfel, G.A., Ogden, R.W. (eds.) Biomechanical Modeling at the Molecular, Cellular and Tissue Levels, pp. 83–107. Springer, Wien (2009)
90. Frank, J.S., Langer, G.A.: The myocardial interstitium: its structure and its role in ionic exchange. J. Cell Biol. **60**(3), 586–601 (1974)
91. Caspari, P.G., Newcomb, M., Gibson, K., Harris, P.: Collagen in the normal and hypertrophied human ventricle. Cardiovasc. Res. **11**(6), 554–558 (1977)
92. Weber, K.T.: Cardiac interstitium in health and disease: the fibrillar collagen network. J. Am. Coll. Cardiol. **13**(7), 1637–1652 (1989)
93. Streeter, D.D. Jr., Spotnitz, H.M., Patel, D.P., Ross, J. Jr., Sonnenblick, E.H.: Fiber orientation in the canine left ventricle during diastole and systole. Circ. Res. **24**(3), 339–347 (1969)
94. Borg, T.K., Caulfield, J.B.: The collagen matrix of the heart. Fed. Proc. **40**(7), 2037–2041 (1981)
95. Caulfield, J.B., Borg, T.K.: The collagen network of the heart. Labor Invest. **40**(3), 364–372 (1979)
96. Robinson, T.F., Cohen-Gould, L., Factor, S.M.: Skeletal framework of mammalian heart muscle: arrangement of inter- and pericellular connective tissue structures. Labor Invest. **49**(4), 482–498 (1983)
97. Robinson, T.F., Factor, S.M., Capasso, J.M., Wittenberg, B.A., Blumenfeld, O.O., Seifter, S.: Morphology, composition, and function of struts between cardiac myocytes of rat and hamster. Cell Tissue Res. **249**(2), 247–255 (1987)
98. Caulfield, J.B., Borg, T.K.: Collagen network of the heart. Lab. Invest. **40**(3), 364–372 (1979)
99. Horowitz, A., Lanir, Y., Yin, F.C.P., Perl, M., Sheinman, I., Strumpf, R.K.: Structural 3-dimensional constitutive law for the passive myocardium. J. Biomech. Eng. **110**(3), 200–207 (1988)
100. Nevo, E., Lanir, Y.: Structural finite deformation model of the left-ventricle during diastole and systole. J. Biomech. Eng. **111**(4), 342–349 (1989)

101. Labeit, S., Kolmerer, B.: Titins: giant proteins in charge of muscle ultrastructure and elasticity. Science **270**(5234), 293–296 (1995)
102. Hill, A.V.: The heat of shortening and the dynamic constants of muscle. Proc. R. Soc. Edinb. Sect. B. Biol. **126**, 136–195 (1938)
103. Jockenhoevel, S., Zund, G., Hoerstrup, S.P., Schnell, A., Turina, M.: Cardiovascular tissue engineering: a new laminar flow chamber for in vitro improvement of mechanical tissue properties. ASAIO J. **48**(1), 8–11 (2002)
104. Nerem, R.M., Seliktar, D.: Vascular tissue engineering. Annu. Rev. Biomed. Eng. **3**, 225–243 (2001)
105. Niklason, L.E., Yeh, A.T., Calle, E.A., Bai, Y., Valentin, A., Humphrey, J.D.: Enabling tools for engineering collagenous tissues integrating bioreactors, intravital imaging, and biomechanical modeling. Proc. Natl. Acad. Sci. USA **107**(8), 3335–3339 (2010)
106. Jackson, Z.S., Gotlieb, A.I., Langille, B.L.: Wall tissue remodeling regulates longitudinal tension in arteries. Circ. Res. **90**(8), 918–925 (2002)
107. Kamiya, A., Togawa, T.: Adaptive regulation of wall shear stress to flow change in the canine carotid artery. Am. J. Physiol. **239**(1), H14–H21 (1980)
108. Langille, B.L., Bendeck, M.P., Keeley, F.W.: Adaptations of carotid arteries of young and mature rabbits to reduced carotid blood flow. Am. J. Physiol. **256**(4 Pt 2), H931–H939 (1989)
109. Wayman, B.H., Taylor, W.R., Rachev, A., Vito, R.P.: Arteries respond to independent control of circumferential and shear stress in organ culture. Ann. Biomed. Eng. **36**(5), 673–684 (2008)
110. Humphrey, J.D., Rajagopal, K.R.: A constrained mixture model for growth and remodeling of soft tissues. Math. Models Methods Appl. Sci. **12**(3), 407–430 (2002)
111. Cowin, S.C.: Tissue growth and remodeling. Annu. Rev. Biomed. Eng. **6**, 77–107 (2004)
112. Lanir, Y.: Mechanistic micro-structural theory of soft tissues growth and remodeling: tissues with unidirectional fibers. Biomech. Model. Mechanobiol. **14**(2), 245–266 (2015)
113. Gleason, R.L., Jr., Humphrey, J.D.: A 2D constrained mixture model for arterial adaptations to large changes in flow, pressure and axial stretch. Math. Med. Biol. **22**(4), 347–369 (2005)
114. Cowin, S.C.: Continuum kinematical modeling of mass increasing biological growth. Int. J. Eng. Sci. **48**(11), 1137–1145 (2010)
115. Chen, H., Liu, Y., Slipchenko, M.N., Zhao, X., Cheng, J.X., Kassab, G.S.: The layered structure of coronary adventitia under mechanical load. Biophys. J. **101**(11), 2555–2562 (2011)
116. Hill, M.R., Duan, X., Gibson, G.A., Watkins, S., Robertson, A.M.: A theoretical and non-destructive experimental approach for direct inclusion of measured collagen orientation and recruitment into mechanical models of the artery wall. J. Biomech. **45**(5), 762–771 (2012)
117. Horny, L., Chlup, H., Zitny, R.: Collagen orientation and waviness within the vein wall. In: XI International Conference on Computational Plasticity: Fundamentals and Applications, Spain. International Center for Numerical Methods in Engineering (CIMNE), Barcelona (2011)
118. Zeinali-Davarani, S., Chow, M.J., Turcotte, R., Zhang, Y.: Characterization of biaxial mechanical behavior of porcine aorta under gradual elastin degradation. Ann. Biomed. Eng. **41**(7), 1528–1538 (2013)
119. Chen, H., Liu, Y., Zhao, X.F., Lanir, Y., Kassab, G.S.: A micromechanics finite-strain constitutive model of fibrous tissue. J. Mech. Phys. Solids **59**(9), 1823–1837 (2011)
120. Imanuel, O.: Stress analysis in the left ventricle of the heart. PhD thesis, Faculty of Biomedical Engineering, Technion–I.I.T.: Haifa (1996)
121. Chen, H., Liu, Y., Slipchenko, M., Zhao, X., Cheng, J.-X., Kassab, G.: The layered structure of coronary adventitia under mechanical load. Biophys. J. **101**, 2555–2562 (2011). doi:10.1016/j.bpj.2011.10.043
122. Zoumi, A., Lu, X., Kassab, G.S., Tromberg, B.J.: Imaging coronary artery microstructure using second-harmonic and two-photon fluorescence microscopy. Biophys. J. **87**(4), 2778–2786 (2004)
123. Freed, A.D., Einstein, D.R., Vesely, I.: Invariant formulation for dispersed transverse isotropy in aortic heart valves: an efficient means for modeling fiber splay. Biomech. Model. Mechanobiol. **4**(2–3), 100–117 (2005)
124. Gasser, T.C., Ogden, R.W., Holzapfel, G.A.: Hyperelastic modelling of arterial layers with distributed collagen fibre orientations. J. Roy. Soc. Interface **3**, 15–35 (2006)
125. Cortes, D.H., Lake, S.P., Kadlowec, J.A., Soslowsky, L.J., Elliott, D.M.: Characterizing the mechanical contribution of fiber angular distribution in connective tissue: comparison of two modeling approaches. Biomech. Model. Mechanobiol. **9**(5), 651–658 (2010)
126. Federico, S., Herzog, W.: Towards an analytical model of soft biological tissues. J. Biomech. **41**(16), 3309–3313 (2008)
127. Bischoff, J.E., Arruda, E.A., Grosh, K.: A microstructurally based orthotropic hyperelastic constitutive law. J. Appl. Mech. **69**, 570–579 (2002)
128. Elata, D., Rubin, M.: Isotropy of strain energy functions which depend only on a finite number of directional strain measures. J. Appl. Mech. **61**(2), 284–289 (1994)

129. Bazant, Z.P.: Efficient numerical integration on the surface of a sphere. ZAMM Z. Angew. Math. Mech. **66**(1), 37–49 (1986)
130. Ehret, A.E., Itskov, M., Schmid, H.: Numerical integration on the sphere and its effect on the material symmetry of constitutive equations—a comparative study. Internat. J. Numer. Methods Engrg. **81**(2), 189–206 (2010)
131. Delsarte, P., Goethals, J.M., Seidel, J.J.: Spherical codes and designs. Geom. Dedicata **6**, 363–388 (1977)
132. Hardin, R.H., Sloane, N.J.A.: McLaren's improved snub cube and other new spherical designs in three dimensions. Discrete Comput. Geom. **15**, 429–441 (1996)
133. Federico, S., Gasser, T.C.: Nonlinear elasticity of biological tissues with statistical fiber orientation. J. Roy. Soc. Interface **7**(47), 955–966 (2010)
134. Martufi, G., Gasser, T.C.: A constitutive model for vascular tissue that integrates fibril, fiber and continuum levels with application to the isotropic and passive properties of the infrarenal aorta. J. Biomech. **44**(14), 2544–2550 (2011)
135. Gray, M.L., Pizzanelli, A.M., Grodzinsky, A.J., Lee, R.C.: Mechanical and physiochemical determinants of the chondrocyte biosynthetic response. J. Orthop. Res. **6**(6), 777–792 (1988)
136. Palmoski, M.J., Brandt, K.D.: Effects of static and cyclic compressive loading on articular cartilage plugs in vitro. Arthritis Rheum. **27**(6), 675–681 (1984)
137. Sah, R.L., Kim, Y.J., Doong, J.Y., Grodzinsky, A.J., Plaas, A.H., Sandy, J.D.: Biosynthetic response of cartilage explants to dynamic compression. J. Orthop. Res. **7**(5), 619–636 (1989)
138. Torzilli, P.A., Grigiene, R., Huang, C., Friedman, S.M., Doty, S.B., Boskey, A.L., Lust, G.: Characterization of cartilage metabolic response to static and dynamic stress using a mechanical explant test system. J. Biomech. **30**(1), 1–9 (1997)
139. Lee, D.A., Bader, D.L.: Compressive strains at physiological frequencies influence the metabolism of chondrocytes seeded in agarose. J. Orthop. Res. **15**(2), 181–188 (1997)
140. Davies, P.F.: Flow-mediated endothelial mechanotransduction. Physiol. Rev. **75**(3), 519–560 (1995)
141. Akhyari, P., Fedak, P.W., Weisel, R.D., Lee, T.Y., Verma, S., Mickle, D.A., Li, R.K.: Mechanical stretch regimen enhances the formation of bioengineered autologous cardiac muscle grafts. Circulation **106**(12 Suppl 1), I137–I142 (2002)
142. Kim, B.S., Nikolovski, J., Bonadio, J., Mooney, D.J.: Cyclic mechanical strain regulates the development of engineered smooth muscle tissue. Nat. Biotechnol. **17**(10), 979–983 (1999)
143. Langer, R., Vacanti, J.P.: Tissue engineering. Science **260**(5110), 920–926 (1993)
144. Fung, Y.C.: Biorheology of soft tissues. Biorheology **10**(2), 139–155 (1973)

J Elast (2017) 129:49–68
DOI 10.1007/s10659-016-9605-2

On Fiber Dispersion Models: Exclusion of Compressed Fibers and Spurious Model Comparisons

Gerhard A. Holzapfel[1,2] · **Ray W. Ogden**[3]

Received: 29 May 2016 / Published online: 25 October 2016

Abstract Fiber dispersion in collagenous soft tissues has an important influence on the mechanical response, and the modeling of the collagen fiber architecture and its mechanics has developed significantly over the last few years. The purpose of this paper is twofold, first to develop a method for excluding compressed fibers within a dispersion for the generalized structure tensor (GST) model, which several times in the literature has been claimed not to be possible, and second to draw attention to several erroneous and misleading statements in the literature concerning the relative values of the GST and the angular integration (AI) models. For the GST model we develop a rather simple method involving a deformation dependent dispersion parameter that allows the mechanical influence of compressed fibers within a dispersion to be excluded. The theory is illustrated by application to simple extension and simple shear in order to highlight the effect of exclusion. By means of two examples we also show that the GST and the AI models have equivalent predictive power, contrary to some claims in the literature. We conclude that from the theoretical point of view neither of these two models is superior to the other. However, as is well known and as we now emphasize, the GST model has proved to be very successful in modeling the data from experiments on a wide range of tissues, and it is easier to analyze and simpler to implement than the AI approach, and the related computational effort is much lower.

Keywords Fiber dispersion model · Generalized structure tensor · Angular integration model · Fibrous tissue · Exclusion of compressed fibers

Mathematics Subject Classification 74B20 · 74E10 · 74L15

✉ G.A. Holzapfel
holzapfel@tugraz.at

1 Institute of Biomechanics, Graz University of Technology, Stremayrgasse 16-II, 8010 Graz, Austria

2 Faculty of Engineering Science and Technology, Norwegian University of Science and Technology (NTNU), 7491 Trondheim, Norway

3 School of Mathematics and Statistics, University of Glasgow, University Gardens, Glasgow G12 8QW, Scotland, UK

1 Introduction

Collagen fiber dispersion in fibrous tissues has been recognized as being important in the mechanical response of the tissue. In recent years continuum mechanical models of this response for tissues such as arteries, the myocardium, heart valves, corneas and articular cartilage have been developed to accommodate the effect of collagen fiber dispersion embedded within a non-collagenous matrix material. There are now many imaging modalities available that can identify fiber orientations within tissues; in particular, second-harmonic generation, see, e.g., [1], and ultra-high field diffusion tensor magnetic resonance imaging, see, e.g., [2]. These modalities are able to capture the 3D distribution of collagen fiber orientations without damage to the tissue, in contrast to histological investigations. This allows a detailed geometrical reconstruction of the fiber architecture which then serves as a basis for more advanced modeling including finite element analysis. It is often considered that in the modeling of such dispersions only those fibers which are extended should contribute to the mechanical response. In other words compression is supported only by the non-collagenous matrix material in which the fibers are embedded, and this consideration forms part of the motivation of the present work.

Basically there exist two different approaches for modeling fiber dispersion. First, the 'angular integration' (AI) approach, which is due to Lanir [3]. In this approach a single collagen fiber with strain energy $w(\lambda)$ is considered as a function of the fiber stretch λ. The strain energy is then integrated over a unit sphere Ω to obtain the strain-energy function Ψ of the aggregate of the fibers per unit reference volume, i.e.

$$\Psi = n \int_{\Omega} \rho(\mathbf{N}) w(\lambda) \mathrm{d}\Omega, \tag{1}$$

where $\mathbf{N}$ is a unit vector pointing in the direction of an arbitrary fiber, n is the number of fibers per unit reference volume, and ρ is the relative angular density of fibers normalized according to

$$\frac{1}{4\pi} \int_{\Omega} \rho(\mathbf{N}) \mathrm{d}\Omega = 1. \tag{2}$$

In (1) it has been assumed that the elastic properties of all the fibers are characterized by the same strain energy w, which was not the case in the original approach by Lanir [3].

Second, the 'generalized structure tensor' (GST) approach [4] considers an energy function Ψ_{f} associated with the fibers given by

$$\Psi_{\mathrm{f}} = \Psi_{\mathrm{f}}(\mathbf{C}, \mathbf{H}), \tag{3}$$

which involves the right Cauchy–Green tensor $\mathbf{C}$, and a generalized structure tensor $\mathbf{H}$, defined by

$$\mathbf{H} = \frac{1}{4\pi} \int_{\Omega} \rho(\mathbf{N}) \mathbf{N} \otimes \mathbf{N} \mathrm{d}\Omega, \tag{4}$$

with the condition $\mathrm{tr}\mathbf{H} = 1$ following from the normalization (2). A list of relevant contributions to these two modeling approaches is provided in the introduction of [5].

Unfortunately, several errors concerning the relationships between the GST and the AI models have been promulgated in the literature, as in, e.g., [6], repeated in [7–9] and other studies. We therefore highlight some of the errors in order to discourage further repetitions.

First, according to [6], the GST model '... *gives excellent results for two reasons: (i) the fiber directional dispersion is small and (ii) all fibers are loaded in tension*'. Both these conclusions are false. The conclusion (i) is false because the argument is based on the very specialized assumption that the strain-energy function Ψ_f in (3) is the same as the strain-energy function w of an individual fiber, and the argument in [6] is also based on a Taylor expansion approximation of w around the mean fiber direction, which is itself only valid for small dispersions. The conclusion (ii) is false because in the application to arteries considered in [4] not all the fibers within the dispersion are loaded in tension but still the results are excellent. Indeed, because of incompressibility, at least one stretch always has to be less than one so that some fibers in a dispersion are always under compression. One implication of conclusion (ii) is that the GST model cannot exclude fibers in compression, which is certainly not the case. Indeed a simple procedure for excluding fibers in compression in the GST approach forms the first part of the content of the present paper. This also complements our analysis of the exclusion procedure in the AI approach contained in [10] and its implementation described in [11]. It should also be mentioned that recently an approach to exclusion of fibers in the GST model based on the use of a Heaviside function was developed in [12].

Second, there was also the suggestion in [13], on the basis of the analysis in [6], that the GST model, in particular the model in [4], is a '... *first order approximation*' of the AI model, again a false conclusion. Third, the errors in [6] have been propagated in subsequent papers without question. For example, in [7], with reference to [6], it is stated that '... *structure tensors can only be used when all the fibers are in tension and the angular distribution is small*'.

Fourth, in [7] further errors have been produced because the authors used the same set of parameters in the exponential forms of the strain-energy function in the AI and the GST approaches (this is a particular example of the assumption used in [6]), which leads to incorrect conclusions concerning the relative predictions of the two models. To elaborate on this point, in the AI model the authors of [7] used the exponential form

$$w(\lambda) = \frac{c_1}{2c_2}\{\exp[c_2(\lambda^2 - 1)^2] - 1\}, \tag{5}$$

and for the GST model, for comparison, they used the strain-energy function

$$\Psi_f(I_1, I_4) = \frac{k_1}{2k_2}\{\exp[k_2(\kappa I_1 + (1 - 3\kappa)I_4)^2] - 1\}, \tag{6}$$

where c_1, c_2, k_1, k_2 are material constants, while I_1 and I_4 are standard invariants and κ is the dispersion parameter. The use of (5) in the integration (1) is legitimate. However, in [7] the predictions of the models were compared by setting $k_1 = c_1$ and $k_2 = c_2$, which is a very restrictive assumption, and leads to erroneous comparisons, a point on which we will elaborate in Sect. 3.

The errors from both studies [6] and [7] have been accepted and further developed with additional errors in [8], and subsequent papers. However, the basis of the comparisons therein is invalid since in each case the same material parameters have been used for two different models. An example of an unfounded statement which is contained in [8] is: '*Numerical tests show that for all the loading cases for which GST models introduce large errors, such as uniaxial loading, shear and biaxial loading, the proposed model has a better performance, in the sense that it provides results closer to the ones furnished by an exact angle integration of the fiber orientation distribution*'. This statement is based on the incorrect comparison in [7]. The GST model does not introduce large errors.

The main justification for selecting a particular model is that it captures very well the underlying structure and physics and the data from experiments. It has to be emphasized that the GST approach has proved to be very successful in modeling the data from experiments on a wide range of tissues, including arteries, the myocardium, heart valves, corneas, and articular cartilage, while comparisons of the AI model with experimental data have been conducted to a much lesser extent. The basis of the comparative study in [7], which suggests that the AI model is superior to the GST model, is unfounded. There is no justification for such a conclusion on the basis of the analysis in [7] or for referring to the AI approach as the 'gold standard'. The purpose of the second part of the present paper is to demonstrate by an appropriate choice of parameters that the AI and the GST modeling approaches are equivalent in regard to their predictive capabilities. Nevertheless, it should be emphasized that the GST approach is more amenable to theoretical analysis than the AI approach. It is also simpler to implement and the related computational effort is much lower because it does not involve integrations.

The outline of the paper is as follows. In Sect. 2 we provide a summary of the equations governing the GST model with rotational symmetry and show how a simple (deformation dependent) modification of the dispersion parameter is used to exclude compressed fibers in a dispersion. The resulting model, based on a generalized structure tensor, is then used to illustrate the consequences of the exclusion of compressed fibers in simple tension and in simple shear. It is shown, in particular, that for simple tension in the mean fiber direction the stress response with the fibers excluded is much stiffer than when they are not excluded when there is significant dispersion. On the other hand, depending on the mean fiber orientation, the simple shear stress response is either stiffer or softer with fiber exclusion than without the exclusion, again when there is significant dispersion.

In order to quantify our comments in the foregoing discussion relating to the studies in, e.g., [6–8], we provide in Sect. 3 an analysis of the comparison between the GST and AI models for small strains. In particular it is shown that the choice $k_1 = c_1$ and $k_2 = c_2$ in the equations (5) and (6), as in [7], is only valid in the two limiting cases of no dispersion or isotropic dispersion. In fact, to properly compare the predictions of the two models on the basis of the exponential models (5) and (6) the parameters c_1 and c_2 must depend not only on k_1 and k_2 but also on the dispersion parameter κ. We then go on to compare the predictions of the AI and the GST models for simple tension and simple shear for three different values of dispersion and in each case, with an appropriate choice of parameters, it can be seen that the predictions of the two models essentially coincide. This is in sharp contrast to the results in [7] for the same deformations.

2 GST Model with Excluded Compressed Fibers

We consider a deformation described in terms of the deformation gradient $\mathbf{F}$. The associated left and right Cauchy–Green tensors are $\mathbf{B} = \mathbf{F}\mathbf{F}^{\mathrm{T}}$ and $\mathbf{C} = \mathbf{F}^{\mathrm{T}}\mathbf{F}$, respectively. In Fig. 1 the unit vector $\mathbf{M}$ represents the mean fiber direction in the reference configuration of a fiber dispersion with an arbitrary fiber direction denoted $\mathbf{N}$, together with the cone within which the fibers are stretched in the deformed configuration, i.e., for which $(\mathbf{CN}) \cdot \mathbf{N} > 1$. Note that $\mathbf{M}$ is included in the cone but if $\mathbf{M}$ is compressed rather than extended then $\mathbf{CN} \cdot \mathbf{N} > 1$ holds in the complement of the cone.

With respect to an orthonormal basis $\mathbf{E}_1, \mathbf{E}_2, \mathbf{E}_3$ with $\mathbf{M} = \mathbf{E}_3$, the unit vector $\mathbf{N}$ is given by

$$\mathbf{N} = \sin\Theta \cos\Phi \mathbf{E}_1 + \sin\Theta \sin\Phi \mathbf{E}_2 + \cos\Theta \mathbf{E}_3, \tag{7}$$

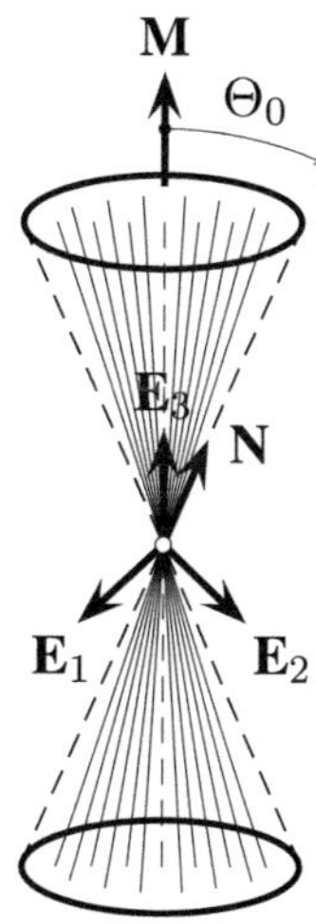

Fig. 1 Unit vector **N** which defines the direction of an arbitrary fiber within a fiber dispersion, mean fiber direction **M** in the reference configuration and cone of fibers which are stretched in the deformed configuration, i.e., $(\mathbf{CN}) \cdot \mathbf{N} > 1$. An orthonormal basis $\mathbf{E}_1, \mathbf{E}_2, \mathbf{E}_3$ with $\mathbf{M} = \mathbf{E}_3$. Note that in general the cone is not symmetric about **M**, and in this example **M** satisfies $(\mathbf{CM}) \cdot \mathbf{M} > 1$

where $\Theta \in [0, \pi]$ and $\Phi \in [0, 2\pi]$ are spherical polar angles, and $\mathbf{M} \cdot \mathbf{N} = \cos\Theta$. Then the equation $(\mathbf{CN}) \cdot \mathbf{N} = 1$ describes the boundary of the region in Fig. 1, and, with (7), the explicit expression reads

$$(\mathbf{CN}) \cdot \mathbf{N} = \sin^2\Theta\big(C_{11}\cos^2\Phi + 2C_{12}\sin\Phi\cos\Phi + C_{22}\sin^2\Phi\big) + 2\sin\Theta\cos\Theta(C_{13}\cos\Phi + C_{23}\sin\Phi) + C_{33}\cos^2\Theta = 1, \tag{8}$$

which, of course, is satisfied identically in the reference configuration, C_{ij} being the components of **C**. When $\Theta \neq \pi/2$ Eq. (8) may be rearranged as

$$\tan^2\Theta\big(C_{11}\cos^2\Phi + 2C_{12}\sin\Phi\cos\Phi + C_{22}\sin^2\Phi - 1\big) + 2\tan\Theta(C_{13}\cos\Phi + C_{23}\sin\Phi) + C_{33} - 1 = 0, \tag{9}$$

which is a quadratic in $\tan\Theta$ and shows that Θ depends on Φ and **C**. If $\Theta = \pi/2$ Eq. (8) reduces to

$$C_{11}\cos^2\Phi + 2C_{12}\sin\Phi\cos\Phi + C_{22}\sin^2\Phi = 1, \tag{10}$$

which governs unstretched fibers perpendicular to **M**.

Under the deformation, the mean fiber direction **M** maps into **FM** in the deformed configuration, and we use the notation $\mathbf{m} = \mathbf{FM}$ to represent this. In general, **m** is neither a unit vector nor the mean fiber direction in the deformed configuration. Correspondingly, for an arbitrary fiber we adopt the notation **n** defined by $\mathbf{n} = \mathbf{FN}$. Next we introduce the angle θ in the current configuration, which, for a given **C**, is related to Θ and Φ by

$$\cos\theta = \frac{\mathbf{m} \cdot \mathbf{n}}{|\mathbf{m}||\mathbf{n}|} = \frac{(C_{13}\cos\Phi + C_{23}\sin\Phi)\sin\Theta + C_{33}\cos\Theta}{C_{33}^{1/2}|\mathbf{n}|} \tag{11}$$

since $|\mathbf{m}| = C_{33}^{1/2}$. The GST model uses a generalized structure tensor, as defined in Eq. (4). For the case in which the dispersion is rotationally symmetric about a mean direction **M**, following [4], the generalized structure tensor has the form $\mathbf{H} = \kappa\mathbf{I} + (1 - 3\kappa)\mathbf{M}\otimes\mathbf{M}$, and the (constant) dispersion parameter is defined by [4]

$$\kappa = \frac{1}{4}\int_0^\pi \rho(\Theta)\sin^3\Theta \mathrm{d}\Theta, \tag{12}$$

where $\rho(\Theta)$ is the fiber orientation density, rotationally symmetric about the mean fiber direction. It satisfies the normalization condition

$$\frac{1}{2}\int_0^\pi \rho(\Theta)\sin\Theta \mathrm{d}\Theta = 1 \tag{13}$$

and enjoys the symmetry properties

$$\rho(\pi + \Theta) = \rho(\pi - \Theta) = \rho(\Theta). \tag{14}$$

A generalized invariant denoted $I_4^\star$ associated with $\mathbf{H}$ is

$$I_4^\star = \mathrm{tr}(\mathbf{HC}) = \kappa I_1 + (1 - 3\kappa)I_4, \tag{15}$$

where

$$I_1 = \mathrm{tr}\,\mathbf{C}, \qquad I_4 = \mathbf{M}\cdot(\mathbf{CM}). \tag{16}$$

We emphasize that (12) includes *all* the fibers in the dispersion whether they are in tension or compression. If we wish to exclude compressed fibers then an appropriate range of angles needs to be omitted from the integral in (12). Towards this goal we define a modified dispersion parameter, denoted $\kappa_1(\Theta)$, by

$$\kappa_1(\Theta) = \frac{1}{4}\int_0^\Theta \rho(\xi)\sin^3\xi \mathrm{d}\xi + \frac{1}{4}\int_{\pi-\Theta}^\pi \rho(\xi)\sin^3\xi \mathrm{d}\xi, \tag{17}$$

where $\Theta \in [0, \pi/2]$. The combination of integrals in (17) is appropriate for the case in which the mean fiber direction is extended, i.e., $(\mathbf{CM})\cdot\mathbf{M} > 1$. However, because of the symmetry properties (14) the two integrals are the same and hence (17) simplifies to

$$\kappa_1(\Theta) = \frac{1}{2}\int_0^\Theta \rho(\xi)\sin^3\xi \mathrm{d}\xi. \tag{18}$$

In particular, if $\Theta = \Theta_0$, which corresponds to the boundary of the cone shown in Fig. 1, then by (9) with Θ replaced by Θ_0 it is clear that Θ_0 depends on Φ and $\mathbf{C}$. Then $\kappa_1(\Theta_0)$ is the dispersion parameter appropriate for the case in which the compressed fibers are excluded, and in this case $\kappa_1(\Theta_0)$ can be obtained from (12) by setting $\rho(\Theta) = 0$ for $\Theta_0 \le \Theta \le \pi - \Theta_0$. To illustrate its behavior a typical plot of $\kappa_1(\Theta)$ is shown in Fig. 2 for Θ between 0 and $\pi/2$. This shows that $\kappa_1(\Theta)$ increases monotonically from 0 to Θ_0 and has the value $\kappa_1(\Theta_0)$ for Θ between Θ_0 and $\pi/2$. From (18) it follows that

$$\kappa_1'(\Theta) = \frac{1}{2}\rho(\Theta)\sin^3\Theta \tag{19}$$

Fig. 2 Typical plot of κ_1 vs Θ contained in $[0, \pi/2]$. For $0 \le \Theta \le \Theta_0$ the curve is determined by (18), while for $\Theta_0 \le \Theta \le \pi/2$, $\kappa_1(\Theta) = \kappa_1(\Theta_0)$, which we denote by $\bar{\kappa}$

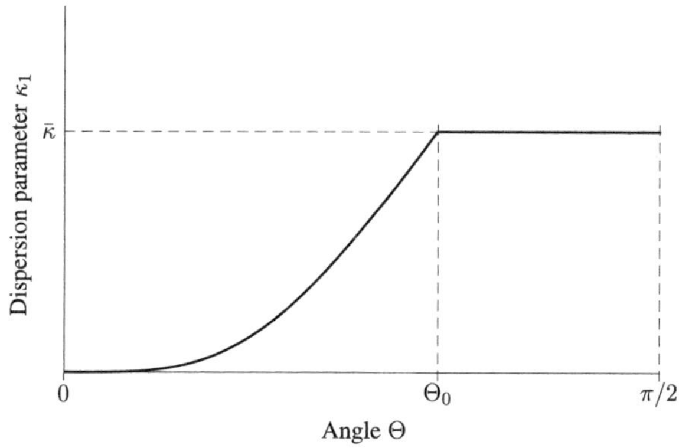

at any point where ρ is continuous but since we have set $\rho(\Theta_0) = 0$, κ_1' is discontinuous at $\Theta = \Theta_0$, with $\kappa_1'(\Theta) = 0$ for $\Theta_0^+ \le \Theta \le \pi/2$.

For convenience we now denote $\kappa_1(\Theta_0)$ by $\bar{\kappa}$ and consider the material to be incompressible with an energy function $\Psi(I_1, I_4, \bar{\kappa})$. Since $\kappa_1'(\Theta_0^+) = 0$ it follows that $\bar{\kappa}$ is stationary with respect to $\mathbf{C}$, and hence the second Piola–Kirchhoff stress tensor $\mathbf{S}$ is given by

$$\mathbf{S} = 2\frac{\partial \Psi}{\partial \mathbf{C}} = 2\psi_1 \mathbf{I} + 2\psi_4 \mathbf{M} \otimes \mathbf{M} - p\mathbf{C}^{-1}, \tag{20}$$

where the abbreviations $\psi_1 = \partial\Psi/\partial I_1$ and $\psi_4 = \partial\Psi/\partial I_4$ have been used, and p is the Lagrange multiplier associated with the constraint $\det \mathbf{C} = 1$. Thus, while its derivative with respect to $\mathbf{C}$ does not appear in (20), $\bar{\kappa} = \kappa_1(\Theta_0)$ does depend on $\mathbf{C}$ and is incorporated in the expression for $\mathbf{S}$ in exactly the same way as if it were a constant. For example, with a typical energy function of the form

$$\Psi = \Psi_{\text{iso}}(I_1) + \Psi_{\text{f}}(I_4^\star), \tag{21}$$

we obtain the stress relation

$$\mathbf{S} = 2\Psi_{\text{iso}}'(I_1)\mathbf{I} + 2\Psi_{\text{f}}'(I_4^\star)\mathbf{H} - p\mathbf{C}^{-1}, \tag{22}$$

where $\Psi_{\text{iso}}(I_1)$ represents the isotropic contribution related to the non-collagenous matrix material and $\Psi_{\text{f}}(I_4^\star)$ is the contribution of the energy stored in the collagen fibers. In addition, $I_4^\star$ is now given by $I_4^\star = \bar{\kappa} I_1 + (1 - 3\bar{\kappa}) I_4$, which is Eq. $(15)_2$ with κ replaced by $\bar{\kappa}$, and we emphasize that $\mathbf{H} = \bar{\kappa}\mathbf{I} + (1 - 3\bar{\kappa})\mathbf{M} \otimes \mathbf{M}$ now involves $\bar{\kappa}$ rather than κ.

If we consider the model adopted in [4] then, for a single family of fibers, we have

$$\Psi_{\text{iso}}(I_1) = \frac{\mu}{2}(I_1 - 3), \qquad \Psi_{\text{f}}(I_4^\star) = \frac{k_1}{2k_2}\{\exp[k_2(I_4^\star - 1)^2] - 1\}, \tag{23}$$

where μ and k_1 are parameters with dimensions of stress and k_2 is a dimensionless parameter. Hence, from (22), the second Piola–Kirchhoff stress tensor becomes

$$\mathbf{S} = \mu\mathbf{I} + 2k_1(I_4^\star - 1)\exp[k_2(I_4^\star - 1)^2]\mathbf{H} - p\mathbf{C}^{-1}. \tag{24}$$

Similarly, if the mean fiber direction is not extended $(\mathbf{CM})\cdot\mathbf{M}\leq 1$ then $(\mathbf{CN})\cdot\mathbf{N}>1$ for $\Theta\in(\Theta_0,\pi-\Theta_0)$, and in this case we define

$$\kappa_2(\Theta_0)=\frac{1}{4}\int_{\Theta_0}^{\pi-\Theta_0}\rho(\Theta)\sin^3\Theta \mathrm{d}\Theta, \tag{25}$$

which is complementary to $\kappa_1(\Theta_0)$ and can be obtained from (12) by setting $\rho(\Theta)=0$ for $\Theta\in[0,\Theta_0]\cup[\pi-\Theta_0,\pi]$. Note that

$$\kappa_1(\Theta_0)+\kappa_2(\Theta_0)=\kappa. \tag{26}$$

2.1 Example: Simple Tension

As a first example we consider a uniaxial stretch $\lambda\geq 1$ in the direction $\mathbf{M}=\mathbf{E}_3$ with a dispersion that is rotationally symmetric about $\mathbf{M}$. Then, by symmetry, the lateral stretches $\lambda_2=\lambda_3$ are $\lambda^{-1/2}$, and the components of $\mathbf{F}$ are diag$[\lambda,\lambda^{-1/2},\lambda^{-1/2}]$, so that Eq. (11) reduces to

$$\cos\theta=\frac{\lambda\cos\Theta}{\sqrt{\lambda^2\cos^2\Theta+\lambda^{-1}\sin^2\Theta}}. \tag{27}$$

Note, in particular, that on the boundary of the cone Eq. (8) reduces to

$$I_4=\lambda^2\cos^2\Theta+\lambda^{-1}\sin^2\Theta=1, \tag{28}$$

which, for $\lambda\neq 1$, can be rearranged with $\Theta=\Theta_0$ as

$$\tan\Theta_0=\sqrt{\lambda(\lambda+1)}. \tag{29}$$

It follows that $\cos\theta_0=\lambda\cos\Theta_0$, where θ_0 is the value θ obtained from (27) when (28) holds. If $0\leq\Theta<\Theta_0$ and $\pi-\Theta_0<\Theta\leq\pi$ then the fibers are extended. Thus, as λ increases more and more fibers are recruited into extension, and Θ_0 becomes larger and approaches $\pi/2$, while θ_0 tends to zero.

We now consider the energy function $\Psi^\star(I_1,I_4^\star)=\Psi(I_1,I_4,\bar{\kappa})$ with $\bar{\kappa}=\kappa_1(\Theta_0)$. Then the Cauchy stress tensor $\boldsymbol{\sigma}=\mathbf{FSF}^{\mathrm{T}}$ is given by

$$\boldsymbol{\sigma}=2\psi_1^\star\mathbf{B}+2\psi_4^\star\mathbf{h}-p\mathbf{I}, \tag{30}$$

where $\psi_1^\star=\partial\Psi^\star/\partial I_1$ and $\psi_4^\star=\partial\Psi^\star/\partial I_4^\star$ and

$$\mathbf{h}=\bar{\kappa}\mathbf{B}+(1-3\bar{\kappa})\mathbf{m}\otimes\mathbf{m},\qquad I_4^\star=\bar{\kappa}I_1+(1-3\bar{\kappa})I_4=\lambda^2-2\bar{\kappa}\left(\lambda^2-\lambda^{-1}\right), \tag{31}$$

$\mathbf{h}=\mathbf{FHF}^{\mathrm{T}}$ being the spatial generalized structure tensor. For the model (23) we have

$$\psi_1^\star=\mu/2,\qquad \psi_4^\star=k_1\left(I_4^\star-1\right)\exp\left[k_2\left(I_4^\star-1\right)^2\right]. \tag{32}$$

For simple tension let σ be the component of the Cauchy stress in the direction of stretch. Then, by specializing (30), we obtain

$$\sigma=\mu\lambda^2+2k_1\left(I_4^\star-1\right)\exp\left[k_2\left(I_4^\star-1\right)^2\right]\left[\bar{\kappa}\lambda^2+(1-3\bar{\kappa})\lambda^2\right]-p, \tag{33}$$

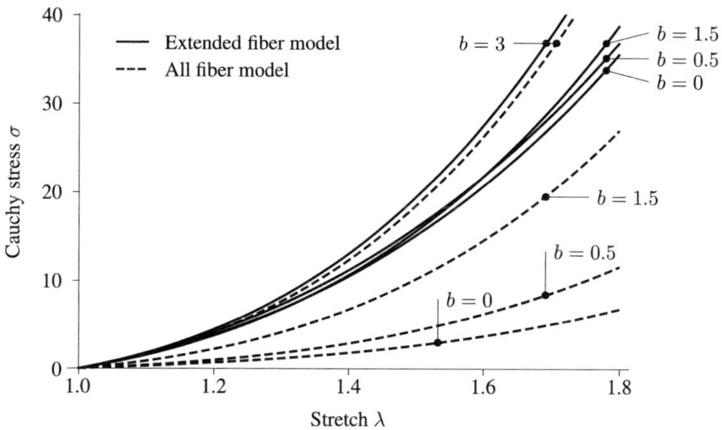

Fig. 3 Uniaxial Cauchy stress σ against the corresponding stretch λ for the four values $b = 0, 0.5, 1.5, 3$ (corresponding to $\kappa = 1/3, 0.285, 0.187, 0.096$) and for $k_1/\mu = 5$ and $k_2 = 0.01$

$$0 = \mu\lambda^{-1} + 2k_1\left(I_4^\star - 1\right)\exp\left[k_2\left(I_4^\star - 1\right)^2\right]\bar{\kappa}\lambda^{-1} - p, \tag{34}$$

and hence, on elimination of p,

$$\begin{aligned}\sigma &= \mu\left(\lambda^2 - \lambda^{-1}\right) + 2k_1\left[\lambda^2 - 1 - 2\bar{\kappa}\left(\lambda^2 - \lambda^{-1}\right)\right]\left(\lambda^2 - 2\bar{\kappa}\lambda^2 - \bar{\kappa}\lambda^{-1}\right)\\ &\quad\times \exp\left\{k_2\left[\lambda^2 - 1 - 2\bar{\kappa}\left(\lambda^2 - \lambda^{-1}\right)\right]^2\right\}.\end{aligned} \tag{35}$$

We emphasize here that $\bar{\kappa} = \kappa_1(\Theta_0)$ depends on λ via (29).

In order to compare the behavior of the model where compressed fibers are excluded with the model where they are not excluded, i.e. with a constant κ, we adopt the von Mises distribution used in [4]. This is given by

$$\rho(\Theta) = 4\sqrt{\frac{b}{2\pi}}\frac{\exp(2b\cos^2\Theta)}{\mathrm{erfi}(\sqrt{2b})}, \tag{36}$$

where erfi is the imaginary error function and b is the concentration parameter. In fact, a specific formula can be obtained for κ by using (36) in the definition (12) with the help of Mathematica [14]. This is

$$\kappa = \frac{1}{2} + \frac{1}{8b} - \frac{1}{4}\sqrt{\frac{2}{\pi b}}\frac{\exp(2b)}{\mathrm{erfi}(\sqrt{2b})}, \tag{37}$$

as given in [15], and it can also be obtained from the formula for the out-of-plane dispersion parameter κ_{op} given in equation (2.28) of [5] by replacing b by $-b$.

We now consider the energy function Ψ specified in (23) and use $k_1/\mu = 5$ and $k_2 = 0.01$, and the values of $b = 0, 0.5, 1.5, 3$ which correspond to $\kappa = 1/3, 0.285, 0.187, 0.096$, respectively. Figure 3 shows plots of the uniaxial Cauchy stress σ versus the corresponding stretch λ based on (35) for the four values of b. The solid curves refer to the model described here while the dashed curves correspond to the case with no fibers excluded and constant κ. As can be seen, as the value of b increases a smaller and smaller proportion of fibers are compressed and hence the difference between the two models decreases; most of the fibers

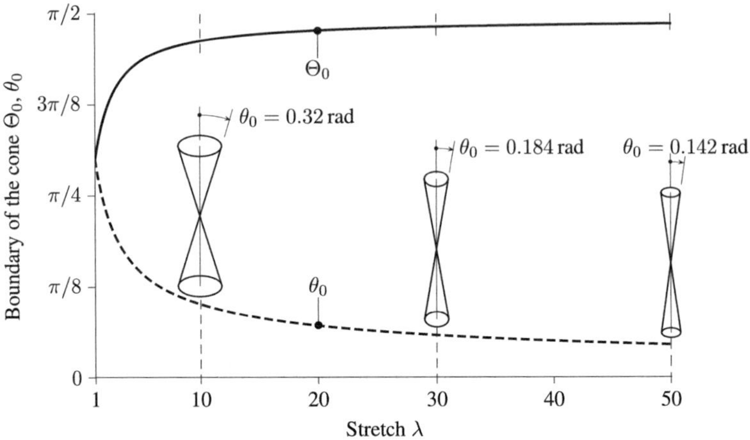

Fig. 4 Evolution of Θ_0 and θ_0 with stretch λ. Cone $\theta_0 = 0.32, 0.184, 0.142$ rad of stretched fibers in the deformed configuration for three representative values $\lambda = 10, 30, 50$

are then aligned with the direction of tension, and the response is stiffest, and if $b \to \infty$, then the curves are identical. For smaller values of b and hence more dispersion there is a significant difference between the two models, and in each case the response is stiffest for the new model. This can be seen from the analytical expression (35) since $\bar{\kappa}$ is smaller than κ. As can be seen from Fig. 3 the continuous curve for $b = 0.5$ crosses over that for $b = 1.5$. This is because for $b = 0.5$, $\bar{\kappa}$ grows faster with λ initially than for $b = 1.5$.

Figure 4 shows the evolution of Θ_0 and θ_0 with stretch λ. As can be seen from the plots the cone of stretched fibers in the deformed configuration is shown for three representative values of λ (10, 30, 50), and we note that the angle of the cone decreases as the value of λ increases (for $\lambda = 10, 30, 50$, $\theta_0 = 0.32, 0.184, 0.142$ rad, respectively). The corresponding cones for the fibers in the reference configuration which are stretched in the deformed configuration are not shown, but we record that the angles for these cones increase with λ (for $\lambda = 10, 30, 50$, $\Theta_0 = 1.476, 1.538, 1.551$ rad, respectively).

2.2 Example: Simple Shear in the (1, 2)-Plane

For the second example we consider simple shear in the $\mathbf{E}_1$-direction in the $(\mathbf{E}_1, \mathbf{E}_2)$-plane, with dispersed fibers lying in this plane. The components of the deformation gradient and the right Cauchy–Green tensors are

$$[\mathbf{F}] = \begin{bmatrix} 1 & \gamma & 0 \\ 0 & 1 & 0 \\ 0 & 0 & 1 \end{bmatrix}, \qquad [\mathbf{C}] = \begin{bmatrix} 1 & \gamma & 0 \\ \gamma & 1+\gamma^2 & 0 \\ 0 & 0 & 1 \end{bmatrix}, \tag{38}$$

respectively, where $\gamma \geq 0$ is the amount of shear, and we take an arbitrary fiber direction $\mathbf{N}$ to have the form

$$\mathbf{N} = \cos\Theta \mathbf{E}_1 + \sin\Theta \mathbf{E}_2 \tag{39}$$

within a dispersion in the (1, 2)-plane which has mean fiber direction $\mathbf{M}$ in that plane given by

$$\mathbf{M} = \cos\Theta_{\mathrm{M}} \mathbf{E}_1 + \sin\Theta_{\mathrm{M}} \mathbf{E}_2. \tag{40}$$

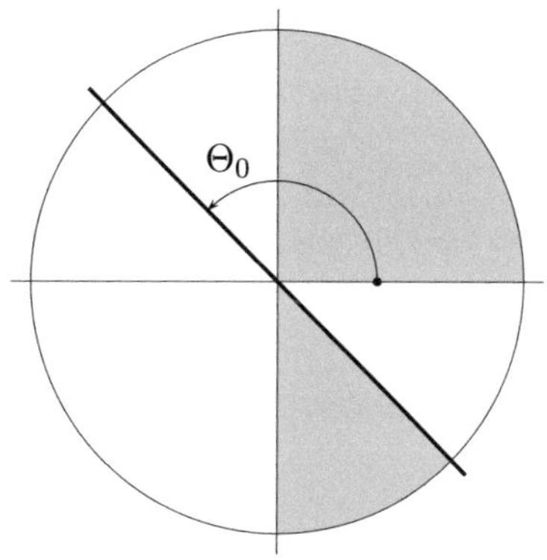

Fig. 5 Depiction of the region in the (1, 2) space traced out by the unit vector **N** in (39) for $0 \leq \Theta \leq 2\pi$. Within the half-circle $-\pi/2 \leq \Theta \leq \pi/2$ the *shaded regions* $-\pi/2 \leq \Theta < -(\pi - \Theta_0)$ and $0 < \Theta \leq \pi/2$ are defined by $(\mathbf{CN}) \cdot \mathbf{N} > 1$. The *angle* Θ_0 satisfies (43) and ranges from $\pi/2$ for $\gamma = 0$ to π for $\gamma \to \infty$

Then

$$(\mathbf{CN}) \cdot \mathbf{N} = 1 + \gamma \sin\Theta(\gamma \sin\Theta + 2\cos\Theta). \tag{41}$$

Since $\gamma \geq 0$, it follows that $(\mathbf{CN}) \cdot \mathbf{N} = 1$ for $\gamma = 0$ or $\Theta = 0$ and, provided $\Theta \neq 0$, $(\mathbf{CN}) \cdot \mathbf{N} > 1$ for

$$\sin\Theta(\gamma \sin\Theta + 2\cos\Theta) > 0. \tag{42}$$

Moreover, $(\mathbf{CN}) \cdot \mathbf{N} = 1$ for $\Theta = \Theta_0$ given by

$$\tan\Theta_0 = -\frac{2}{\gamma}. \tag{43}$$

If we restrict the attention to the range of values $\Theta \in [-\pi/2, \pi/2]$, then $(\mathbf{CN}) \cdot \mathbf{N} > 1$ for the shaded region in Fig. 5, i.e., for $-\pi/2 \leq \Theta < -(\pi - \Theta_0)$ and $0 < \Theta \leq \pi/2$. Equivalently, $(\mathbf{CN}) \cdot \mathbf{N} > 1$ for the range $0 < \Theta < \Theta_0$. Note that this is independent of whether the mean fiber direction **M** is compressed or extended.

Assuming the symmetry $\rho(\Theta + \pi) = \rho(\Theta)$ for the considered 2D dispersion with orientation distribution $\rho(\Theta)$, the normalization condition is

$$\frac{1}{\pi} \int_{-\pi/2}^{\pi/2} \rho(\Theta) \mathrm{d}\Theta = 1. \tag{44}$$

For this case the von Mises distribution $\rho(\Theta)$ symmetric about $\Theta = 0$ is given by

$$\rho(\Theta) = \frac{\exp(b \cos 2\Theta)}{I_0(b)}, \tag{45}$$

where I_0 is the modified Bessel function of the first kind of order 0. Without omitting any directions the dispersion parameter is defined by

$$\kappa = \frac{1}{\pi} \int_{-\pi/2}^{\pi/2} \rho(\Theta) \sin^2\Theta \mathrm{d}\Theta = \frac{1}{2}\left(1 - \frac{I_1(b)}{I_0(b)}\right), \tag{46}$$

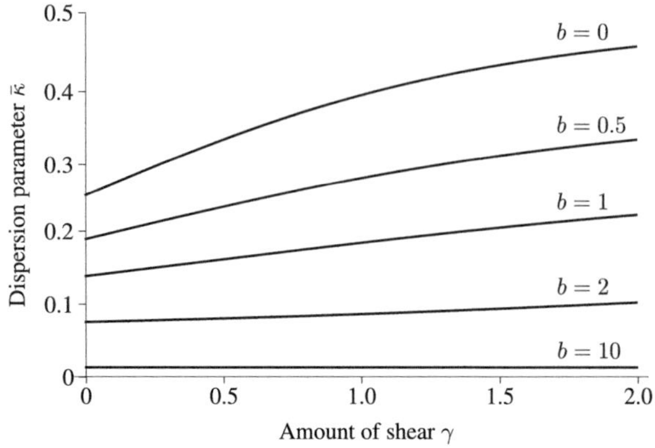

Fig. 6 Example for simple shear showing the dependence of the dispersion parameter $\bar{\kappa} = \kappa_1(\Theta_0)$ on the amount of shear γ for different values of the concentration parameter b, when compressed fibers are excluded

where $I_1(b)$ is the modified Bessel function of the first kind of order 1. The corresponding (generalized) structure tensor is

$$\hat{\mathbf{H}} = \kappa \hat{\mathbf{I}} + (1 - 2\kappa)\mathbf{M} \otimes \mathbf{M} \tag{47}$$

with $\mathbf{M} = \mathbf{E}_1$, where the hat indicates restriction in the $(1, 2)$ plane. For a general mean fiber direction $\mathbf{M}$, given in (40), the structure tensor again has the form (47).

Similarly to the three-dimensional case we define $\kappa_1(\Theta)$ according to

$$\kappa_1(\Theta) = \frac{1}{\pi} \int_{-\pi/2}^{-(\pi-\Theta)} \rho(\xi) \sin^2 \xi \mathrm{d}\xi + \frac{1}{\pi} \int_0^{\pi/2} \rho(\xi) \sin^2 \xi \mathrm{d}\xi. \tag{48}$$

Equivalently, because of the symmetry of ρ, this can be expressed as

$$\kappa_1(\Theta) = \frac{1}{\pi} \int_0^{\Theta} \rho(\xi) \sin^2 \xi \mathrm{d}\xi, \tag{49}$$

with $\Theta \in [0, \pi]$, and hence

$$\kappa_1'(\Theta) = \frac{1}{\pi} \rho(\Theta) \sin^2(\Theta). \tag{50}$$

Similarly to the three-dimensional case Eq. (49) can be obtained by setting $\rho(\Theta) = 0$ for $\Theta \in [\Theta_0, \pi]$, so $\rho(\Theta_0) = 0$ and hence $\kappa_1'(\Theta) = 0$ for $\Theta_0 \le \Theta \le \pi$.

As in the three-dimensional case we use the notation $\bar{\kappa} = \kappa_1(\Theta_0)$. Representative plots of $\bar{\kappa} = \kappa_1(\Theta_0)$ as a function of γ for different values of the concentration parameter b are shown in Fig. 6 based on Eq. (49) with $\Theta = \Theta_0$ given by (43).

Consider again the model $\Psi^\star(I_1, I_4^\star)$ but now with the restriction to plane strain. Then, similarly to the three-dimensional case, the two-dimensional Cauchy stress is obtained as the specialization of (30), written as

$$\hat{\sigma} = 2\psi_1^\star \hat{\mathbf{B}} + 2\psi_4^\star \hat{\mathbf{h}} - p\hat{\mathbf{I}}, \tag{51}$$

where a hat again signifies the restriction to two dimensions, so that $\hat{\mathbf{h}} = \bar{\kappa}\hat{\mathbf{B}} + (1-2\bar{\kappa})\mathbf{m}\otimes\mathbf{m}$. For simple shear we have

$$[\mathbf{m}] = [\mathbf{F}][\mathbf{M}] = \begin{bmatrix} M_1 + \gamma M_2 \\ M_2 \\ 0 \end{bmatrix}, \qquad [\mathbf{B}] = \begin{bmatrix} 1+\gamma^2 & \gamma & 0 \\ \gamma & 1 & 0 \\ 0 & 0 & 1 \end{bmatrix}, \tag{52}$$

where $[\mathbf{M}] = [M_1, M_2, 0] = [\cos\Theta_{\mathrm{M}}, \sin\Theta_{\mathrm{M}}, 0]$, and

$$\hat{I}_1 = \operatorname{tr}\hat{\mathbf{B}} = 2+\gamma^2, \qquad \hat{I}_4^\star = \bar{\kappa}(2+\gamma^2) + (1-2\bar{\kappa})(1+2\gamma M_1 M_2 + \gamma^2 M_2^2). \tag{53}$$

For the considered plane strain the strain-energy function (23) takes on the form

$$\Psi = \frac{\mu}{2}(\hat{I}_1 - 2) + \frac{k_1}{2k_2}\{\exp[k_2(\hat{I}_4^\star - 1)^2] - 1\}, \tag{54}$$

and the in-plane Cauchy stress components are derived as

$$\hat{\sigma}_{11} = \mu(1+\gamma^2) + 2\psi_4^\star[\bar{\kappa}(1+\gamma^2) + (1-2\bar{\kappa})(M_1+\gamma M_2)^2] - p, \tag{55}$$

$$\hat{\sigma}_{22} = \mu + 2\psi_4^\star[\bar{\kappa} + (1-2\bar{\kappa})M_2^2] - p, \tag{56}$$

$$\hat{\sigma}_{12} = \mu\gamma + 2\psi_4^\star[\bar{\kappa}\gamma + (1-2\bar{\kappa})(M_1+\gamma M_2)M_2]. \tag{57}$$

Of particular interest is the dependence of the shear stress $\hat{\sigma}_{12}$ on the amount of shear γ, and henceforth we focus on this relationship. First, however, we note that for a general direction (39) we have $\hat{I}_4 = 1 + \gamma\sin 2\Theta + \gamma^2\sin^2\Theta$, which has its maximum value at $\Theta = \Theta_{\max}$ given by

$$\tan(2\Theta_{\max}) = -\frac{2}{\gamma}, \tag{58}$$

an equation which should be compared with (43). It follows that $\Theta_{\max} = \Theta_0/2$ and this ranges from $\pi/4$ at $\gamma = 0$ to $\pi/2$ as $\gamma \to \infty$. Thus, the direction in which the stretch is largest is close to $\Theta = \pi/4$ for moderate values of γ and this has a strong influence on the stiffness of the shear response. In particular, for the range $\gamma \in [0,2]$ considered in the subsequent examples $\Theta_{\max} \in [\pi/4, 3\pi/8]$.

The following four illustrative examples are now considered.

Example 1 $\Theta_{\mathrm{M}} = 0$, $\hat{I}_4^\star - 1 = \bar{\kappa}\gamma^2$, with

$$\hat{\sigma}_{12} = \mu\gamma + 2k_1\bar{\kappa}^2\gamma^3\exp(k_2\bar{\kappa}^2\gamma^4). \tag{59}$$

Example 2 $\Theta_{\mathrm{M}} = \pi/4$, $\hat{I}_4^\star - 1 = \gamma + \frac{1}{2}\gamma^2 - 2\bar{\kappa}\gamma$, with

$$\hat{\sigma}_{12} = \mu\gamma + k_1(\hat{I}_4^\star - 1)\exp[k_2(\hat{I}_4^\star - 1)^2](\gamma + 1 - 2\bar{\kappa}). \tag{60}$$

Example 3 $\Theta_{\mathrm{M}} = \pi/2$, $\hat{I}_4^\star - 1 = (1-\bar{\kappa})\gamma^2$, with

$$\hat{\sigma}_{12} = \mu\gamma + 2k_1(1-\bar{\kappa})^2\gamma^3\exp[k_2(1-\bar{\kappa})^2\gamma^4]. \tag{61}$$

Example 4 $\Theta_{\mathrm{M}} = 3\pi/4$, $\hat{I}_4^\star - 1 = -\gamma + \frac{1}{2}\gamma^2 + 2\bar{\kappa}\gamma$, with

$$\hat{\sigma}_{12} = \mu\gamma + k_1(\hat{I}_4^\star - 1)\exp[k_2(\hat{I}_4^\star - 1)^2](\gamma - 1 + 2\bar{\kappa}). \tag{62}$$

We now compare the shear stress $\hat{\sigma}_{12}$ versus the amount of shear γ behavior for the model in which compressed fibers are omitted with that in which all fibers are included for these four representative examples. The results are shown in Fig. 7 for $b = 0.1, 1, 10$, which correspond to $\kappa = 0.475, 0.277, 0.026$, respectively. We choose $k_1/\mu = 5$ and $k_2 = 0.01$ and take these values for purposes of illustration but the qualitative nature of the plots is very similar for other values.

First, for $b = 10$, we note that there is no essential difference between the plots with or without compressed fibers excluded (and the slight difference disappears altogether as b becomes larger and larger). This is because there is very little dispersion and the fibers are concentrated close to the mean fiber direction. For $\Theta_{\mathrm{M}} = 3\pi/4$, in particular, we note that the mean fiber direction is outside the range of integration for small values of γ so that the fibers do not contribute to the strain energy or stress initially, which is supported only by the isotropic term. As γ increases the fiber contribution to the stress is slightly negative for a small range of values of γ and leads to a maximum in the shear stress versus amount of shear behavior. Thereafter the fiber contribution becomes positive and increases rapidly with $\gamma > 2$ (not shown). Note that for $\Theta_{\mathrm{M}} = 0$ the stress is relatively small since for simple shear there is no stretch in the direction of shear. On the other hand, for $\Theta_{\mathrm{M}} = \pi/4$ and $\Theta_{\mathrm{M}} = \pi/2$, which are close to the direction of maximum stretch, the stress is much larger.

The pattern is similar for $b = 1$ in respect of the stress magnitudes although they are larger for $\Theta_{\mathrm{M}} = 0$ and $\Theta_{\mathrm{M}} = 3\pi/4$ because of the dispersion and there is a difference between the two models. For these two angles the 'all fiber' model has a higher stress than the excluded fiber model, but for $\Theta_{\mathrm{M}} = \pi/4$ and $\Theta_{\mathrm{M}} = \pi/2$ it is the other way around. And the situation for $b = 0.1$ is similar to that for $b = 1$ except that the stiffnesses for all four mean fiber directions are very similar, because this is very close to the isotropic situation. These differences can be seen in the Eqs. (59)–(62) since for Examples 1 and 4 the $\bar{\kappa}$ terms are positive, while in the Examples 2 and 3 they are negative. Thus, in contrast to uniaxial extension, for which the excluded fiber model gives a stiffer response than the all fiber model, in simple shear the stress can be either stiffer or softer depending on the mean fiber direction.

3 Some Comparisons with the AI Model

We now compare the GST model with the AI model due to Lanir [3] but without excluding compressed fibers. For the AI model the energy function is an integral of the form

$$n\int_{\Omega} \rho(\Theta, \Phi) w(\lambda) \sin\Theta \, \mathrm{d}\Theta \, \mathrm{d}\Phi, \tag{63}$$

where Ω is the unit sphere $\{(\Theta, \Phi) \mid \Theta \in [0, \pi], \Phi \in [0, 2\pi]\}$, the orientation density ρ depends on Θ and Φ in general, as does λ, which is defined as $\sqrt{I_4}$, where $I_4 = \mathbf{N} \cdot (\mathbf{CN})$, with $\mathbf{N}$ given by (7), for a general Cauchy–Green tensor $\mathbf{C}$, and $w(\lambda)$ is the strain energy of an individual fiber in the direction $\mathbf{N}$. Note that (63) does not include an isotropic term associated with the non-collagenous matrix material.

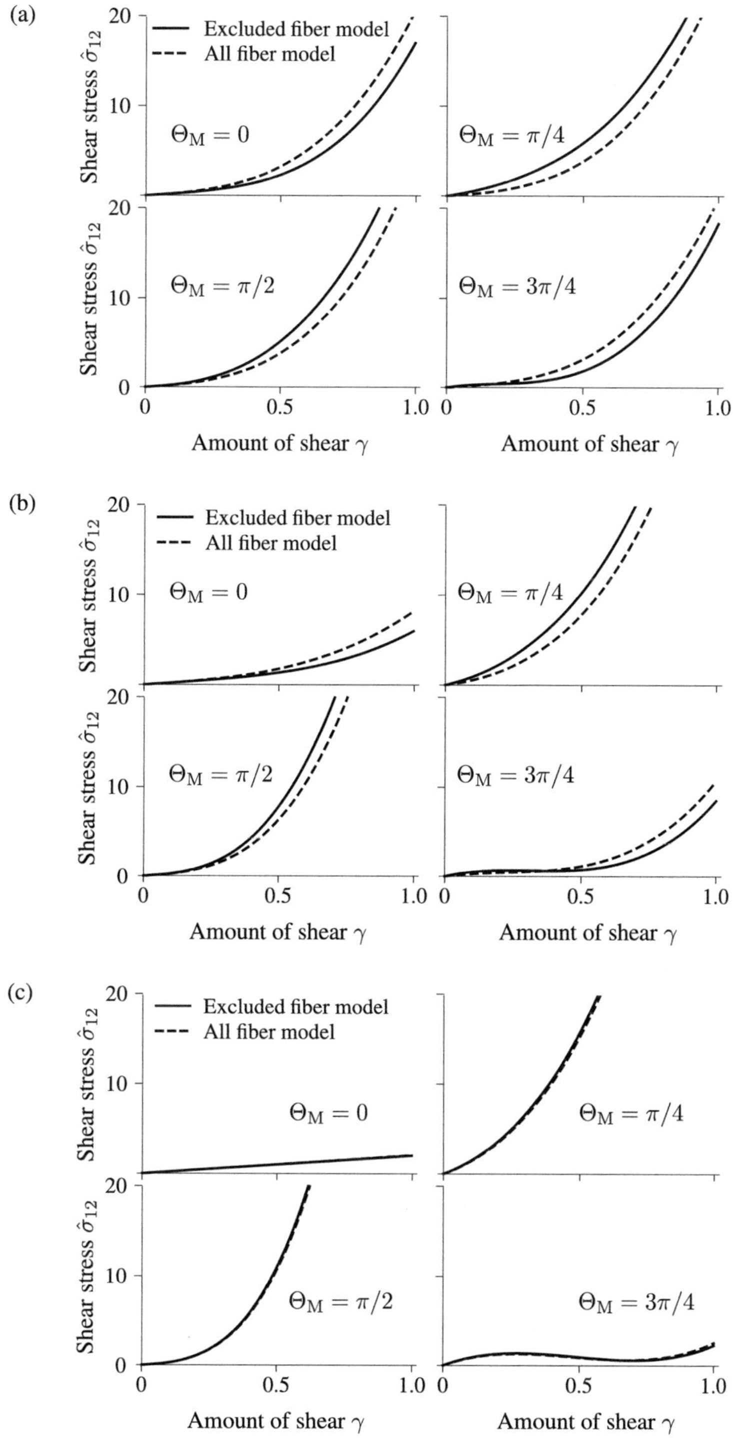

Fig. 7 Shear stress $\hat{\sigma}_{12}$ versus amount of shear γ for both the excluded fiber model (*continuous curves*) and the 'all fiber' model (*dashed curves*) for $\Theta_M = 0, \pi/4, \pi/2, 3\pi/4$: (**a**) $b = 0.1$ ($\kappa = 0.475$); (**b**) $b = 1$ ($\kappa = 0.277$); (**c**) $b = 10$ ($\kappa = 0.026$). For each case $k_1/\mu = 5$ and $k_2 = 0.01$

For purposes of comparison we include a neo-Hookean isotropic term and characterize the AI model by the strain-energy function Ψ_{AI} as

$$\Psi_{\mathrm{AI}} = \frac{1}{2}\mu(I_1 - 3) + n\int_{\Omega} \rho(\Theta, \Phi) w(\lambda) \sin\Theta \, \mathrm{d}\Theta \, \mathrm{d}\Phi. \tag{64}$$

The corresponding GST model has the form of (21) and (23), which we now write as

$$\Psi_{\mathrm{GST}} = \frac{1}{2}\mu(I_1 - 3) + \Psi_{\mathrm{f}}\big(I_4^\star\big), \tag{65}$$

where Ψ_{f} represents the strain energy of the dispersed fibers, and, for a rotationally symmetric dispersion, $I_4^\star$ is given by (15). Assuming that no energy or stress is associated with the fibers in the reference configuration, we have

$$w(1) = w'(1) = 0, \qquad \Psi_{\mathrm{f}}(1) = \Psi_{\mathrm{f}}'(1) = 0. \tag{66}$$

In the paper [7] it was attempted to compare the predictions of the above GST model with the AI model having the same exponential forms (5) and (6) and the same material constants $c_1 = k_1$ and $c_2 = k_2$. The associated Cauchy stress relations are

$$\boldsymbol{\sigma}_{\mathrm{AI}} = \mu\mathbf{B} + 2nc_1\int_{\Omega} \rho(\Theta, \Phi)(I_4 - 1)\exp\big[c_2(I_4 - 1)^2\big]\mathbf{n}\otimes\mathbf{n} \sin\Theta \, \mathrm{d}\Theta \, \mathrm{d}\Phi - p\mathbf{I} \tag{67}$$

and

$$\boldsymbol{\sigma}_{\mathrm{GST}} = \mu\mathbf{B} + 2k_1\big(I_4^\star - 1\big)\exp\big[k_2\big(I_4^\star - 1\big)^2\big]\big[\kappa\mathbf{B} + (1 - 3\kappa)\mathbf{m}\otimes\mathbf{m}\big] - p\mathbf{I}, \tag{68}$$

where we recall that $\mathbf{n} = \mathbf{FN}$ and $\mathbf{m} = \mathbf{FM}$.

We now elaborate on this particular point by considering the case of small strains. The Green–Lagrange strain tensor $\mathbf{E}$ is defined by $\mathbf{E} = (\mathbf{C} - \mathbf{I})/2$, so that $I_1 = 3 + 2\,\mathrm{tr}\,\mathbf{E}$ and $I_4 = 1 + 2\mathbf{N}\cdot(\mathbf{EN})$. These expressions for the invariants are exact but when we expand $w(\lambda)$ and $\Psi_{\mathrm{f}}(I_4^\star)$ to the second order in $\mathbf{E}$ we obtain

$$w(\lambda) \approx \frac{1}{2}(\mathbf{N}\cdot\mathbf{EN})^2 w''(1), \qquad \Psi_{\mathrm{f}}\big(I_4^\star\big) \approx 2\Psi_{\mathrm{f}}''(1)(1 - 3\kappa)^2\big[\mathbf{M}\cdot(\mathbf{EM})\big]^2, \tag{69}$$

where the properties (66) have been used. By substituting these approximations into (64) and (65) we obtain

$$\Psi_{\mathrm{AI}} = \mu\,\mathrm{tr}\,\mathbf{E} + \frac{n}{2}w''(1)\int_{\Omega} \rho(\Theta, \Phi)\big[\mathbf{N}\cdot(\mathbf{EN})\big]^2 \sin\Theta \, \mathrm{d}\Theta \, \mathrm{d}\Phi, \tag{70}$$

and

$$\Psi_{\mathrm{GST}} = \mu\,\mathrm{tr}\,\mathbf{E} + 2\Psi_{\mathrm{f}}''(1)(1 - 3\kappa)^2\big[\mathbf{M}\cdot(\mathbf{EM})\big]^2, \tag{71}$$

respectively. Note that, since we are considering incompressible materials, expansion of the incompressibility condition $\det\mathbf{C} = 1$ to the second order gives $\mathrm{tr}\,\mathbf{E} = \mathrm{tr}(\mathbf{E}^2)$, which can be used in (70) and (71). The corresponding linearized Cauchy stresses are deduced from (70) and (71) according to

$$\boldsymbol{\sigma}_{\mathrm{AI}} = 2\mu\mathbf{E} + nw''(1)\int_{\Omega} \rho(\Theta, \Phi)\big[\mathbf{N}\cdot(\mathbf{EN})\big]\mathbf{N}\otimes\mathbf{N} \sin\Theta \, \mathrm{d}\Theta \, \mathrm{d}\Phi - p\mathbf{I} \tag{72}$$

and

$$\sigma_{\mathrm{GST}} = 2\mu\mathbf{E} + 4\Psi_{\mathrm{f}}''(1)(1-3\kappa)^2\big[\mathbf{M}\cdot(\mathbf{EM})\big]\mathbf{M}\otimes\mathbf{M} - p\mathbf{I}, \tag{73}$$

respectively.

For definiteness we now consider a uniaxial strain with component ε in the direction $\mathbf{M} = \mathbf{E}_3$, with Θ measured from $\mathbf{E}_3$ and with ρ independent of Φ. Then, the dispersion is rotationally symmetric about $\mathbf{E}_3$ and, by symmetry and the incompressibility condition, the lateral strains are, to the first order, each equal to $-\varepsilon/2$. By using the definition (7) and $\mathbf{M} = \mathbf{E}_3$ it follows that $\mathbf{M}\cdot(\mathbf{EM}) = \varepsilon$ and $\mathbf{N}\cdot(\mathbf{EN}) = (3\cos^2\Theta - 1)\varepsilon/2$. The corresponding uniaxial stresses after elimination of p are obtained as

$$\sigma_{\mathrm{AI}} = 3\mu\varepsilon + \frac{n\pi}{2}w''(1)f(b)\varepsilon, \qquad \sigma_{\mathrm{GST}} = 3\mu\varepsilon + 4\Psi_{\mathrm{f}}''(1)(1-3\kappa)^2\varepsilon, \tag{74}$$

where

$$f(b) = \int_0^\pi \rho(\Theta)\big(3\cos^2\Theta - 1\big)^2 \sin\Theta \,\mathrm{d}\Theta, \tag{75}$$

has been introduced, which, on use of Mathematica [14], can be integrated to give

$$f(b) = 2 + \frac{27}{8b^2} + \frac{3}{b} + \frac{3}{2b^2}\sqrt{\frac{b}{2\pi}}\frac{(4b-9)\exp(2b)}{\mathrm{erfi}(\sqrt{2b})}, \tag{76}$$

and then, on use of (37), this can be simplified to

$$f(b) = 4 + (1-3\kappa)\big(4 - 9b^{-1}\big). \tag{77}$$

If the two models are to predict the same stress then from (74) we must have

$$n\pi w''(1)f(b) = 8(1-3\kappa)^2\Psi_{\mathrm{f}}''(1). \tag{78}$$

For the exponential model used in (23) we obtain $\Psi_{\mathrm{f}}''(1) = k_1$. If we use the exponential model in Eq. (5) we obtain $w''(1) = 4c_1$, and hence the constants c_1 and k_1 must be related by

$$n\pi c_1\big[4 + (1-3\kappa)\big(4 - 9b^{-1}\big)\big] = 2k_1(1-3\kappa)^2. \tag{79}$$

In the paper [7] the exponential model with $k_1 = c_1$ was adopted for the GST and AI models in order to compare the predictions of the models for several modes of deformation for a range of values of κ. In view of the formula (37), the above relation would then give two equations relating b and κ. The only solutions of these simultaneous equations are $\kappa = 0$ and $\kappa = 1/3$, as can easily be verified, so the use of the same exponential models with the same material constants for other values of κ in [7] is inappropriate, and the resulting comparisons between the predictions of the GST and the AI models are therefore misleading. Moreover, for each different deformation considered a formula relating c_1 and k_1 different from (79) would be obtained. This means that even in the small deformation range the two models are not equivalent.

Thus, to compare the predictions of the two models, for each separate κ (or equivalently b) in uniaxial extension, for example, and for a given k_1 the formula (79) needs to be used to obtain c_1 for each b separately for the corresponding AI model. To illustrate the point the plots in Fig. 8 provide comparisons of the predictions of the two models for several values of b for both simple tension and simple shear. For this purpose it suffices to

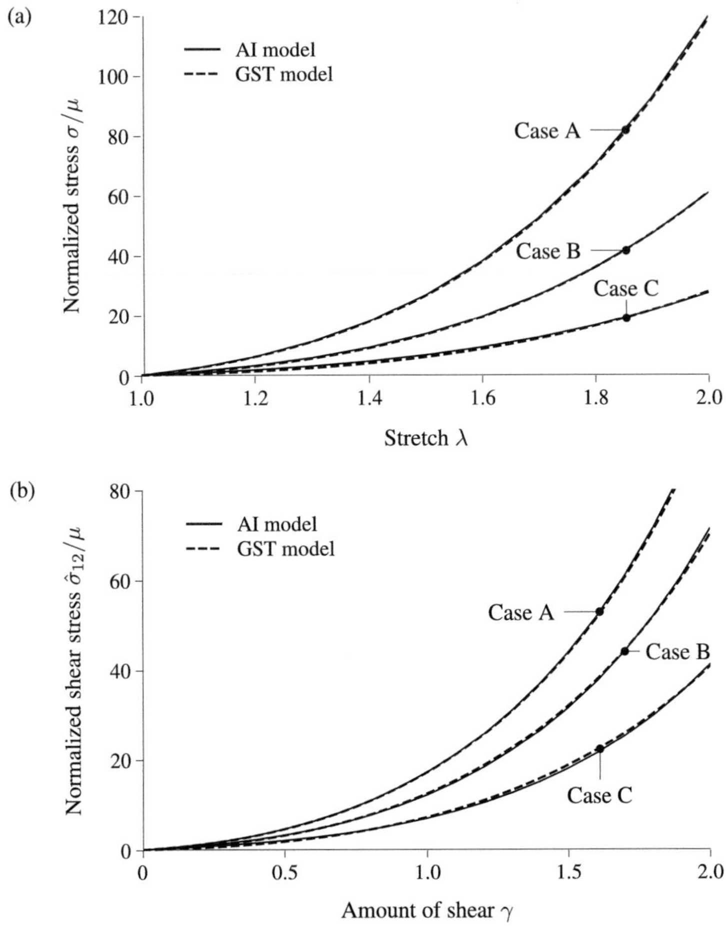

Fig. 8 AI model (*continuous curve*) and GST model (*dashed curve*) for Case A (very little dispersion), Case B (intermediate dispersion) and Case C (large dispersion, close to isotropy): (**a**) simple tension; (**b**) simple shear for $\Theta_M = 60°$. Parameters are for Case A (AI: $b = 10$, $c_1 = 5$; GST: $\kappa = 0.026$, $k_1 = 5$), Case B (AI: $b = 1.5$, $c_1 = 5$; GST: $\kappa = 0.15$, $k_1 = 5.2$), Case C (AI: $b = 0.1$, $c_1 = 5$; GST for (**a**): $\kappa = 0.26$, $k_1 = 5.7$; GST for (**b**): $\kappa = 0.26$, $k_1 = 4.1$), while for all cases and the two models $c_2 = k_2 = 0.01$. Note that the factor n in (67) has been absorbed in the parameter c_1

take $c_2 = k_2$ for all the considered values of b. The continuous curves correspond to the AI model and the dashed curves to the GST model. The parameter values have been set so that the continuous and dashed curves can be seen separately but by refining these parameters it is possible to arrange for the plots to be indistinguishable. It is quite clear that the predictions of the two models essentially coincide. Note that in Case A, where there is very little dispersion, $c_1 = k_1$. In Case B, where there is more dispersion, c_1 and k_1 are slightly different, while in Case C, where the dispersion is larger, close to isotropy, the values for c_1 and k_1 are quite different for both simple extension and simple shear. Note that for Case C the values of k_1 are different for simple extension and simple shear because of the role of the mean fiber direction and the different dispersions used in the two cases. We emphasize that for the plots in Fig. 8 fibers under compression were not excluded, but similar results can be obtained when compressed fibers are excluded.

4 Summary and Concluding Remarks

In contrast to the claims in the literature that compressed fibers cannot be excluded in the generalized structure tensor model, in the present paper we have shown that this is not the case by developing a rather simple and efficient method for excluding fibers under compression for the GST model. The consequences of this have been illustrated for the cases of simple tension and simple shear, and the differences between the stress responses under exclusion and non-exclusion have been highlighted. The theory presented herein can be used as a basis for a general implementation into a finite element program in which the dispersion parameter needs to be evaluated at each Gauss point; details of this will be provided in a subsequent paper. The present model just constitutes a small extension to the dispersion model developed in [4]. The analysis can also be extended to the case of a non-symmetric fiber dispersion involving two dispersion parameters, as introduced in [5], and will result in two deformation dependent dispersion parameters. The GST model has also been used in considering inelastic effects such as damage in, e.g., [16] and [17]. We emphasize that we have already provided a detailed analysis of the exclusion of compressed fibers for the AI approach in [10, 11]; however, it seems that the AI approach has not been implemented in commercially available finite element programs.

In the second part of the present paper we have drawn attention to misleading statements in the literature concerning the relative values of the two approaches. Unfortunately, statements which suggest that the AI approach is superior by referring to it as the 'gold standard', are based on erroneous considerations, as we have highlighted at several locations in the preceding sections. Indeed we have shown that both models have equivalent predictive power in two examples for which the study [7] and subsequent studies have indicated significant differences which are referred to as 'errors' therein. As the above examples demonstrate, careful analysis of the literature is needed, and one should not just accept that what is published is correct.

References

1. Schriefl, A.J., Wolinski, H., Regitnig, P., Kohlwein, S.D., Holzapfel, G.A.: An automated approach for 3D quantification of fibrillar structures in optically cleared soft biological tissues. J. R. Soc. Interface **10**, 20120760 (2013)
2. Pierce, D.M., Trobin, W., Raya, J.G., Trattnig, S., Bischof, H., Glaser, Ch., Holzapfel, G.A.: DT-MRI based computation of collagen fiber deformation in human articular cartilage: a feasibility study. Ann. Biomed. Eng. **38**, 2447–2463 (2010)
3. Lanir, Y.: Constitutive equations for fibrous connective tissues. J. Biomech. **16**, 1–12 (1983)
4. Gasser, T.C., Ogden, R.W., Holzapfel, G.A.: Hyperelastic modelling of arterial layers with distributed collagen fibre orientations. J. R. Soc. Interface **3**, 15–35 (2006)
5. Holzapfel, G.A., Niestrawska, J.A., Ogden, R.W., Reinisch, A.J., Schriefl, A.J.: Modelling non-symmetric collagen fibre dispersion in arterial walls. J. R. Soc. Interface **12**, 20150188 (2015)
6. Federico, S., Herzog, W.: Towards an analytical model of soft biological tissues. J. Biomech. **41**, 3309–3313 (2008)
7. Cortes, D.H., Lake, S.P., Kadlowec, J.A., Soslowsky, L.J., Elliot, D.M.: Characterizing the mechanical contribution of fiber angular distribution in connective tissue: comparison of two modeling approaches. Biomech. Model. Mechanobiol. **9**, 651–658 (2010)
8. Pandolfi, A., Vasta, M.: Fiber distributed hyperelastic modeling of biological tissues. Mech. Mater. **44**, 151–162 (2012)
9. Lanir, Y., Namani, R.: Reliability of structure tensors in representing soft tissues structure. J. Mech. Behav. Biomed. Mater. **46**, 222–228 (2015)
10. Holzapfel, G.A., Ogden, R.W.: On the tension–compression switch in soft fibrous solids. Eur. J. Mech. A, Solids **49**, 561–569 (2015)

11. Li, K., Ogden, R.W., Holzapfel, G.A.: Computational method for excluding fibers under compression in modeling soft fibrous solids. Eur. J. Mech. A, Solids **57**, 178–193 (2016)
12. Melnik, A.V., Borja Da Rocha, H., Goriely, A.: On the modeling of fiber dispersion in fiber-reinforced elastic materials. Int. J. Non-Linear Mech. **75**, 92–106 (2015)
13. Gizzi, A., Pandolfi, A., Vasta, M.: Statistical characterization of the anisotropic strain energy in soft materials with distributed fibers. Mech. Mater. **92**, 119–138 (2016)
14. Wolfram Research, Inc. Mathematica, Version 10.4. Champaign, Illinois (2016)
15. Ogden, R.W.: Nonlinear continuum mechanics and modelling the elasticity of soft biological tissues with a focus on artery walls. In: Holzapfel, G.A., Ogden, R.W. (eds.) Lecture Notes from the Summer School "Biomechanics: Trends in Modeling and Simulation", Graz, Austria, September, 2014. Springer, Heidelberg (2016)
16. Weisbecker, H., Pierce, D.M., Regitnig, P., Holzapfel, G.A.: Layer-specific damage experiments and modeling of human thoracic and abdominal aortas with non-atherosclerotic intimal thickening. J. Mech. Behav. Biomed. Mater. **12**, 93–106 (2012)
17. Fereidoonnezhad, B., Naghdabadi, R., Holzapfel, G.A.: Stress softening and permanent deformation in human aortas: continuum and computational modeling with application to arterial clamping. J. Mech. Behav. Biomed. Mater. **61**, 600–616 (2016)

J Elast (2017) 129:69–105
DOI 10.1007/s10659-017-9630-9

Reactive Constrained Mixtures for Modeling the Solid Matrix of Biological Tissues

Robert J. Nims[1] · **Gerard A. Ateshian**[1]

Received: 27 July 2016 / Published online: 1 March 2017

Abstract This article illustrates our approach for modeling the solid matrix of biological tissues using reactive constrained mixtures. Several examples are presented to highlight the potential benefits of this approach, showing that seemingly disparate fields of mechanics and chemical kinetics are actually closely interrelated and may be elegantly expressed in a unified framework. Thus, constrained mixture models recover classical theories for fibrous materials with bundles oriented in different directions or having different reference configurations, that produce characteristic fiber recruitment patterns under loading. Reactions that exchange mass among various constituents of a mixture may be used to describe tissue growth and remodeling, which may also alter the material's anisotropy. Similarly, reactions that describe the breaking and reforming of bonds may be used to model free energy dissipation in a viscoelastic material. Therefore, this framework is particularly well suited for modeling biological tissues.

Keywords Mixture theory · Reactive mixtures · Biomechanics · Growth and remodeling · Viscoelasticity

Mathematics Subject Classification (2010) 74L15 · 92C10

1 Introduction

Biological tissues are living structures that grow and remodel in response to mechanical, chemical and electrical stimuli. Growth and remodeling are reactive processes that transform reactants into products, thus altering the composition, properties, and ultrastructure of those tissues over time. A framework that models these reactive processes provides the necessary foundation for describing the behavior of such tissues. Mixture theory represents a natural framework for modeling reactive continua since it accounts for the presence of any number of constituents within an elemental volume of the continuum, allowing mass

✉ G.A. Ateshian
ateshian@columbia.edu

[1] Columbia University, 500 West 120th St, MC4703, New York, NY 10027, USA

exchanges among those constituents. These reactions may describe phenomena occurring at a molecular level, that influence the macroscopic mixture response, thereby providing a fundamental multi-scale modeling framework.

The general framework of reactive mixtures has been described in the prior literature [4, 12, 13, 21]. A characteristic aspect of this general framework is that each constituent may have its own independent motion. The relative motion between constituents introduces independent diffusive terms in the governing equations; such terms account for well-known fundamental mechanisms, such as the diffusive drag between a porous solid matrix and its interstitial fluid, as embodied in Darcy's law for porous media, and the diffusive drag between a solute and its surrounding solvent, as embodied in Fick's law for mass transport. In the general framework of unconstrained mixtures, these diffusive terms also produce less familiar contributions to the mixture stress tensor, internal and free energy, heat flux, heat supply, etc. Consequently, formulations of unconstrained reactive mixture models generally exhibit a level of complexity that often deters the adoption of this framework or limits it to highly specialized cases, such as non-reactive binary mixtures of intrinsically incompressible solid and fluid constituents for modeling porous media [34, 41]. In this study we focus on a highly useful specialization of this general framework.

Humphrey and Rajagopal [30, 31] proposed to specialize the mixture framework to the case where all solid constituents are constrained to move together. They used this constrained solid mixture approach to model vascular tissue consisting of collagen, elastin, and smooth muscle cells. This concept of constrained solid mixtures was later applied to model residual stresses in soft tissue growth and remodeling, in the arterial wall as well as in cells and other tissues [9, 19, 24, 25, 33, 47, 49–51, 58–61, 66]. The theoretical foundation for constrained reactive mixtures adopted in this presentation has been presented by Ateshian and Ricken [9], where it was specialized from the general framework of unconstrained reactive mixtures. In subsequent studies, it was shown that this framework could also be used to model viscoelasticity, by modeling strong and weak bonds as mixture constituents, with weak bonds breaking and reforming in response to loading [5]; and damage mechanics, by modeling bonds that break permanently [42].

The primary benefit of using a reactive mixture approach for modeling growth, viscoelasticity and damage mechanics is that the evolution of the tissue response with time is explicitly tied to the evolving composition (the concentration of each constituent, such as intact, broken and reformed bonds), which is governed by the axiom of mass balance [18]. Therefore, growth mechanics, viscoelasticity and damage mechanics may be modeled using only observable variables, without the need to introduce internal (hidden) variables and associated evolution laws, as commonly done in classical approaches [17].

In this paper we review these recent developments by presenting them in a unified context and illustrating them with simple examples, mostly using infinitesimal strain theory. Section 2 summarizes basic definitions for mixtures and examines the axiom of mass balance for reactive mixture constituents, which provides the fundamental basis for this reactive framework. This introductory section reviews concepts of chemical kinetics that are generally familiar to chemistry and chemical engineering students, but less familiar to students of solid mechanics.

In Sect. 3, we briefly show that familiar classical models for fibrous tissue mechanics illustrate examples of non-reactive constrained mixtures. This is followed by examples of the application of constrained mixtures to the analysis of soft tissue remodeling and multi-generational growth mechanics. Finally, in Sect. 4, we derive the classical theory of linear viscoelasticity using the framework of constrained reactive mixtures. All of these illustrations serve to demonstrate that the framework of reactive constrained mixtures provides a common framework for a broad range of modeling tasks relevant to living biological tissues.

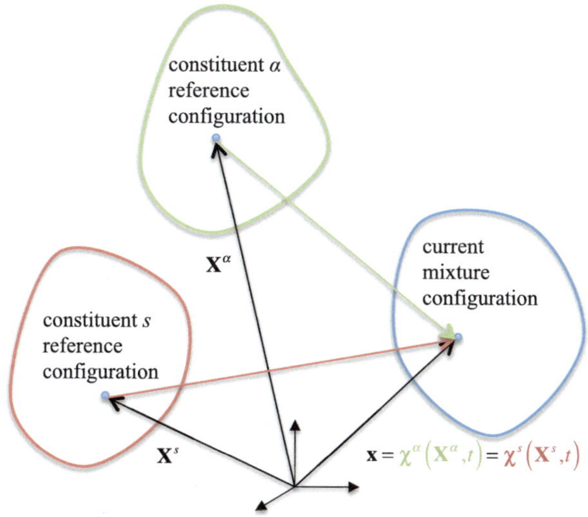

Fig. 1 Motion of a mixture and its constituents

2 Mass Balance in Reactive Mixtures

2.1 Motion

A mixture may consist of any number of constituents, each denoted by α. The motion of constituent α is given by $\chi^\alpha(\mathbf{X}^\alpha, t)$, where $\mathbf{X}^\alpha$ denotes material points in their referential configuration. At the current time t, all constituents that occupy an elemental region at the spatial position $\mathbf{x}$ may thus have originated from different referential positions $\mathbf{X}^\alpha$, such that $\mathbf{x} = \chi^\alpha(\mathbf{X}^\alpha, t)$. The positions $\mathbf{X}^\alpha$ of material points of constituent α represent the *reference configuration* for that constituent at some initial time. In a reactive mixture, constituents may form as products of a reaction at various time points. Therefore, the initial time t^α for the reference configuration of constituent α may differ for each constituent. The positions $\mathbf{x} = \chi^\alpha(\mathbf{X}^\alpha, t)$ of the same material points at the current time t represent the *current configuration* for that constituent. All constituents α of the mixture share the same current configuration (Fig. 1).

The velocity $\mathbf{V}^\alpha(\mathbf{X}^\alpha, t)$ of constituent α in its material frame is defined as

$$\mathbf{V}^\alpha(\mathbf{X}^\alpha, t) = \frac{\partial \chi^\alpha(\mathbf{X}^\alpha, t)}{\partial t}, \tag{2.1}$$

whereas the deformation gradient $\mathbf{F}^\alpha(\mathbf{X}^\alpha, t)$ is defined as

$$\mathbf{F}^\alpha(\mathbf{X}^\alpha, t) = \frac{\partial \chi^\alpha(\mathbf{X}^\alpha, t)}{\partial \mathbf{X}^\alpha}. \tag{2.2}$$

From these relations, the material time derivative of $\mathbf{F}^\alpha$ in the material frame is

$$\frac{\partial \mathbf{F}^\alpha(\mathbf{X}^\alpha, t)}{\partial t} = \frac{\partial \mathbf{V}^\alpha(\mathbf{X}^\alpha, t)}{\partial \mathbf{X}^\alpha} = \mathbf{L}^\alpha\left(\chi^\alpha(\mathbf{X}^\alpha, t), t\right) \cdot \mathbf{F}^\alpha(\mathbf{X}^\alpha, t), \tag{2.3}$$

where

$$\mathbf{L}^\alpha(\mathbf{x}, t) = \operatorname{grad} \mathbf{v}^\alpha(\mathbf{x}, t) \tag{2.4}$$

is the spatial gradient of the velocity $\mathbf{v}^\alpha(\mathbf{x},t)$ of α. We may rewrite this expression as $\dot{\mathbf{F}}^\alpha = \mathbf{L}^\alpha \cdot \mathbf{F}^\alpha$.

A *constrained mixture* represents the special case when all solid constituents share the same spatial velocity [30]; this constraint is expressed in the spatial frame as

$$\mathbf{v}^\alpha(\mathbf{x},t) = \mathbf{v}^s(\mathbf{x},t), \tag{2.5}$$

where s represents one of the constrained mixture constituents. This relation implies that $\mathbf{L}^\alpha(\mathbf{x},t) = \mathbf{L}^s(\mathbf{x},t)$ for all constrained mixture constituents α. Note that the respective referential configurations $\mathbf{X}^\alpha$ of the various constituents α need not be the same, since each constituent may have formed (come into existence as the product of a reaction) at a different time point. For the pair of constituents α and s, it follows that $d\mathbf{x} = \mathbf{F}^\alpha(\mathbf{X}^\alpha,t)\cdot d\mathbf{X}^\alpha = \mathbf{F}^s(\mathbf{X}^s,t)\cdot d\mathbf{X}^s$, explicitly establishing the relation between constituent deformation gradients. In this expression, $d\mathbf{X}^\alpha$ is a differential line element at $\mathbf{X}^\alpha$ which transforms into $d\mathbf{x}$; $d\mathbf{X}^s$ is the line element at $\mathbf{X}^s$ that also transforms into $d\mathbf{x}$, where $\mathbf{X}^\alpha$ and $\mathbf{X}^s$ represent those material points of constituents α and s that meet at $\mathbf{x} = \chi^\alpha(\mathbf{X}^\alpha,t) = \chi^s(\mathbf{X}^s,t)$ at the current time t. In particular, it may be convenient to identify the reference configuration s as the *master reference configuration*, $\mathbf{X}^s$, so that

$$\mathbf{F}^s\left(\mathbf{X}^s,t\right) = \mathbf{F}^\alpha\left(\mathbf{X}^\alpha,t\right)\cdot\frac{\partial\mathbf{X}^\alpha(\mathbf{X}^s)}{\partial\mathbf{X}^s} \equiv \mathbf{F}^\alpha\left(\mathbf{X}^\alpha,t\right)\cdot\mathbf{F}^{\alpha s}\left(\mathbf{X}^s\right), \tag{2.6}$$

where $\mathbf{F}^{\alpha s}(\mathbf{X}^s) = \partial\mathbf{X}^\alpha(\mathbf{X}^s)/\partial\mathbf{X}^s$ represents the deformation gradient of α relative to s. For example, in a reactive mixture, s may represent the oldest solid constituent. In a constrained mixture, the relation between $\mathbf{X}^\alpha$ and $\mathbf{X}^s$ is *invariant* in time and $\mathbf{F}^{\alpha s}(\mathbf{X}^s)$ is a pre-determined spatial mapping that defines the function $\mathbf{X}^\alpha(\mathbf{X}^s)$. In the remainder of this treatment, the time-invariant dependence of $\mathbf{X}^\alpha$ on $\mathbf{X}^s$ is implied for any expression using $\mathbf{X}^\alpha$. Therefore, the material time derivative of $\mathbf{F}^s(\mathbf{X}^s,t)$ in (2.6) is $\dot{\mathbf{F}}^s(\mathbf{X}^s,t) = \dot{\mathbf{F}}^\alpha(\mathbf{X}^\alpha,t)\cdot\mathbf{F}^{\alpha s}(\mathbf{X}^s)$ and this relation remains consistent with the kinematic relations $\dot{\mathbf{F}}^s = \mathbf{L}^s\cdot\mathbf{F}^s$ and $\dot{\mathbf{F}}^\alpha = \mathbf{L}^s\cdot\mathbf{F}^\alpha$. The volume ratios $J^\alpha(\mathbf{X}^\alpha,t) = \det\mathbf{F}^\alpha(\mathbf{X}^\alpha,t)$ similarly satisfy

$$J^s\left(\mathbf{X}^s,t\right) = J^\alpha\left(\mathbf{X}^\alpha,t\right)J^{\alpha s}\left(\mathbf{X}^s\right), \tag{2.7}$$

where $J^s(\mathbf{X}^s,t) = \det\mathbf{F}^s(\mathbf{X}^s,t)$ and $J^{\alpha s}(\mathbf{X}^s) = \det\mathbf{F}^{\alpha s}(\mathbf{X}^s)$. As discussed in greater detail below, the mapping $\mathbf{F}^{\alpha s}(\mathbf{X}^s)$ in a constrained mixture must be postulated by constitutive assumption.

Under infinitesimal strain analyses, we note that

$$\mathbf{F}^\alpha \approx \mathbf{I} + \boldsymbol{\varepsilon}^\alpha + \boldsymbol{\omega}^\alpha, \tag{2.8}$$

where $\boldsymbol{\varepsilon}^\alpha$ is the symmetric infinitesimal strain tensor and $\mathbf{I} + \boldsymbol{\omega}^\alpha$ is the infinitesimal rotation tensor for constituent α. Here, $\boldsymbol{\omega}^\alpha$ is antisymmetric and its dual vector points along the axis of rotation and its magnitude is the infinitesimal angle of rotation. From a Taylor series expansion of $\det\mathbf{F}^\alpha$, it can be shown that

$$J^\alpha \approx 1 + \operatorname{tr}\boldsymbol{\varepsilon}^\alpha. \tag{2.9}$$

Under these conditions, the relation of (2.6) may be rewritten as

$$\mathbf{I} + \boldsymbol{\varepsilon}^s + \boldsymbol{\omega}^s \approx \left(\mathbf{I} + \boldsymbol{\varepsilon}^\alpha + \boldsymbol{\omega}^\alpha\right)\cdot\left(\mathbf{I} + \boldsymbol{\varepsilon}^{\alpha s} + \boldsymbol{\omega}^{\alpha s}\right). \tag{2.10}$$

Equating symmetric and antisymmetric parts of this relation and neglecting higher order products produces

$$\begin{aligned}\boldsymbol{\varepsilon}^s &= \boldsymbol{\varepsilon}^{\alpha s} + \boldsymbol{\varepsilon}^\alpha ,\\ \boldsymbol{\omega}^s &= \boldsymbol{\omega}^{\alpha s} + \boldsymbol{\omega}^\alpha .\end{aligned} \tag{2.11}$$

Just like $\mathbf{F}^{\alpha s}$, the relative strain $\boldsymbol{\varepsilon}^{\alpha s}$ and relative rotation $\boldsymbol{\omega}^{\alpha s}$ are temporally invariant measures which are prescribed based on constitutive assumptions. In particular, when all constituents α are assumed to share a common reference configuration with the master constituent s, it follows that $\mathbf{F}^{\alpha s} = \mathbf{I}$ and $\boldsymbol{\varepsilon}^{\alpha s} = \mathbf{0}$.

2.2 Apparent Density

The *apparent density* ρ^α of mixture constituent α is defined as

$$\rho^\alpha(\mathbf{x}, t) = \frac{dm^\alpha}{dV}, \tag{2.12}$$

where dm^α is the mass of constituent α in the mixture elemental volume dV, both given in the current configuration. The apparent density of α may also be called the *mass concentration* of α. In a general analysis, both dm^α and dV may evolve with time.

We may also define the *referential apparent density* $\rho_r(\mathbf{X}^\alpha, t)$ of constituent α as

$$\rho_r^\alpha(\mathbf{X}^\alpha, t) = \frac{dm^\alpha}{dV_r}, \tag{2.13}$$

where dV_r is the mixture volume in the reference configuration of the master constituent s (Fig. 1). This definition is consistent with the adoption of the reference configuration of the master constituent as that of the mixture domain. For example, in a finite element model of a reactive mixture, the undeformed mesh represents the stress-free configuration of constituent s. Since $dV/dV_r = J^s$, the relation between $\rho^\alpha(\mathbf{x}, t)$ and $\rho_r^\alpha(\mathbf{X}^\alpha, t)$ is

$$\rho_r^\alpha(\mathbf{X}^\alpha, t) = \rho^\alpha(\chi^s(\mathbf{X}^s, t), t) J^s(\mathbf{X}^s, t). \tag{2.14}$$

Example 1 Consider a soft hydrated tissue such as articular cartilage, which may be modeled as a mixture of a collagen fibrillar matrix, proteoglycans, water, sodium and chloride ions. The collagen-proteoglycan mixture may be modeled as a constrained solid mixture, since the proteoglycan macromolecules are typically enmeshed within, and bound to the collagen. The remaining constituents are fluids that may transport through the porous, permeable collagen-proteoglycan solid matrix. The macroscopic boundaries of the fibrillar collagen matrix define the boundaries of a cartilage tissue specimen, as collagen is the most abundant solid constituent; in contrast, proteoglycan macromolecules are interspersed within the collagen matrix. Therefore, the collagen constituent is the natural choice of master constituent in a cartilage mixture model. Since fluids do not have a natural stress-free configuration, they represent less convenient choices when defining a master constituent.

Other common definitions of concentration exist, most notably the molar concentration c^α of constituent α, given by

$$c^\alpha = \frac{dn^\alpha}{dV}, \tag{2.15}$$

where dn^α is the number of moles of α in the elemental volume dV. We may relate mass and molar concentrations using

$$\rho^\alpha = M^\alpha c^\alpha, \tag{2.16}$$

where $M^\alpha = dm^\alpha/dn^\alpha$ is the *molar mass* of constituent α, which is an invariant material property of constituent α.

2.3 Axiom of Mass Balance for Each Constituent

The axiom of mass balance may be derived for constituent α in integral form using a control volume V. The time rate of change of mass of α in the control volume is equal to the rate at which the mass of α is convected into V across its control surface ∂V and the rate at which mass is supplied to α from *reactions* with other mixture constituents,

$$\frac{d}{dt}\int_V \rho^\alpha dV = -\int_{\partial V} \rho^\alpha (\mathbf{v}^\alpha \cdot \mathbf{n}) dS + \int_V \hat{\rho}^\alpha \, dV. \tag{2.17}$$

Here, $\hat{\rho}^\alpha(\mathbf{x}, t)$ is the *mass density supply* (with units of mass per volume, per time) to α from all other constituents resulting from reactions. The mass supply is a function of state that needs to be defined with a suitable constitutive relation. Using the divergence theorem, this integral form may be converted to the differential form

$$\frac{\partial \rho^\alpha}{\partial t} + \operatorname{div}(\rho^\alpha \mathbf{v}^\alpha) = \hat{\rho}^\alpha. \tag{2.18}$$

Expanding the expression in the divergence operator and defining the material time derivative in the spatial frame, following constituent α, as

$$\frac{D^\alpha(\cdot)}{Dt} = \frac{\partial(\cdot)}{\partial t} + \operatorname{grad}(\cdot) \cdot \mathbf{v}^\alpha, \tag{2.19}$$

we may rewrite the mass balance for constituent α as

$$\frac{D^\alpha \rho^\alpha}{Dt} + \rho^\alpha \operatorname{div} \mathbf{v}^\alpha = \hat{\rho}^\alpha. \tag{2.20}$$

For the constituents α of a constrained mixture, since there is no relative motion between α and s, we may express the integral form of the mass balance in the material frame of the master constituent s as

$$\frac{d}{dt}\int_{V_r} \rho_r^\alpha(\mathbf{X}^\alpha, t)\, dV_r = \int_{V_r} \hat{\rho}_r^\alpha(\mathbf{X}^\alpha, t)\, dV_r, \tag{2.21}$$

where

$$\hat{\rho}_r^\alpha(\mathbf{X}^\alpha, t) = J^s(\mathbf{X}^s, t)\hat{\rho}^\alpha(\boldsymbol{\chi}^s(\mathbf{X}^s, t), t) \tag{2.22}$$

may be called the *referential mass density supply* to constituent α. Thus, the differential form of the mass balance in the material frame is given by

$$\dot{\rho}_r^\alpha \equiv \frac{\partial \rho_r^\alpha(\mathbf{X}^\alpha, t)}{\partial t} = \hat{\rho}_r^\alpha(\mathbf{X}^\alpha, t). \tag{2.23}$$

Equation (2.23) shows that ρ_r^α remains time-invariant for non-reactive constituents α in a mixture, since $\hat{\rho}_r^\alpha = 0$ for those constituents. From the definition given in (2.13), it becomes evident that ρ_r^α evolves only as a result of changes in dm^α, since dV_r is constant by definition. In contrast, looking at the definition of ρ^α in (2.12), we find that ρ^α of a constrained mixture constituent may evolve as a result of reactions (evolving dm^α) and deformations (evolving dV). Therefore, in contrast to ρ^α, ρ_r^α is a more convenient choice of state variable when modeling reactive constrained mixtures, since it can track the evolution of the mass content of α occurring exclusively in response to reactions, whereas the complete state of deformation of that constituent may be tracked using $\mathbf{F}^\alpha$.

The mass balance for the mixture is obtained by summing the mass balance in Eq. (2.18) over all constituents. The *axiom of mixtures* states that the governing equations for the mixture should have the same form as those of a single, pure substance (a single, non-reactive constituent α). As shown previously [4, 12, 13, 21], it follows that the mass supplies must satisfy

$$\sum_\alpha \hat{\rho}^\alpha = 0, \tag{2.24}$$

implying that the rate at which mass is lost from reactants must be matched by the rate at which mass is added to products, in all reactions occurring in the mixture.

2.4 Reactive Mixtures

Reactions may occur among the constituents of a mixture which result in a temporal evolution of the mass content of reactants and products. A forward reaction between mixture constituents may be written down as

$$\sum_\alpha \nu_R^\alpha \mathcal{E}^\alpha \rightarrow \sum_\alpha \nu_P^\alpha \mathcal{E}^\alpha, \tag{2.25}$$

where $\mathcal{E}^\alpha$ is the chemical (molecular) species associated with constituent α, ν_R^α represents the stoichiometric coefficient of reactant α and ν_P^α is that of the corresponding product. Similarly, a reversible reaction may be written as

$$\sum_\alpha \nu_R^\alpha \mathcal{E}^\alpha \rightleftharpoons \sum_\alpha \nu_P^\alpha \mathcal{E}^\alpha. \tag{2.26}$$

The summations are taken over all mixture constituents, though constituents that are not reactants in that particular reaction will have $\nu_R^\alpha = 0$, and those that are not products will have $\nu_P^\alpha = 0$. The stoichiometry of the reaction imposes constraints on the mass supplies such that we may define a molar production rate $\hat{\zeta}$ (units of moles per volume, per time),

$$\hat{\rho}^\alpha = M^\alpha \nu^\alpha \hat{\zeta}, \tag{2.27}$$

where

$$\nu^\alpha = \nu_P^\alpha - \nu_R^\alpha \tag{2.28}$$

is the net stoichiometric coefficient of α in the reaction. The relation (2.27) may be substituted into the constraint on mass supplies, Eq. (2.24) to produce

$$\sum_\alpha M^\alpha \nu^\alpha = 0. \tag{2.29}$$

This relation may be recognized as the classical requirement to balance the molar mass of reactants and products in a reaction. Thus, the constraint resulting from the axiom of mixtures produces a classical outcome when applied to the mass balance. For constrained mixture constituents, the relation (2.27) may also be written as

$$\hat{\rho}_r^\alpha = M^\alpha \nu^\alpha \hat{\zeta}_r, \tag{2.30}$$

where $\hat{\zeta}_r = J^s \hat{\zeta}$.

Example 2 In the reaction

$$H_2SO_4 \to 2H^+ + SO_4^{2-},$$

we have a total of three constituents, $\alpha = H_2SO_4, H^+, SO_4^{2-}$. Their stoichiometric coefficients are given by

$$\begin{aligned} \nu_R^{H_2SO_4} &= 1, & \nu_P^{H_2SO_4} &= 0, & \nu^{H_2SO_4} &= -1, \\ \nu_R^{H^+} &= 0, & \nu_P^{H^+} &= 2, & \nu^{H^+} &= 2, \\ \nu_R^{SO_4^{2-}} &= 0, & \nu_P^{SO_4^{2-}} &= 1, & \nu^{SO_4^{2-}} &= 1. \end{aligned}$$

Therefore, based on Eq. (2.27), we find that

$$\hat{\rho}^{H_2SO_4} = -M^{H_2SO_4}\hat{\zeta}, \quad \hat{\rho}^{H^+} = 2M^{H^+}\hat{\zeta}, \quad \hat{\rho}^{SO_4^{2-}} = M^{SO_4^{2-}}\hat{\zeta},$$

and the constraint of Eq. (2.29) reduces to

$$-M^{H_2SO_4} + 2M^{H^+} + M^{SO_4^{2-}} = 0.$$

2.5 Chemical Kinetics

The mass balance equations for mixture constituents require the specification of constitutive relations for the mass supplies $\hat{\rho}^\alpha$. Equation (2.27) shows that it suffices to provide a single constitutive relation for the molar production rate $\hat{\zeta}$. Constitutive relations are generally formulated based on experimental observations. For chemical reactions, a commonly used constitutive relation is the *law of mass action* [46]. In this section we adopt the form of these constitutive relations which employs referential mass densities and density supplies, convenient for the analysis of constrained mixtures.

For forward reactions, the law of mass action may take the form

$$\hat{\zeta}_r = k \prod_\alpha \left(c_r^\alpha\right)^{\nu_R^\alpha}, \quad c_r^\alpha \equiv \frac{\rho_r^\alpha}{M^\alpha}, \tag{2.31}$$

where k is the *specific reaction rate*. For reversible reactions, the net molar production rate is

$$\hat{\zeta}_r = \hat{\zeta}_F - \hat{\zeta}_R, \tag{2.32}$$

where

$$\hat{\zeta}_F = k_F \prod_\alpha \left(c_r^\alpha\right)^{\nu_R^\alpha},$$
$$\hat{\zeta}_R = k_R \prod_\alpha \left(c_r^\alpha\right)^{\nu_P^\alpha}, \tag{2.33}$$

represent forward and reverse production rates. We may combine these relations as

$$\hat{\zeta}_r = \hat{\zeta}_F \left[1 - \frac{k_R}{k_F} \prod_\alpha \left(c_r^\alpha\right)^{\nu^\alpha}\right], \tag{2.34}$$

where the ratio of forward and reversible specific reaction rates reduces to the equilibrium constant K_c when the reaction has reached steady state,

$$\left.\frac{k_F}{k_R}\right|_{\hat{\zeta}_r=0} \equiv K_c. \tag{2.35}$$

Example 3 We reprise the reaction of Example 2 and use the law of mass action in Eq. (2.31),

$$\hat{\zeta}_r = k \frac{\rho_r^{H_2SO_4}}{M^{H_2SO_4}}.$$

This expression may be substituted into (2.30) to produce $\hat{\rho}_r^{H_2SO_4}$, which may then be substituted into the mass balance relation (2.23)

$$\frac{\partial \rho_r^{H_2SO_4}}{\partial t} = -M^{H_2SO_4} \hat{\zeta}_r = -k\, \rho_r^{H_2SO_4}.$$

This linear differential equation may be integrated subject to the homogeneous initial condition $\rho_r^{H_2SO_4}(\mathbf{X}^s, 0) = \rho_0^{H_2SO_4}$ to produce

$$\rho_r^{H_2SO_4}\left(\mathbf{X}^s, t\right) = \rho_0^{H_2SO_4} e^{-kt},$$

showing that the concentration of the reactant decays exponentially with a time constant $1/k$. Accordingly, this type of response is called a *first-order reaction* and we find that the specific reaction rate controls the temporal response. Given this solution, it follows that

$$\hat{\zeta}_r = k \frac{\rho_0^{H_2SO_4}}{M^{H_2SO_4}} e^{-kt},$$

which may be used to similarly solve for the concentrations of the products,

$$\frac{\rho_r^{H^+}(\mathbf{X}^s, t)}{M^{H^+}} = \frac{\rho_0^{H^+}}{M^{H^+}} + 2\frac{\rho_0^{H_2SO_4}}{M^{H_2SO_4}}\left(1 - e^{-kt}\right),$$
$$\frac{\rho_r^{SO_4^{2-}}(\mathbf{X}^s, t)}{M^{SO_4^{2-}}} = \frac{\rho_0^{SO_4^{2-}}}{M^{SO_4^{2-}}} + \frac{\rho_0^{H_2SO_4}}{M^{H_2SO_4}}\left(1 - e^{-kt}\right).$$

Example 4 Receptor-ligand kinetics is a classical example of a reversible reaction in biochemistry, where a ligand binds to a receptor to produce a receptor-ligand complex. The reaction is reversible, implying that the complex may also dissociate back into its constituents. The forward and reverse reactions may proceed at different rates. Let the receptor be denoted with R and the ligand with L, whereas the receptor-ligand complex is C. The reaction is given by

$$\mathrm{R} + \mathrm{L} \rightleftharpoons \mathrm{C}.$$

The initial concentrations are $c_r^{\mathrm{R}}(\mathbf{X}^s, 0) = c_0^{\mathrm{R}}$, $c_r^{\mathrm{L}}(\mathbf{X}^s, 0) = c_0^{\mathrm{L}}$, and $c_r^{\mathrm{C}}(\mathbf{X}^s, 0) = 0$. The molar production rate is

$$\hat{\zeta}_r = k_F c_r^{\mathrm{R}} c_r^{\mathrm{L}} - k_R c_r^{\mathrm{C}}.$$

From mass balance it follows that $c_r^{\mathrm{R}} = c_0^R - c_r^{\mathrm{C}}$ and $c_r^{\mathrm{L}} = c_0^{\mathrm{L}} - c_r^{\mathrm{C}}$. Therefore,

$$\hat{\zeta}_r = k_F \left(c_0^R - c_r^{\mathrm{C}}\right)\left(c_0^{\mathrm{L}} - c_r^{\mathrm{C}}\right) - k_R c_r^{\mathrm{C}},$$

and the mass balance for the complex becomes

$$\frac{\partial c_r^{\mathrm{C}}}{\partial t} = k_F \left(c_0^R - c_r^{\mathrm{C}}\right)\left(c_0^{\mathrm{L}} - c_r^{\mathrm{C}}\right) - k_R c_r^{\mathrm{C}}.$$

The solution to this nonlinear ordinary differential equation is

$$c_r^{\mathrm{C}}\left(\mathbf{X}^s, t\right) = c_0 - \chi \tanh\left(k_F \chi t + \tanh^{-1} \frac{c_0}{\chi}\right),$$

where

$$c_0 = \frac{1}{2}\left(c_0^{\mathrm{L}} + c_0^{\mathrm{R}} + K_d\right), \qquad K_d = \frac{k_R}{k_F}, \qquad \chi = \sqrt{c_0^2 - c_0^{\mathrm{R}} c_0^{\mathrm{L}}}.$$

At steady state the solution is

$$c_r^{\mathrm{C}}\left(\mathbf{X}^s, t \to \infty\right) = c_0 - \chi = c_0 - c_0 \sqrt{1 - \frac{c_0^{\mathrm{R}} c_0^{\mathrm{L}}}{c_0^2}}.$$

In the limit when $c_0^{\mathrm{R}} c_0^{\mathrm{L}} / c_0^2 \ll 1$, i.e., when the initial ligand concentration or the initial receptor concentration is much smaller than the dissociation constant K_d, this solution reduces to

$$c_r^{\mathrm{C}}\left(\mathbf{X}^s, t \to \infty\right) \approx \frac{1}{2} \frac{c_0^{\mathrm{R}} c_0^{\mathrm{L}}}{c_0} = \frac{c_0^{\mathrm{R}} c_0^{\mathrm{L}}}{c_0^{\mathrm{L}} + c_0^{\mathrm{R}} + K_d}.$$

2.6 Interstitial Solid Growth and Remodeling

Interstitial growth implies solid mass deposition within a porous mixture. This type of growth is not necessarily biological. For example, a metallic porous filter may be imbibed with a fluid that causes metal oxidation, so that the mass of metal decreases while that of the metal oxide increases throughout the entire filter domain. In the case of biological tissues, biochemical reactions may occur that are not necessarily driven directly by cell activity, such as binding of exogenous growth factors to the extracellular matrix. Conversely, cells may directly process soluble reactants to produce extracellular matrix products in a process variably called growth or remodeling. In a generic sense, this type of reaction may be written as

$$\text{cells} + \text{nutrients} \rightarrow \text{cells} + \text{solid matrix}.$$

For such growth processes, the law of mass action may or may not be directly applicable, especially when a multitude of reactions are taking place simultaneously, whose aggregate response on solid matrix production is not reducible to the functional form of the law of mass action. Furthermore, when dealing with biological tissues where solid deformation occurs, the growth process may be significantly influenced by such deformations and their history over an extended period of intermittent loading. Therefore, constitutive modeling of such growth processes may be reformulated to reflect experimental observations at a more global level, rather than the intricate details of individual chemical reactions.

In the sections below we examine basic problems in tissue growth and remodeling. Interstitial growth implies that the referential apparent density ρ_r^α of a solid constituent increases over time; this type of growth leads to an increase in the mass of constituent α but does not necessarily produce changes to the overall tissue dimensions. For example, in tissue engineering where cells are seeded within a scaffold, the cells may synthesize solid matrix that fills the scaffold interstitial space, without altering the overall dimensions of the engineered construct. If the scaffold is degradable, its loss of mass (e.g., solubilization) represents negative growth.

In contrast to growth, tissue remodeling may be viewed as the effect of reactions that alter the mass content of various solid constituents, without an obligatory net change in total solid mass content. This distinction between growth and remodeling is semantic; effectively, both phenomena involve reactions that alter the mass content of various solid constituents.

In biomechanics, it is common to model biological soft tissues as being nearly incompressible. This choice of material behavior is sometimes convenient for simplifying analyses of tissue responses, as in studies where two-dimensional strain measurements are performed on the surface of a tissue layer and the strain through the layer thickness is deduced from the assumption that the response is isochoric. It is useful to understand when this assumption may be valid, especially in the context of reactive mixtures where growth and resorption of solid constituents alter the tissue porosity.

The best evidence available to us is arguably from studies of articular cartilage, which has been alternatively modeled as an incompressible or nearly incompressible elastic solid [1, 20] or a porous-permeable biphasic material (a solid-fluid mixture or poroelastic material) [2, 41]. Cartilage has a porosity ranging from 90 to 65 percent, decreasing with depth from the articular surface, and with age [56]. The interstitial fluid of cartilage consists primarily of water and inorganic salts; this fluid may flow within cartilage since its pores communicate, as evidenced by direct measurements of its hydraulic permeability [39]. Experiments have shown that loading of articular cartilage causes its interstitial fluid to pressurize; this pressure subsides when the loading remains static for a sustained period of time, as the pressurized fluid flows away from the loaded region [43]. Excellent agreement has been found between direct measurements of the interstitial fluid pressure and predictions from the biphasic theory [44, 52, 53], which assumes that each mixture constituent (solid and fluid) is intrinsically incompressible, though the mixture (defined over the domain of the porous solid matrix) may change in volume with time as the fluid flows through it [41]. The assumption of intrinsic incompressibility of each constituent was validated from direct experimental measurements of cartilage subjected to hydrostatic pressures up to 12 MPa [11].

The biphasic theory predicts that the instantaneous response of cartilage to loading is isochoric, since the interstitial fluid cannot instantaneously flow out of the mixture due its frictional interactions with the solid [2, 6, 38]. This behavior has been verified from experimental measurements of Poisson's ratio in cartilage subjected to rapid loading under small strains, which produced values approaching 0.5 [28, 65]. Moreover, the biphasic theory predicts that the equilibrium response of cartilage, achieved when the interstitial fluid pressure and flow have subsided, is equivalent to that of a compressible elastic solid [2, 38, 41]. Experimental measurements of Poisson's ratio under equilibrium conditions have yielded values as low as 0.02 [16]. This very low value implies that the porous solid matrix has negligible lateral expansion upon axial compression, with the reduction in volume occurring in the pore space. In contrast, a biphasic material with zero porosity (no fluid content), or with non-communicating pores, is intrinsically incompressible since no fluid may enter or leave the tissue.

In reality, interstitial fluid may also be exchanged with hydrated biological tissues via osmotic mechanisms. Experiments on cartilage, intervertebral disc, arterial wall and cornea have shown that they swell or shrink in response to osmotic loading, when the salt concentration of the surrounding fluid bath is altered [10, 14, 16, 40, 57]. Therefore, the assumption that these tissues may be modeled as incompressible solids that cannot change in volume should be adopted carefully, as it is valid only under special circumstances: As long as the interstitial fluid has not had sufficient time to flow into or out of the porous solid matrix in response to mechanical or osmotic loading, or growth and remodeling processes that alter the porosity, it is reasonable to assume that the response to mechanical loading will be isochoric.

Conversely, since growth and remodeling may alter the solid matrix porosity of a biological tissue, as also illustrated in our prior theoretical study of cell growth [7], it would be unreasonable to illustrate this mechanism with examples that use a value of 0.5 for Poisson's ratio. Therefore, in the examples that follow, a value of zero is selected for simplicity. This value does not preclude isochoric responses under rapid loading and unloading when the pore space is filled with interstitial fluid [2, 6, 38].

2.6.1 Bone Remodeling in Response to Loading

Bone is known to respond to loading according to Wolff's law, whereby bone mass increases or decreases when loading deviates from a set point. We may generically represent these reactions as

$$\text{cells} + \text{nutrients} \overset{\text{loading}}{\rightarrow} \text{cells} + \text{solid matrix}, \tag{2.36}$$

$$\text{cells} + \text{solid matrix} \overset{\text{unloading}}{\rightarrow} \text{cells} + \text{soluble waste products}. \tag{2.37}$$

In this case the first reaction is only triggered when the loading exceeds the set point, whereas the second reaction is exclusively triggered when the loading falls below the threshold. The cells involved in these processes may even be different; in the case of bone, matrix synthesis is driven by osteoblasts whereas matrix degradation is performed by osteoclasts.

For trabecular bone remodeling, Huiskes and co-workers proposed the following constitutive relation to embody Wolff's law,

$$\hat{\rho}_r^s = B\left(\frac{\Psi_r}{\rho_r^s} - \psi_0\right), \tag{2.38}$$

where Ψ_r is the strain energy density in the bone, ψ_0 is the specific strain energy at the set point, and B controls the remodeling rate [29, 63]. This constitutive relation, which responds indiscriminately to tensile, compressive and shear loading, may be substituted into the mass balance (2.23) to solve for the evolving apparent bone density ρ_r^s,

$$\frac{\partial \rho_r^s}{\partial t} = B\left(\frac{\Psi_r}{\rho_r^s} - \psi_0\right). \tag{2.39}$$

An increase in ρ_r^s implies that existing trabeculae become thicker, or more trabeculae form, within a given trabecular bone region, thus decreasing the corresponding pore volume. Conversely, a decrease in ρ_r^s implies thinning or resorbing trabeculae, leading to increased porosity.

For a well-posed problem it is necessary to propose a constitutive relation for the free energy density of the solid. For example, we may idealize bone as a linear elastic solid. For a one-dimensional analysis, the stress-strain relation is $\sigma = E\varepsilon$ where E is Young's modulus, so that

$$\Psi_r = \frac{1}{2}\sigma\varepsilon = \frac{1}{2}E\varepsilon^2 = \frac{\sigma^2}{2E}. \tag{2.40}$$

Experimentally, Carter and Hayes [15] found that the strength of trabecular bone varied as a power-law of its apparent density, and a similar relation has been adopted for its Young's modulus [22, 63],

$$E = c\left(\rho_r^s\right)^\gamma, \tag{2.41}$$

where c and γ are material parameters obtained empirically. We may now analyze the problem of uniaxial loading of a bar as a canonical problem in this remodeling framework, which may be representative of a long bone such as the femur or humerus. When the bar is subjected to a prescribed normal traction σ, we use the form $\Psi_r = \sigma^2/2E$ in (2.40), so that the only unknown is the density ρ_r^s appearing in the expression for E in (2.41). Substituting these relations above now produces a differential equation that depends on the known state of stress as well as the unknown bone density ρ_r^s,

$$\frac{\partial \rho_r^s}{\partial t} = B\left(\frac{\sigma^2}{2c(\rho_r^s)^{\gamma+1}} - \psi_0\right). \tag{2.42}$$

This equation governs the remodeling of bone in response to loading. Conversely, had the bar been subjected to a prescribed normal strain ε, we would have used $\Psi_r = E\varepsilon^2/2$ in (2.40), so that (2.39) would instead reduce to

$$\frac{\partial \rho_r^s}{\partial t} = B\left(\frac{1}{2}c\left(\rho_r^s\right)^{\gamma-1}\varepsilon^2 - \psi_0\right). \tag{2.43}$$

Example 5 Consider a bar with initial uniform density $\rho_r^s(x,t) = \rho_0^s$, subjected to a constant traction $\sigma(t) = \sigma_0$ on both ends.[1] The transient solution to the above differential equation is most conveniently obtained numerically. However, the response at steady state is easily evaluated by letting $\partial\rho_r^s/\partial t = 0$ to produce

$$\rho_\infty^s = \lim_{t\to\infty} \rho_r^s(x,t) = \sqrt[\gamma+1]{\frac{\sigma_0^2}{2c\psi_0}}. \tag{2.44}$$

[1] For an infinitesimal strain analysis we may assume that the material and spatial configurations are nearly identical, $x \approx X^s$.

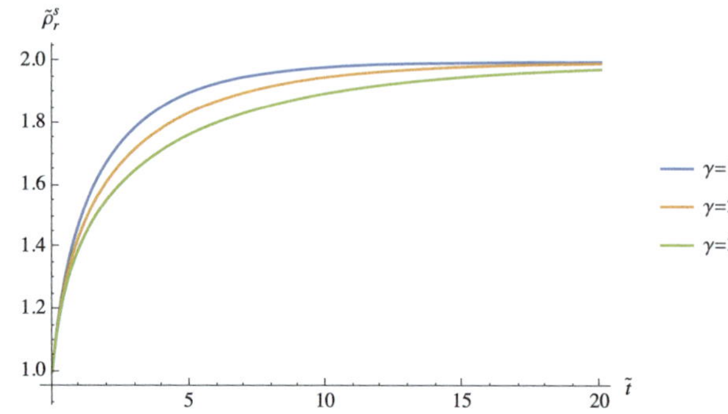

Fig. 2 Transient response of $\tilde{\rho}_r^s$ versus $\tilde{t}$ in Example 5, using three different values of γ, with $\rho_0^s/\rho_\infty^s = 1/2$

We conclude that the apparent density remodels from ρ_0^s to the new equilibrium value given above, determined by the material parameters c, γ, and ψ_0, and the prescribed traction σ_0. This formula shows that the bone density increases with increasing applied stress σ_0; though the formula suggests no limit on its upper bound, in reality ρ_r^s reaches a maximum value when all the pore space has been filled with bone matrix, producing a value essentially equivalent to the density of cortical bone. This solution also shows that bone density decreases with increasing set point value ψ_0. This type of prediction suggests that a disease such as osteoporosis, a complex metabolic bone disease that leads to loss of bone density over time, may be simulated in a growth and remodeling analysis by altering the set point ψ_0 with disease progression. Effectively, this would indicate that bone cells have become less sensitive to mechanical stimuli, requiring a higher threshold for producing a response comparable to healthy bone.

As a special case, the transient response when $\sigma_0 = 0$ reduces to $\rho_r^s(x,t) = \rho_0^s - B\psi_0 t$, a linear decrease which remains valid until ρ_r^s reduces to 0. In general, the rate of remodeling is given directly by the mass balance relation (2.42), which shows that it is proportional to B but evolves with time; at $t = 0$ the characteristic time constant τ for remodeling is

$$\tau = \frac{\rho_0^s}{B\left(\frac{\sigma_0^2}{2c(\rho_0^s)^{\gamma+1}} - \psi_0\right)}. \tag{2.45}$$

Letting $\tilde{\rho}_r^s = \rho_r^s/\rho_0^s$ and $\tilde{t} = t/\tau$, we may non-dimensionalize the differential equation (2.42) as

$$\frac{\partial \tilde{\rho}_r^s}{\partial \tilde{t}} = \frac{1}{1-r}\left(\frac{1}{(\tilde{\rho}_r^s)^{\gamma+1}} - r\right), \quad r = \left(\frac{\rho_0^s}{\rho_\infty^s}\right)^{\gamma+1}. \tag{2.46}$$

The initial condition is $\tilde{\rho}_r^s(x,0) = 1$. This form is more suitable for numerical implementation and select responses are illustrated in Fig. 2.

Example 6 If we allow the traction σ to vary cyclically over time with a period T, e.g., $\sigma(t) = \sqrt{2}\sigma_0 \sin(2\pi t/T)$, the temporal evolution of ρ_r^s will also vary cyclically. In a long bone for example, this type of cyclical loading may arise from locomotion and T may be on the order of one second. The non-dimensional form of the differential equation (2.42) becomes

$$\frac{\partial \tilde{\rho}_r^s}{\partial \tilde{t}} = \frac{1}{1-r}\left(\frac{2}{(\tilde{\rho}_r^s)^{\gamma+1}} \sin^2\left(2\pi \frac{\tau}{T}\tilde{t}\right) - r\right).$$

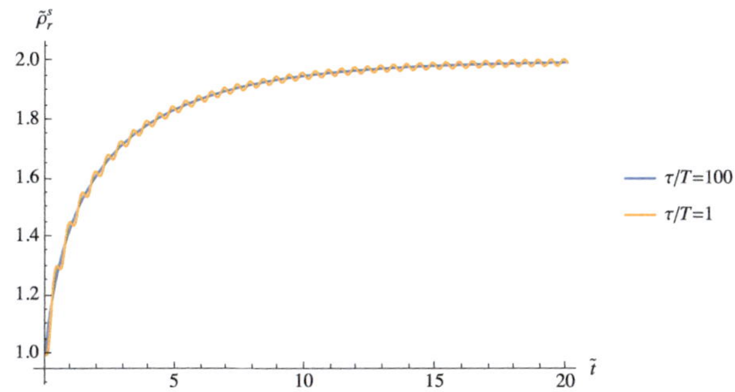

Fig. 3 Transient response of $\tilde{\rho}_r^s$ versus $\tilde{t}$ in Example 6, using two different values of τ/T, with $\rho_0^s/\rho_\infty^s = 1/2$ and $\gamma = 2$. The effect of cyclical loading is negligible when $\tau/T \gg 1$

An examination of this equation shows that the influence of cyclical loading will be negligible when the period T is much shorter than the characteristic remodeling time τ, which is typical of biological tissue remodeling. Numerical examples are illustrated in Fig. 3.

Example 7 Finally, consider a thick-walled cylindrical tube with inner radius a and outer radius b, subjected to an internal pressure p_0. This type of geometry may represent an artery for example. In a cylindrical coordinate system, the state of stress is [54]

$$[\boldsymbol{\sigma}] = \begin{bmatrix} \sigma_{rr} & 0 & 0 \\ 0 & \sigma_{\theta\theta} & 0 \\ 0 & 0 & 0 \end{bmatrix}, \tag{2.47a}$$

where

$$\sigma_{rr} = \frac{a^2 p_0}{b^2 - a^2}\left(1 - \frac{b^2}{r^2}\right), \qquad \sigma_{\theta\theta} = \frac{a^2 p_0}{b^2 - a^2}\left(1 + \frac{b^2}{r^2}\right), \quad a \le r \le b. \tag{2.47b}$$

For a linear isotropic elastic solid, assuming that Poisson's ratio is zero for simplicity, the free energy density is given by

$$\Psi_r = \frac{1}{2E(\rho_r^s)}\left(\sigma_{rr}^2 + \sigma_{\theta\theta}^2\right) = \frac{p_0^2}{E(\rho_r^s)}\frac{a^4}{(b^2 - a^2)^2}\left(1 + \frac{b^4}{r^4}\right). \tag{2.47c}$$

Now (2.42) reduces to

$$\frac{\partial \rho_r^s}{\partial t} = B\left(\frac{p^2(t)}{c(\rho_r^s)^{\gamma+1}}\frac{a^4}{(b^2 - a^2)^2}\left(1 + \frac{b^4}{r^4}\right) - \psi_0\right). \tag{2.47d}$$

At steady state, when $\partial \rho_r^s / \partial t = 0$, we find that

$$\lim_{t \to \infty} \rho_r^s(r,t) = \rho_\infty^s \sqrt[\gamma+1]{\frac{a^4}{(b^2 - a^2)^2}\left(1 + \frac{b^4}{r^4}\right)}, \quad \rho_\infty^s = \sqrt[\gamma+1]{\frac{p_0^2}{c\psi_0}}.$$

Unlike the previous examples, this result shows that the steady-state remodeling response need not produce a homogeneous distribution for ρ_r^s. In this analysis, the referential apparent density ρ_r^s is highest at the inner wall and decreases monotonically toward the outer wall (Fig. 4).

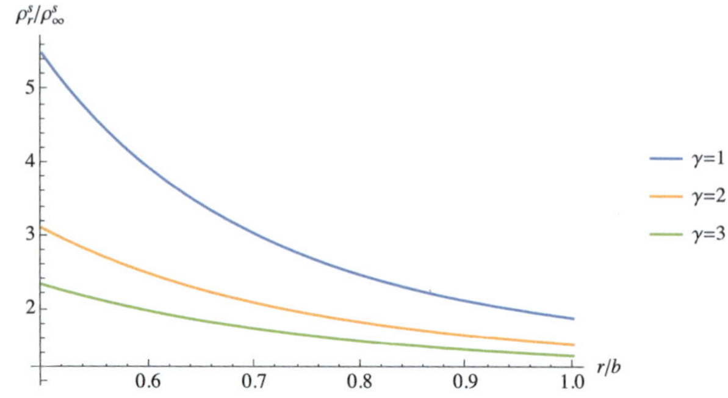

Fig. 4 Steady-state response of ρ_r^s/ρ_∞^s in Example 7, along the radius r of a thick-walled cylindrical tube subjected to an internal pressure p_0. The inner wall radius is a and the outer wall radius is b. Results are presented for $b/a = 2$ and various values of γ

3 Mechanics of Constrained Mixtures

The equations of momentum balance for constrained mixture constituents were presented previously [9]. A notable finding from that earlier study is that the state of stress in individual mixture constituents remains indeterminate. This indeterminacy occurs because the internal forces resulting from momentum exchanges between constrained constituents remain unobservable due to lack of relative motion between them. The only observable stress measure is the mixture stress $\boldsymbol{\sigma}$, which may be evaluated from the mixture momentum balance. Under quasi-static conditions, in the absence of external body forces, this equation reduces to the familiar relation

$$\operatorname{div}\boldsymbol{\sigma} = \mathbf{0}. \tag{3.1}$$

To solve for the function of state $\boldsymbol{\sigma}$, it is necessary to first select the desired list of state variables for describing the response of biological tissues. Under isothermal processes, when analyzing reactive solid constituents in a constrained mixtures, we choose $(\theta, \mathbf{F}^\beta, \rho_r^\beta)$ as our list of state variables, where β spans all constituents α in the mixture [9]. The absolute temperature θ is included since material properties appearing in the constitutive relations for $\boldsymbol{\sigma}$ and $\hat{\rho}^\alpha$, such as reaction rates, may vary with temperature. The deformation gradients $\mathbf{F}^\beta$ are included to account for the role of solid matrix strain. The referential apparent densities ρ_r^β are included since they may evolve in time and space in reactive processes, thereby altering material properties over time.

Given these state variables, the axiom of entropy inequality imposes the following thermodynamic constraint on $\boldsymbol{\sigma}$,

$$\boldsymbol{\sigma} = \frac{1}{J^s}\sum_\alpha \frac{\partial \Psi_r}{\partial \mathbf{F}^\alpha}\cdot\left(\mathbf{F}^\alpha\right)^T, \tag{3.2}$$

where Ψ_r is the referential mixture free energy density, which may be evaluated from the specific free energy ψ^α of each constituent α according to

$$\Psi_r = \sum_\alpha \rho_r^\alpha \psi^\alpha. \tag{3.3}$$

Other constraints imposed by the axiom of entropy inequality are $H_r = -\partial\Psi_r/\partial\theta$, where H_r is the referential mixture entropy density, and $\mathbf{q} = \mathbf{0}$, where $\mathbf{q}$ is the heat flux in the mixture, leaving the residual dissipation

$$\sum_\alpha \hat{\rho}^\alpha \mu^\alpha \le 0, \tag{3.4}$$

where

$$\mu^\alpha = \frac{\partial \Psi_r}{\partial \rho_r^\alpha} \tag{3.5}$$

represents the *chemical potential* of constituent α.

The residual dissipation effectively places a thermodynamic constraint on the constitutive relations for mass supplies $\hat{\rho}^\alpha$. When the inequality in (3.4) is violated, the reaction embodied in $\hat{\rho}^\alpha$ may not proceed. Students of chemistry may recognize this inequality constraint as equivalent to the requirement that chemical reactions may only proceed when they produce a net reduction in the Gibbs energy [55].[2]

In summary, the mixture momentum balance (3.1) and constituent mass balances (2.23) may be solved simultaneously for the unknowns $\mathbf{F}^s$ and ρ_r^α, given constitutive relations for Ψ_r and $\hat{\rho}^\alpha$, and constitutive assumptions for $\mathbf{F}^{\alpha s}$.

Example 8 Under infinitesimal strain analyses, the relation (3.2) for the mixture stress simplifies to

$$\boldsymbol{\sigma} = \sum_\alpha \frac{\partial \Psi_r}{\partial \boldsymbol{\varepsilon}^\alpha}. \tag{3.6}$$

For example, assuming linear elasticity, the mixture strain energy density may be given by

$$\Psi_r = \sum_\alpha \frac{1}{2} \boldsymbol{\varepsilon}^\alpha : \boldsymbol{\mathcal{C}}^\alpha \left(\rho_r^\beta\right) : \boldsymbol{\varepsilon}^\alpha, \tag{3.7}$$

where $\boldsymbol{\mathcal{C}}^\alpha(\rho_r^\beta)$ is the fourth-order elasticity tensor for constituent α. In this example, to maintain some generality, we have assumed that the elasticity of constituent α may depend on the referential apparent density of all constituents β in the mixture. Based on this relation, the mixture stress simplifies to

$$\boldsymbol{\sigma} = \sum_\alpha \boldsymbol{\mathcal{C}}^\alpha \left(\rho_r^\beta\right) : \boldsymbol{\varepsilon}^\alpha = \sum_\alpha \boldsymbol{\mathcal{C}}\left(\rho_r^\beta\right) : \left(\boldsymbol{\varepsilon}^s - \boldsymbol{\varepsilon}^{\alpha s}\right), \tag{3.8}$$

where we have used the general relation (2.11) between the strain tensor of constituent α relative to the master constituent s.

3.1 Non-reactive Elasticity

3.1.1 Tissues with Distinct Fiber Orientations

The simplest application of constrained mixtures of solid constituents is the classical field of non-reactive elasticity ($\hat{\rho}^\alpha = 0, \forall \alpha$), when all constituents α share the same reference configuration $\mathbf{X}^s$. In that case, the mixture stress in (3.2) simplifies to

$$\boldsymbol{\sigma}\left(\theta, \mathbf{F}^s\right) = \frac{1}{J^s} \frac{\partial \Psi_r}{\partial \mathbf{F}^s} \cdot \left(\mathbf{F}^s\right)^T. \tag{3.9}$$

[2] Chemistry textbooks typically adopt temperature and pressure as state variables, so that the Gibbs energy emerges as the natural scalar potential for examining the thermodynamics of reactions. In contrast, our treatment employs temperature and strain (thus, volume) as state variables, so that the Helmholtz free energy emerges as the natural choice of scalar potential.

Example 9 Consider a fibrous tissue consisting of multiple fiber bundles, initially oriented along distinct directions in the reference configuration. We may associate each bundle with a distinct constituent α, so that the referential orientation of that bundle is given by the unit vector $\mathbf{n}_r^\alpha$; similarly, we can characterize that bundle's mass content using ρ_r^α (which is time-invariant in this non-reactive example). For simplicity, we use infinitesimal strain theory and linear elasticity, though the same presentation may be given using finite deformation with nonlinear elasticity. Therefore, the stress response may be specialized from (3.8) with $\boldsymbol{\varepsilon}^{\alpha s} = \mathbf{0}$, since all constituents share the same reference configuration,

$$\boldsymbol{\sigma} = \sum_\alpha \boldsymbol{\mathcal{C}}^\alpha : \boldsymbol{\varepsilon}^s. \tag{3.10}$$

The normal strain ε_n^α in bundle α is given by $\varepsilon_n^\alpha = \mathbf{n}_r^\alpha \cdot \boldsymbol{\varepsilon}^s \cdot \mathbf{n}_r^\alpha$. Assuming that each fiber bundle behaves as a one-dimensional structure, and that its material properties only depend on its own density, the elasticity tensor may be expressed as

$$\boldsymbol{\mathcal{C}}^\alpha = E\left(\rho_r^\alpha\right)\mathbf{n}_r^\alpha \otimes \mathbf{n}_r^\alpha \otimes \mathbf{n}_r^\alpha \otimes \mathbf{n}_r^\alpha, \tag{3.11}$$

where $E(\rho_r^\alpha)$ is Young's modulus for fiber bundle α. Combining these expressions, and further assuming that fiber bundles sustain stress only in tension, we get

$$\boldsymbol{\sigma} = \sum_\alpha H\left(\varepsilon_n^\alpha\right) E\left(\rho_r^\alpha\right)\varepsilon_n^\alpha \mathbf{n}_r^\alpha \otimes \mathbf{n}_r^\alpha, \tag{3.12}$$

where $H(\cdot)$ is the Heaviside unit step function. This relation shows that the contribution of each fiber bundle to the mixture stress is weighted by the bundle's modulus $E(\rho_r^\alpha)$, which may generally be a nonlinear function of ρ_r^α. This formulation is consistent with a fibrous cartilage model reported in an earlier study, without explicitly appealing to the concept of constrained mixtures [8]. In the case of a linear constitutive model, $E(\rho_r^\alpha) = c\rho_r^\alpha$, where c is a constant, we effectively propose that the weight factor of each fiber bundle is directly proportional to its mass content ρ_r^α. Then, the expression (3.12) becomes consistent with the classical treatment of fibrous tissues consisting of multiple fiber bundles; it may be easily generalized to continuous fiber distributions by suitably substituting the summation with an integration over all directions $\mathbf{n}_r^\alpha$ [36, 37]. The densities ρ_r^α are observable (measurable) state variables. They may be directly related to the fiber angular distribution function $R(\mathbf{n}_r^\alpha)$, which represents the fraction of total fibers oriented along $\mathbf{n}_r^\alpha$ in the reference configuration, and is routinely reported in the soft tissue mechanics literature [3, 23, 35, 45, 48].

Importantly, this example illustrates that the tissue ultrastructure can be directly related to its composition (mass content), since each constituent α represents a fiber bundle oriented in a different direction. More complex relations between ultrastructure and composition may be similarly formulated. For example, the ultrastructure of trabecular bone may be characterized by the concentrations ρ_r^β of various critical sub-structures, such as rods and plates with particular dimensional characteristics, and their connectivities. Thus, the evolution of tissue ultrastructure during tissue remodeling may be entirely described by the axiom of mass balance, using scalar observable variables ρ_r^β.

3.1.2 Tissues with Distinct Fiber Reference Configurations

In our second example, consider multiple fiber bundles α which are all oriented along the same direction $\mathbf{n}_r$ in the master reference configuration. Each bundle α has a distinct refer-

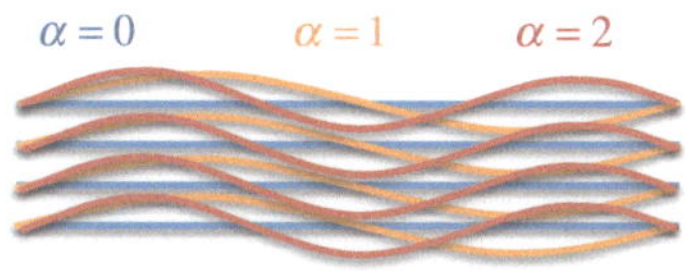

Fig. 5 Schematic illustration of a constrained mixture of three fiber bundles α, all oriented along the same direction $\mathbf{n}_r$, with distinct reference configurations. Each bundle is crimped by a different amount, such that the transition between compression and tension differs for each constituent α. Let $\alpha=0$ correspond to the master constituent s, and let ε_n^s denote the normal strain along $\mathbf{n}_r$ in this bundle. We may define the relative normal strain $\varepsilon_n^{\alpha s}$ along $\mathbf{n}_r$ as the value of ε_n^s at which bundle α transitions between compression and tension

ence configuration $\mathbf{X}^\alpha$, so that the relative strain tensors $\boldsymbol{\varepsilon}^{\alpha s}$, which are prescribed by constitutive assumption, are not zero. Assuming that the elasticity tensor has the same form as (3.11), with $\mathbf{n}_r^\alpha=\mathbf{n}_r$ for all α, and accounting for the tension-only response of fiber bundles, the stress-strain response of (3.8) may be rearranged as

$$\boldsymbol{\sigma}=\sum_\alpha H\left(\varepsilon_n^s-\varepsilon_n^{\alpha s}\right)E\left(\rho_r^\alpha\right)\left(\varepsilon_n^s-\varepsilon_n^{\alpha s}\right)\mathbf{n}_r\otimes\mathbf{n}_r, \tag{3.13}$$

where $\varepsilon_n^s=\mathbf{n}_r\cdot\boldsymbol{\varepsilon}^s\cdot\mathbf{n}_r$, and similarly for $\varepsilon_n^{\alpha s}$. Note that $\varepsilon_n^{\alpha s}$ represents the value of ε_n^s when bundle α transitions between compression and tension. In this example, fiber bundles are incrementally recruited as the normal strain ε_n^s exceeds the threshold $\varepsilon_n^{\alpha s}$ of each bundle α (Fig. 5). This incremental recruitment may thus produce the typical toe region observed in experimental measurements on fibrous tissues, as illustrated in the next example. The shape of the toe region may be modulated by the specific values of $\varepsilon_n^{\alpha s}$ and the function $E(\rho_r^\alpha)$ used for each bundle.

Example 10 Consider a mixture of n constituents ($\alpha=0$ to $n-1$, where s corresponds to $\alpha=0$) such that

$$\varepsilon^{\alpha s}=\frac{\alpha}{n}\varepsilon_0, \tag{3.14a}$$

and

$$E\left(\rho_r^\alpha\right)=\frac{E_0}{n}. \tag{3.14b}$$

Here, ε_0/n represents the uniform strain increment at which fiber bundles get consecutively recruited, whereas E_0/n is the modulus of each fiber bundle.[3] Under uniaxial loading along $\mathbf{n}_r$, with normal strain $\varepsilon_n^s\equiv\varepsilon$, we find that $\boldsymbol{\sigma}=\sigma\,\mathbf{n}_r\otimes\mathbf{n}_r$ where

$$\sigma=\frac{E_0}{n}\sum_{\alpha=0}^{n-1}H\left(\varepsilon-\frac{\alpha}{n}\varepsilon_0\right)\left(\varepsilon-\frac{\alpha}{n}\varepsilon_0\right). \tag{3.14c}$$

A plot of σ/E_0 versus ε illustrates the fiber bundle recruitment within the toe region, and the linear response beyond it, for select values of n (Fig. 6). This discrete fiber model may

[3]An explicit dependence of E on ρ_r^α may also be adopted, e.g., assuming that each bundle has the same density $\rho_r^\alpha=\rho_0/n$, where ρ_0 is the mixture density; then, letting $E(\rho_r^\alpha)=(\rho_r^\alpha/\rho_0)E_0$ would reproduce the relation adopted in (3.14b).

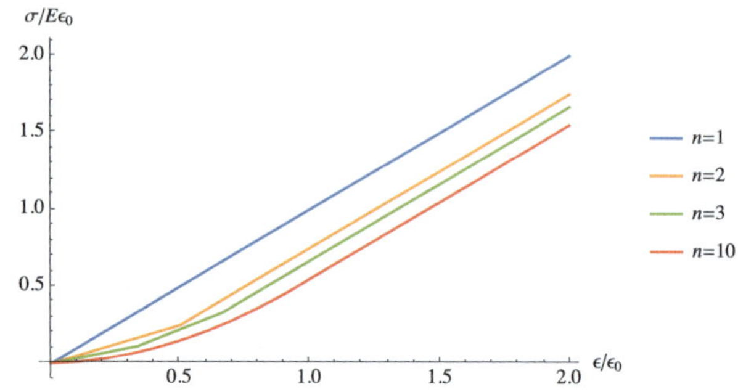

Fig. 6 Stress-strain response for the fiber recruitment model described in Example 10, with increasing number of fiber bundles n and uniform increment in recruitment strain

be generalized to a continuous fiber model ($n \to \infty$) by converting the above summation to an integral,

$$\sigma = \frac{E_0}{\varepsilon_0} \int_0^{\varepsilon_0} H(\varepsilon - \eta)(\varepsilon - \eta)\, d\eta, \tag{3.14d}$$

which produces

$$\sigma = E_0 \begin{cases} 0 & \varepsilon < 0 \\ \frac{\varepsilon^2}{2\varepsilon_0} & 0 \le \varepsilon < \varepsilon_0 \\ \varepsilon - \frac{\varepsilon_0}{2} & \varepsilon_0 \le \varepsilon \end{cases}. \tag{3.14e}$$

We conclude from this result that our specific constitutive assumptions for $\varepsilon_n^{\alpha s}$ and $E(\rho_r^\alpha)$ produce a quadratic toe region in the limit of a continuous fiber distribution, with ε_0 representing the width of the toe region and E_0 representing the modulus of the linear region.

This modeling approach for progressive recruitment of fiber bundles has been used in many prior investigations [27, 32, 36, 67]; this example illustrates that this classical approach may be reproduced within the context of constrained, non-reactive mixtures. In the examples that follow, we illustrate how a reactive mixture framework easily extends these types of formulations to fibrous tissue growth and remodeling.

3.2 Reactive Elasticity

A reactive elasticity framework extends the previous examples by letting the apparent densities ρ_r^α evolve in response to various signals. In a biological tissue, cells may synthesize or degrade extracellular matrix in response to the state of loading. This type of response can be embodied in a suitable constitutive model for the mass supply $\hat{\rho}_r^\alpha$ for each constituent, which is then used in the mass balance equation (2.23). We have already illustrated this approach for a single solid constituent s in Sect. 2.6.1. In the sections below we examine additional examples that illustrate common features in biological tissue growth and remodeling.

Example 11 We revisit the tissue model of Example 9, with distinct fiber orientations $\mathbf{n}_r^\alpha$, but common reference configuration $\mathbf{X}^s$. All fibers bundles start with a uniform density $\rho_r^\alpha = \rho_0$ and lie in the x_1–x_2 plane, with bundle α oriented at an angle α from the x_1-axis (Fig. 7),

$$\mathbf{n}_r^\alpha = \cos\alpha\, \mathbf{e}_1 + \sin\alpha\, \mathbf{e}_2. \tag{3.15a}$$

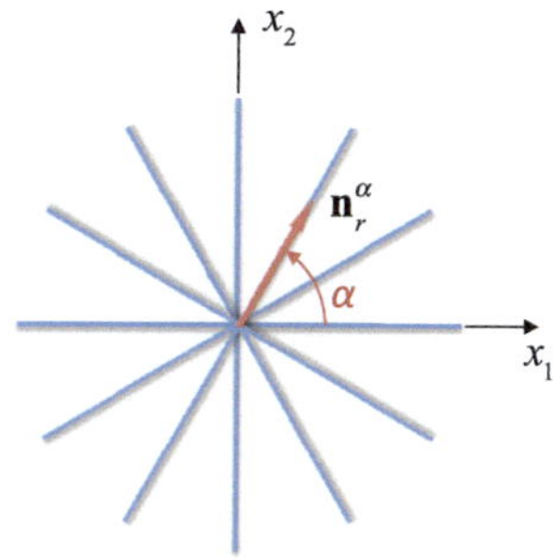

Fig. 7 Planar fiber distribution in the x_1–x_2 plane. Constituent α in this constrained mixture represents the fiber bundle making an angle α with the x_1-axis

The fibrous matrix is initially isotropic in the x_1–x_2 plane. Consider that the tissue is now subjected to biaxial tension, such that the matrix of the stress $\boldsymbol{\sigma}$ in the basis $\{\mathbf{e}_1, \mathbf{e}_2, \mathbf{e}_3\}$ is

$$[\boldsymbol{\sigma}] = \begin{bmatrix} \sigma_1 & 0 & 0 \\ 0 & \sigma_2 & 0 \\ 0 & 0 & 0 \end{bmatrix}, \quad \sigma_1 \geq \sigma_2 \geq 0. \tag{3.15b}$$

Under infinitesimal strains the fiber bundle directions $\mathbf{n}_r^\alpha$ remain nearly unchanged, so that the normal stress along bundle α is

$$\sigma = \mathbf{n}_r^\alpha \cdot \boldsymbol{\sigma} \cdot \mathbf{n}_r^\alpha = \sigma_1 \cos^2 \alpha + \sigma_2 \sin^2 \alpha. \tag{3.15c}$$

We assume that the fibers remodel in response to loading according to the constitutive relation (2.38),

$$\frac{\partial \rho_r^\alpha}{\partial t} = B\left(\frac{\Psi_r^\alpha}{\rho_r^\alpha} - \psi_0^\alpha\right), \quad \Psi_r^\alpha = \frac{\sigma^2}{2E(\rho_r^\alpha)}. \tag{3.15d}$$

For simplicity, we assume a linear dependence of the fiber modulus $E(\rho_r^\alpha)$ on referential apparent density,

$$E(\rho_r^\alpha) = c\rho_r^\alpha. \tag{3.15e}$$

If we assume that $\psi_0^\alpha = \psi_0$ is constant for all fibers, the steady-state remodeling response becomes

$$\rho_\infty^\alpha = \lim_{t\to\infty} \rho_r^\alpha = \rho_\infty\left(\cos^2 \alpha + \frac{\sigma_2}{\sigma_1} \sin^2 \alpha\right), \quad \rho_\infty = \sqrt{\frac{\sigma_1^2}{2c\psi_0}}. \tag{3.15f}$$

If we assume that the fiber distribution is continuous, we may integrate the expression for ρ_∞^α in (f) with respect to α from 0 to 2π to get the mixture density,

$$\rho = \rho_\infty\left(1 + \frac{\sigma_2}{\sigma_1}\right)\pi. \tag{3.15g}$$

Then, the mass fraction of fibers along α may be evaluated from ρ_∞^α/ρ. A plot of this fiber angular density distribution for various values of the ratio σ_2/σ_1 shows that fibers become preferentially denser in the direction of loading (Fig. 8). In the limiting case when $\sigma_2/\sigma_1 = 1$, the density distribution remains isotropic.

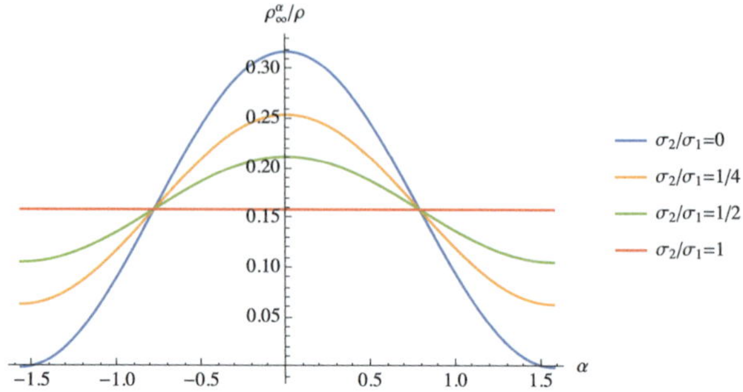

Fig. 8 Steady-state remodeling response of fibers when $\psi_0^\alpha = \psi_0$ is constant, as described in Example 11. *Curves* show the steady-state mass fraction $\rho_\infty^\alpha / \rho$ of fiber bundles oriented at angle α, for various values of the biaxial stress ratio σ_2/σ_1, as summarized in Eq. (3.15f)

Example 12 Alternatively, we may assume that the metabolic activity of cells is very sensitive to their orientation relative to the principal directions of stress. In that case, we may assume that the threshold value of the specific free energy ψ_0^α is nonlinearly dependent on the fiber orientation relative to the principal direction of stress,

$$\psi_0^\alpha = \psi_0 e^{-2b\cos 2\alpha}, \quad b = \ln \frac{\sigma_1}{\sigma_2}. \tag{3.16a}$$

In this model, the non-dimensional parameter b controls the sensitivity of matrix remodeling to the ratio of principal stresses. The steady-state remodeling response of ρ_r^α under this state of stress is

$$\rho_\infty^\alpha = \lim_{t\to\infty} \rho_r^\alpha = \rho_\infty \left(\cos^2\alpha + e^{-b}\sin^2\alpha\right) e^{b\cos 2\alpha}, \quad \rho_\infty = \sqrt{\frac{\sigma_1^2}{2c\psi_0}}. \tag{3.16b}$$

For a continuous fiber distribution, integrating this expression over the range $0 \le \alpha < 2\pi$ produces the mixture density

$$\rho = \rho_\infty \left[\left(1 + e^{-b}\right) I_0(b) + \left(1 - e^{-b}\right) I_1(b) \right] \pi, \tag{3.16c}$$

where I_0 and I_1 are modified Bessel functions of the first kind, of order 0 and 1, respectively. A plot of $\rho_\infty^\alpha / \rho$ versus α, for various choices of b, shows that fibers are recruited preferentially along the dominant loading direction ($\alpha = 0$), with a fiber angular distribution that narrows considerably with increasing b (Fig. 9). Once again, in the special case when $\sigma_2 = \sigma_1$ ($b = 0$), we note that the fiber angular distribution remains isotropic ($\rho_\infty^\alpha / \rho = 1/2\pi$ for all α). In contrast to the previous example, the remodeling response produces a much narrower fiber distribution for this choice of constitutive model for ψ_0^α. In fact, the constitutive relation given in (3.16a) was selected specifically to reproduce the π-periodic von Mises fiber angular distribution, commonly used to model fibrous biological tissues [23, 48].

3.3 Multi-generational Growth

When growth occurs in a biological tissue, the newly deposited solid material may not necessarily be in the same state of stress as the existing substrate. Therefore, we can treat each new generation of deposited solid material as a new constituent α within a constrained mixture, having its own distinct reference configuration $\mathbf{X}^\alpha$. The reference configuration of each generation becomes a constitutive assumption in the modeling of a growing tissue. A more comprehensive treatment of multi-generational growth mechanics has been presented previously [9].

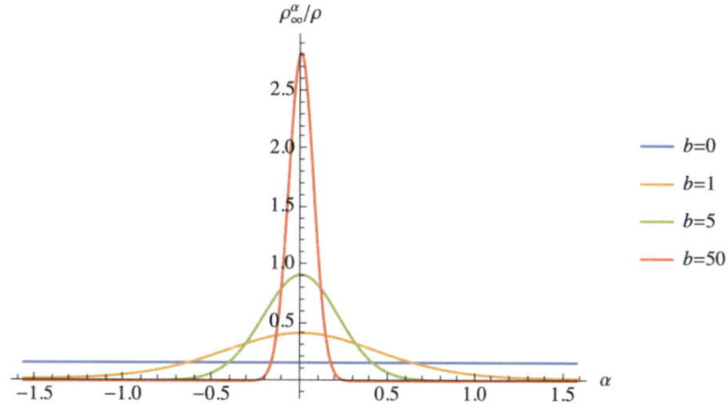

Fig. 9 Steady-state remodeling response of fibers as described in Example 12, showing the mass fraction $\rho^\alpha_\infty/\rho_\infty$ of fiber bundles oriented along α, for various values of the parameter b

A critical question that arises in this context is whether the newly deposited matrix should be in a stress-free state, or stressed. If it is stressed, should its state of stress match that of the underlying substrate or should it be distinct from it? The answer to any of these questions depends very much on the specifics of a given growth process, including environmental conditions at the time of matrix deposition, and the potential need to account for microscopic-level phenomena (such as molecular forces that stretch or coil a long-chain molecule as it binds to its substrate). For example, looking at Fig. 5, we need to hypothesize the manner by which cells synthesized fiber bundles at different time points, under different loading conditions, such that they were deposited within the extracellular matrix with different amounts of crimping.

One simple constitutive assumption is that the newly-deposited matrix is in the same state of strain as the underlying substrate, implying that $\mathbf{X}^\alpha = \mathbf{X}^s$, $\mathbf{F}^\alpha = \mathbf{F}^s$ and $\mathbf{F}^{\alpha s} = \mathbf{I}$ (or $\boldsymbol{\varepsilon}^\alpha = \boldsymbol{\varepsilon}^s$ and $\boldsymbol{\varepsilon}^{\alpha s} = \mathbf{0}$ under infinitesimal strains) for all generations α. This assumption for multi-generational growth may be hard to justify physically, since it is difficult to explain how the newly-deposited matrix would bind onto the existing substrate while reproducing its current state of strain exactly. Therefore, we have to view that constitutive assumption as a simplifying idealization, satisfied approximately in an average sense, as might occur for example when the substrate is being loaded cyclically in tension and compression, while new material is continuously being added. In that case, the net macroscopic effect would be to produce a tissue with no apparent residual stresses, consistent with the idealization that the new matrix is being deposited in the same state of strain as the substrate. This type of growth is believed to occur in bone, since bone does not typically exhibit residual stresses, despite the fact that it remodels continually under load.

Another simple constitutive assumption is that the matrix of generation α, newly deposited at time t^α, is in a stress-free state, so that its reference configuration matches the current configuration of the tissue, $\mathbf{X}^\alpha = \boldsymbol{\chi}^s(\mathbf{X}^s, t^\alpha)$. This assumption implies that $\mathbf{F}^{\alpha s}(\mathbf{X}^s) = \mathbf{F}^s(\mathbf{X}^s, t^\alpha)$ (or $\boldsymbol{\varepsilon}^{\alpha s}(\mathbf{X}^s) = \boldsymbol{\varepsilon}^s(\mathbf{X}^s, t^\alpha)$ under infinitesimal strains), producing $\mathbf{F}^\alpha(\mathbf{X}^s, t^\alpha) = \mathbf{I}$ (equivalently, $\boldsymbol{\varepsilon}^\alpha(\mathbf{X}^s, t^\alpha) = \mathbf{0}$). This assumption is easier to understand physically, since matrix products released from a cell are initially in soluble form, especially if we ignore molecular forces or assume that their effects average out to produce a stress-free state at the time of binding. This modeling assumption may account for the evolution of residual stresses in a tissue, when consecutive generations are deposited in different stress-free configurations, because the current configuration of the underlying substrate varies over time. This type of growth is consistent with observations of residual stresses in many soft tissues, such as articular cartilage, intervertebral disc, or arterial wall.

Importantly, both of these constitutive assumptions lead to observable measures for $\mathbf{F}^{\alpha s}$, since these measures are based on the observable deformation $\mathbf{F}^s$ of the master constituent.

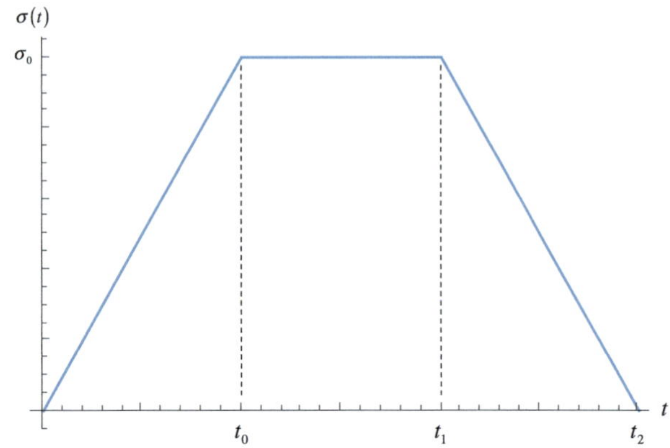

Fig. 10 Prescribed uniaxial stress history $\sigma(t)$ for multi-generational growth problem in Example 13

Thus, observation of the kinematics, $\mathbf{F}^s(\mathbf{X}^s,t)$, and composition $\rho_r^\alpha(\mathbf{X}^s,t)$, provides a complete determination of the state variables, whereas constitutive relations for $\Psi_r(\theta,\mathbf{F}^s,\rho_r^\alpha)$ and $\hat{\rho}_r^\alpha(\theta,\mathbf{F}^s,\rho_r^\alpha)$, and constitutive assumptions for $\mathbf{F}^{\alpha s}$, provide the functions of state necessary for this framework.

Example 13 Consider uniaxial loading of a bar having a solid matrix whose first generation is the master constituent s. The prescribed uniaxial stress $\sigma(t)$ in the bar is given by

$$\sigma(t) = \begin{cases} \sigma_0 \frac{t}{t_0}, & 0 \le t < t_0, \\ \sigma_0, & t_0 \le t < t_1, \\ \sigma_0 \frac{t_2 - t}{t_2 - t_1}, & t_1 \le t \le t_2. \end{cases} \tag{3.17a}$$

This profile is illustrated in Fig. 10. We assume that a new generation u is deposited interstitially, and homogeneously along the bar, over the time interval $t_0 \le t < t_1$. The moduli of generations s and u are respectively $E^s = E(\rho_r^s)$ and $E^u = E(\rho_r^u)$, where E is a given function of density. What is the history of the state of strain $\varepsilon^s(t)$?

We solve this problem using the infinitesimal strain relation (3.8), specialized to this one-dimensional analysis of a bar. The state of stress is already given, so our only challenge is to identify ε^{us} by constitutive assumption, and determine the history of ε^s. We adopt the assumption that generation u is deposited in a stress-free state over the time interval $t_0 \le t < t_1$. The mixture stress may be evaluated from (3.8) as

$$\sigma(t) = \sum_{\alpha=s,u} E(\rho_r^\alpha)\left(\varepsilon^s(t) - \varepsilon^{\alpha s}\right) = \begin{cases} E^s \varepsilon^s(t), & 0 \le t < t_0, \\ E^s \varepsilon^s(t) + E(\rho_r^u)(\varepsilon^s(t) - \varepsilon^{us}), & t_0 \le t < t_1, \\ E^s \varepsilon^s(t) + E^u(\varepsilon^s(t) - \varepsilon^{us}), & t_1 \le t \le t_2. \end{cases} \tag{3.17b}$$

Note that generation u does not exist prior to $t = t_0$, therefore it does not contribute to the mixture stress prior to that time. Since generation s does not grow or degrade, its density ρ_r^s and its modulus E^s are constants. The density ρ_r^u of generation u increases from 0 to its final value over the time interval $t_0 \le t < t_1$, during which the modulus $E(\rho_r^u)$ may evolve; however, from t_1 onward, the modulus $E^u = E(\rho_r^u)$ remains constant. Equating (3.17a) and (3.17b) over the time interval $0 \le t < t_0$, we can solve for $\varepsilon^s(t)$ and conclude that

$$\varepsilon^{us} = \varepsilon^s(t_0) = \frac{\sigma_0}{E^s}. \tag{3.17c}$$

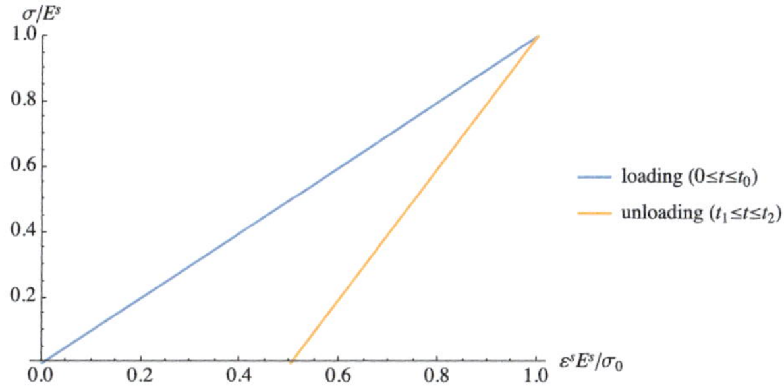

Fig. 11 Stress-strain response for multigenerational growth of a uniaxially loaded bar. The response to loading of the first generation ($0 \leq t \leq t_0$) has a slope E^s. After the growth of the second generation from t_0 to t_1, the modulus has risen to $E^s + E^u$ and the unload from t_1 to t_2 produces a residual strain, as given by Eq. (3.17e) in Example 10. This illustration uses $E^u = E^s$

Substituting this relation into (3.17b) and equating the resulting expressions with corresponding time intervals in (3.17a), we obtain the complete history

$$\varepsilon^s(t) = \begin{cases} \frac{\sigma_0}{E^s}\frac{t}{t_0}, & 0 \leq t < t_0, \\ \frac{\sigma_0}{E^s}, & t_0 \leq t < t_1, \\ \frac{\sigma_0}{E^s+E^u}\left(\frac{t_2-t}{t_2-t_1} + \frac{E^u}{E^s}\right), & t_1 \leq t \leq t_2. \end{cases} \tag{3.17d}$$

In particular, when the bar is unloaded at $t = t_2$, we find that the strain does not return to zero,

$$\varepsilon^s(t_2) = \frac{\sigma_0}{E^s + E^u}\frac{E^u}{E^s}. \tag{3.17e}$$

In other words, the bar does not return to the reference configuration of the master generation s, as the growth process has changed the traction-free configuration of the constrained mixture. The tissue has become stiffer due to the addition of mass in generation u. Because of the homogeneous growth and homogeneous state of stress throughout the growth history, the final state of stress is zero and the residual strain $\varepsilon^s(t_2)$ is homogeneous along the entire length of the bar. Importantly, to an observer who only examines the bar starting at $t > t_2$, it is theoretically impossible to detect this residual strain because of its homogeneous distribution, nor is it possible to reconstruct the true growth history, regardless of the measurements performed. Thus, unobserved past growth history may only be hypothesized, but not fully reconstructed.

Since we are using linear infinitesimal strain theory, we can use the superposition principle as a quick way of reproducing the result of (3.17e). We recognize that the bar has a modulus of E^e before growth, so that a stress σ_0 produces a strain σ_0/E^e. After growth, the bar's modulus has increased to $E^s + E^u$, therefore a reversal of the prescribed stress, $-\sigma_0$, produces a strain $-\sigma_0/(E^s + E^u)$. Superposing these two prescribed stresses and resulting strains, the net stress becomes zero and the net strain is identical to the formula in (3.17e), as also illustrated in Fig. 11.

The above example illustrates homogeneous growth under a homogeneous state of stress, which produces homogeneous residual strain (but no residual stress) in the final, traction-free configuration of the constrained mixture. In the next example, we examine homogeneous interstitial growth (homogeneous deposition of ρ_r^u) under an inhomogeneous state of stress, using an internally pressurized thick-walled cylinder.

Example 14 Consider the thick-walled cylindrical tube analyzed in Example 7. Consider that the tube is subjected to an internal pressure $p(t)$. The resulting state of stress is given in that previous example. Consider that the pressure $p(t)$ varies over time according to

$$p(t) = \begin{cases} p_0 \frac{t}{t_0}, & 0 \le t < t_0, \\ p_0, & t_0 \le t < t_1, \\ p_0 \frac{t_2 - t}{t_2 - t_1}, & t_1 \le t \le t_2, \end{cases} \tag{3.18a}$$

and growth occurs over the time interval $t_0 \le t < t_1$, in analogy to Example 13. Assuming that generations s and u behave as linear isotropic elastic materials with Poisson's ratio equal to zero for both generations, what is the state of stress and strain in the traction-free state at $t = t_2$?

We solve this problem in the same manner as in Example 13, using the quick superposition method summarized at the end of that example. Based on the constitutive relation of a linear isotropic elastic solid, when Poisson's ratio is zero, the non-zero normal strains of generation s at time t_0 are given by

$$\begin{aligned} \varepsilon_{rr}^s(r, t_0) = \varepsilon_{rr}^{us}(r) &= \frac{\sigma_{rr}}{E^s} = \frac{a^2}{b^2 - a^2}\left(1 - \frac{b^2}{r^2}\right)\frac{p_0}{E^s}, \\ \varepsilon_{\theta\theta}^s(r, t_0) = \varepsilon_{\theta\theta}^{us}(r) &= \frac{\sigma_{\theta\theta}}{E^s} = \frac{a^2}{b^2 - a^2}\left(1 + \frac{b^2}{r^2}\right)\frac{p_0}{E^s}. \end{aligned} \tag{3.18b}$$

After the growth of generation u, the modulus has increased to $E^s + E^u$. Therefore, the strains resulting from a negative internal pressure $-p_0$ have a similar form as (3.18b), with p_0 replaced by $-p_0$ and E^s replaced by $E^s + E^u$. Superposing these two loading states produces a net state of strain at t_2 given by

$$\begin{aligned} \varepsilon_{rr}^s(r, t_2) &= \frac{a^2}{b^2 - a^2}\left(1 - \frac{b^2}{r^2}\right)\frac{p_0}{E^s + E^u}\frac{E^u}{E^s}, \\ \varepsilon_{\theta\theta}^s(r, t_2) &= \frac{a^2}{b^2 - a^2}\left(1 + \frac{b^2}{r^2}\right)\frac{p_0}{E^s + E^u}\frac{E^u}{E^s}. \end{aligned} \tag{3.18c}$$

The state of stress at t_2 is evaluated from (3.8), using the stress-strain relation specialized to the case when Poisson's ratio is zero, $\boldsymbol{\sigma} = E\boldsymbol{\varepsilon}$, and we confirm that the stresses return to zero upon unloading, $\sigma_{rr}(r, t_2) = \sigma_{\theta\theta}(r, t_2) = 0$. The development of inhomogeneous residual strains, summarized in (3.18c) and plotted in Fig. 12, is a consequence of the multi-generational growth process under an inhomogeneous state of stress. Observing the material at time $t > t_2$ does not make it possible to reconstruct the growth history; however, the inhomogeneous strain state provides a hint that such a growth process has occurred. Cutting the tube radially at any location around its circumference will cause a deformation that breaks the axisymmetric geometry; this type of deformation serves as evidence that the tube was residually strained. After cutting, some residual stresses may develop in the tube.

In conclusion, we find that a constrained reactive mixture framework allows us to account for tissue growth and remodeling, the evolution of fibrous tissue anisotropy in response to

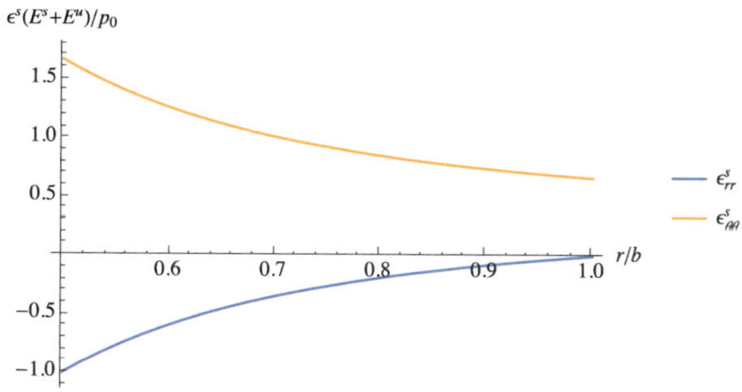

Fig. 12 Residual radial and circumferential strains in multi-generational growth of a pressurized thick-walled cylindrical tube, as described in Example 14. In this example, $a = b/2$ and $E^u = E^s$

loading, and the evolution of residual strains and stresses. As shown in various examples, the tissue microstructure may be described entirely with compositional measures, namely, the concentrations ρ_r^α of various constituents, whose evolution is governed by the axiom of mass balance. No internal variables are needed to describe these processes, and this approach neatly recovers many classical formulations for biological tissues within a unified framework.

4 Reactive Linear Viscoelasticity

4.1 Strong and Weak Bonds

In this section we apply the framework of constrained reactive mixtures to formulate the theory of linear viscoelasticity for solids. A more comprehensive treatment of reactive nonlinear viscoelasticity has been presented in an earlier study [5]. Here, we treat viscoelastic solids as a mixture of two bond families: Strong bonds ($\alpha = e$) that do not break in response to loading and produce the equilibrium elastic response; and weak bonds $\alpha \neq e$, which break in response to loading and immediately reform into a stress-free configuration. Weak bonds are therefore reactive constituents in this mixture framework. Weak bonds may belong to different bond families b that exhibit different responses to loading; in this introductory presentation, we assume that a single weak bond family b is present. This microscopic description of viscoelasticity was originally proposed by Green and Tobolsky [26] and this concept was reprized in more recent studies [64], though the present treatment differs from those prior articles.

Weak bonds that reform at time $t = u$ are denoted by $\alpha = u$ and called u-generation bonds. Similarly, weak bonds that break and reform at a subsequent time $t = v$ are denoted by $\alpha = v$. The reaction describing bond breaking and reforming is simply

$$\mathcal{E}^u \to \mathcal{E}^v, \tag{4.1}$$

indicating that loaded bonds belonging to the u-generation break and reform into stress-free bonds at time $t = v$, forming v-generation bonds. The earliest generation is $\alpha = s$ where $s \to -\infty$ and it represents the master constituent whose reference configuration is the stress-free configuration $\mathbf{X}^s$ of the weak bonds. We assume that $\mathbf{X}^s$ also represents the stress-free reference configuration of strong bonds. The motion of the master constituent is given by $\boldsymbol{\chi}^s(\mathbf{X}^s, t)$.

We adopt the constitutive assumption that bonds belonging to the u-generation have a stress-free reference configuration $\mathbf{X}^u$ that coincides with the current configuration at the time of bond reformation,

$$\mathbf{X}^u = \chi^s(\mathbf{X}^s, u). \tag{4.2}$$

Based on the relation (2.11), the infinitesimal strain for this generation is

$$\boldsymbol{\varepsilon}^u(\mathbf{X}^s, t) = \boldsymbol{\varepsilon}^s(\mathbf{X}^s, t) - \boldsymbol{\varepsilon}^{us}(\mathbf{X}^s). \tag{4.3}$$

This relation indicates that $\boldsymbol{\varepsilon}^u(\mathbf{X}^s, t)$ is the strain relative to time u. The constitutive assumption of (4.2) implies that $\boldsymbol{\varepsilon}^{us}$ may be evaluated as

$$\boldsymbol{\varepsilon}^{us}(\mathbf{X}^s) = \boldsymbol{\varepsilon}^s(\mathbf{X}^s, u), \tag{4.4}$$

which remains invariant for all $t \geq u$. In other words, the strain $\boldsymbol{\varepsilon}^s$ of the master constituent uniquely determines the relative strain $\boldsymbol{\varepsilon}^u$ of all subsequent generations u.

4.2 Constitutive Model for Free Energy Density

We now propose a constitutive form for the mixture strain energy density Ψ_r,

$$\Psi_r(\theta, \boldsymbol{\varepsilon}^s, \rho_r^\beta) = \Psi_r^e(\theta, \boldsymbol{\varepsilon}^s) + \sum_u \Psi_r^u(\theta, \rho_r^u, \boldsymbol{\varepsilon}^u), \tag{4.5}$$

where Ψ_r^e is the strain energy density of strong bonds and Ψ_r^u is the strain energy density of u-generation bonds. The simplifying constitutive assumptions adopted here are that (a) Ψ_r^e is independent of the weak bond concentrations, and (b) Ψ_r^u only depends on the concentration ρ_r^u of that generation. We make a further simplifying assumption that the free energy density of each generation has the same functional form for all generations,

$$\Psi_r^u(\theta, \rho_r^u, \boldsymbol{\varepsilon}^u) = \rho_r^u \psi^b(\theta, \boldsymbol{\varepsilon}^u), \tag{4.6}$$

where ψ^b is the specific free energy of any generation in the bond family b, and is independent of the concentration of that generation. All these simplifications represent constitutive assumptions that may be validated against experimental measurements of specific viscoelastic solids.

Based on this constitutive model we find that the chemical potential of each bond generation, given by (3.5), reduces to

$$\mu^u = \frac{\partial \Psi_r^u}{\partial \rho_r^u} = \psi^b(\theta, \boldsymbol{\varepsilon}^u), \tag{4.7}$$

which is the specific free energy of that bond. The chemical potential of strong bonds (which are non-reactive) is zero, $\mu^e = 0$. The mixture stress may be evaluated by substituting (4.5) into (3.6),

$$\boldsymbol{\sigma} = \frac{\partial \Psi_r^e}{\partial \boldsymbol{\varepsilon}^s} + \sum_u \rho_r^u \frac{\partial \psi^b(\theta, \boldsymbol{\varepsilon}^u)}{\partial \boldsymbol{\varepsilon}^u}. \tag{4.8}$$

This relation may also be rewritten as $\boldsymbol{\sigma} = \boldsymbol{\sigma}^e + \boldsymbol{\sigma}^{\mathrm{v}}$, where

$$\boldsymbol{\sigma}^e = \frac{\partial \Psi_r^e}{\partial \boldsymbol{\varepsilon}^s} \tag{4.9}$$

is the elastic response of strong bonds, while

$$\boldsymbol{\sigma}^{\mathrm{v}} = \sum_{u} \rho_r^u \frac{\partial \psi^b(\theta, \boldsymbol{\varepsilon}^u)}{\partial \boldsymbol{\varepsilon}^u} \tag{4.10}$$

is the stress in weak bonds, which is effectively the viscous response since ρ_r^u eventually decays to zero for all loaded bond generations, as shown below.

4.3 Constitutive Model for Mass Supply

The mass balance for each constituent of a constrained mixture is given in (2.23), subject to a suitable initial condition. Based on Sect. 2.4, the initial bond reaction $\mathcal{E}^s \to \mathcal{E}^u$ is constrained stoichiometrically by

$$\hat{\rho}_r^u = -\hat{\rho}_r^s = M^s \hat{\zeta}_r, \tag{4.11}$$

where $\hat{\zeta}_r$ is the molar production rate and $M^s = M^u$ is the molar mass for this weak bond species. We now make the constitutive assumption that $\hat{\zeta}_r$ describes a first-order forward reaction, according to the law of mass action,

$$M^s \hat{\zeta}_r = \frac{1}{\tau} \rho_r^s, \tag{4.12}$$

where $1/\tau$ is the reaction rate and τ is the time constant of the reaction. Using a material frame to solve for $\rho_r^s(\mathbf{X}^s, t)$, the mass balance for the s-generation becomes

$$\frac{\partial \rho_r^s}{\partial t} = -\frac{1}{\tau} \rho_r^s, \tag{4.13}$$

whose solution is

$$\rho_r^s(\mathbf{X}^s, t) = \rho_0 e^{-(t-u)/\tau}, \quad t \geq u. \tag{4.14}$$

Here, ρ_0 is the initial concentration of s-generation bonds of bond family b, which represents the total density of weak bonds in the mixture. The corresponding equation for ρ_r^u is

$$\frac{\partial \rho_r^u}{\partial t} = \frac{1}{\tau} \rho_r^s = \frac{1}{\tau} \rho_0 e^{-(t-u)/\tau}, \tag{4.15}$$

with initial condition $\rho_r^u(\mathbf{X}^s, u) = 0$ at $t = u$. Therefore, the solution for ρ_r^u is

$$\rho_r^u(\mathbf{X}^s, t) = \rho_0 \left(1 - e^{-(t-u)/\tau}\right), \quad t \geq u. \tag{4.16}$$

Next, consider the subsequent bond reaction $\mathcal{E}^u \to \mathcal{E}^v$, which starts at time v and is assumed to exhibit the same reaction kinetics. Following the approach above, the governing equation for u-generation bonds is

$$\frac{\partial \rho_r^u}{\partial t} = -\frac{1}{\tau} \rho_r^u, \tag{4.17}$$

subject to the initial condition $\rho_r^u(\mathbf{X}^s, v) = \rho_0(1 - e^{-(v-u)/\tau})$ obtained from (4.16). The solution to this equation is

$$\rho_r^u(\mathbf{X}^s, t) = \rho_0 \left(e^{-(t-v)/\tau} - e^{-(t-u)/\tau}\right), \quad t \geq v. \tag{4.18}$$

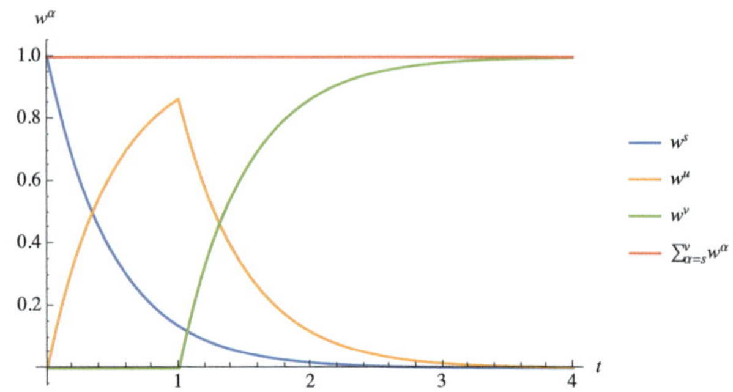

Fig. 13 Temporal response of weak bond mass fractions w^α for three consecutive bond generations, where α denotes the time when that generation started forming: $s=-\infty$, $u=0$, $v=1$, with $\tau=1/2$. Based on the constraint of (2.24) and the definition (4.20), the bond mass fractions sum up to unity, $\sum_\alpha w^\alpha = 1$

Therefore,

$$\rho_r^u(\mathbf{X}^s, t) = \begin{cases} 0, & t < u, \\ \rho_0(1 - e^{-(t-u)/\tau}), & u \le t < v, \\ \rho_0(e^{-(t-v)/\tau} - e^{-(t-u)/\tau}), & v \le t. \end{cases} \tag{4.19}$$

We may define the bond mass fraction

$$w^u(\mathbf{X}^s, t) \equiv \frac{\rho_r^u(\mathbf{X}^s, t)}{\rho_0}, \tag{4.20}$$

such that w^u is now given by

$$w^u(\mathbf{X}^s, t) = \begin{cases} 0, & t < u, \\ 1 - e^{-(t-u)/\tau}, & u \le t < v, \\ e^{-(t-v)/\tau} - e^{-(t-u)/\tau}, & v \le t. \end{cases} \tag{4.21}$$

While s- and u-generation bonds are breaking over time $t \ge v$, v-generation bonds are growing from the reformation of bonds from both prior generations, with ρ_r^v satisfying

$$\frac{\partial \rho_r^v}{\partial t} = \frac{1}{\tau}(\rho_r^s + \rho_r^u) = \frac{\rho_0}{\tau} e^{-(t-v)/\tau}, \tag{4.22}$$

subject to $\rho_r^v(\mathbf{X}^s, v) = 0$. Thus, the solution for ρ_r^v is

$$\rho_r^v(\mathbf{X}^s, t) = \rho_0(1 - e^{-(t-v)/\tau}), \quad t \ge v, \tag{4.23}$$

such that $w^v = \rho_r^v/\rho_0 = 1 - e^{-(t-v)/\tau}$. Comparing this relation to (4.16), it becomes apparent that the relation for w^u in (4.21) is recursive and applies to all subsequent generations (Fig. 13).

The mixture strain energy density and stress may be evaluated using these mass fractions as

$$\Psi_r = \Psi_r^e(\theta, \boldsymbol{\varepsilon}^s) + \sum_u w^u \Psi_r^b(\theta, \boldsymbol{\varepsilon}^u), \tag{4.24}$$

and

$$\boldsymbol{\sigma} = \frac{\partial \Psi_r^e}{\partial \boldsymbol{\varepsilon}^s} + \sum_u w^u \frac{\partial \Psi_r^b(\theta, \boldsymbol{\varepsilon}^u)}{\partial \boldsymbol{\varepsilon}^u}, \tag{4.25}$$

where $\Psi_r^b \equiv \rho_0 \psi^b$.

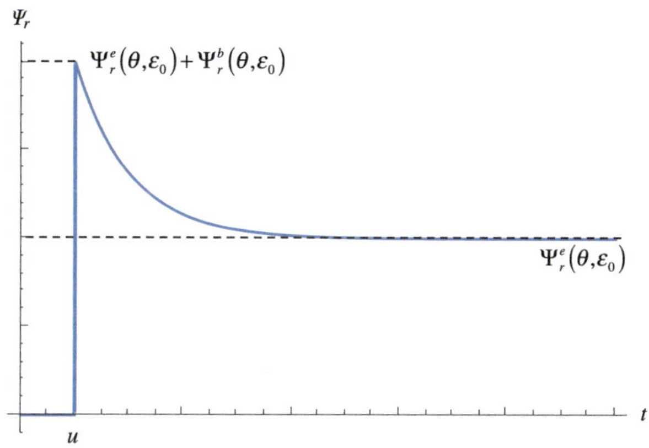

Fig. 14 Time-dependent response of the free energy density Ψ_r in a reactive viscoelastic material subjected to a step strain $\boldsymbol{\varepsilon}_0$ at time $t = u$, according to Example 15

Example 15 The stress-relaxation response to a single step strain at $t = u$ may be evaluated from these expressions as follows. Let $\boldsymbol{\varepsilon}^s(\mathbf{X}^s, t) = H(t-u)\boldsymbol{\varepsilon}_0$. According to (4.4), we find that $\boldsymbol{\varepsilon}^{us} = \boldsymbol{\varepsilon}_0$ so that $\boldsymbol{\varepsilon}^u(\mathbf{X}^s, t) = \mathbf{0}$ for $t \geq u$ according to (4.3). In other words, generation u comes into existence at the instant of loading at $t = u$, and its reference configuration is the current configuration of the s-generation at $t = u$. Since $\boldsymbol{\varepsilon}^u(\mathbf{X}^s, t)$ is zero, the u-generation strain energy density is equal to zero, $\Psi_r^b(\theta, \boldsymbol{\varepsilon}^u) = 0$. Therefore, the only contributions to the mixture strain energy density Ψ_r are the strong bonds and the s-generation weak bonds, whose concentration is decaying over time according to (4.14),

$$\Psi_r = \Psi_r^e(\theta, \boldsymbol{\varepsilon}_0) + e^{-(t-u)/\tau}\Psi_r^b(\theta, \boldsymbol{\varepsilon}_0), \quad t \geq u. \tag{4.26a}$$

Accordingly, at $t = u$, the strain energy density in the mixture is $\Psi_r = \Psi_r^e(\theta, \boldsymbol{\varepsilon}_0) + \Psi_r^b(\theta, \boldsymbol{\varepsilon}_0)$, and as $t \to \infty$, it decays to $\Psi_r = \Psi_r^e(\theta, \boldsymbol{\varepsilon}_0)$ (Fig. 14). Thus, instantaneously upon loading, both strong and weak bonds can store strain energy; however, the progressive breaking and reforming of weak bonds causes a concomitant dissipation of mixture strain energy density, eventually reducing to the contribution of strong bonds only. The stress response shows the identical behavior,

$$\boldsymbol{\sigma} = \boldsymbol{\sigma}^e(\theta, \boldsymbol{\varepsilon}_0) + e^{-(t-u)/\tau}\boldsymbol{\sigma}^b(\theta, \boldsymbol{\varepsilon}_0), \quad t \geq u, \tag{4.26b}$$

where the strong and weak bond stresses are evaluated from their respective strain energy density functions as $\boldsymbol{\sigma}^e = \partial\Psi_r^e/\partial\boldsymbol{\varepsilon}$ and $\boldsymbol{\sigma}^b = \partial\Psi_r^b/\partial\boldsymbol{\varepsilon}$. Comparing the expression of (4.26b) to (4.8), we conclude that the viscous stress is $\boldsymbol{\sigma}^{\text{v}} = e^{-(t-u)/\tau}\boldsymbol{\sigma}^b$. This example shows that the stress-strain response of strong and weak bonds need not be the same, since the constitutive relations for Ψ_r^e and Ψ_r^b may be different. In most classical, introductory presentations of linear viscoelasticity, it is common to select these functions to differ only by a scale factor. In particular, $\Psi_r^b = 0$ produces an elastic solid; $\Psi_r^e = 0$ produces a Maxwell fluid; and $\Psi_r^b = \beta\Psi_r^e$ produces the standard linear solid, where β is a positive scalar.

For continuous time increments between consecutive generations, let $v = u + du$ in the recursive relation for w^u in (4.21), and perform a Taylor series expansion of the third entry about u to show that

$$w^u = \frac{1}{\tau}e^{-(t-u)/\tau}du, \quad u \leq t. \tag{4.27}$$

(The second entry in (4.21) reduces to zero as $v \to u$ and $t \to u$). Now, the summation in the evaluation of $\boldsymbol{\sigma}$ may be converted to an integral,

$$\boldsymbol{\sigma} = \frac{\partial \Psi_r^e}{\partial \boldsymbol{\varepsilon}^s} + \int_{-\infty}^{t} \frac{1}{\tau} e^{-(t-u)/\tau} \frac{\partial \Psi_r^b(\theta, \boldsymbol{\varepsilon}^u)}{\partial \boldsymbol{\varepsilon}^u}\, du. \tag{4.28}$$

We may define the function

$$\boldsymbol{\sigma}^b(\theta, \boldsymbol{\varepsilon}^u) = \frac{\partial \Psi_r^b(\theta, \boldsymbol{\varepsilon}^u)}{\partial \boldsymbol{\varepsilon}^u}, \tag{4.29}$$

and use the definition of (4.9), so that the above expression for the stress may be rewritten as

$$\boldsymbol{\sigma} = \boldsymbol{\sigma}^e(\theta, \boldsymbol{\varepsilon}^s) + \frac{1}{\tau} \int_{-\infty}^{t} e^{-(t-u)/\tau} \boldsymbol{\sigma}^b(\theta, \boldsymbol{\varepsilon}^u)\, du. \tag{4.30}$$

Similarly, the mixture free energy density is

$$\Psi_r(\theta, \boldsymbol{\varepsilon}^s, \rho_r^\gamma) = \Psi_r^e(\theta, \boldsymbol{\varepsilon}^s) + \sum_u w^u \Psi_r^b(\theta, \boldsymbol{\varepsilon}^u) \tag{4.31}$$

in discrete form, and

$$\Psi_r(\theta, \boldsymbol{\varepsilon}^s, \rho_r^\gamma) = \Psi_r^e(\theta, \boldsymbol{\varepsilon}^s) + \frac{1}{\tau} \int_{-\infty}^{t} e^{-(t-u)/\tau} \Psi_r^b(\theta, \boldsymbol{\varepsilon}^u)\, du \tag{4.32}$$

in continuous form.

For a linear stress-strain relation for $\boldsymbol{\sigma}^b$ versus $\boldsymbol{\varepsilon}^u$, the relation (4.3) between $\boldsymbol{\varepsilon}^u$ and $\boldsymbol{\varepsilon}^s$ implies that

$$\boldsymbol{\sigma}^b(\theta, \boldsymbol{\varepsilon}^u(\mathbf{X}^s, t)) = \boldsymbol{\sigma}^b(\theta, \boldsymbol{\varepsilon}^s(\mathbf{X}^s, t)) - \boldsymbol{\sigma}^b(\boldsymbol{\varepsilon}^s(\mathbf{X}^s, u)). \tag{4.33}$$

This expression may be substituted into (4.30) to produce

$$\boldsymbol{\sigma} = \boldsymbol{\sigma}^e(\boldsymbol{\varepsilon}^s(\mathbf{X}^s, t)) + \boldsymbol{\sigma}^b(\boldsymbol{\varepsilon}^s(\mathbf{X}^s, t)) - \frac{1}{\tau} \int_{-\infty}^{t} e^{-(t-u)/\tau} \boldsymbol{\sigma}^b(\boldsymbol{\varepsilon}^s(\mathbf{X}^s, u))\, du, \tag{4.34}$$

which is the classical relation for infinitesimal strain viscoelasticity in isotropic materials.

In summary, we were able to derive the theory of linear viscoelasticity without an *ad hoc* appeal to Boltzmann's superposition principle. Similarly, we did not propose the existence of internal variables, such as the deformation of springs and dashpots placed in various parallel and series configurations, to produce the desired material response. Instead, we formulated a viscoelasticity theory based on the existence of microscopic weak bonds that break and reform in response to loading, which is the fundamental physical mechanism believed to describe viscoelasticity [26, 64]. This formulation shows that stress may be evaluated from the derivative of the strain energy density with respect to strain, as shown in (4.25) and (4.28), in analogy to elasticity theory.

Example 16 Consider the one-dimensional analysis of a bar subjected to uniaxial loading. Let the strong bond modulus be E^e and the weak bond modulus be E^b, so that $\sigma^e = E^e \varepsilon^s$ and $\sigma^b = E^b \varepsilon^s$. Under a prescribed strain $\varepsilon^s = \varepsilon_0 H(t)$, the stress response is obtained from (4.34),

$$\sigma(t) = E^e \varepsilon_0 H(t) + E^b \varepsilon_0 H(t) - \frac{E^b \varepsilon_0}{\tau} \int_0^t e^{-(t-u)/\tau}\, du, \tag{4.35a}$$

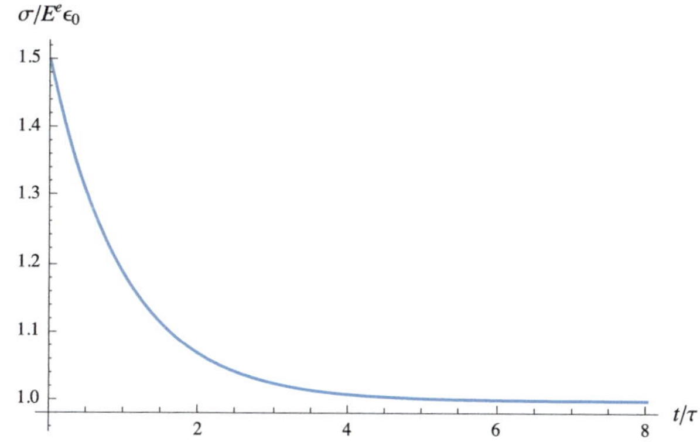

Fig. 15 One-dimensional stress-relaxation response as described in Example 16, with $E^b/E^e = 1/2$

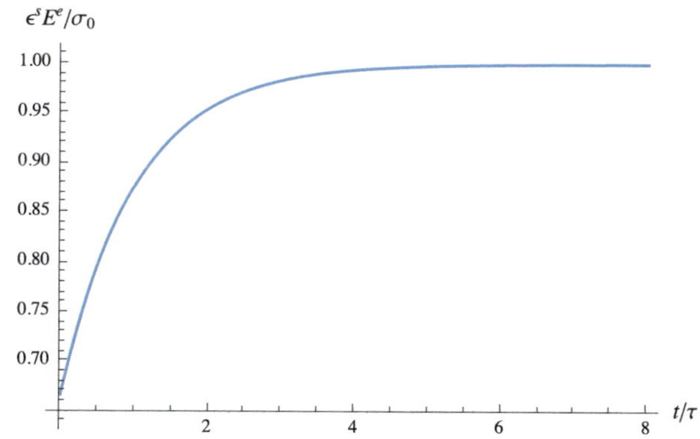

Fig. 16 One-dimensional creep response as described in Example 16, with $E^b/E^e = 1/2$

which evaluates to

$$\sigma(t) = E^e \varepsilon_0 H(t) \left(1 + \frac{E^b}{E^e} e^{-t/\tau} \right). \tag{4.35b}$$

Thus, the stress relaxes exponentially from a peak value $(E^e + E^b)\varepsilon_0$ at $t = 0$ to an equilibrium value $E^e \varepsilon_0$ (Fig. 15).

Under a prescribed stress $\sigma = \sigma_0 H(t)$, we solve for $\varepsilon^s(t)$ using

$$\sigma_0 H(t) = \left(E^e + E^b\right)\varepsilon^s(t) - \frac{E^b}{\tau} \int_{-\infty}^{t} e^{-(t-u)/\tau} \varepsilon^s(u)\, du. \tag{4.35c}$$

This equation may be solved using the method of Laplace transforms to produce

$$\varepsilon^s(t) = \frac{\sigma_0}{E^e}\left(1 - \beta e^{-\beta \frac{t}{\tau}}\right), \quad \beta = \frac{E^b}{E^e + E^b}. \tag{4.35d}$$

This creep response shows that the strain is initially $\sigma_0/(E^e + E^b)$, and rises exponentially to an equilibrium value σ_0/E^e (Fig. 16).

4.4 Thermodynamic Constraints

The reaction $\mathcal{E}^s \to \mathcal{E}^u$ involves two constituents and their referential mass supplies are given in (4.11), whereas the chemical potentials are given in (4.7). Substituting these relations into the residual dissipation (3.4) produces

$$\hat{\rho}_r^s \mu^s + \hat{\rho}_r^u \mu^u = M^s \hat{\zeta}_r \big(\psi^b(\theta, \boldsymbol{\varepsilon}^u) - \psi^b(\theta, \boldsymbol{\varepsilon}^s)\big) \leq 0. \tag{4.36}$$

Note that $M^s \hat{\zeta}_r$ is always positive since the reaction proceeds forward (thus $\tau > 0$), as seen from (4.12). We conclude that the thermodynamic constraint placed by the residual dissipation is that the specific free energy $\psi^b(\theta, \boldsymbol{\varepsilon}^u)$ of broken and reformed bonds in the u-generation must be less than the specific free energy $\psi^b(\theta, \boldsymbol{\varepsilon}^s)$ of intact bonds s. This constraint is automatically satisfied based on our modeling assumptions, since $\boldsymbol{\varepsilon}^u = \mathbf{0}$ at time u, implying that $\psi^b(\theta, \boldsymbol{\varepsilon}^u) = 0$ at the time of u-generation bond reformation.

5 Conclusion

This article reviews some recent advances in modeling biological tissues using reactive constrained mixture theory. Many simple illustrations have been presented to highlight the potential benefits of this approach. The main message conveyed in this presentation is that seemingly disparate fields of mechanics and chemical kinetics are actually closely interrelated and may be elegantly expressed in a unified framework. Thus, constrained mixture models recover classical theories for fibrous materials with bundles oriented in different directions or having different reference configurations, that produce characteristic fiber recruitment patterns under loading. Reactions that exchange mass among various constituents of a mixture may be used to describe tissue growth and remodeling. Similarly, reactions that describe the breaking and reforming of bonds may be used to model free energy dissipation in a viscoelastic material.

In contrast to general (unconstrained) mixtures, constrained mixture only require the solution of a single momentum balance equation. No constitutive relations are needed to describe the momentum exchange among various constituents, since there is no relative motion between them within a constrained mixture framework. Therefore, just like elasticity theory, there is only one unknown motion $\boldsymbol{\chi}^s(\mathbf{X}^s, t)$ that needs to be obtained from the analysis of a constrained mixture, by solving the momentum balance for the mixture. When the mixture is reactive, the axiom of mass balance may be used to solve for the evolving concentration $\rho_r^\alpha(\mathbf{X}^s, t)$ of mixture constituents α. Much freedom exists in the formulation of constitutive relations for the mixture free energy density $\Psi_r(\theta, \mathbf{F}^s, \rho_r^\alpha)$, constituent mass density supplies $\hat{\rho}_r^\alpha(\theta, \mathbf{F}^s, \rho_r^\alpha)$, and constitutive assumptions about the mapping $\mathbf{F}^{\alpha s}$ between the reference configuration of constituent α and that of the master constituent s. Contrary to theories that employ internal variables, which necessarily evolve in time and require additional evolution equations [17], $\mathbf{F}^{\alpha s}$ is time-independent. It is a property of the mixture which is postulated by constitutive assumption when solid constituent α comes into existence. This approach allowed us to model the evolution of material anisotropy using ρ_r^α as the evolving measure of constituents with preferred spatial orientations, as illustrated with fibrous tissues in Sect. 3.2; a similar approach could be used with other quantitative measures of material structure, such as rods and plates in trabecular bone [62]. Importantly, the state variables $(\theta, \mathbf{F}^s, \rho_r^\alpha)$ are all observable and governed by fundamental axioms of mass, momentum and energy balance, such that this framework does not require the introduction of additional evolution equations.

Acknowledgements Research reported in this publication was supported by the National Institutes of Health under Award Number R01 GM083925. The content is solely the responsibility of the author and does not necessarily represent the official views of the National Institutes of Health.

References

1. Armstrong, C.G.: An analysis of the stresses in a thin layer of articular cartilage in a synovial joint. Eng. Med. **15**(2), 55–61 (1986)
2. Armstrong, C.G., Lai, W.M., Mow, V.C.: An analysis of the unconfined compression of articular cartilage. J. Biomech. Eng. **106**(2), 165–173 (1984)
3. Aspden, R.M., Hukins, D.W.: Collagen organization in articular cartilage, determined by X-ray diffraction, and its relationship to tissue function. Proc. R. Soc. Lond. B, Biol. Sci. **212**(1188), 299–304 (1981)
4. Ateshian, G.A.: On the theory of reactive mixtures for modeling biological growth. Biomech. Model. Mechanobiol. **6**(6), 423–445 (2007)
5. Ateshian, G.A.: Viscoelasticity using reactive constrained solid mixtures. J. Biomech. **48**(6), 941–947 (2015)
6. Ateshian, G.A., Ellis, B.J., Weiss, J.A.: Equivalence between short-time biphasic and incompressible elastic material responses. J. Biomech. Eng. **129**(3), 405–412 (2007)
7. Ateshian, G.A., Morrison, B. III, Holmes, J.W., Hung, C.T.: Mechanics of cell growth. Mech. Res. Commun. **42**, 118–125 (2012)
8. Ateshian, G.A., Rajan, V., Chahine, N.O., Canal, C.E., Hung, C.T.: Modeling the matrix of articular cartilage using a continuous fiber angular distribution predicts many observed phenomena. J. Biomech. Eng. **131**(6), 061003 (2009)
9. Ateshian, G.A., Ricken, T.: Multigenerational interstitial growth of biological tissues. Biomech. Model. Mechanobiol. **9**(6), 689–702 (2010)
10. Azeloglu, E.U., Albro, M.B., Thimmappa, V.A., Ateshian, G.A., Costa, K.D.: Heterogeneous transmural proteoglycan distribution provides a mechanism for regulating residual stresses in the aorta. Am. J. Physiol., Heart Circ. Physiol. **294**(3), H1197–H1205 (2008)
11. Bachrach, N.M., Mow, V.C., Guilak, F.: Incompressibility of the solid matrix of articular cartilage under high hydrostatic pressures. J. Biomech. **31**(5), 445–451 (1998)
12. Bedford, A., Drumheller, D.S.: Theories of immiscible and structured mixtures. Int. J. Eng. Sci. **21**(8), 863–960 (1983)
13. Bowen, R.M.: Thermochemistry of reacting materials. J. Chem. Phys. **49**(4), 1625–1637 (1968)
14. Bryant, M.R., McDonnell, P.J.: A triphasic analysis of corneal swelling and hydration control. J. Biomech. Eng. **120**(3), 370–381 (1998)
15. Carter, D.R., Hayes, W.C.: Bone compressive strength: the influence of density and strain rate. Science **194**(4270), 1174–1176 (1976)
16. Chahine, N.O., Wang, C.C.-B., Hung, C.T., Ateshian, G.A.: Anisotropic strain-dependent material properties of bovine articular cartilage in the transitional range from tension to compression. J. Biomech. **37**(8), 1251–1261 (2004)
17. Coleman, B.D., Gurtin, M.E.: Thermodynamics with internal state variables. J. Chem. Phys. **47**(2), 597–613 (1967)
18. Cowin, S., Hegedus, D.: Bone remodeling. I: Theory of adaptive elasticity. J. Elast. **6**(3), 313–326 (1976)
19. Cyron, C.J., Aydin, R.C., Humphrey, J.D.: A homogenized constrained mixture (and mechanical analog) model for growth and remodeling of soft tissue. Biomech. Model. Mechanobiol. **15**(6), 1389–1403 (2016)
20. Eberhardt, A.W., Keer, L.M., Lewis, J.L., Vithoontien, V.: An analytical model of joint contact. J. Biomech. Eng. **112**(4), 407–413 (1990)
21. Eringen, A.C., Ingram, J.D.: A continuum theory of chemically reacting media—I. Int. J. Eng. Sci. **3**(2), 197–212 (1965)
22. Fyhrie, D.P., Carter, D.R.: A unifying principle relating stress to trabecular bone morphology. J. Orthop. Res. **4**(3), 304–317 (1986)
23. Gasser, T.C., Ogden, R.W., Holzapfel, G.A.: Hyperelastic modelling of arterial layers with distributed collagen fibre orientations. J. R. Soc. Interface **3**(6), 15–35 (2006)
24. Gleason, R.L., Humphrey, J.D.: Effects of a sustained extension on arterial growth and remodeling: a theoretical study. J. Biomech. **38**(6), 1255–1261 (2005)
25. Gleason, R.L., Taber, L.A., Humphrey, J.D.: A 2-d model of flow-induced alterations in the geometry, structure, and properties of carotid arteries. J. Biomech. Eng. **126**(3), 371–381 (2004)
26. Green, M., Tobolsky, A.: A new approach to the theory of relaxing polymeric media. J. Chem. Phys. **14**(2), 80–92 (1946)
27. Hill, M.R., Duan, X., Gibson, G.A., Watkins, S., Robertson, A.M.: A theoretical and non-destructive experimental approach for direct inclusion of measured collagen orientation and recruitment into mechanical models of the artery wall. J. Biomech. **45**(5), 762–771 (2012)
28. Hori, R.Y., Mockros, L.F.: Indentation tests of human articular cartilage. J. Biomech. **9**(4), 259–268 (1976)

29. Huiskes, R., Weinans, H., Grootenboer, H.J., Dalstra, M., Fudala, B., Slooff, T.J.: Adaptive bone-remodeling theory applied to prosthetic-design analysis. J. Biomech. **20**(11–12), 1135–1150 (1987)
30. Humphrey, J., Rajagopal, K.: A constrained mixture model for growth and remodeling of soft tissues. Math. Models Methods Appl. Sci. **12**(03), 407–430 (2002)
31. Humphrey, J.D., Rajagopal, K.R.: A constrained mixture model for arterial adaptations to a sustained step change in blood flow. Biomech. Model. Mechanobiol. **2**(2), 109–126 (2003)
32. Hurschler, C., Loitz-Ramage, B., Vanderby, R. Jr.: A structurally based stress-stretch relationship for tendon and ligament. J. Biomech. Eng. **119**(4), 392–399 (1997)
33. Karsaj, I., Humphrey, J.D.: A mathematical model of evolving mechanical properties of intraluminal thrombus. Biorheology **46**(6), 509–527 (2009)
34. Kenyon, D.E.: The theory of an incompressible solid-fluid mixture. Arch. Ration. Mech. Anal. **62**(2), 131–147 (1976)
35. Lake, S.P., Miller, K.S., Elliott, D.M., Soslowsky, L.J.: Effect of fiber distribution and realignment on the nonlinear and inhomogeneous mechanical properties of human supraspinatus tendon under longitudinal tensile loading. J. Orthop. Res. **27**(12), 1596–1602 (2009)
36. Lanir, Y.: A structural theory for the homogeneous biaxial stress-strain relationships in flat collagenous tissues. J. Biomech. **12**(6), 423–436 (1979)
37. Lanir, Y.: Constitutive equations for fibrous connective tissues. J. Biomech. **16**(1), 1–12 (1983)
38. Mak, A.F., Lai, W.M., Mow, V.C.: Biphasic indentation of articular cartilage—I: Theoretical analysis. J. Biomech. **20**(7), 703–714 (1987)
39. Mansour, J.M., Mow, V.C.: The permeability of articular cartilage under compressive strain and at high pressures. J. Bone Jt. Surg., Am. Vol. **58**(4), 509–516 (1976)
40. Maroudas, A., Bannon, C.: Measurement of swelling pressure in cartilage and comparison with the osmotic pressure of constituent proteoglycans. Biorheology **18**(3–6), 619–632 (1981)
41. Mow, V., Kuei, S., Lai, W., Armstrong, C.: Biphasic creep and stress relaxation of articular cartilage in compression: theory and experiments. J. Biomech. Eng. **102**(1), 73–84 (1980)
42. Nims, R.J., Durney, K.M., Cigan, A.D., Dusséaux, A., Hung, C.T., Ateshian, G.A.: Continuum theory of fibrous tissue damage mechanics using bond kinetics: application to cartilage tissue engineering. Interface Focus **6**(1), 20150063 (2016)
43. Oloyede, A., Broom, N.D.: Is classical consolidation theory applicable to articular cartilage deformation? Clin. Biomech. **6**(4), 206–212 (1991)
44. Park, S., Krishnan, R., Nicoll, S.B., Ateshian, G.A.: Cartilage interstitial fluid load support in unconfined compression. J. Biomech. **36**(12), 1785–1796 (2003)
45. Pierce, D.M., Ricken, T., Holzapfel, G.A.: Modeling sample/patient-specific structural and diffusional responses of cartilage using dt-mri. Int. J. Numer. Methods Biomed. Eng. **29**(8), 807–821 (2013)
46. Prud'homme, R.: Flows of Reactive Fluids. Fluid Mechanics and Its Applications, vol. 94. Springer, New York (2010)
47. Rachev, A., Gleason, R.L. Jr.: Theoretical study on the effects of pressure-induced remodeling on geometry and mechanical non-homogeneity of conduit arteries. Biomech. Model. Mechanobiol. **10**(1), 79–93 (2011)
48. Sacks, M.S.: Incorporation of experimentally-derived fiber orientation into a structural constitutive model for planar collagenous tissues. J. Biomech. Eng. **125**(2), 280–287 (2003)
49. Satha, G., Lindström, S.B., Klarbring, A.: A goal function approach to remodeling of arteries uncovers mechanisms for growth instability. Biomech. Model. Mechanobiol. **13**(6), 1243–1259 (2014)
50. Seyedsalehi, S., Zhang, L., Choi, J., Baek, S.: Prior distributions of material parameters for Bayesian calibration of growth and remodeling computational model of abdominal aortic wall. J. Biomech. Eng. **137**(10), 101001 (2015)
51. Soares, J.S., Sacks, M.S.: A triphasic constrained mixture model of engineered tissue formation under in vitro dynamic mechanical conditioning. Biomech. Model. Mechanobiol. **15**(2), 293–316 (2016)
52. Soltz, M.A., Ateshian, G.A.: Experimental verification and theoretical prediction of cartilage interstitial fluid pressurization at an impermeable contact interface in confined compression. J. Biomech. **31**(10), 927–934 (1998)
53. Soltz, M.A., Ateshian, G.A.: Interstitial fluid pressurization during confined compression cyclical loading of articular cartilage. Ann. Biomed. Eng. **28**(2), 150–159 (2000)
54. Timoshenko, S., Goodier, J.N.: Theory of Elasticity, 3rd edn. McGraw-Hill, New York (1970)
55. Tinoco, I., Sauer, K., Wang, J.C.: Physical Chemistry: Principles and Applications in Biological Sciences, 3rd edn. Prentice-Hall, Englewood Cliffs (1995)
56. Torzilli, P., Askari, E., Jenkins, J.: Water content and solute diffusion properties in articular cartilage. In: Biomechanics of Diarthrodial Joints, vol. I, pp. 363–390. Springer, Berlin (1990)
57. Urban, J.P., Maroudas, A.: Swelling of the intervertebral disc in vitro. Connect. Tissue Res. **9**(1), 1–10 (1981)

58. Valentín, A., Holzapfel, G.A.: Constrained mixture models as tools for testing competing hypotheses in arterial biomechanics: a brief survey. Mech. Res. Commun. **42**, 126–133 (2012)
59. Vernerey, F.J., Farsad, M.: A constrained mixture approach to mechano-sensing and force generation in contractile cells. J. Mech. Behav. Biomed. Mater. **4**(8), 1683–1699 (2011)
60. Wagenseil, J.E.: A constrained mixture model for developing mouse aorta. Biomech. Model. Mechanobiol. **10**(5), 671–687 (2011)
61. Wan, W., Hansen, L., Gleason, R.L. Jr.: A 3-d constrained mixture model for mechanically mediated vascular growth and remodeling. Biomech. Model. Mechanobiol. **9**(4), 403–419 (2010)
62. Wang, J., Zhou, B., Liu, X.S., Fields, A.J., Sanyal, A., Shi, X., Adams, M., Keaveny, T.M., Guo, X.E.: Trabecular plates and rods determine elastic modulus and yield strength of human trabecular bone. Bone **72**, 71–80 (2015)
63. Weinans, H., Huiskes, R., Grootenboer, H.J.: The behavior of adaptive bone-remodeling simulation models. J. Biomech. **25**(12), 1425–1441 (1992)
64. Wineman, A.: On the mechanics of elastomers undergoing scission and cross-linking. Int. J. Adv. Eng. Sci. Appl. Math. **1**(2–3), 123–131 (2009)
65. Wong, M., Ponticiello, M., Kovanen, V., Jurvelin, J.S.: Volumetric changes of articular cartilage during stress relaxation in unconfined compression. J. Biomech. **33**(9), 1049–1054 (2000)
66. Wu, J., Shadden, S.C.: Stability analysis of a continuum-based constrained mixture model for vascular growth and remodeling. Biomech. Model. Mechanobiol. **15**(6), 1669–1684 (2016)
67. Zeinali-Davarani, S., Wang, Y., Chow, M.-J., Turcotte, R., Zhang, Y.: Contribution of collagen fiber undulation to regional biomechanical properties along porcine thoracic aorta. J. Biomech. Eng. **137**(5), 051001 (2015)

J Elast (2017) 129:107–124
DOI 10.1007/s10659-016-9619-9

Solid Tumors Are Poroelastic Solids with a Chemo-mechanical Feedback on Growth

D. Ambrosi[1] · S. Pezzuto[2] · D. Riccobelli[1] · T. Stylianopoulos[3] · P. Ciarletta[1]

Received: 14 May 2016 / Published online: 9 January 2017

Abstract The experimental evidence that a feedback exists between growth and stress in tumors poses challenging questions. First, the rheological properties (the "constitutive equations") of aggregates of malignant cells are still a matter of debate. Secondly, the feedback law (the "growth law") that relates stress and mitotic–apoptotic rate is far to be identified. We address these questions on the basis of a theoretical analysis of *in vitro* and *in vivo* experiments that involve the growth of tumor spheroids. We show that solid tumors exhibit several mechanical features of a poroelastic material, where the cellular component behaves like an elastic solid. When the solid component of the spheroid is loaded at the boundary, the cellular aggregate grows up to an asymptotic volume that depends on the exerted compression. Residual stress shows up when solid tumors are radially cut, highlighting a peculiar tensional pattern. By a novel numerical approach we correlate the measured opening angle and the underlying residual stress in a sphere. The features of the mechanobiological system can be explained in terms of a feedback of mechanics on the cell proliferation rate as modulated by the availability of nutrient, that is radially damped by the balance between diffusion and consumption. The volumetric growth profiles and the pattern of residual stress can be theoretically reproduced assuming a dependence of the target stress on the concentration of nutrient which is specific of the malignant tissue.

Keywords Solid tumor · Poroelasticity · Growth · Inhomogeneity · Osmotic pressure

Mathematics Subject Classification 74F10 · 74L15

✉ D. Ambrosi
davide.ambrosi@polimi.it

[1] MOX—Dipartimento di Matematica, Politecnico di Milano, piazza Leonardo da Vinci 32, 20133 Milano, Italy

[2] Faculty of Informatics, Institute of Computational Science, Università della Svizzera italiana, Via Giuseppe Buffi 13, 6900 Lugano, Switzerland

[3] Cancer Biophysics Laboratory, Department of Mechanical and Manufacturing Engineering, University of Cyprus, P.O. Box 20537, Nicosia 1678, Cyprus

1 Introduction

After Folkman and Hochberg [1], the *multicellular spheroid* is a standard *in vitro* system used to evaluate the uncontrolled duplication rate of a tumor cell aggregate. A tumor spheroid is a cluster of cells floating in a culture medium, it is an ensemble of cells freely proliferating in an environment with large availability of nutrient. The malignant cells have lost the ability to self-regulate their own number through a normal apoptosis mechanism, regulated by the homeostasis with the environment; they duplicate in an uncontrolled manner, isotropically, producing a nearly spherical shape.

In the standard free-growth case, a plot of the diameter of the tumor vs. time typically exhibits an early stage of exponential growth, followed by a linear one. The transition from one regime to the other is mainly regulated by the availability of nutrient, that is driven by diffusion through the intercellular space. In fact, when the size of the tumor $R_o(t)$ is smaller than the typical diffusion length, the nutrient is everywhere available in the spheroid and the growth is volumetric [2]:

$$\frac{dR_o^3}{dt} \simeq {R_o}^3, \tag{1}$$

so that $R_o \simeq e^t$. Conversely, when the diameter of the spheroid is much larger than the penetration length of the nutrient, one obtains surface growth, that is

$$\frac{dR_o^3}{dt} \simeq {R_o}^2, \tag{2}$$

and $R_o \simeq t$. In a realistic intermediate regime, the concentration of nutrients decays exponentially with the radius [3], favoring the external proliferation vs the internal one.

This work is motivated by a number of recent experiments that demonstrate the dependence of the growth rate of a tumor spheroid on the mechanical load at the boundary. Some papers report a reduced apoptosis, with no significant changes in proliferation [4]. According to others, the cell division, rather than the cell death rate, is affected by stress [5]. To disentangle the puzzle of the biological feedback of stress on growth, we discuss first the rheology of the cellular aggregate as a living material, to point out its constitutive properties. We illustrate a number of arguments that support the hypothesis that a solid tumor is a poroelastic material, where the cells and the extra-cellular matrix represent the solid elastic component. A mathematical model based on such an assumption is able to predict inhomogeneities that can not be justified by fluid-like assumptions.

In the last section we address the numerical simulation of the growth of a murine tumor. In vivo tumors reach a larger size, they can be partially vascularized, they have a more complex internal composition and exhibit release of residual stress. We test the ability of our mathematical model comparing the observed and the predicted opening angle after excision.

2 Background: Elementary Rheology and Growth Theory

The simplest distinction among fluid and solid materials can be based on an elementary ideal experiment: under a pure shear load fluids flow, while solids do not, at the time scale of interest. This draconian categorization encompasses also viscoelastic materials, as they typically exhibit fluid-like or solid-like properties depending on the relaxation time scales. As an example, a "Maxwell fluid" behaves as a solid if observed at a time scale much smaller than its relaxation time. Analogously, a "Kelvin solid" flows like a fluid when observed on short enough time scales. Things become a little bit more complex when flow is prompted only above a yield stress, but the distinction persists when loads are neatly below or beyond the threshold.

Many biological materials are composed by a mixture of several components: interstitial fluid, different species of cells, collagen fibers, and so on. For these microscopically heterogeneous materials the overall mechanical behavior is represented, at the macroscale, by the superposition of single phase contributions, proportionally to the volume fraction occupied by each component. The archetypical example of a mixture is a porous elastic material permeated by a fluid: the stress in a poroelastic medium is the sum of the interstitial pressure of the fluid plus the solid stress, which is proportional to the solid volume fraction.

Fluids and solids behave in a very different manner when internal stresses arise not because of external loads, but as due to the inner material reorganization (growth and remodeling). The simplest example are thermal stresses in inhomogeneously heated materials with temperature-dependent density: residual stresses relax in fluids, not in solids. The persistence of residual stress is therefore the signature of solid-like behavior which has to be properly addressed in a modelling framework. In case of small strains, linearized elasticity applies and stress (and strains) can be superimposed. In case of large strains, as it is often the case with soft matter, a multiplicative decomposition of the tensor gradient of deformation has to be introduced.

For our purposes, we represent the motion of every material point of a continuous body as a smooth invertible map $\chi(\mathbf{X})$ with Jacobian $\mathsf{F} = \frac{\partial \chi}{\partial \mathbf{X}}$. For a nonlinear elastic material the strain energy is $W(\mathsf{F})$; when the body grows and residual stress is present, the strain energy rewrites

$$W\left(\mathsf{F}\mathsf{G}^{-1}\right), \tag{3}$$

where G is usually called "growth tensor".

3 Are Solid Tumors Fluids?

While the availability of nutrients is the major factor affecting tumor growth, other external agents can play a role. The mechanical influence of external loading on tumor growth has been first demonstrated by Helmlinger et al. [4]. They designed an experimental setup in order to control the load applied at the boundary of tumor cell spheroids *in vitro* in agarose gels, and checked the influence of such a stressed state on the growth rate of the multicell spheroid. They compared the free growth of a floating multicell spheroid with the size of cell aggregates placed into the agarose gel. The gel is produced at a given (known) stiffness by suitably tuning the concentration of the solid phase. As the spheroid grows, it displaces the surrounding gel, which then exerts a compressive force at the surface of the tumor spheroid. An *a priori* mechanical characterization of the gel allows to calculate the pressure exerted by the gel on the spheroid, depending on its radius.

The main result of the experiments carried out by the group of Rakesh Jain [4] is that the stress field reduces the final size of the spheroids, with a decreased apoptosis and non significant changes in proliferation. It is therefore clear that a precise determination of the constitutive laws that characterize the mechanical behavior of a tumor spheroid is a prerequisite in order to assess a reliable stress-growth relationship.

Early attempts in this respect assumed that a cell conglomerate behaves like a viscoelastic fluid, able to bear a static load because of its surface tension [6]. At equilibrium, measurements of the curvature radius of a loaded sample provide the surface tension of the "fluid".

According to the Laplace formula, the pressure jump across a curved interface between two fluids is inversely proportional to the radius of the curvature. If the spheroid is loaded with the force F acting on a contact surface A, by continuity of the stress, the inner pressure is F/A and therefore

$$\frac{F}{A} = \sigma\left(\frac{1}{R_1} + \frac{1}{R_2}\right), \tag{4}$$

where σ is the surface tension and R_1, R_2 are the curvature radii of the free surface. According to the experiments, the surface tension of a cell aggregate ranges in $1\text{–}22 \times 10^{-3}$ N/m (as a reference value, the surface tension of the water is about 72×10^{-3} N/m). Relaxation times range between 1 and 50 seconds [6].

The opposite approach is to describe a solid tumor as a viscoelastic solid. In this case, at equilibrium the external load should be balanced by the stress in the body, depending on the strain of its material points. Assuming an homogeneous deformation and using the same data provided by the experiments above, one can estimate the Young modulus E according to the following rule:

$$\frac{F}{A} = E\frac{h - h_0}{h_0}, \tag{5}$$

where h, h_0 are the height of the loaded and unloaded sample, respectively. In this case one finds $E \simeq 4$ kPa, a typical soft-range value for living cells [7].

A second argument supporting the assumption of solid-like constitutive equations is based on the spatial correlation between stress and apoptosis–mitosis in loaded ellipsoidal spheroids [8]. The non-homogeneous proliferation pattern can be produced only by a solid-like material: a hydrostatic stress generates a pressure independent on the position in any symmetric geometry, while in a solid material, high stress concentrates around the tips.

Furthermore, the work by Netti et al. [9] support the view that tumors behave as solid-like materials. In their study, stress-relaxation experiments of various tumor types in confined compression were performed and at the end of the experiment all tumors equilibrated in a constant, non-zero stress, typical of viscoelastic solids.

Finally, evidence of residual stress in murine and human tumors is reported by Stylianopoulos et al [10]. They cut the tumor azimuthally and observed an opening angle, which is the signature of a solid like behavior. Residual stress is likely produced by an inhomogeneous duplication rate of the cells as well as by mechanical interactions between the cells and extracellular matrix components, particularly collagen and hyaluronic acid, that strain the tumor microenvironment. Only solids can contain residual stress, due to the evolution of their relaxed configuration produced by incompatible growth [11]: energy can be elastically stored in the unloaded body only if it is a solid. In particular, Stylianopoulos et al. observed a compressive residual stress (i.e., negative opening angle) in the kernel and a tensile residual stress (positive opening angle) in the outer shell of the tumor. This behavior is paradoxical in terms of availability of nutrients: their concentration is larger near the boundary, thus favoring proliferation and eventually producing compressive stress. In a solid tumor *in vivo*, this intuitive explanation does not work and we address such a puzzle in the next sections.

4 Growth and Stress

An evocative definition of a tumor is "a living system that has lost its self-regulating ability towards homeostasis". In other words, tumor cells do not correctly detect or elaborate the external signals that should regulate its proliferation and apoptosis, and duplicate without control. When the stress state of the system is not in homeostatic mechanical equilibrium, it remodels (growing or resorbing matter) until the target tensional state is recovered. In this respect, all the genetic information that details the shape and function of organs is encoded in the target stress. A suggestive mechanical interpretation of a tumor therefore naturally arises: a tumor is an open system (in terms of mass and energy) with a damaged inner mechano-biological control inducing a disregulation of tensional homeostasis, i.e. the feedback that normally self-regulates growth in terms of stress-modulated control does not properly work. In other words, tumor cells regulate production and consumption not according to a benefit of the whole organism but only in view of maximum invasion of malignant cells: the control on growth does indeed exist, also as a function of available nutrients, but the corresponding duplication/apoptosis strategy has a different aim.

The experiments illustrated in the section above do not only demonstrate the existence of residual stress in tumors, but they also show that the inhomogeneous proliferation and apoptosis, triggered by the differential availability of nutrients, is enhanced in a mechanically loaded spheroid. Their main result is that mechanical stress affects proliferation and apoptosis inside the spheroid in a non-homogeneous way, a correlation existing between strong apoptosis and high stress.

In another series of experiments, the compression of the spheroid is controlled by the concentration of a large molecule (Dextran) soluted in the bath [5, 12]. As Dextran molecules cannot enter neither the cell membrane nor the interstitial (intracellular) space, an imbalance of osmotic pressure at the boundary loads the cellular aggregate. It is reported that for larger concentrations of Dextran the diameter of the spheroid grows slower and reaches a plateau at smaller radius, in agreement with the results of Helmlinger et al [4]. While a single cell is almost incompressible with respect to the pressure due to the concentration of Dextran, the volume of the cell aggregate strongly depends on the osmotic pressure [13]. The reduction in volume in the cellular aggregate therefore mainly occurs because of reduction of the intercellular space, in the inner region of the spheroid.

The large number of available data suggests that a cell aggregate behaves as a poroelastic material. The mathematical modelling of solid tumors as porous deformable media has been addressed in a number of papers [14–16]; it is suitable mechanical framework to account for the coupled dynamics of cells and extracellular matrix (the solid matrix) and interstitial fluid. The interstitial flow is typically represented by a Darcy-type equation, while the mass exchange among phases which allows a prediction of the growth of the mass.

In the experimental setup by Montel et al. [5, 12] the porous media theory offers a transparent explanation for the interplay between the pressure of the fluid, the chemical potential of the Dextran and the stress in the solid matrix. The external load at the boundary is the sum of two terms: the pressure of the fluid plus the chemical potential of the Dextran. Observing that the diameter of the macromolecules is typically larger than the size of the intracellular pores, we split the fluid load into two contributions: one that balances the interstitial pressure, the other one loading the solid (cellular) component. Formally, we assume that the global balance at the boundary

$$(-p - p_D)_{\text{out}}\mathbf{n} = (\mathsf{T} - p\mathsf{I})_{\text{in}}\mathbf{n}, \tag{6}$$

splits into

$$-p_{\text{out}} = -p_{\text{in}}, \tag{7}$$

$$-p_D\mathbf{n} = \mathsf{T}\mathbf{n}, \tag{8}$$

where p is the pressure of the interstitial fluid, p_D is the osmotic pressure contribution due to the concentration of Dextran, T is the Cauchy stress tensor in the cellular aggregate, I is the identity tensor and $\mathbf{n}$ is the outgoing normal (radially directed) vector. This assumption is in agreement with the observation that the solid stress is not affected by the interstitial fluid pressure [10].

On the basis of this hypothesis, the stress state in the loaded spheroid can be determined solving the force balance equations for the solid component only. Assuming spherical symmetry, the tensor gradient of deformation and the growth tensor read

$$\mathsf{F} = \mathrm{diag}\left(r', \frac{r}{R}, \frac{r}{R}\right), \qquad \mathsf{G} = g(r)\mathsf{I}, \tag{9}$$

where $r(R,t)$ is the radial coordinate of the material point that was in R at time $t = 0$, I is the identity tensor and the prime $'$ denotes differentiation in R. The solid component of the poroelastic spheroid must satisfy the force balance equation

$$\frac{d}{dr} T_{rr} + \frac{2}{r}(T_{rr} - T_{\theta\theta}) = 0, \tag{10}$$

with boundary conditions

$$r(0,t) = 0, \qquad T_{rr}(r_o) = -p_D. \tag{11}$$

where $r_o = r(R_o, t)$. A simple representation of an hyperelastic compressible material is provided by the strain energy

$$\hat{W}(\mathsf{F}) = \frac{\mu}{2}\big(\mathsf{F} \cdot \mathsf{F} - 2\log(\det \mathsf{F}) - 3\big). \tag{12}$$

If the material grows, the strain energy depends on the growth tensor too, through a classical multiplicative decomposition

$$W\big(\mathsf{FG}^{-1}\big) = \frac{\mu}{2}\big(\mathsf{FG}^{-1} \cdot \mathsf{FG}^{-1} - 2\log\big(\det\big(\mathsf{FG}^{-1}\big)\big) - 3\big), \tag{13}$$

where μ is the shear elastic modulus. First variation and pull back to the reference configuration yields the first Piola-Kirchhoff stress

$$\mathsf{P} = \mu \det(\mathsf{G})\big(\mathsf{FG}^{-1}\mathsf{G}^{-T} - \mathsf{F}^{-T}\big), \tag{14}$$

where, explicitly,

$$P_{RR} = \mu g^3\left(\frac{r'}{g^2} - \frac{1}{r'}\right), \qquad P_{\Theta\Theta} = \mu g^3\left(\frac{r}{Rg^2} - \frac{R}{r}\right). \tag{15}$$

The force balance equation (10) in material coordinates reads

$$\frac{d}{dR} P_{RR} + \frac{2}{R}(P_{RR} - P_{\Theta\Theta}) = 0, \tag{16}$$

or, explicitly,

$$\frac{d}{dR}\left(r'g - \frac{g^3}{r'}\right) = 2\left(g\frac{r}{R^2} - \frac{g^3}{r} - g\frac{r'}{R} + \frac{g^3}{r'R}\right), \tag{17}$$

to be supplemented by boundary conditions (11) rewritten in material coordinates

$$P_{RR}(R_o) = (\det \mathsf{F}) T_{rr} F_{rr}^{-1}\big|_{R_0} = -(\det \mathsf{F}) p_D F_{rr}^{-1}\big|_{R_0}, \tag{18}$$

or, explicitly

$$\mu g^3 \left(\frac{r'}{g^2} - \frac{1}{r'} \right)\bigg|_{R_0} = -p_D \frac{r^2}{R^2}\bigg|_{R_0}. \tag{19}$$

For constant g the force balance equation (17) with boundary conditions (19) has solution

$$r(R; g) = \gamma g R, \tag{20}$$

where γ is the positive root of the third order polynomial

$$f(\gamma) = p_D \gamma^3 + \mu \gamma^2 - \mu = 0. \tag{21}$$

One may notice that $f(0) = -\mu < 0$ while f' is always positive, therefore the root is unique. Moreover $f(1) = p_D > 0$, so that it must be $0 < \gamma < 1$.

Remark. One could observe that poroelasticity has been advocated for the model above, but its use is apparently very limited: there is no interstitial fluid flow, and the porosity, the volume fraction of solid vs. liquid component, is not even mentioned. There is a rationale behind such a minimal choice. Fluid flow is so slow that it carries no contribution in the stress balance equation; of course, mass exchange among species is the true physical mechanism for the growth of the tumor mass, however here it is directly incorporated in the growth tensor G. Secondly, the porosity of the matrix should contribute to the stress tensor T with a multiplicative factor depending on the determinant of the gradient of deformation; as a matter of fact, we incorporate such a contribution in the compressibility of the strain energy function (12). The numerical results to be illustrated in the next sections will confirm that good predictions can be obtained even with such a simple constitutive law, thus confirming that the theory weakly depends on the specific constitutive equation for the strain energy density of the solid matrix. The crucial ingredient of the model is the multiphase split of the load at the boundary into a fluid and a solid component (7) and (8).

5 Mechanobiological Feedback and Equilibrium

In the general case, a growth law for G is to be supplemented to close the differential equation (17)–(19). We consider first the case of growth controlled by a mechanical feedback only. If nutrients are largely available everywhere, the growth in time is expected to depend on the stress only. In finite elasticity, the growth must depend on an invariant measure of the stress. A thermodynamically consistent choice is to adopt the dependence on the Eshelby stress [17]. To minimize the calculations, while preserving the essential biophysical features, we chose here to measure the stress in terms of the second Piola-Kirchhoff tensor S, which reads

$$\mathsf{S} = \mathsf{F}^{-1} \mathsf{P}. \tag{22}$$

The mitotic rate of single tumor cells is known to be inhibited by compression [4], and promoted by tension [18], and a very simple growth law that can account for such a behavior is

$$\dot{g} = \frac{g}{\tau}\left(1 + \frac{\operatorname{tr} S}{3\kappa\mu} - \frac{g}{\alpha}\right), \tag{23}$$

where $1/\tau$ is the mitotic rate in absence of external stimuli, $\kappa\mu$ is a threshold stress and the last term in brackets accounts for apoptosis, the natural cellular death rate. We highlight that the assumption of an isotropic growth tensor allows to set a functional dependence on the trace of S. For a general anisotropic growth a more complex dependence on the principal stresses would be needed, guided by thermo-mechanical requirements.

Consider the unloaded case first: $-p_D = 0$ and at time $t = 0$ the solid component has $g = 1$. As $\mathsf{S} = 0$, the evolution in time of G is autonomous and independent of the radial position, so that $g(t)$ is constant in space and its evolution in time initially follows the well known exponential growth in size of the cell aggregate up to a saturation dictated by the value of α. The solid component of the poroelastic spheroid is therefore relaxed, exactly as a sponge in the deep ocean, where the interstitial pressure balances the water head.

If Dextran is present, the extra pressure compresses the cellular phase and triggers the mechanobiological feedback via Eq. (23). The growth $g(t)$ is given by the solution of the first order ordinary differential equation

$$\dot{g} = \frac{g}{\tau}\left(1 - g\left(\frac{\gamma p_D}{\kappa\mu} + \frac{1}{\alpha}\right)\right). \tag{24}$$

Equation (24) has two equilibrium points: $g = 0$, always unstable, and

$$g_e = \frac{\alpha\kappa\mu}{\alpha\gamma p_D + \kappa\mu}, \tag{25}$$

which is always stable (for the fixed $r(R;t)$ of (20)). The mathematical model therefore predicts the following scenario, corresponding to the observed dynamics. For null osmotic pressure, the system grows exponentially, then it tends to saturation. For sufficiently large osmotic pressure the stable equilibrium depends on the applied pressure p_D. After derivation of Eqs. (21) and (25) we get

$$\frac{dg_e}{dp_D} = -\frac{\alpha^2\kappa\mu}{(\alpha\gamma p_D + \kappa\mu)^2}\,\frac{\gamma(2\gamma p_D + 2\mu)}{3\gamma p_D + 2\mu} < 0. \tag{26}$$

The solution of Eq. (24) explains the plateau in growth vs time reported for loaded spheroids at different Dextran concentrations, but it does not account for the radial density inhomogeneities observed in excised aggregates. Remaining in a purely mechanical setting, an explanation for such a discrepancy between theory and experiments could be provided by the possible onset of an instability for the equilibrium solution (24) of the coupled problem. This question is addressed in the Appendix, where we study the stability of the solution of the nonlinear system (17) and (23) with boundary conditions (19) in order to explain the emergence of inhomogeneity. The result of the analysis is that the small perturbations are always damped in time, so that a purely mechanical framework cannot account for the observed dependency of growth on the radial coordinate. The biophysics of the system needs therefore to be enriched: in the next section we show that the kinetics of nutrients can trigger dependence of the asymptotic state on the radial coordinate.

6 Dynamics of the Nutrient and Inhomogeneity of Growth

In an avascular tumor, nutrients are provided to malignant cells by diffusion through the boundary of the spheroid. The balance between diffusion and uptake is fast with respect to the growth times (one hour vs. days) and obeys a linear reaction–diffusion equation:

$$-\frac{1}{r^2}\frac{d}{dr}\left(r^2\frac{dc}{dr}\right) = -\frac{c}{\lambda^2}, \tag{27}$$

with boundary conditions

$$\left.\frac{dc}{dr}\right|_{r=0} = 0, \qquad c(r_0) = c_0, \tag{28}$$

where the decay length λ is on the order of 100–200 micrometers and c_0 is the external (constant) concentration. We remind that the boundary value problems refers to the avascular phase of tumor growth. At later stages, neovascularization can be triggered after the diffusion-limited radius is reached. In such a case, a distributed nutrient supply from the tumor vascular network should also be taken into consideration.

During the avascular growth phase, the concentration profile can be calculated by direct integration of the equation in spatial coordinates [3], yielding an exponential decay of the concentration of nutrient going from the boundary to the center of the spheroid:

$$c(r) = c_0 \frac{r_0}{\sinh(r_0/\lambda)} \frac{\sinh(r/\lambda)}{r}. \tag{29}$$

To account for the combined action of stress and nutrient pattern, we propose to rephrase Eq. (24) to the following growth law

$$\dot{g} = \frac{g}{\tau} c\left(1 + c\left(\frac{\operatorname{tr} S}{3\kappa\mu} - \frac{g}{\alpha}\right)\right). \tag{30}$$

According to Eq. (30), the proliferation of the malignant cells is enhanced by the availability of nutrient, as it is usually assumed in mathematical models that do not specifically account for mechanics. In the same way as in Eq. (23), it is expected that the system reaches an equilibrium when the term in brackets vanishes: a plateau in size is observed for large enough times. The novelty of this growth law is that the equilibrium does not correspond to a homogeneous growth tensor $G = g\mathsf{I}$, but it depends on the radial position through the concentration of nutrient, thus originating an inhomogeneous residual stress. Using numerical simulations, in the next sections we are able to show that the predicted residual stress is in agreement with the reported opening angles from cutting experiments.

7 Numerical Simulations

Numerical integration of Eqs. (17) and (30) with boundary conditions (19) and initial conditions

$$r(R, t = 0) = R, \qquad g(R, t = 0) = 1, \tag{31}$$

is performed using a finite difference scheme with centered discretization in space and a fourth order Runge-Kutta scheme in time. The parameters used in the numerical simulations

Fig. 1 Volume of a spheroid vs. time for different values of the osmotic pressure

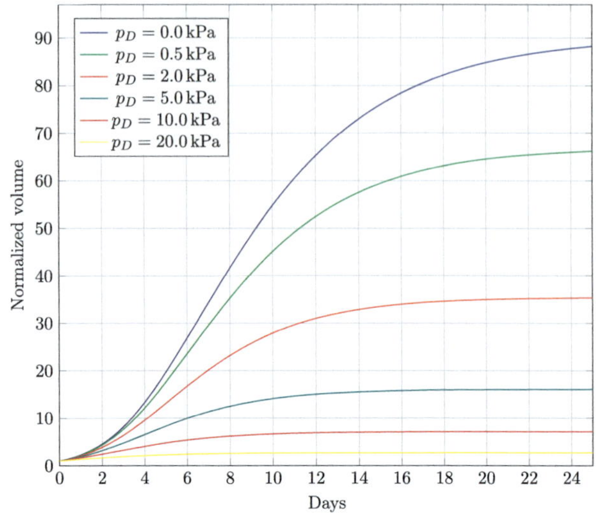

Fig. 2 Radial residual stress versus the radial position at final time for different values of the osmotic pressure

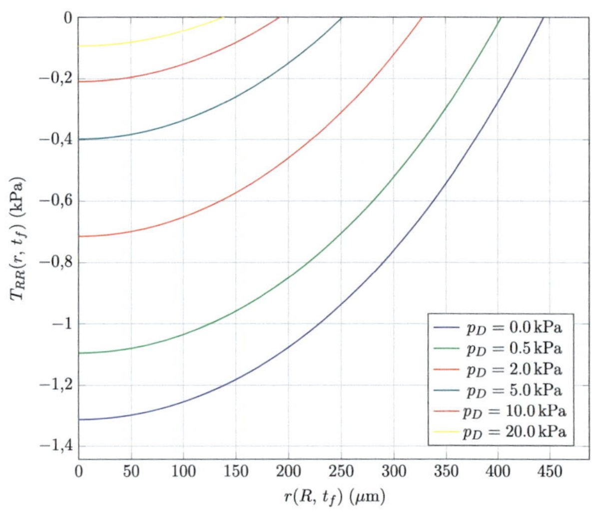

are $\tau = 2.5$ days, $\kappa = 2.9$ kPa, $\alpha = 3.7$, $\lambda = 250$ μm and $\mu = 10$ kPa. The initial radius is 100 μm, the final simulation time is $t_f = 25$ days and the boundary condition of (27) is $c_0 = 1$.

The volume of the spheroid initially grows very rapidly for all values of p_D. At large times, for null or small values of osmotic pressure the slope of the curve becomes very small, and it becomes horizontal for large p_D (Fig. 1).

As expected, the non-uniform pattern of nutrients triggers a weak inhomogeneity in growth. While the predicted growth pattern cannot be directly compared with data, it is indirectly supported by the residual stress that it produces by the relation (3). The radial and hoop component of the residual stress are plotted in Figs. 2 and 3 versus the radial coordinate at equilibrium.

As expected, the radial stress vanishes on the boundary of the spheroid, while it is internally compressive. Conversely, the hoop stress changes sign, being compressive in the

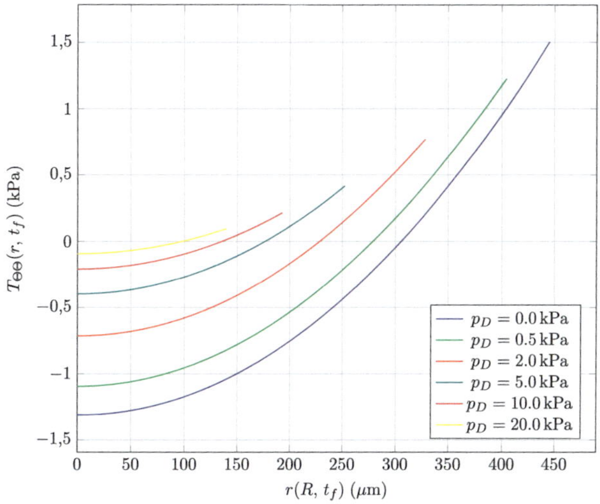

Fig. 3 Hoop residual stress versus the radial coordinate at final time for different values of the osmotic pressure

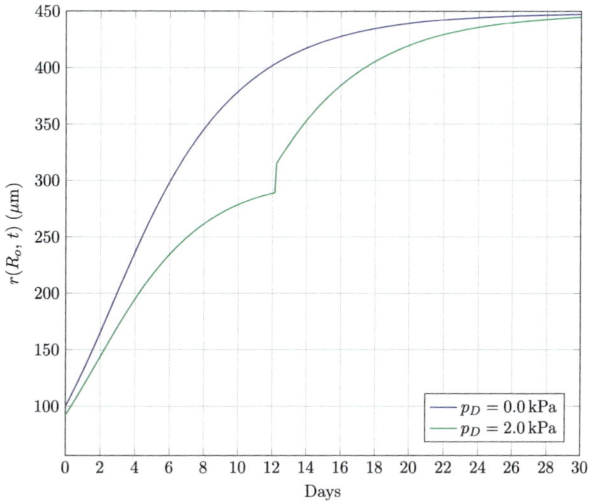

Fig. 4 Radius of the spheroid vs. time for $p_D = 0$ (*blue line*) and $p_D = 2000$ Pa (*green line*). After 12 days the external pressure is removed and the system returns the curve corresponding to the unloaded state

core and tensional in the outer layer. Such a residual stress distribution is stable against both circumferential and azimuthal perturbations of the tumor boundary, as investigated in [19].

Data on residual stress of *in vitro* tumor spheroids are not available, probably because they are too soft and do not reach a size such that a mechanical manipulation and a precise cut can be operated. However the pattern reported in Figs. 2 and 3 is in qualitative agreement with experiments on (much bigger) human tumors implanted in mice [10]. Stylianopoulos et al. observe that cells at the periphery of the spheroid are restricted by the surrounding tissue and thus, during radial tumor growth they develop tensile circumferential forces. Surrounding tissues in vivo would then produce on the tumor a compressive hydrostatic pressure increasing with the tumor growth. Furthermore, Figs. 2 and 3 depict that the magnitude of stress—either compressive or tensile—increases as the osmotic pressure exerted on the cells, p_D, decreases. This is explained by the fact that for low osmotic pressures the tumor becomes larger in size and the stresses increase.

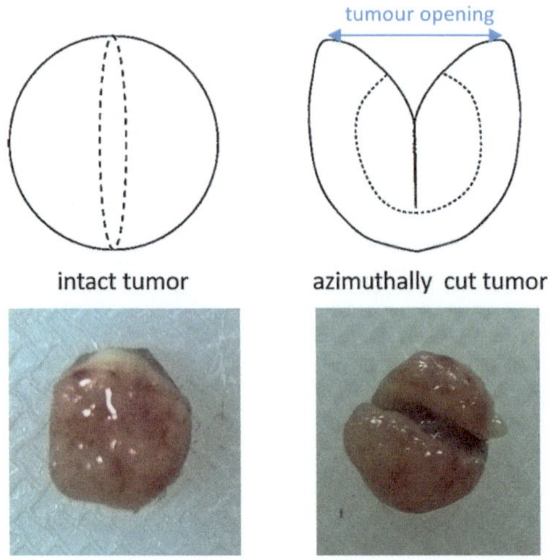

Fig. 5 An intact, residually stressed spheroid (*left*) is cut along an azimuthal plane for 80 % of its diameter. The outer region opens up while the inner one swells (*right*), thus distribution of residual stress that goes from compressive to tensile along the radial coordinate

Results for MCF10CA1a tumors

Tumor number	Volume (mm³)	Tumor Opening (mm)	Circumferential Stress (mmHg)	Radial Stress (mmHg)
1	140.8651	3.37	-37.795	-40.654
2	109.6826	3.63	-41.906	-52.061
3	131.0658	4.85	-50.932	-55.19
4	82.71174	3.61	-78.583	-114.04
5	74.0102	3.2	-72.043	-102.003
6	85.65257	3.39	-70.495	-99.263
7	96.40847	3.12	-37.079	-45.605

Results for MiaPaCa2 tumors

Tumor number	Volume (mm³)	Tumor Opening (mm)	Circumferential Stress (mmHg)	Radial Stress (mmHg)
1	52.27961	2.23	-25.67	-25.67
2	65.41667	2.49	-28.42	-28.42
3	27.0857	1.83	-21.32	-21.32
4	176.6973	1.3	-15.4	-15.4
5	41.60866	2.13	-24.53	-24.53

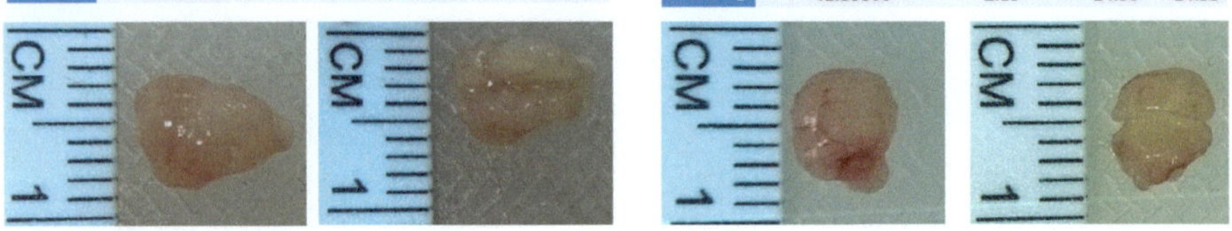

Fig. 6 Tables with experimental measures (*top*) of cutting experiments (*bottom*) for MCF10CA1a breast tumor cells (*left*) and for MiaPaCa2 pancreatic tumor cells (*right*) implanted orthotopically in the mammary fat pad of nude mice

If the external pressure is removed, the radius quickly grows and reaches the same value of the free-growth case (Fig. 4), in agreement with the experimental results [12].

7.1 Stress Release in a Cut Spheroid

A quantitative comparison among observed and predicted residual stress can be obtained on the basis of the opening angle of cut specimens. To this aim, tumor spheroids have been grown in mice and then they have been cut along their azimuthal plane for about 80 % of their diameter. The spheroids then partially relax their residual stress: the cut surface opens up at the periphery while the inner region swells (see Fig. 5). Figure 6 depicts the cutting experiments for breast and pancreatic tumors implanted in nude mice, also reporting the tumor opening length and the maximum residual stresses within the tumor specimens.

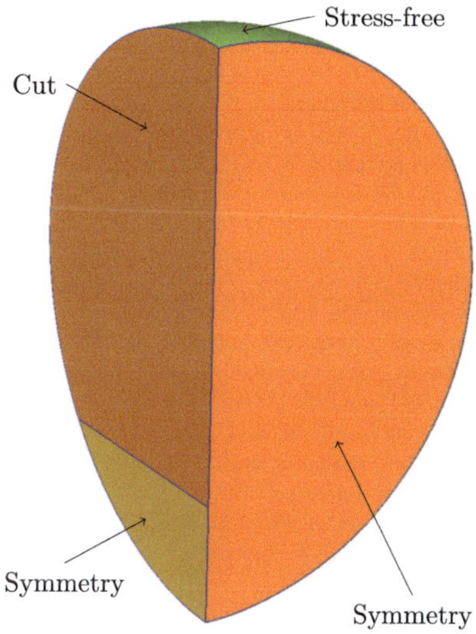

Fig. 7 Computational domain and boundary conditions for the 3d numerical experiment of the opening angle. The "symmetry" label indicates that the surface is fixed in the normal direction. The vertical cut is 80 % of the diameter deep

The observed behavior, which is in qualitative agreement with our predictions in the stress pattern in small *in vitro* spheroids, can be quantitatively compared with opening angles data on the basis of a three dimensional numerical simulation only. As a matter of fact, an axial cut of a ring preserves the cylindrical symmetry of the problem [20], while an azimuthal cut of a sphere breaks it.

Numerical simulations are obtained using a finite element code that solves the equation of finite elasticity on a spherical wedge. The computation reproduces the physical observations: the spheroid grows under spherical symmetry which is eventually broken by the cut. We therefore use the growth tensor computed under radial symmetry assumption and we evaluate the opening angle that it produces.

The 3D numerical problem is based on FEniCS [21]. The computational domain is discretized with quadratic tetrahedral elements, with an average diameter of 10 μm. Since we expect near to singular stresses around the edge of the cut, the mesh is gradually refined nearby this edge to one twentieth of the original size. The mesh contains roughly 28143 elements and it has been produced by Gmsh [22]. The non-linear variational problem is discretized with quadratic isoparametric finite elements, and the final problem has 137949 degrees of freedom. The solver for non-linear problem is based on a modified Newton's method specifically designed for variational inequalities, and implemented in the PETSc framework [23]. The solver can deal with inequality constraints, as we have on the cut boundary surface to avoid self-contact during the swelling. The solver for the linear system is MUMPS [24].

The simulation is performed in two steps: first, we apply a homogeneous growth tensor obtained by averaging the target one, while keeping the cut sealed; then, we release the cut and we enforce the final growth tensor. This strategy facilitates the convergence of the non-linear solver, which performs 40 iterations at most. A relative error below 1 % on the opening angle is observed when the mesh is uniformly refined, certificating the numerical convergence.

Boundary conditions, reported in Fig. 7, apply as follows: the outer boundary is stress-free, the cut surface is enforced to have non-negative displacement in the normal direction to avoid self-contact, and on the internal, intact, portion of the boundary symmetry arguments yield null normal displacement and while the other components of the displacement must

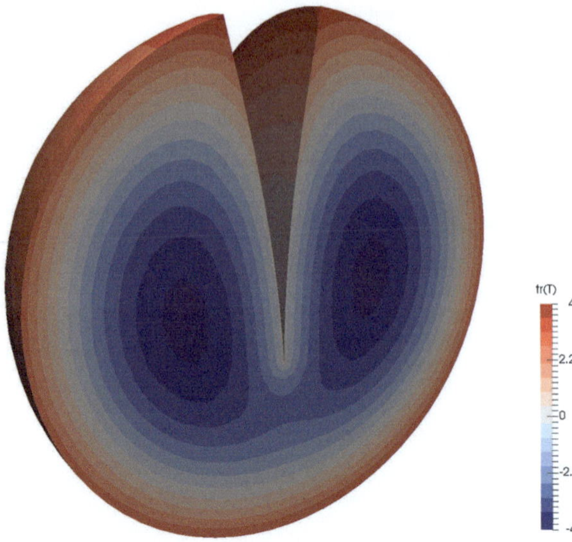

Fig. 8 Current configuration of an unloaded grown sphere with initial radius of 100 μm. The sphere is azimuthally cut at the final time. The parameters in the simulation are: $\tau = 2.5$ days, $\kappa = 33.35$, $\alpha = 37$, $\lambda = 2.5$ mm, $\mu = 27.0$ kPa and $p_D = 5.0$ kPa. The domain of the numerical simulation is a quarter of a sphere. For the sake of graphical clarity of the opening angle, the spherical wedge is combined with its symmetric counterpart. The body is unloaded but not stress free: the *color map* represents the trace of the Cauchy stress tensor. The lowest part of the cut is partially resew by the swelling

Table 1 Numerical result of a stepwise increase of the difference in growth between the center and the boundary of the tumor. The growth function $g(R)$ is linear in the radial component

$g\ (R=0)$	45.0	44.0	43.0	42.0	41.0	40.0
$g\ (R=R_0)$	45.0	48.0	51.0	54.0	57.0	60.0
Opening length [mm]	0.0	0.540	1.12	1.70	2.30	2.92
Opening angle [deg]	0.0	3.43	6.94	10.5	14.2	18.0
volume [mm^3]	94.7	94.8	95.0	95.4	95.9	96.6

have null derivative with respect to the normal direction. In Fig. 8 the deformed configuration obtained after a vertical cut of 80 % of the diameter is shown.

The parameters used in the numerical simulations are $\tau = 2.5$ days, $\kappa = 33.35$, $\alpha = 37$, $\lambda = 2.5$ mm, $\mu = 27.0$ kPa and $p_D = 5.0$ kPa. The initial radius is 100 μm, the simulation ends at $t = 50$ days and the boundary condition of (27) is $c_0 = 1$. In the numerical experiment the tumor opens with an angle of 11.70°, corresponding to 1.41 mm of opening length. The final volume is 169.84 mm^3. The inner-most part of the cut, for about 20 % of the diameter, is in self-contact, certificating that this portion of the tumor tends to swell outward after the cut. It is to be remarked that the diffusion length assumed here is larger than the one used for small in vitro spheroids.

These results are in agreement with the *ex-vivo* experiments: the opening length, the final volume and the hoop stress are very close to the reported ones for the MiaPaCa2 tumor number 4 (see Fig. 6) [10].

In order to investigate the relationship between the heterogeneity in the growth tensor and the opening angle, we have performed a numerical experiment where the difference between the growth at the center and the boundary of the tumor is stepwise increased from zero to a value of 20. Table 1 summarizes the result of the simulation. As expected from the theory, a uniform growth yields no residual stress and the tumor does not open after the cut (first column of the table). On the other hand, the greater the difference in growth between the center and the boundary, the larger the opening length and consequently the opening angle (from the second column of the table). The volume is mostly affected by the average value of the growth over the entire domain, and not by the heterogeneity. The numerical experiment also shows that the angle linearly increases with the difference in growth of about 8° every 10 units per mm of growth.

8 Final Remarks

The growth of a tumor spheroid can be controlled using mechanical stress: when an osmotic pressure is applied at the boundary, the radius of the aggregate grows in time until it reaches an equilibrium volume which inversely depends on the load. The size control is fully reversible: when traction is released, the cellular matrix relaxes and returns the original growth curve. The observation that the intercellular space forms a pore-like structure, that macromolecules cannot enter, suggests to represent mechanically the cellular aggregate as a poroelastic material [25]. The evidence of a residual stress leads to the assumption that the solid phase is hyperelastic: the large compliance of the cell aggregate is due to the squeezing of the intracellular fluid and the corresponding reduction of the intracellular space, while the single cells are much stiffer [13]. Boundary conditions are split accordingly: the osmotic pressure generated by the Dextran solution of the surrounding fluid loads the solid phase only.

The exponential decay in the pattern of nutrients makes the proliferation process of a sufficiently large loaded spheroid inhomogeneous and the generated residual stress depends on the radial position: it is compressive near the center and tensional at the periphery [10]. This feature is paradoxical when compared with the usual scenario: large availability of nutrients at the periphery of the spheroid is expected to favor the proliferation and, therefore, emergence of compressive residual stress. The dynamics of tumor growth is apparently different: tumor cells duplicate (or control their apoptosis) on the basis of the available nutrients, but their *target stress* modulates so as to produce compression in regions with small concentration of nutrients. In other words, the reported radial distribution of residual stress can be explained only admitting that in the inner regions, where the concentration of nutrients is very small, malignant cells slow down their apoptotic rate, in agreement with the observations of Helmlinger et al [4].

After a standard multiplicative decomposition of the tensor gradient of deformation to account for growth, we introduce a simple law of biomechanical feedback, and we are eventually able to explain the observed dynamics. On the basis of such a conjecture, we are able to reproduce growth profiles of tumor spheroids for different values of the applied load and the open angles of mice bearing breast tumors. Particularly for the second result, the solid phase of mice bearing tumors apart from cells consists also of the extra-cellular matrix, which can contribute to the development of residual stress. Here the mechanical contribution of all components of the tumor are resumed in the solid phase, and the growth tensor G accounts also for the possible tensional contribution due to the elongation of the collagen fibers.

Our mechanobiological model explains the observed smaller asymptotic volume as a function of increasing osmotic load on the basis of a stress-growth coupling. At later stages, not covered by the present model, when the radial inhomogeneity is fully developed, the solid (cellular) component of the spheroid undergoes a stress per volume fraction larger than a threshold that takes it into the plastic regime [26]; then cells start flowing centripetally, producing an internalization of the cells from the periphery to the center of the tumor [27, 28].

Acknowledgements The work has been partially supported by the My First AIRC Grant—MFAG 2015—code 17412 titled "A Mathematical insights of glioblastoma growth: a mechano-biology approach for patient-specific clinical tools" and by "Progetto Giovani GNFM 2016" funded by the National Group of Mathematical Physics (GNFM—INdAM). TS is supported by European Research Council (ERC-2013-StG-336839).

Appendix: Stability of the Homogeneous Solution

The integration of the "stress-modulated growth" illustrated in the previous section predicts a spatially homogeneous solution, parametrically depending on the time-dependent growth rate. The growth $g(R;t)$ is independent of the radial position because the stress is the same everywhere. In such a purely mechanical setting we now study the stability of the homogeneous solution (20) and (25). In other words, the question is whether the spatial inhomogeneity observed in grown spheroids could be produced by the mechanobiological feedback, thus amplifying the spatial perturbations of the stress to yield inhomogeneous growth.

To investigate this hypothesis we consider the following perturbation of the homogeneous solution:

$$r(R,t) = \gamma g_0(t)R + \rho(R,t), \quad \gamma g_0(t)R \gg \rho(R,t), \tag{32}$$

$$g(R,t) = g_0(t) + \delta(R,t), \quad g_0(t) \gg \delta(R,t), \tag{33}$$

where γ and $g_0(t)$ are solutions of Eqs. (21) and (24), respectively, and $g_0(0) = 1$.

When the perturbed solutions are plugged in Eqs. (21) and (24) and only first order terms are retained, the following linear equations are found

$$\left(\rho' + 2\frac{\rho}{R}\right)' = \gamma \frac{3-\gamma^2}{\gamma^2+1}\delta', \tag{34}$$

$$\dot{\delta} = \left(1 + 2\frac{g_0}{\kappa} - 2\frac{g_0}{\alpha} - 4\frac{g_0}{\kappa\gamma^2}\right)\frac{\delta}{\tau} + \frac{2}{3}\frac{g_0}{\kappa\tau\gamma^3}\left(\rho' + 2\frac{\rho}{R}\right). \tag{35}$$

Derivation of the former equation in space, derivation of the latter in time and cross substitution yields

$$\dot{\delta}' = \left(1 + 2\frac{g_0}{\kappa} - 2\frac{g_0}{\alpha} - 4\frac{g_0}{\kappa\gamma^2} + \frac{2}{3}\frac{g_0}{\kappa\gamma^2}\frac{3-\gamma^2}{\gamma^2+1}\right)\frac{\delta'}{\tau}, \tag{36}$$

which determines the evolution in time of the spatial perturbation in the growth $g(t)$. Instability shows up if

$$\alpha\kappa\gamma^2(\gamma^2+1) > 2g_0\left(\alpha\left(1 + \frac{4}{3}\gamma^2 - \gamma^4\right) + \kappa\gamma^2(\gamma^2+1)\right), \tag{37}$$

for some $1 < g_0(t) < g_e$, where $0 < \gamma(p_D) < 1$.

The result (37) is negative versus our conjecture: it predicts a stabilization of the system for large enough growth g_0 which is not in agreement with experiments. If the purely mechanical system is stable, the reported inhomogeneity (large proliferation near the boundary, smaller internally) should instead be explained accounting for the role of nutrients.

References

1. Folkman, J., Hochberg, M.: Self-regulation of growth in three dimensions. J. Exp. Med. **138**(4), 745–753 (1973)
2. Ambrosi, D., Mollica, F.: The role of stress in the growth of a multicell spheroid. J. Math. Biol. **48**(5), 477–499 (2004)
3. Greenspan, H.: Models for the growth of a solid tumor by diffusion. Stud. Appl. Math. **51**(4), 317–340 (1972)
4. Helmlinger, G., Netti, P.A., Lichtenbeld, H.C., Melder, R.J., Jain, R.K.: Solid stress inhibits the growth of multicellular tumor spheroids. Nat. Biotechnol. **15**(8), 778–783 (1997)
5. Montel, F., Delarue, M., Elgeti, J., Vignjevic, D., Cappello, G., Prost, J.: Isotropic stress reduces cell proliferation in tumor spheroids. New J. Phys. **14**(5), 055008 (2012)
6. Forgacs, G., Foty, R.A., Shafrir, Y., Steinberg, M.S.: Viscoelastic properties of living embryonic tissues: a quantitative study. Biophys. J. **74**(5), 2227–2234 (1998)
7. Kuznetsova, T.G., Starodubtseva, M.N., Yegorenkov, N.I., Chizhik, S.A., Zhdanov, R.I.: Atomic force microscopy probing of cell elasticity. Micron **38**(8), 824–833 (2007)
8. Cheng, G., Tse, J., Jain, R.K., Munn, L.L.: Micro-environmental mechanical stress controls tumor spheroid size and morphology by suppressing proliferation and inducing apoptosis in cancer cells. PLoS ONE **4**(2), e4632 (2009)
9. Netti, P.A., Berk, D.A., Swartz, M.A., Grodzinsky, A.J., Jain, R.K.: Role of extracellular matrix assembly in interstitial transport in solid tumors. Cancer Res. **60**(9), 2497–2503 (2000)
10. Stylianopoulos, T., Martin, J.D., Chauhan, V.P., Jain, S.R., Diop-Frimpong, B., Bardeesy, N., Smith, B.L., Ferrone, C.R., Hornicek, F.J., Boucher, Y., et al.: Causes, consequences, and remedies for growth-induced solid stress in murine and human tumors. Proc. Natl. Acad. Sci. **109**(38), 15101–15108 (2012)
11. Ambrosi, D., Mollica, F.: On the mechanics of a growing tumor. Int. J. Eng. Sci. **40**(12), 1297–1316 (2002)
12. Montel, F., Delarue, M., Elgeti, J., Malaquin, L., Basan, M., Risler, T., Cabane, B., Vignjevic, D., Prost, J., Cappello, G., et al.: Stress clamp experiments on multicellular tumor spheroids. Phys. Rev. Lett. **107**(18), 188102 (2011)
13. Monnier, S., Delarue, M., Brunel, B., Dolega, M.E., Delon, A., Cappello, G.: Effect of an osmotic stress on multicellular aggregates. Methods **94**, 114–119 (2016)
14. Byrne, H., Preziosi, L.: Modelling solid tumor growth using the theory of mixtures. Math. Med. Biol. **20**(4), 341–366 (2003)
15. Ambrosi, D., Preziosi, L.: On the closure of mass balance models for tumor growth. Math. Models Methods Appl. Sci. **12**(05), 737–754 (2002)
16. Roose, T., Netti, P.A., Munn, L.L., Boucher, Y., Jain, R.K.: Solid stress generated by spheroid growth estimated using a linear poroelasticity model. Microvasc. Res. **66**(3), 204–212 (2003)
17. Ambrosi, D., Guana, F.: Stress-modulated growth. Math. Mech. Solids **12**(3), 319–342 (2007)
18. Chanet, S., Martin, A.C.: Mechanical force sensing in tissues. Prog. Mol. Biol. Transl. Sci. **126**, 317 (2014)
19. Ciarletta, P.: Buckling instability in growing tumor spheroids. Phys. Rev. Lett. **110**(15), 158102 (2013)
20. Destrade, M., Murphy, J.G., Ogden, R.W.: On deforming a sector of a circular cylindrical tube into an intact tube: existence, uniqueness, and stability. Int. J. Eng. Sci. **48**(11), 1212–1224 (2010)
21. Alnæs, M., Blechta, J., Hake, J., Johansson, A., Kehlet, B., Logg, A., Richardson, C., Ring, J., Rognes, M.E., Wells, G.N.: The FEniCS project version 1.5. Arch. Numer. Softw. **3**(100) (2015)
22. Geuzaine, C., Remacle, J.-F.: Gmsh: A 3-D finite element mesh generator with built-in pre- and post-processing facilities. Int. J. Numer. Meth. Engng. **79**, 1309–1331 (2009)
23. Balay, S., Abhyankar, S., Adams, M.F., Brown, J., Brune, P., Buschelman, K., Dalcin, L., Eijkhout, V., Gropp, W.D., Kaushik, D., et al.: PETSc users manual. Technical Report ANL-95/11—Revision 3.6. Argonne National Laboratory (2015)
24. Amestoy, P.R., Duff, I.S., L'Excellent, J.Y., Koster, J.: A fully asynchronous multifrontal solver using distributed dynamic scheduling. SIAM J. Matrix Anal. Appl. **23**(1), 15–41 (2001)
25. Mascheroni, P., Stigliano, C., Carfagna, M., Boso, D.P., Preziosi, L., Decuzzi, P., Schrefler, B.A.: Predicting the growth of glioblastoma multiforme spheroids using a multiphase porous media model. Biomech. Model. Mechanobiol. **15**(5), 1215–1228 (2016)
26. Ambrosi, D., Preziosi, L.: Cell adhesion mechanisms and stress relaxation in the mechanics of tumours. Biomech. Model. Mechanobiol. **8**(5), 397–413 (2009)

27. Dorie, M.J., Kallman, R.F., Rapacchietta, D.F., Van Antwerp, D., Huang, Y.R.: Migration and internalization of cells and polystyrene microspheres in tumor cell spheroids. Exp. Cell Res. **141**(1), 201–209 (1982)
28. Delarue, M., Montel, F., Caen, O., Elgeti, J., Siaugue, J.M., Vignjevic, D., Prost, J., Joanny, J.F., Cappello, G.: Mechanical control of cell flow in multicellular spheroids. Phys. Rev. Lett. **110**(13), 138103 (2013)

J Elast (2017) 129:125–144
DOI 10.1007/s10659-017-9626-5

A Computational Model of the Biochemomechanics of an Evolving Occlusive Thrombus

Manuel K. Rausch[1,2] · **Jay D. Humphrey**[1]

Received: 9 May 2016 / Published online: 8 March 2017

Abstract Blood clots are fundamental to preventing excessive blood loss in cases of vascular injury and to promoting subsequent wound healing, but also to many disease conditions. The biomechanical properties of clots play important roles in dictating the natural history in health and disease. In this paper, we present a novel multiscale computational model of biological, chemical, and mechanical contributions to the maturation of an occlusive clot from a fibrin-dominated to a collagen-dominated tissue. For an occlusive venous thrombus, simulations show the potential coupling between mechanical deformations and chemical processes, which can result in competing phenomena. This mixture-based framework also promises to guide future experimentation, which is vitally needed to increase our understanding of the complex and important biochemomechanics of clots.

Keywords Growth and remodeling · Deep vein thrombosis · Microsphere · Venous thrombosis · Multiscale

Mathematics Subject Classification 74F10

1 Introduction

Thrombosis, that is the formation of a blood clot, is essential for controlling hemorrhage in cases of tissue injury. Yet, many devastating disease conditions can also result from a thrombus—heart attack, stroke, and pulmonary embolism to name a few [23, 33, 36].

✉ M.K. Rausch
manuel.rausch@utexas.edu

J.D. Humphrey
jay.humphrey@yale.edu

1 Department of Biomedical Engineering, Yale University, 55 Prospect Street, New Haven, CT 06520, USA

2 Department of Aerospace Engineering & Engineering Mechanics, University of Texas at Austin, 210 E 24th St, Austin, TX 78712, USA

Notwithstanding the critical importance of the biochemistry involved in the formation, lysis, or maturation of a clot, quantification of the evolving biomechanical properties is equally important for understanding the role of thrombus in both physiology and pathophysiology. Whereas many prior biomechanical studies have focused on the initial few minutes of the blood clotting cascade [51, 52], here we introduce a new model within the framework of a reactive/constrained mixture theory that allows studying the possible longer term evolution of a thrombus. Specifically, we are interested in the competition between thrombus degradation or lysis and the deposition of structural constituents during maturation, which may determine whether thrombus embolizes, that is, sheds fragments into the flowing blood, or whether it persists intact [20].

The blood clotting cascade is well studied [15] and can be captured via complex systems of differential equations that describe reactions among the many different biomolecules [17, 34]. However, given that the initial clotting process occurs within minutes and our interest is in longer term maturation of the clot (on the order of weeks), we focus on the latter by prescribing the biomechanical properties of a fresh fibrin network in our initial model. During its evolution, we consider thrombus to be a reactive mixture, consisting of a constrained mixture of two structurally significant solid constituents, ubiquitous interstitial fluid, a generic family of migrating synthetic cells, and one key biomolecule. This biomolecule, plasmin(ogen), which represents both the zymogen plasminogen and its active form, the serine protease plasmin, facilitates the degradation of one of the solid constituents (fibrin) while the cells, ostensibly (myo)fibroblasts, deposit the other solid constituent (collagen fibers). We use the term (myo)fibroblasts for it is the fibroblast that is recruited and then changes phenotypically to increase its production of collagen and better organize the matrix; we do not model the phenotypic modulation however. To further focus our attention on the longer term maturation of a fibrin-based clot into a more collagenous tissue, we assume that local plasmin levels correspond directly to those for plasminogen. For conceptual purposes, we also consider plasmin(ogen) and (myo)fibroblasts as non-charged "solutes" that diffuse and advect within the interstitial fluid that saturates the pores of the solid. Because local deformations affect the transport of important thrombotic and thrombolytic biomolecules, the migration of relevant synthetic cell species, critical reactions between constituents and biomolecules, biological processes such as collagen deposition, our model, for the first time, additionally includes mechanically and directionally sensitive constitutive equations for transport and migration as well as mechanically and directionally sensitive reaction equations [2, 12, 13, 49].

The primary goal of this work is to develop a first framework to model the nonlinear, anisotropic mechanical behavior of evolving thrombus by capturing the roles of specific cells and biomolecules within a multiscale framework that largely melds concepts from Ateshian [6, 7] and Humphrey and Rajagopal [27]. The secondary goal of our work is to use our newly introduced approach to investigate mechanically-mediated feedback mechanisms that may predict different fates of a thrombus, including resolution, maturation, or embolization. This capability may be critical from both clinical and basic scientific standpoints as the location of advanced lysis and/or reduced collagen deposition may determine whether thrombus embolizes partially and fragments or dislodges in whole, thus increasing the risk of a fatal pulmonary embolus for example. For purposes of illustration, our computational model focuses on the evolution of occlusive venous thrombus and is based on data available from a well-accepted mouse model [18, 32]. The basic framework is much more general, however.

2 Modeling Framework

2.1 Solid Constituents—Fibrin and Collagen

The solid portion of an evolving thrombus, denoted by superscript s, is assumed to consist of a constrained mixture [27] of three constituents ξ: an amorphous ground substance matrix ($\xi = g$), a fibrin network ($\xi = f$), and fibrillar collagen ($\xi = c$). Their kinematics are constrained such that $\boldsymbol{v}^\xi = \boldsymbol{v}^s$, where $\boldsymbol{v}^\xi$ and $\boldsymbol{v}^s$ denote the velocity of each solid constituent ξ and the total solid s, respectively. We account for growth and remodeling of thrombus by modeling chemical reactions between the "solutes" plasmin(ogen) and (myo)fibroblasts and the solid constituents fibrin and collagen. The addition/subtraction of mass of an individual structural constituent may thus be expressed via a mass exchange term in the statement of mass balance, namely

$$\frac{\mathrm{D}\rho_r^\xi}{\mathrm{D}t} = \hat{\rho}_r^\xi, \quad \text{where } \xi = g, f, c, \tag{1}$$

where $\mathrm{D}\{\circ\}/\mathrm{D}t$ denotes the material time derivative, ρ_r^ξ is the current apparent mass density with respect to a reference volume, and $\hat{\rho}_r^\xi$ denotes the apparent mass density supply, also with respect to the reference volume. In our model, two mechanisms contribute to $\hat{\rho}_r^\xi$: first, we consider a decrease in mass density of fibrin due to degradation by plasmin(ogen) and, second, an increase in mass density of collagen by (myo)fibroblast deposition. These reactions will be discussed in more detail later.

2.2 Plasmin(ogen) and (Myo)fibroblasts

We assume that plasmin(ogen) ($\iota = pl$) is capable of transport via diffusion and advection; it can also react with fibrin. We further assume that (myo)fibroblasts ($\iota = fb$) are capable of migration and deposit collagen. For ease of computation, we treat both as "solutes" denoted by superscript ι. Mass balance for these constituents can be written

$$J^{-1}\frac{\mathrm{D}(J[1-\varphi^s]c^\iota)}{\mathrm{D}t} + \operatorname{div} \boldsymbol{j}^\iota = \left[1-\varphi^s\right]\hat{c}^\iota, \quad \text{with } \iota = pl, fb, \tag{2}$$

where $J = \det \boldsymbol{F}$ tracks changes in the volume of the mixture relative to its reference volume, φ^s is again the solid volume fraction, $\hat{c}^\iota$ is the mass exchange (i.e., molar supply or sink due to chemical reactions), and finally $\boldsymbol{j}^\iota$ is the molar flux relative to the solid.

2.3 Chemical Reactions

A key feature of our model is the ability of constituents to interact via chemical reactions, which contribute to lysis and maturation. In contrast to the common treatment of chemical kinetics, we introduce the deformation gradient $\boldsymbol{F}$ in addition to the standard state variables, which are concentrations c^ι of solutes and apparent referential densities ρ_r^ξ of solid constituents. It proves convenient to express apparent referential mass densities as concentrations through $c^\xi = \rho_r^\xi/[J - \varphi_r^s]M^\xi$, with φ_r^s the referential solid volume fraction and M^ξ the molar mass of constituents ξ. During lysis and maturation of thrombus, we consider all reactions to follow the law of forward mass action, viz. $\sum_\alpha \nu_R^\alpha \mathcal{E}^\alpha \to \sum_\alpha \nu_P^\alpha \mathcal{E}^\alpha$, where $\mathcal{E}^\alpha$ are the chemical species that correspond to constituents α ($= \xi$ or ι) and ν_R^α and ν_P^α are the stoichiometric coefficients of reactant and product, respectively. Therefore, the molar supply $\hat{c}^\alpha$ of solid constituents and solutes may be expressed as $\hat{c}^\alpha = [\nu_P^\alpha - \nu_R^\alpha]\hat{\zeta}$, with

$$\hat{\zeta} = k(\boldsymbol{F}, c^\alpha) \prod_\alpha [c^\alpha]^{\nu_R^\alpha}, \quad \text{with } \alpha = \xi, \iota, \tag{3}$$

where $k(\boldsymbol{F}, c^\alpha)$ is the specific reaction rate. The balance of mass of the solid constituents, solutes, and the mixture may now be rewritten in terms of (3), see [7].

2.4 Interstitial Fluid

Interstitial fluid (mainly water, denoted by superscript w) is modeled as incompressible and inviscid, while frictional losses between the fluid and the solid matrix are included via a momentum exchange term within the statement of linear momentum balance [5]. By neglecting the volume fraction of the solutes, we may express the flux of interstitial fluid as $\boldsymbol{w} = [1 - \varphi^s][\boldsymbol{v}^w - \boldsymbol{v}^s]$, where φ^s is the solid volume fraction.

With the introduction of $\boldsymbol{w}$, the statement of mass balance for the entire mixture can be written

$$\operatorname{div}(\boldsymbol{v}^s + \boldsymbol{w}) = \sum_\alpha \frac{\hat{\rho}^\alpha}{\rho_T^\alpha}, \quad \text{with } \alpha = \xi, \iota, \tag{4}$$

where ρ_T^α is the true density of constituent α (current mass of constituent α divided by current volume of constituent α).

2.5 Balance of Linear Momentum

Given the expression for the total Cauchy stress, $\boldsymbol{\sigma} = -p\boldsymbol{I} + \boldsymbol{\sigma}^s$, where $\boldsymbol{\sigma}^s$ is the solid matrix stress, we can further write the spatial balance of linear momentum of the total mixture as

$$\operatorname{div} \boldsymbol{\sigma} = -\operatorname{grad} p + \operatorname{div} \boldsymbol{\sigma}^s = \boldsymbol{0}, \tag{5}$$

and where we assume quasi-static conditions.

2.6 Finite Element Implementation

The governing equations of the reactive mixture are solved using a nonlinear finite element method [35]. To this end, we express the initial-boundary value problem in weak form, which can be solved monolithically and fully implicitly using Newton's method. The first two equations that enter the statement of virtual work are the mass balance for each solute, (2), and mass balance of the mixture, (4). Adding the balance of linear momentum for the solid-fluid mixture, (5), complements the set of governing differential equations that lead to the solution of our variables of interest, the nodal displacement for the solids $\boldsymbol{u}$, the interstitial fluid pressure p, and the solute concentrations c^ι. Because the interstitial fluid pressure p is not guaranteed to be continuous across element boundaries, we instead solve for an effective fluid pressure $\tilde{p}$ that is continuous across element boundaries, where $\tilde{p} = p - R\theta\Phi \sum_\iota c^\iota$, R the universal gas constant, θ the absolute temperature, and Φ the osmotic coefficient. Importantly, c^ι is only continuous across element boundaries in the case of ideal solubility, which we assume here. Because the referential apparent mass densities of the solid constituents fibrin and collagen can be obtained by simple integration of (1) they enter the statement of virtual work directly. With all this said, the total statement of virtual work

$$\delta W = \sum_{\iota} \int_b \delta c^{\iota} \big[J^{-1} \mathrm{D}^s \big(J(1-\varphi^s) c^{\iota} \big) / \mathrm{D}t + \operatorname{div} \boldsymbol{j}^{\alpha} \big] \mathrm{d}v$$
$$- \bar{\mathcal{V}} \int_b \delta \tilde{p} \big[1 - \varphi^s \big] \hat{\zeta} \,\mathrm{d}v - \sum_{\iota} \nu^{\iota} \int_b \delta c \big[1 - \varphi^s \big] \hat{\zeta} \,\mathrm{d}v$$
$$+ \int_b \delta \boldsymbol{v} \cdot \operatorname{div} \boldsymbol{\sigma} \,\mathrm{d}v + \int_b \delta \tilde{p} \operatorname{div} \big(\boldsymbol{v}^s + \boldsymbol{w} \big) \mathrm{d}v, \tag{6}$$

where $\delta \boldsymbol{v}$ is a virtual velocity vector, $\delta \tilde{p}$ is a virtual effective interstitial pressure, δc^{ι} is a virtual molar energy, and $\bar{\mathcal{V}} = \sum_{\alpha} \nu^{\alpha} \mathcal{V}^{\alpha}$ with $\mathcal{V}^{\alpha} = M^{\alpha} / \rho_T^{\alpha}$ the molar volume. Integrations are over current mixture volume, v.

Expressions for the volumetric fluid flux $\boldsymbol{w}$ and solute molar fluxes $\boldsymbol{j}^{\iota}$ follow from inversion of the balances of linear momentum for the interstitial fluid and the "solutes" and result in a generalized Darcy's law and Fick's law, respectively. In addition, constitutive relations are required for the solid portion of the thrombus, mass forward reaction equations, and mechanically dependent permeability and diffusivity tensors, which we define in the following section.

3 Illustrative Constitutive Relations

3.1 Solid Matrix

We postulate the existence of a strain energy function $\mathcal{W}$ that determines the constitutive behavior of the solid constituents. Specifically, we describe the solid matrix as a constrained mixture of an isotropic ground substance $\mathcal{W}^g$ and an anisotropic fibrous matrix that may be populated with fibrin $\mathcal{W}^f$ or collagen $\mathcal{W}^c$, namely

$$\mathcal{W} = \mathcal{W}^g + \mathcal{W}^f + \mathcal{W}^c. \tag{7}$$

$\mathcal{W}^f$ and $\mathcal{W}^c$ are sensitive to decreases and increases in fiber density due to degradation and deposition, respectively. Fibers are deposited into extant matrix at a physiological prestretch. Hence, the total stretch of a fiber at the current time results multiplicatively from the stretch of the mixture in the direction of that fiber, the inverse of the mixture stretch at the time of deposition, and the deposition stretch value [8]. Thus, the total strain energy of the matrix follows from the integral over all active fibers based on their respective deposition times and individual reference configurations.

Specifically, we model the ground substance as neo-Hookean parameterized with Lamé parameters μ and λ, namely

$$\mathcal{W}^g = \frac{\mu}{2} [I - 3] - \mu \ln J + \frac{\lambda}{2} [\ln J]^2, \tag{8}$$

where $I = \boldsymbol{C} : \boldsymbol{I}$ is the first invariant of the right Cauchy-Green tensor, with $\boldsymbol{I}$ the second order identity tensor, and $J = \det \boldsymbol{F}$ again. We consider the fibrous matrix using continuous distributions functions integrated over the unit sphere having surface S [22, 30],

$$\mathcal{W}^m = \frac{1}{4\pi} \int_S \int_0^s \frac{\hat{\rho}_r^m(\phi, \psi, \tau)}{\rho_0^m} \bar{\mathcal{W}}^m(\phi, \psi, \tau) \,\mathrm{d}\tau \,\mathrm{d}\theta \,\mathrm{d}\Phi, \quad \text{where } m = f, c, \tag{9}$$

with s being the current thrombus evolution time and $\tau \in [0, s]$ the collagen deposition time.

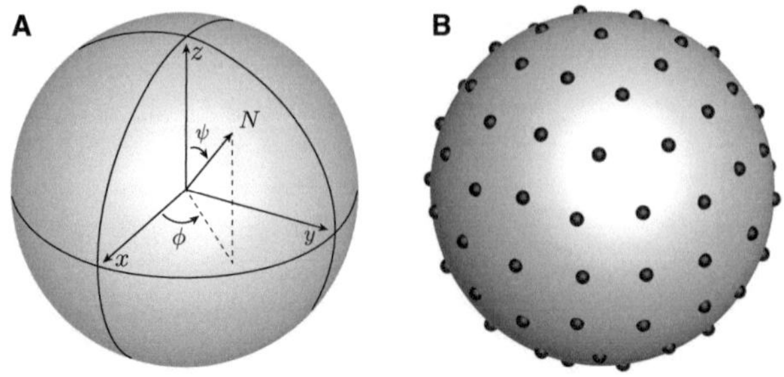

Fig. 1 (**A**) Parametrization of the fiber orientation vector $N(\phi, \psi)$ within the unit sphere. (**B**) Discretization of the continuous spherical surface according to the fully symmetric cubature formula by Heo and Yuan [24] with 45 integration points over the half sphere. Adapted from [22]

Hence, each fiber contributes as a function of the mass densities of either fibrin or collagen, associated with the mass density production rate for that fiber, $\hat{\rho}_r{}^m(\phi, \psi)$, relative to a reference value ρ_0^m. In turn, the constitutive behavior of the unmodified fiber is described by a Fung-type strain energy function [25],

$$\bar{\mathcal{W}}^m(\phi, \psi, \tau) = \frac{k_1^m}{4k_2^m}\big[\exp\big(k_2^m[\bar{IV}^m(\phi, \psi, \tau) - 1]^2\big) - 1\big], \tag{10}$$

with the fiber direction-dependent invariant $\bar{IV}^m(\phi, \psi, \tau) = IV(\phi, \psi) IV(\phi, \psi, \tau)^{-1} IV_p^m$, where $IV(\phi, \psi) = \boldsymbol{C} : \boldsymbol{N}(\phi, \psi) \otimes \boldsymbol{N}(\phi, \psi)$ is the square of the current fiber stretch, $IV(\phi, \psi, \tau) = \boldsymbol{C}(\tau) : \boldsymbol{N}(\phi, \psi) \otimes \boldsymbol{N}(\phi, \psi)$ is the square of the fiber stretch at the time of deposition τ, and $IV_p^m = [\lambda_p^m]^2$ the deposition stretch squared. Note that $\boldsymbol{C} = \boldsymbol{F}^T \boldsymbol{F}$ and $\boldsymbol{C}(\boldsymbol{\tau}) = \boldsymbol{F}(\boldsymbol{\tau})^T \boldsymbol{F}(\boldsymbol{\tau})$ is the right Cauchy-Green tensor at time τ and $\boldsymbol{N}(\phi, \psi) = \{\cos\phi \sin\psi, \sin\phi \sin\psi, \cos\psi\}$ is the referential fiber direction vector.

Thus, as the referential apparent mass densities $\rho_r^m(\phi, \psi, s) = \int_0^s \hat{\rho}_r^m(\phi, \psi, \tau)\mathrm{d}\tau$, associated with individual fibrin or collagen fibers, evolve with time due to chemical reactions between fibrin and plasmin(ogen) or fibroblasts and collagen, the relative weights of the fiber contributions decrease for fibrin and increase for collagen. For practical purposes, the expression in (9) is integrated numerically and the time integral approximated, hence

$$\mathcal{W}^m = \sum_\beta \sum_t w^\beta \frac{\Delta\rho_r^{m,\beta,t}}{\rho_0^m} \bar{\mathcal{W}}^m\big(\boldsymbol{N}^\beta, t\big), \tag{11}$$

where $\Delta\rho_r^{m,\beta,t} = [\rho_r^{m,\beta,t} - \rho_r^{m,\beta,t-1}]$ and the directions $\boldsymbol{N}^\beta = \boldsymbol{N}(\phi^\beta, \psi^\beta)$ and weights w^β follow from a fully symmetric cubature formula by Heo and Yuan [24]. Specifically, we approximate the definite integral in (9) by 45 discrete directions and weights over the half sphere (Fig. 1), where the discrete times t coincide with converged time steps.

These constitutive relations were implemented in FEBio [35] via its plugin interface. To this end, our subroutine provides the Cauchy stress for the solid matrix $\boldsymbol{\sigma}^s = 2J^{-1}\boldsymbol{F}\partial\mathcal{W}/\partial\boldsymbol{C}\boldsymbol{F}^T$

$$\begin{aligned}\boldsymbol{\sigma}^s &= J^{-1}\mu[\boldsymbol{b} - \boldsymbol{I}] + J^{-1}\lambda \ln J\boldsymbol{I} \\ &\quad + 2J^{-1}\sum_m\sum_\beta\sum_t w^\beta \frac{\Delta\rho_r^{m,\beta,t}}{\rho_0^m} k_1^m\big[\bar{IV}^{m,\beta,t} - 1\big]\frac{IV_p^m}{IV^{\beta,t}}\exp\big(k_2^m\big[\bar{IV}^{m,\beta,t} - 1\big]^2\big)\boldsymbol{n}^\beta \otimes \boldsymbol{n}^\beta.\end{aligned} \tag{12}$$

Here, $\boldsymbol{b}$ is the left Cauchy-Green tensor and $\boldsymbol{n}^\beta$ are direction vectors in the spatial configuration, $\boldsymbol{n}^\beta = \boldsymbol{F}\boldsymbol{N}^\beta$. The spatial tangent $\mathbb{c}^s = 4J^{-1}[\boldsymbol{F} \overline{\otimes} \boldsymbol{F}] : \partial^2\mathcal{W}/\partial\boldsymbol{C} \otimes \partial\boldsymbol{C} : [\boldsymbol{F}^T \overline{\otimes} \boldsymbol{F}^T]$, also implemented in the plugin, becomes

$$\mathbb{c}^s = J^{-1}\lambda \boldsymbol{I} \otimes \boldsymbol{I} + J^{-1}[\mu - \lambda \ln J]\mathbb{I}$$
$$+ 4J^{-1} \sum_m \sum_\beta \sum_t w^\beta \frac{\Delta\rho_r^{m,\beta,t}}{\rho_0^m} k_1^m \big[1 + 2k_2^m \big[\bar{IV}^{m,\beta,t} - 1\big]^2\big] \left[\frac{IV_p^m}{IV^{\beta,t}}\right]^2$$
$$\times \exp\big(k_2^m \big[\bar{IV}^{m,\beta,t} - 1\big]^2\big)\mathbb{N} \tag{13}$$

with the fourth order identity tensors $\mathbb{I} = [\boldsymbol{I} \overline{\otimes} \boldsymbol{I} + \boldsymbol{I} \underline{\otimes} \boldsymbol{I}]/2$ and the structural tensor $\mathbb{N} = \boldsymbol{n}^\beta \otimes \boldsymbol{n}^\beta \otimes \boldsymbol{n}^\beta \otimes \boldsymbol{n}^\beta$, where we used the short-hand $\{\circ \overline{\otimes} \bullet\}_{ijkl} = \{\circ\}_{ik}\{\bullet\}_{jl}$ as well as $\{\circ \underline{\otimes} \bullet\}_{ijkl} = \{\circ\}_{il}\{\bullet\}_{jk}$.

3.2 Reaction Rate Constitutive Behavior

Thrombus lysis and maturation are accounted for by considering chemical reactions involving the "solutes" plasmin(ogen) and (myo)fibroblasts and the solid constituents fibrin and collagen. Specifically, we consider two separate mass forward reactions. First, we model a reaction between fibrin and plasmin(ogen) to yield a reaction by-product. Because plasmin(ogen) is assumed to be constant (is not depleted), this is a pseudo first order reaction. Second, we consider a reaction between (myo)fibroblasts and nutrients to generate collagen. Here, nutrients and (myo)fibroblasts are also assumed to remain constant, making this equation a pseudo zeroth order reaction equation. For both types of reactions, we model the deformation-dependent specific reaction rate as

$$k^{m,\beta} = \epsilon k_0^m \big[IV^\beta\big]^\gamma, \quad \text{where } m = f, c, \tag{14}$$

where ϵ equals 1 in the case of fibrin and $[1 - \rho_r^{c,\beta}/\rho_0^c]$ in the case of collagen, thus limiting the apparent referential mass density of the latter to a maximum allowable value ρ_0^c. Further, k_0^m is the specific reaction rate, IV^β the square of the fiber stretch ($IV^\beta = \boldsymbol{C} : \boldsymbol{N}^\beta \otimes \boldsymbol{N}^\beta$), and $\gamma \in \mathbb{W}$. Fibers under tension or compression may thus be degraded/deposited faster or slower depending on the value of γ [1, 11, 43].

This constitutive behavior was similarly implemented as a plugin in FEBio. To this end we provide $k^{m,\beta}$ and its derivative $\boldsymbol{k}^{m,\beta} = 2J^{-1}\boldsymbol{F}\partial k^{m,\beta}/\partial \boldsymbol{C}\boldsymbol{F}^T$, with

$$\boldsymbol{k}^{m,\beta} = 2J^{-1}\epsilon k_0^m \gamma \big[IV^\beta\big]^{\gamma-1} \boldsymbol{n}^\beta \otimes \boldsymbol{n}^\beta. \tag{15}$$

3.3 Mechanically Sensitive Permeability and Diffusivity

Transport of key biomolecules within the thrombus plays important roles in its lysis and maturation [17]. Such transport depends on the mechanical state of the thrombus, hence we model both plasmin(ogen) diffusivity and thrombus permeability as a function of deformation. Specifically, we assume an initially isotropic permeability tensor $\boldsymbol{p}$ and diffusivity tensor $\boldsymbol{d}^m$ as

$$\boldsymbol{p} = p_0 \boldsymbol{I} + p_1 J^{-2} \boldsymbol{b} \tag{16}$$

and

$$\boldsymbol{d}^m = d_0^m \boldsymbol{I} + d_1^m J^{-2} \boldsymbol{b}, \quad \text{with } m = pl, fb, \tag{17}$$

respectively, where J is the Jacobian, $\boldsymbol{b}$ is the left Cauchy-Green tensor, and p_0, p_1, d_0^m, d_1^m are scalar material parameters. In addition, linearized permeability tensor, $\mathbb{p}$, and diffusivity

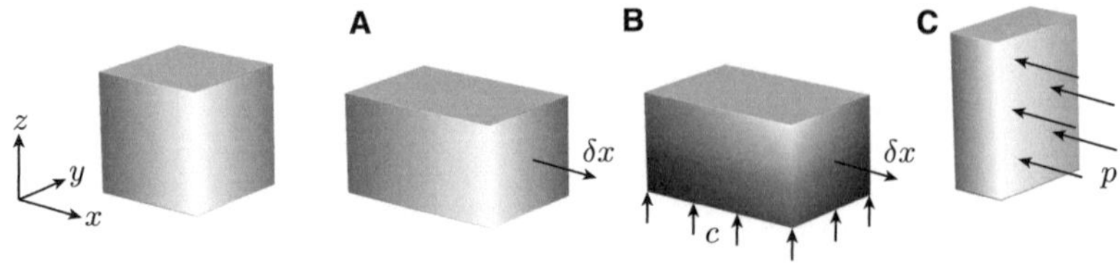

Fig. 2 Three different quarter symmetric loading scenarios for sensitivity study. (**A**) Strip biaxial extension (1.25:1 stretch ratio in x–y plane, traction free $z = 1$ surface) with homogeneous plasmin(ogen) and (myo)fibroblast distribution. (**B**) Same as (**A**) with diffusing plasmin(ogen) and migrating (myo)fibroblasts through the $z = 0$ surface. (**C**) Strip biaxial pressure (pressure applied to $x = 1$ surface, $y = 0$ surface is fixed in y-direction, traction free $z = 1$ surface) with homogeneous plasmin(ogen) and (myo)fibroblast distribution

tensor, $\mathbb{d}^m$, are

$$\mathbb{p} = p_0 \boldsymbol{I} \otimes \boldsymbol{I} - 2p_0 \mathbb{I} - p_1 J^{-2} \boldsymbol{b} \otimes \boldsymbol{I} \tag{18}$$

and

$$\mathbb{d}^m = d_0 \boldsymbol{I} \otimes \boldsymbol{I} - 2d_0^m \mathbb{I} - d_1^m J^{-2} \boldsymbol{b} \otimes \boldsymbol{I}, \quad \text{where } m = pl, fb. \tag{19}$$

Here, $\mathbb{I}$ is again the fourth order identity tensor, $\mathbb{I} = [\boldsymbol{I} \overline{\otimes} \boldsymbol{I} + \boldsymbol{I} \underline{\otimes} \boldsymbol{I}]/2$.

3.4 Sensitivity Study

Our model considers the mechanically mediated interplay among solid and fluid constituents as well as cellular and molecular "solutes". In a set of simple numerical experiments we demonstrate the sensitivity of our model to critical parameters and constitutive relations. Specifically, we consider three scenarios. First, we study separately fibrin-dominated and collagen-dominated unit cubes under displacement-driven strip biaxial extension (Fig. 2A). In this set of simulations, both cubes are instantaneously saturated with plasmin(ogen) or (myo)fibroblasts that homogeneously degrade fibrin and deposit collagen, respectively. During a second set of experiments, we apply the same mechanical boundary conditions, but study fibrin degradation and collagen deposition in response to plasmin(ogen) diffusion and (myo)fibroblast migration through the $z = 0$ surface of the cube (Fig. 2B). In the third set of experiments, we model the same unit cubes but apply biaxial stresses instead of displacement-driven biaxial extension (Fig. 2C). For all simulations, free fluid flow is allowed through the $z = 1$ surface of the material cube.

Because degradation follows a pseudo first-order reaction equation (i.e. plasmin(ogen) is not consumed during the reaction), we report the evolution of fibrin apparent referential density as a function of normalized time, using the time constant $\tau^f = [k_0^f \cdot c^{pl}]^{-1}$, where, k_0^f is the specific degradation reaction rate and c^{pl} is the concentration of plasmin(ogen). Similarly, we present the deposition of collagen referential apparent density as a function of normalized time, with time constant $\tau^c = [k_0^c \cdot c^{fb}/c^c]^{-1}$, where k_0^c is the specific deposition reaction rate, c^{fb} is the (myo)fibroblast "concentration", and c_0^c is the maximum allowable collagen concentration. All remaining parameter values are listed in Table 1.

Using our first model problem, Figs. 3A, B illustrate the sensitivity of the apparent referential fibrin and collagen density to parameter γ (recall equation (14)). Increasing γ increases the degradation rate since the reaction rate parameter k^f is amplified through its

Table 1 Parameters in sensitivity study and simulation of occlusive venous thrombus. Values were estimated from references or chosen when no data was available

Symbol	Description	Value	Unit	Reference
Thrombus/Wall geometry				
	Thrombus length	4.0	mm	[32]
	Thrombus diameter	1.2	mm	[32]
	Venous wall length	8.0	mm	[–]
	Venous wall thickness	0.1	mm	[–]
Thrombus material				
φ_0	Initial solid volume fraction	0.2	–	[17]
μ	Ground substance shear modulus	0.0038	MPa	[–]
λ	Ground substance Lamé parameter	0.058	MPa	[–]
k_1^f	Fibrin material parameter	0.01	MPa	[–]
k_2^f	Fibrin material parameter	0.0	–	[–]
k_1^c	Collagen parameter	1.031	MPa	[–]
k_2^f	Collagen parameter	6.559	–	[–]
T^a	Platelet tension parameter	$1.6\mathrm{e}^{-4}$	MPa	[14]
λ_p^f	Fibrin deposition stretch	1.00	–	[–]
λ_p^c	Collagen deposition stretch	1.04	–	[10]
Venous wall material				
	Wall shear modulus	0.1	MPa	[31]
	Wall bulk modulus	1000	MPa	[–]
λ^z	Axial prestretch	1.7	–	[31]
Solid and solute constituents				
M^f	Fibrin molecular mass	340	kDa	[19]
M^c	Collagen molecular mass	285	kDa	[26]
M^{pl}	Plasminogen molecular mass	80	kDa	[9]
M^{fb}	(Myo)fibroblast molecular mass	80	kDa	[–]
c_0^{pl}	Initial plasmin(ogen) concentration	$1.875\mathrm{e}^{-4}$	nmol/mm^3	[42]
c^{pl}	Blood plasmin(ogen) concentration	$2.5\mathrm{e}^{-3}$	nmol/mm^3	[42]
c^{fb}	Wall (myo)fibroblast concentration	$1.0\mathrm{e}^{-8}$	nmol/mm^3	[–]
Transport				
p_0	Permeability parameter	0.0	mm^4/N/s	[–]
p_1	Permeability parameter	0.91	mm^4/N/s	[3]
d_0^{pl}	Plasminogen diffusivity parameter	$2.5\mathrm{e}^{-3}$	mm^2/s	[37]
d_1^{pl}	Plasminogen diffusivity parameter	$0.5\mathrm{e}^{-4}$	mm^2/s	[37]
d_0^{fb}	(Myo)fibroblast diffusivity parameter	$5.0\mathrm{e}^{-8}$	mm^2/s	[–]
d_1^{fb}	(Myo)fibroblast diffusivity parameter	0.0	mm^2/s	[–]
Reactions				
k_0^f	Specific fibrin degradation rate	$2.0\mathrm{e}^{-4}$	1/s	[–]
k_0^c	Specific collagen deposition rate	5.0	1/s	[–]
γ^f	Fibrin degradation rate exponent	1.0	–	[–]
γ^c	Collagen deposition rate exponent	1.0	–	[–]

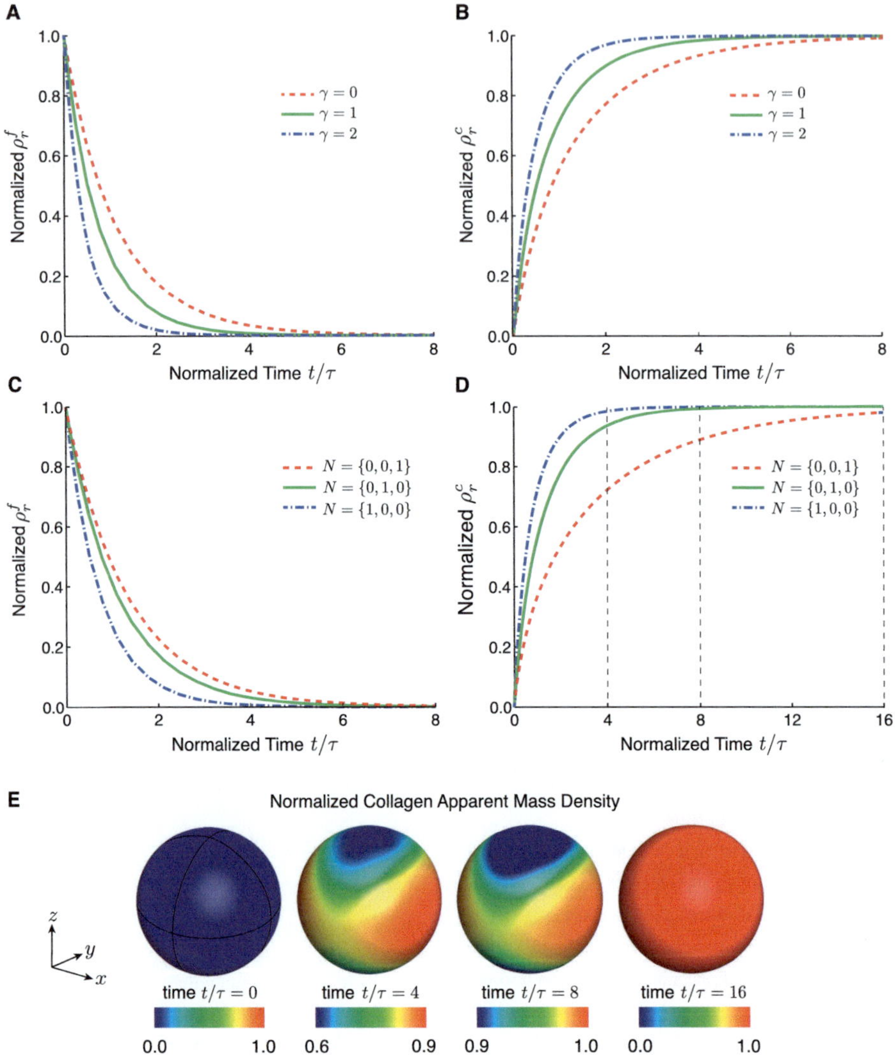

Fig. 3 Fibrin degradation rate and collagen deposition rate depend on the solid deformation in a direction dependent manner. (**A**) Fibrin degradation rate is accelerated by increasing γ through dependence on the pseudo invariant IV^γ. (**B**) Collagen deposition rate is accelerated by increasing γ through dependence on the pseudo invariant IV^γ. (**C**) Fibrin degradation rate varies with direction $\boldsymbol{N}$ and is the highest in first principal loading direction and the lowest in third principal loading direction. (**D**) Collagen deposition rate varies with direction $\boldsymbol{N}$ and is the highest in first principal loading direction and the lowest in third principal loading direction. (**E**) Anisotropic collagen deposition as a function of deformation at four times (see **D** for comparison), where density distributions reflect the principal load directions. Note, continuous distributions are obtained through cubic interpolation of 45 integration points over the half sphere using symmetry conditions. Further note, in (**E**), we chose different color scales to highlight remodeling induced material anisotropy

dependence on the pseudo invariant IV to the power of γ (Fig. 3A). Similarly, collagen deposition accelerates with increasing γ, as shown in Fig. 3B. We further observe that for $\gamma = 1$ the non-equibiaxial loading causes spatial variations in fibrin degradation as well as collagen deposition rates that result in load-induced material anisotropy (Fig. 3C, D). In

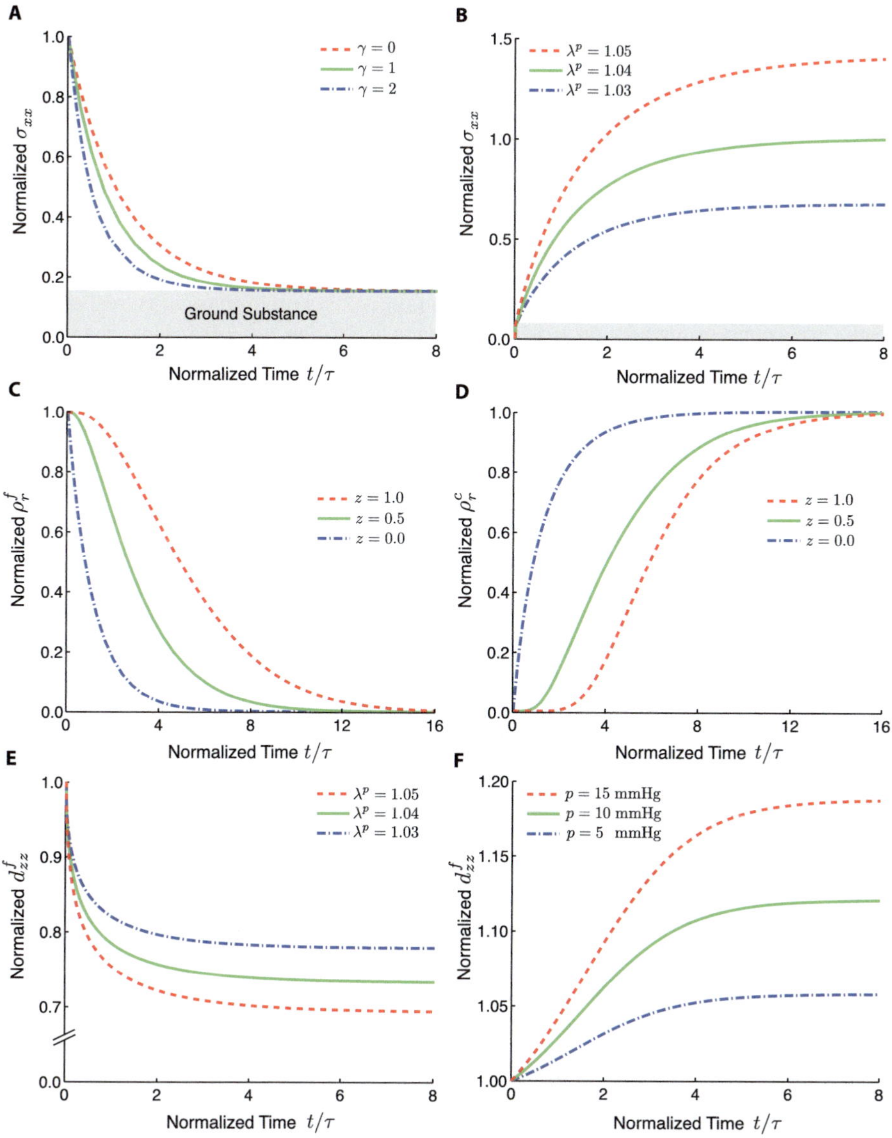

Fig. 4 Fibrin degradation and collagen deposition affect material stiffness: (**A**) Degradation reduces the material stiffness and thus the state of stress via removal of fibrin as a function of γ. (**B**) Collagen deposition increases material stiffness, wherein a larger collagen prestretch increases the equilibrium value. Transport induced material heterogeneity: (**C**) Intra-thrombus transport of plasmin(ogen) delays fibrin degradation and thus results in spatially varying fibrin densities. (**D**) (Myo)fibroblast migration delays collagen deposition and thus results in spatially varying collagen densities. Mechanical sensitivity of plasmin(ogen) diffusivity: (**E**) Plasmin(ogen) diffusivity decreases with thrombus contraction as a function of collagen prestretch. (**F**) Plasmin(ogen) diffusivity increases under pressure and is sensitive to pressure magnitude

Fig. 3E this anisotropy is further illustrated at one representative material point (of the homogenous unit cube). Here, evolving collagen density is interpolated over the unit sphere, based on the 45 discrete fiber directions (Fig. 1B).

As the microstructure of the thrombus evolves during fibrin degradation and collagen deposition, its state of stress also changes. Where fibrin is initially the primary load carrying component, its degradation results in reduced stiffness (Fig. 4A). Once all fibrin is degraded (and in the case of no collagen deposition), the ground substance provides the only remaining resistance to stretch. On the other hand, Fig. 4B illustrates that the state of stress of collagen dominated sample strongly depends on the deposition stretch of the collagen fibers. Specifically, collagen deposited at a higher stretch results in increased equilibrium stress.

In Fig. 4C, D we illustrate our second scenario (Fig. 2B) in which plasmin(ogen) enters and (myo)fibroblasts invade the biaxially stretched sample through the $z = 0$ surface. A diffusivity and migration mediated delay of plasmin(ogen) and (myo)fibroblasts, respectively, results in early fibrin degradation and early collagen deposition at $z = 0$ and late degradation and deposition at $z = 1$. Thus, material inhomogeneities naturally emerge from inclusion of thrombus transport phenomena such as diffusion and migration.

Mechanically altered intra-thrombus transport through diffusion and advection is critical to both thrombus lysis and maturation. We implement these effects via stretch dependence of the diffusivity and permeability tensors, see equations ((16)–(17)). Figure 4E, F demonstrate the model's sensitivity to degradation and deposition by means of the z-component of the plasmin(ogen) diffusivity tensor. Using the first loading scenario (Fig. 2A), we see that diffusivity may vary with collagen deposition stretch (Fig. 4E). As collagen deposition stretch is increased, thrombus compaction in the z-direction results in decreasing diffusivity, thus coupling progressive collagen deposition to reduced plasmin(ogen) transport and thus decreasing fibrin degradation. Similarly, using our third loading scenario (Fig. 2C) we see that decreasing fibrin density results in increased compression under stress in the planar directions and thus increased out-of-plane sample extension. Positive stretch in the z-direction subsequently increases diffusivity and provides a positive feedback for fibrin degradation via an increased plasmin(ogen) supply (Fig. 4F). These examples show that intra-thrombus transport may be increased or decreased by fibrin degradation and collagen deposition, respectively, and thus may be accelerated or decelerated through mechanically mediated feedback mechanisms.

4 Remodeling Venous Thrombus

We now combine all aspects of our proposed model in a three dimensional simulation of an occlusive venous thrombus. Specifically, we begin our simulation once the fibrin network has mostly developed (minutes to hours after initiation). Consequently, we assume that no additional fibrin deposition occurs. Starting from this point, we consider the transport of plasmin(ogen) from the thrombus/blood interface and associated degradation of the existing fibrin network by plasmin(ogen) as well as the migration of (myo)fibroblasts from the thrombus/wall interface and associated deposition of collagen during maturation. We also include the venous wall in our simulations. Specifically, we model the wall as neo-Hookean and use a three-field finite element interpolation scheme to approximate its nearly incompressible material behavior. We also include axial prestretch in our venous wall model. To this end, we assume that our reference configuration is not stress-free, but is rather the result of an isochoric deformation from a fictitious unloaded configuration according to a prestretch tensor $\boldsymbol{F}^z = \lambda^z \boldsymbol{N}^z \otimes \boldsymbol{N}^z + 1/\sqrt{\lambda^z}[\boldsymbol{I} - \boldsymbol{N}^z \otimes \boldsymbol{N}^z]$, where λ^z is the axial prestretch and $\boldsymbol{N}^z$ is the axial direction vector [39]. Lastly, we implement an additive active stress model to include thrombus compaction resulting from platelet activation and contraction. Thus, we add the active stress tensor $\boldsymbol{\sigma}^a$ to the solid stress tensor, where $\boldsymbol{\sigma}^a = T^a \boldsymbol{b}$, with T^a the tension parameter and $\boldsymbol{b}$ is the left Cauchy-Green deformation tensor.

Because there are few specific data on the evolution of human venous thrombus, we base our computational model largely on observations and measurements from a well-accepted stasis deep vein thrombosis mouse model [18]. Toward this end, we surgically ligated the inferior vena cava just below the renal veins to initiate the formation of (relatively) large, cylindrical thrombus samples. At 2 and 4 weeks, we explanted the venous thrombus and used combined histology/immunohistochemistry and micro-uniaxial tensile testing to characterize the specimens [32]. We measured thrombus length and diameter, spatial fibrin and collagen distributions, (myo)fibroblast distribution and count, neutrophil and macrophage distribution and count, and quantified the uniaxial tensile stress-stretch behavior.

Based on these data, we first estimate the material parameters for our ground substance matrix (μ, λ) as well as those for the fibrin and collagen (k_1^f, k_2^f and k_1^c, k_2^c). To this end, we built virtual cylindrical thrombus samples at 2 weeks and 4 weeks based on geometric experimental data as well as relative distributions of fibrin and collagen: 4 mm length and 1.2 mm diameter at 2 weeks and 3 mm length and 0.7 mm diameter at 4 weeks, with relative ground substance:fibrin:collagen distributions of 1:7:2 and 1:0:9, respectively. Thus, at 2 weeks our thrombus material is dominated by fibrin with some early collagen deposition, while at 4 weeks our sample is dominated by collagen. Subsequently, we apply a trust-region algorithm to minimize an objective function that penalizes the sum of the squared differences between experimentally determined load-stretch curves and finite element simulations, simultaneously for the samples at 2 and 4 weeks. Furthermore, during initial simulations we iteratively adjusted the fibrin degradation rate, the collagen deposition rate, and the (myo)fibroblast migration rate to match our histological observations qualitatively. All other parameters are either literature values or were chosen for lack of available data, see Table 1.

Notwithstanding the single ligature, for computational ease, we apply venous blood pressure to the distal and proximal thrombus surfaces. In contrast, we model the distal and proximal surfaces of the venous segment as fixed in space. Without data to the contrary, we also model the thrombus and wall as perfectly adherent. Finally, while this model is fully three dimensional, in the interest of computational time, all simulations are carried out employing axisymmetric boundary conditions. The final geometry and mesh are shown in Fig. 6A.

Before simulating thrombus lysis and maturation, we apply mechanical as well as chemical initial conditions step-wise. In the first load step we apply venous blood pressure and wait for intra-thrombus fluid pressure to equilibriate. Secondly, we activate platelet contraction, and, again, let the thrombus pressure equilibrate. In the last initiating load step, we perfuse the thrombus homogeneously with an initial plasmin(ogen) concentration.

The actual simulation of thrombus evolution begins with the application of a blood-equivalent plasmin(ogen) concentration at the thrombus/blood interface and a count-down for (myo)fibroblast migration from the thrombus/wall interface, which begins at 7 days into the simulation. We end the simulation after 8 weeks of thrombus evolution. In Fig. 5 we show the resulting transport of plasmin(ogen) and the migration of (myo)fibroblasts as well as the evolution of fibrin referential apparent density and collagen referential apparent density at times 0 days, 1/8 days, 7 days, 14 days, 28 days, and 56 days.

The first row in Fig. 5 depicts the transport of plasmin(ogen) from the thrombus/blood interface. Recalling the initial uniform plasmin(ogen) concentration, subsequent transport of plasmin(ogen) from the thrombus/blood interface inward results in a concentration gradient, which we can still observe at 3 hours. After 7 days, this gradient vanishes as the entire thrombus is saturated with plasmin(ogen).

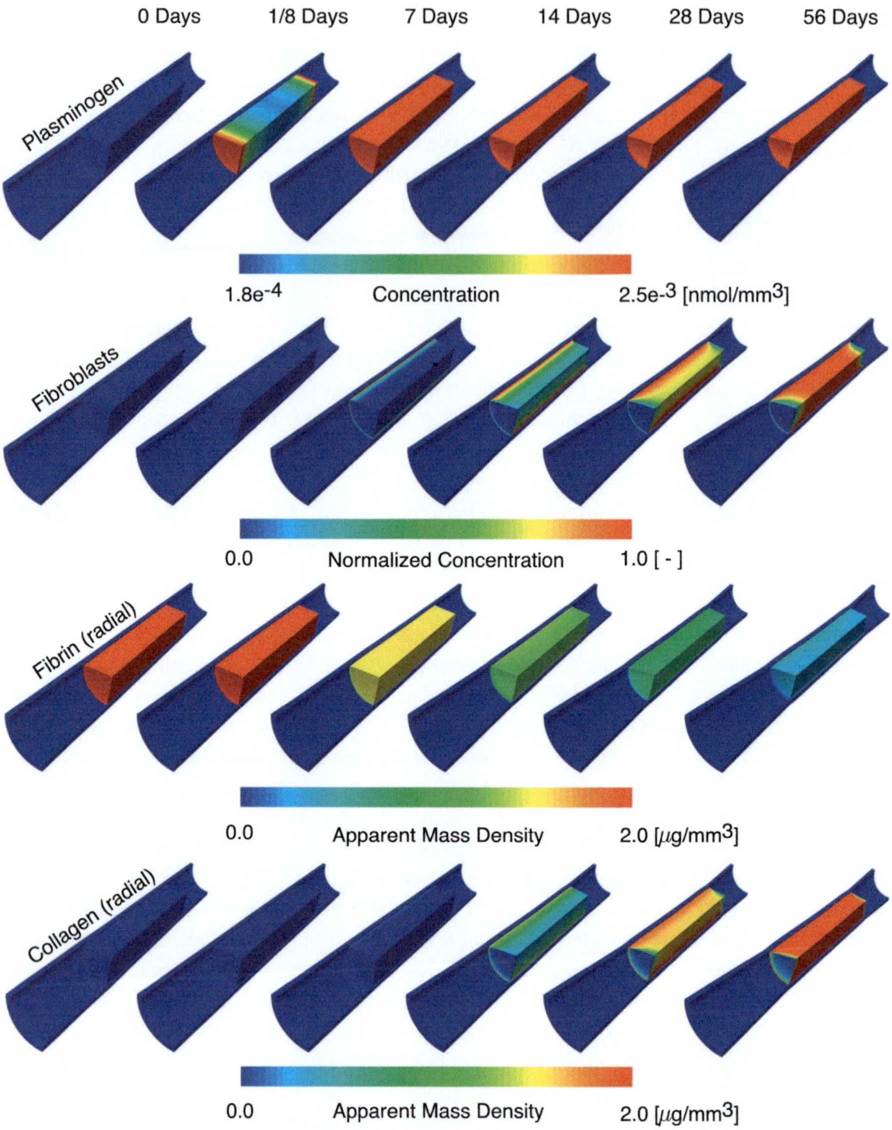

Fig. 5 Simulated occlusive venous thrombus remodeling. Between 0 days and 56 days differing time scales of plasmin(ogen) transport and (myo)fibroblast migration result in nearly homogeneous degradation of fibrin and heterogeneous deposition of collagen (shown are the results in radial direction). The simulated results match qualitatively our prior experimental observations [32]

In contrast, (myo)fibroblasts do not begin to invade the thrombus until day 7. As observed in our experimental data, these cells have made their way into the thrombus volume by day 14, with the highest "concentration" at the thrombus/wall interface and only few (myo)fibroblasts in the thrombus center. This picture changes by day 28 when the cells have substantially invaded the thrombus. By day 56 the entire thrombus is populated with (myo)fibroblasts with the exception of the distal and proximal thrombus surfaces which we exclude from (myo)fibroblast invasion due to its direct contact with blood.

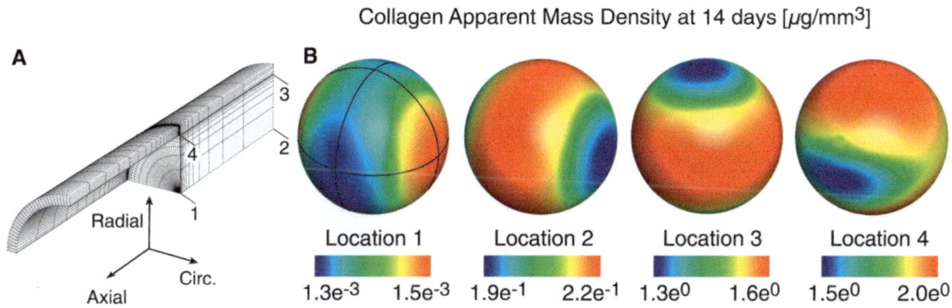

Fig. 6 (**A**) Quarter symmetric segment of thrombus/wall complex and finite element discretization. (**B**) Collagen distribution on microspheres at four different locations after 14 days reflects heterogeneous and anisotropic evolution of thrombus microstructure. Note, continuous distributions are obtained through cubic interpolation of 45 integration points over the half sphere using symmetry conditions. Further note, in (**B**) we chose different color scales to highlight remodeling induced material anisotropy

The different times scales of plasmin(ogen) diffusion and (myo)fibroblast migration are reflected in characteristically different evolution patterns between fibrin and collagen. Because plasmin(ogen) is transported at a high rate into the thrombus center, while the degradation rate of fibrin is relatively low, fibrin changes almost homogeneously throughout the thrombus volume. Fibrin density decreases asymptotically from 2 $\mu g/mm^3$ at 0–7 days to close to 0 $\mu g/mm^3$ at 56 days, see Fig. 5 for fibrin density in the radial direction. On the other hand, because the collagen deposition rate is relatively high in comparison to the (myo)fibroblast migration rate, collagen density mimics (myo)fibroblast density qualitatively. Collagen density further shows a clear gradient from the periphery toward the thrombus center line at 14 days, which vanishes at 28 days and 56 days when collagen has mostly replaced fibrin as the predominant structural constituent of thrombus.

Figure 6B depicts both the heterogeneity as well as the anisotropy of collagen evolution after 2 weeks of maturation. Collagen density varies considerably, as indicated by the different scales at four representative locations. In addition, variations over the unit sphere represent load-induced material anisotropy. At location 3, for example, we find that collagen is deposited predominantly in axial and circumferential directions and not in the radial direction, which reflects the time average of the principal load directions. This locally reduced deposition of collagen in the radial direction may further present a focal material weakness, potentially indicative of a subsequent failure point.

Changes in fibrin and collagen density are also reflected in changes in thrombus geometry. In the current problem, we apply traction boundary conditions to the distal and proximal thrombus surfaces (blood pressure), as well as to the thrombus peripheral boundary (venous wall). Thus, collagen deposition results in thrombus contraction rather than the increased equilibrium stresses that we saw in Fig. 4B. This contraction, consistent with the phenomenon of wound contraction, is depicted in contour plots of total displacement over time in Fig. 7. In this time sequence, we find at 0 days that venous blood pressure applied to the distal and proximal surface of the thrombus and platelet contraction results in compaction at the distal and proximal boundaries but negligible compaction at the radial boundary. Yet, because the thrombus is constrained to move with the venous wall, this displacement does not translate to the thrombus center. We see also slight increases in displacement at day 7 as fibrin degradation reduces thrombus' stiffness. Beginning at day 14 and continuing through day 56, we notice a clear compaction throughout the thrombus as a result of increased collagen deposition.

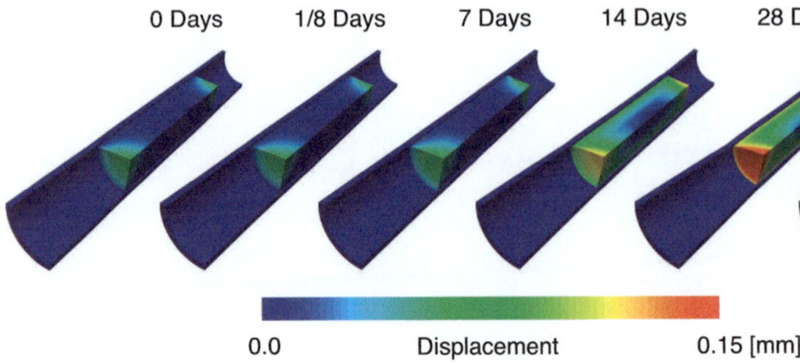

Fig. 7 Occlusive venous thrombus compaction. Between 0 days and 56 days venous blood pressure, platelet contraction, fibrin degradation and collagen deposition result in increased thrombus compaction over time. While blood pressure and platelet contraction dominate early compaction up to 7 days, late compaction is dominated by collagen deposition

5 Discussion

Thrombus plays a critical role in many diseases as well as clinical complications, including deep vein thrombosis [23], stroke [36], aortic dissections [48], abdominal aortic aneurysms [44], coronary disease [33], and medical device failure [45]. The complex multiscale/multiphysical phenomena defining thrombus lysis and maturation largely determine the progression of these diseases as well as their outcomes. Furthermore, thrombus lysis and maturation are targets of current treatments and therapeutic technologies [4]. Computational models may hold the potential to better understand thrombus evolution and thus many of the above diseases. Furthermore, computational models may serve as platforms for personalized treatment planning and therapeutic optimization.

Historically, the study of thrombus has largely focused on the biochemistry of thrombus initiation and formation [17, 51, 52]. The biochemomechanics of thrombus during long-term evolution has received less attention, with few exceptions [29]. Recently, however, it has become clear that biochemomechanics play an important role, not only in thrombus formation [16], but also during thrombus lysis or maturation [12, 13, 38, 50]. We introduced the first finite element framework to combine the nonlinear continuum mechanics of thrombus, intra-thrombus transport, and reactions among solids, interstitial fluid, and diffusing and migrating "solutes". By doing so, we unite biology and chemistry with mechanics on multiple scales. Specifically, we consider spatio-temporal changes in and interactions among biomolecular and cellular species and structural constituents and their consequences on the macroscopic mechanical behavior of thrombus. To the best of our knowledge, this is also the first use of the microsphere approach to model the anisotropically and heterogeneously remodeling microstructure of thrombus. We interpret the discrete quadrature derived directions representing the microspherical surface as fibrin and collagen bundles and allow each to remodel individually. Consequently, the anisotropic degradation of fibrin and deposition of collagen can be resolved, in our case, in 45 discrete directions. Thus, evolving material anisotropy and heterogeneities emerge naturally from our model rather than requiring a priori definition of growth tensors, as, for example, in finite growth theory [21, 40, 41].

The current framework was inspired by our recent report on the evolving histomechanics of venous thrombus in a deep vein thrombosis mouse model [32]. Combined histology/immunohistochemistry and uniaxial tensile test data revealed dramatic changes from 2 weeks to 4 weeks. In short, venous thrombus at 2 weeks is primarily a fibrin-based mesh with relatively low stiffness, with abundant inflammatory cells throughout the thrombus but (myo)fibroblasts near the periphery of thrombus. By 4 weeks, however, (myo)fibroblasts were seen throughout the thrombus, co-localized with fibrillar collagen that likely was responsible for the significantly higher material stiffness at that time. That is, thrombus evolved significantly within a few weeks from a relatively compliant fibrin mesh to a stiff, collagenous pseudo-tissue, largely driven by the degradation of fibrin by plasmin and the invasion of (myo)fibroblasts that deposited collagen. Our numerical example of an occlusive venous thrombus (cf. Sect. 4) was informed by these observations on cell distributions, cell counts, and mechanical properties as a function of relative amounts of the structural constituents fibrin and collagen.

One of the challenges in the study of venous thrombus lysis or maturation lies in delineating the origin of diverging fates; thrombus in vivo may either lyse spontaneously or may mature into a chronic lesion. This outcome is critical from a clinical perspective because a lysing thrombus is more likely to embolize than a chronic thrombus. We hypothesize that thrombus fate may be determined by the competition between fibrin degradation and collagen deposition, with local differences likely critical. In this work we show that this competition may be partially determined through mechanically-mediated feedback mechanisms. For example, collagen deposition and subsequent thrombus contraction may reduce diffusion and advection of important thrombolytic biomolecules such as plasmin(ogen), thus slowing lysis and further promoting maturation.

Thrombus microstructure appears heterogeneous and anisotropic [32, 47]. To date, no model of thrombus remodeling has accounted for local differences in constituent density or spatially varying material properties. We employed the microsphere concept to allow thrombus to remodel freely in response to reactive interactions among solid, fluid, and "solute" constituents as a function of their locally and directionally varying mechanical environment. We believe that material heterogeneity and anisotropy may be critical to the fate of venous thrombus; local differences in collagen and fibrin density, for example, could account for different failure modes such that advanced lysis or reduced maturation at the thrombus wall interface may result in whole thrombus dislodgement, while homogeneous degradation may result in thrombus fragmentation and smaller emboli. In our simulation, an occlusive venous thrombus remodeled in a heterogeneous, anisotropic fashion. For example, collagen deposition at the thrombus wall interface is minimal in the radial direction, which, under the radial forces of thrombus contraction, could increase the risk for thrombus detachment and subsequent embolization. Microstructural studies of occlusive venous thrombus at varying stages of remodeling will be needed to confirm this hypothesis, which likely would need refinement if one accounts for remodeling of the underlying vascular wall as well.

While venous thrombus and arterial thrombus are often treated separately (e.g., white thrombus (arterial) and red thrombus (venous)), it is becoming increasingly clear that they represent two observations from a continuous disease spectrum [28]. Therefore, our model, with appropriate extensions, may also be applicable to arterial thrombus. Our particular interest on the arterial side lies in the study of thrombus in aortic aneurysms and dissections. Here we suspect thrombus may play a dual role. On one hand, thrombus may be protective, shielding the adjacent aortic wall. On the other hand, it may serve as source of biological (proteolytic) activity that could promote the progressive break-down of aortic wall constituents and ultimately contribute to wall rupture. Our computational model of the biochemomechanics of occlusive thrombus (with appropriate extensions) may similarly aid in testing these hypotheses. Lastly, considering the important physiological role of thrombus in hemostasis following vascular injury, it is not surprising that thrombus evolution closely resembles the wound healing response. Our work may, thus, also be relevant to the study of wound healing [46].

Of course our model depends on a number of assumptions, parameters, and constitutive relations that are currently either only partially known or unknown. For example, no data are available on the evolving porosity of venous thrombus, the mechanical dependence of thrombus permeability, or the mechanical dependence of plasmin(ogen) diffusivity. Similarly, we currently do not have a three dimensional model of platelet contraction and do not know the mathematical form of mechanically dependent reaction equations, or the mechanical properties of venous thrombus under compression at various stages of evolution. Other limitations include our choice to neglect interactions among multiple migrating cell species, focus on plasmin(ogen) but not fibrin(ogen), and the potential evolution of the venous wall, all of which may affect thrombus properties. Further, we currently assume perfect adherence between thrombus and venous wall, which is an approximation of the actual interfacial properties at best. Based on the data we have available, however, we consider the present work another step toward a general framework that includes important biological, chemical, and mechanical contributors to thrombus lysis or maturation.

Acknowledgements This work was funded, in part, by NIH grants R01 HL086418, U01 HL116323, and T32 HL007974. We also thank Steve Maas and David Rawlins for their invaluable help with the implementation of our constitutive laws in FEBio and Professor Gerard Ateshian for his input and support.

References

1. Adhikari, A.S., Chai, J., Dunn, A.R.: Mechanical load induces a 100-fold increase in the rate of collagen proteolysis by MMP-1. J. Am. Chem. Soc. **133**(6), 1686–1689 (2011)
2. Adhikari, A.S., Mekhdjian, A.H., Dunn, A.R.: Strain tunes proteolytic degradation and diffusive transport in fibrin networks. Biomacromolecules **13**(2), 499–506 (2012)
3. Adolph, R., Vorp, D.A., Steed, D.L., Webster, M.W., Kameneva, M.V., Watkins, S.C.: Cellular content and permeability of intraluminal thrombus in abdominal aortic aneurysm. J. Vasc. Surg. **25**(5), 916–926 (1997)
4. Alexandrov, A.V., Molina, C.A., Grotta, J.C., Garami, Z., Ford, S.R., Alvarez-Sabin, J., Montaner, J., Saqqur, M., Demchuk, A.M., Moyé, L.A., et al.: Ultrasound-enhanced systemic thrombolysis for acute ischemic stroke. N. Engl. J. Med. **351**(21), 2170–2178 (2004)
5. Ateshian, G.A., Albro, M.B., Maas, S., Weiss, J.A.: Finite element implementation of mechanochemical phenomena in neutral deformable porous media under finite deformation. J. Biomech. Eng. **133**(8), 081005 (2011)
6. Ateshian, G.A., Maas, S., Weiss, J.A.: Multiphasic finite element framework for modeling hydrated mixtures with multiple neutral and charged solutes. J. Biomech. Eng. **135**(11), 111001 (2013)
7. Ateshian, G.A., Nims, R.J., Maas, S., Weiss, J.A.: Computational modeling of chemical reactions and interstitial growth and remodeling involving charged solutes and solid-bound molecules. Biomech. Model. Mechanobiol. **13**(5), 1105–1120 (2014)
8. Baek, S., Rajagopal, K., Humphrey, J.: A theoretical model of enlarging intracranial fusiform aneurysms. J. Biomech. Eng. **128**(1), 142–149 (2006)
9. Barlow, G.H., Summaria, L., Robbins, K.C.: Molecular weight studies on human plasminogen and plasmin at the microgram level. Int. J. Biol. Chem. **244**(5), 1138–1141 (1969)
10. Bersi, M.R., Bellini, C., Wu, J., Montaniel, K.R., Harrison, D.G., Humphrey, J.D.: Excessive adventitial remodeling leads to early aortic maladaptation in angiotensin-induced hypertension. Hypertension **67**(5), 890–896 (2016)
11. Breen, E.C.: Mechanical strain increases type I collagen expression in pulmonary fibroblasts in vitro. J. Appl. Physiol. **88**(1), 203–209 (2000)
12. Brown, A.E., Litvinov, R.I., Discher, D.E., Purohit, P.K., Weisel, J.W.: Multiscale mechanics of fibrin polymer: gel stretching with protein unfolding and loss of water. Science **325**(5941), 741–744 (2009)
13. Bucay, I., O'Brien, E.T. III, Wulfe, S.D., Superfine, R., Wolberg, A.S., Falvo, M.R., Hudson, N.E.: Physical determinants of fibrinolysis in single fibrin fibers. PLoS ONE **10**(2), e0116 (2015)
14. Cohen, I., Gerrard, J.M., White, J.G.: Ultrastructure of clots during isometric contraction. J. Cell Biol. **93**(3), 775–787 (1982)
15. Davie, E.W., Fujikawa, K., Kisiel, W.: The coagulation cascade: initiation, maintenance, and regulation. Biochemistry **30**(43), 10363 (1991)

16. Di Achille, P., Tellides, G., Figueroa, C., Humphrey, J.: A haemodynamic predictor of intraluminal thrombus formation in abdominal aortic aneurysms. Proc. R. Soc. A **470**(2172), 20140 (2014). 163
17. Diamond, S.L., Anand, S.: Inner clot diffusion and permeation during fibrinolysis. Biophys. J. **65**(6), 2622 (1993)
18. Diaz, J.A., Obi, A.T., Myers, D.D., Wrobleski, S.K., Henke, P.K., Mackman, N., Wakefield, T.W.: Critical review of mouse models of venous thrombosis. Arterioscler. Thromb. Vasc. Biol. **32**(3), 556–562 (2012)
19. Doolittle, R.F.: Fibrinogen and fibrin. Encyclopedia of Life Sciences (2001)
20. Emelianov, S., Chen, X., O'Donnell, M., Knipp, B., Myers, D., Wakefield, T., Rubin, J.: Triplex ultrasound: elasticity imaging to age deep venous thrombosis. Ultrasound Med. Biol. **28**(6), 757–767 (2002)
21. Genet, M., Rausch, M., Lee, L., Choy, S., Zhao, X., Kassab, G., Kozerke, S., Guccione, J., Kuhl, E.: Heterogeneous growth-induced prestrain in the heart. J. Biomech. **48**(10), 2080–2089 (2015)
22. Göktepe, S., Miehe, C.: A micro–macro approach to rubber-like materials. Part III: the micro-sphere model of anisotropic mullins-type damage. J. Mech. Phys. Solids **53**(10), 2259–2283 (2005)
23. Goldhaber, S.Z., Bounameaux, H.: Pulmonary embolism and deep vein thrombosis. Lancet **379**(9828), 1835–1846 (2012)
24. Heo, S., Xu, Y.: Constructing fully symmetric cubature formulae for the sphere. Math. Comput. **70**(233), 269–279 (2001)
25. Holzapfel, G.A., Gasser, T.C., Ogden, R.W.: A new constitutive framework for arterial wall mechanics and a comparative study of material models. J. Elast. **61**(1–3), 1–48 (2000)
26. Huang, L., Nagapudi, K., Apkarian, R.P., Chaikof, E.L.: Engineered collagen-PEO nanofibers and fabrics. J. Biomater. Sci., Polym. Ed. **12**(9), 979–993 (2001)
27. Humphrey, J., Rajagopal, K.: A constrained mixture model for growth and remodeling of soft tissues. Math. Models Methods Appl. Sci. **12**(03), 407–430 (2002)
28. Jerjes-Sanchez, C.: Venous and arterial thrombosis: a continuous spectrum of the same disease? Eur. Heart J. **26**(1), 3–4 (2005)
29. Karšaj, I., Humphrey, J.D.: A mathematical model of evolving mechanical properties of intraluminal thrombus. Biorheology **46**(6), 509–527 (2009)
30. Lanir, Y., Namani, R.: Reliability of structure tensors in representing soft tissues structure. J. Mech. Behav. Biomed. Mater. **46**, 222–228 (2015)
31. Lee, Y., Naito, Y., Kurobe, H., Breuer, C., Humphrey, J.: Biaxial mechanical properties of the inferior vena cava in C57BL/6 and CB-17 SCID/bg mice. J. Biomech. **46**(13), 2277–2282 (2013)
32. Lee, Y., Lee, A., Humphrey, J., Rausch, M.: Histological and biomechanical changes in a mouse model of venous thrombus remodeling. Biorheology **52**(3), 235–245 (2015)
33. Libby, P., Theroux, P.: Pathophysiology of coronary artery disease. Circulation **111**(25), 3481–3488 (2005)
34. Lo, K., Denney, W.S., Diamond, S.L.: Stochastic modeling of blood coagulation initiation. Pathophysiol. Haemost. Thromb. **34**(2–3), 80–90 (2005)
35. Maas, S.A., Ellis, B.J., Ateshian, G.A., Weiss, J.A.: FEBio: finite elements for biomechanics. J. Biomech. Eng. **134**(1), 011005 (2012)
36. Mackman, N.: Triggers, targets and treatments for thrombosis. Nature **451**(7181), 914–918 (2008)
37. Matveyev, M.Y., Domogatsky, S.P.: Penetration of macromolecules into contracted blood clot. Biophys. J. **63**(3), 862 (1992)
38. Rausch, M.K., Humphrey, J.D.: A microstructurally inspired damage model for early venous thrombus. J. Mech. Behav. Biomed. Mater. **55**, 12–20 (2015)
39. Rausch, M.K., Kuhl, E.: On the effect of prestrain and residual stress in thin biological membranes. J. Mech. Phys. Solids **61**(9), 1955–1969 (2013)
40. Rausch, M.K., Kuhl, E.: On the mechanics of growing thin biological membranes. J. Mech. Phys. Solids **63**, 128–140 (2014)
41. Rausch, M.K., Zöllner, A.M., Genet, M., Baillargeon, B., Bothe, W., Kuhl, E.: A virtual sizing tool for mitral valve annuloplasty. Int. J. Numer. Methods Biomed. Eng. (2016). doi:10.1002/cnm.2788
42. Robbie, L., Bennett, B., Croll, A., Brown, P., Booth, N.: Proteins of the fibrinolytic system in human thrombi. Thromb. Haemost. **75**(1), 127–133 (1996)
43. Ruberti, J.W., Hallab, N.J.: Strain-controlled enzymatic cleavage of collagen in loaded matrix. Biochem. Biophys. Res. Commun. **336**(2), 483–489 (2005)
44. Sakalihasan, N., Limet, R., Defawe, O.: Abdominal aortic aneurysm. Lancet **365**(9470), 1577–1589 (2005)
45. Starling, R.C., Moazami, N., Silvestry, S.C., Ewald, G., Rogers, J.G., Milano, C.A., Rame, J.E., Acker, M.A., Blackstone, E.H., Ehrlinger, J., et al.: Unexpected abrupt increase in left ventricular assist device thrombosis. N. Engl. J. Med. **370**(1), 33–40 (2014)

46. Tepole, A.B., Kuhl, E.: Computational modeling of chemo-bio-mechanical coupling: a systems-biology approach towards wound healing. Comput. Methods Biomech. Biomed. Eng. **24**, 1–18 (2014)
47. Tong, J., Cohnert, T., Regitnig, P., Holzapfel, G.A.: Effects of age on the elastic properties of the intraluminal thrombus and the thrombus-covered wall in abdominal aortic aneurysms: biaxial extension behaviour and material modelling. Eur. J. Vasc. Endovasc. Surg. **42**(2), 207–219 (2011)
48. Tsai, T.T., Evangelista, A., Nienaber, C.A., Myrmel, T., Meinhardt, G., Cooper, J.V., Smith, D.E., Suzuki, T., Fattori, R., Llovet, A., et al.: Partial thrombosis of the false lumen in patients with acute type b aortic dissection. N. Engl. J. Med. **357**(4), 349–359 (2007)
49. Varjú, I., Sótonyi, P., Machovich, R., Szabó, L., Tenekedjiev, K., Silva, M., Longstaff, C., Kolev, K.: Hindered dissolution of fibrin formed under mechanical stress. J. Thromb. Haemost. **9**(5), 979–986 (2011)
50. Wilson, J.S., Virag, L., Di Achille, P., Karšaj, I., Humphrey, J.D.: Biochemomechanics of intraluminal thrombus in abdominal aortic aneurysms. J. Biomech. Eng. **135**(2), 021011 (2013)
51. Xu, Z., Lioi, J., Mu, J., Kamocka, M.M., Liu, X., Chen, D.Z., Rosen, E.D., Alber, M.: A multiscale model of venous thrombus formation with surface-mediated control of blood coagulation cascade. Biophys. J. **98**(9), 1723–1732 (2010)
52. Xu, Z., Kamocka, M., Alber, M., Rosen, E.D.: Computational approaches to studying thrombus development. Arterioscler. Thromb. Vasc. Biol. **31**(3), 500–505 (2011)

J Elast (2017) 129:145–170
DOI 10.1007/s10659-016-9616-z

A Constitutive Model for Swelling Pressure and Volumetric Behavior of Highly-Hydrated Connective Tissue

Peter M. Pinsky[1] · **Xi Cheng**[1]

Received: 31 May 2016 / Published online: 6 January 2017

Abstract The fluid pressure within water-filled connective tissues such as cartilage, intervertebral disk and cornea facilitates a vital part of their functionality. Cartilage and intervertebral disk must resist compressive loading, and even the cornea uses fluid pressure loading to form its precise refractive geometry. The fluid pressure is composed of hydrostatic and swelling pressure components, with the latter deriving primarily from osmotic forces associated with ion concentrations. The tissues, like electrolyte gels in general, have a strong tendency to absorb water and swell. The goal of this work is to formulate an *in vivo* tissue theory from first principles and to arrive at the simplest possible model which captures the essential features of tissue charge effects and swelling behavior. The Gibbs free energy of the tissue, including elastic, hydrostatic, and electrostatic components, is characterized and equilibrium thermodynamics is employed to find explicit constitutive equations for the tissue osmotic pressure and osmotic compressibility over a unit cell. It is shown that the osmotic compressibility essentially defines the tissue macroscopic pressure-volume relationship at equilibrium. To illustrate the theory in detail, the human cornea is taken as a model tissue, and it is shown how the nanometer features of the glycosaminoglycan charge distribution can be modeled within the proposed theory. The cornea model is further extended to include the effects of the pump-leak hydration control mechanism based on active ion and passive water transport across the corneal endothelium. The model has been implemented in a standard finite element code and is shown to be capable of reproducing fundamental *in vitro* swelling experiments, including massive swelling, and typical *in vivo* swelling observed in disease states such as Fuch's dystophy.

Keywords Connective tissue · Cornea · Electrolyte gel · Osmotic pressure · Swelling · Pump-leak mechanism

✉ P.M. Pinsky
pinsky@stanford.edu

X. Cheng
xichengsu@alumni.stanford.edu

[1] Department of Mechanical Engineering, Stanford University, 230 Building 520, 452 Escondido Mall, Stanford, CA 94305, USA

Mathematics Subject Classification 74B20 · 74F20 · 74G10 · 74L15 · 74Q05 · 74Q15 · 74S05 · 92C05 · 92C10 · 92C50

1 Introduction

In this work we consider the modeling of those connective tissues whose water content is dominant. Such water-filled (i.e. highly-hydrated) tissues, including cartilage, intervertebral disk and cornea, have water content in the range of 70–90 % by weight [1–3]. The major structural constituent of these tissues is collagen but they also contain charged proteoglycans and mobile ions which indirectly to contribute to the tissue structural behavior, and particularly the volumetric behavior, through their role in generating tissue swelling pressure [2, 4, 5]. It is widely appreciated that the mechanical behavior of these tissues depends on their multiphasic constituents and a full treatment of the tissue constitutive behavior requires a multiphasic approach such as poroelasticity, or more generally, some type of mixture theory [6–8]. These theories are necessarily complex because they describe the interactions of the constituent phases as well as the conservation principles that the phases must satisfy and invariably they require the coupled solution of multiple fields, for example, the deformation, fluid pressure, ionic concentrations, and electrostatic potential.

It is perhaps not surprising that the modeling of such tissues is often based more on expediency than rigor in an effort to minimize complexity. One such approach is to regard the tissue as an incompressible or nearly-incompressible hyperelastic material [9, 10]. The elasticity of the tissue is derived from the collagen network and the presence of the interstitial water is assumed to justify the assumption of incompressibility. This approximation might be justified for the case of determining the instantaneous response of the tissue to a load under the assumption that the interstitial water does not have time to flow. However, this is of dubious value in the context of most clinical problems where interest is focussed on the time-course of response to load or on the long-term, steady-state solution [11]. However, far more important than this technical issue, is the fact that hydrated tissues swell and the incompressible elasticity model provides no information about this crucial feature of the tissue.

In highly hydrated tissues the swelling pressure is associated with the tissue osmotic pressure which results from the interaction of mobile ions with the highly negatively charged glycosaminoglycans (GAGs) found distributed thoughout the tissue in essentially fixed positions. The greater the concentration of GAG charge, the greater the concentration of ions and resulting osmotic pressure. In cartilage, the osmotic pressure is measured to be on the order of 3000 mmHg [12], whereas in corneal stroma it is approximately 60 mmHg [13–15]. A modeling capability for predicting tissue swelling pressure and the biomechanical response to that pressure[1] is of considerable clinical significance for many tissues [11, 14, 16, 17].

The cornea presents a special and interesting case because the tissue depends on active transport processes to maintain its hydration level and prevent swelling.[2] Bare stroma will

[1]In tissues, the change of hydration (fluid buildup), described as swelling in this work, can also have a component due to inflammation associated with an immune response, with the overall behavior being called edema. The immune effect is not considered in the current study.

[2]Many tissues do employ epithelial and/or endothelial layers as barriers to separate phases and to actively mediate the exchange of solutes and solvent between those phases. Examples include the lining of blood vessels, the kidney, organs of the gastrointestinal tract, the choroid plexus in the brain, and the cornea.

swell to many times its normal thickness when placed in a water bath [3, 18, 19]. This cannot be tolerated in the living cornea since swelling will affect the regular spacing between collagen fibrils in the stromal lamellae producing increased light scatter and loss of transparency [5, 20]. Swelling also modifies the curvatures of the cornea's bounding surfaces and defeats its focusing function. The corneal endothelium, a cellular layer which separates the stroma and aqueous humor, is responsible for maintaining the hydration of the cornea through a "pump-leak" mechanism [21–23]. The metabolically supported "pump" function is provided by active ion transporters located in the endothelium layer [21]. The "leak" function refers to the passive transport of water across the endothelium through water channels [24] in response to the stromal osmotic pressure. Normal (physiological) hydration is maintained by balancing the endothelial pump rate with the osmotic pressure-driven leak [11, 20, 23]. When ion pumping is compromised, as in Fuch's dystrophy, the leak of water into the stroma will be continuous and the resulting corneal swelling can lead to loss of transparency, impaired vision or corneal blindness.

A full description of the mechanics of the tissue must describe how the collagen network resists (or does not resist) the osmotic and fluid pressures, depending on its organization. For example, in the human cornea, the collagen fibers are aligned with the corneal surfaces and cannot resist swelling in the thickness direction except near the anterior region where they become inclined, possibly to stabilize the optical surface [11, 25, 26]. As a result, the human cornea swells through its thickness and mainly as a posterior bulge [27]. Cartilage exhibits a somewhat analogous collagen arrangement, with collagen unable to resist swelling/deswelling but also exhibiting inclination near the articular surface [28]. In this paper we focus on a description of the swelling driving force and do not consider interaction with collagen.

The goal of this work is to formulate an *in vivo* tissue theory from first principles, including active hydration control mechanisms, and to arrive at the simplest possible model which captures the essential features of charge effects and swelling missed by purely hyperelastic approaches [9, 10, 26, 29]. The work is developed under a thermodynamic framework with two basic assumptions. The first is that the model is restricted to tissue which is highly hydrated. In this case the swelling pressure corresponds to the tissue osmotic pressure—the entropy of mixing described by, for example, Flory-Huggins theory [30] is assumed to be secondary and ignored. The second major assumption is that the tissue system is in steady-state (although not necessarily thermodynamic equilibrium). This limits the model to long-time solutions and does not allow the time-course of the solution to be obtained. However, this assumption provides significant simplifications while retaining the ability to model problems of clinical relevance. Furthermore, in order to make the presentation more concrete, the cornea is adopted as the model tissue; we have more experience with this tissue and a good set of fundamental physical constants.

The Gibbs free energy of the system, including the osmotic and electrostatic contributions, is first characterized and it is shown that stationarity (i.e., thermodynamic equilibrium) yields governing equations that are equivalent to those found using a classical mixture theory. However, the energy approach, unlike the mixture theory approach, provides a very convenient basis for homogenization of the osmotic pressure, taking into account the nanometer-scale details of charge distributions (which vary between tissues), over a unit cell. In addition, the use of Donnan-type analytical approximations for the local (unit cell) electrostatic potential finally provides an explicit constitutive equation for the osmotic pressure and osmotic compressibility (and which are the key results in this simplified theory). Finally, the active ion transport in the corneal endothelium is treated by modeling the endothelium as a simple bounding layer with ions that are transported both passively, with a driving force from the gradient of the electrochemical potential, and actively. The model quantitatively describes how active ion transport across the endothelium reduces the osmotic pressure in the tissue.

The resulting model is rather simple to implement, hardly more complex than a classical compressible hyperelastic model. Constitutive equations for the osmotic pressure and osmotic compressibility are easily implemented; the finite element tangent operator is given in full detail. The accuracy of the model is assessed by comparison to direct experimental measurement of *in vitro* swelling pressure and is used to model swelling in the *in vivo* cornea under normal (healthy) and diseased conditions corresponding to Fuch's dystrophy.

2 Free Energy of a Hydrated Polyelectrolyte Gel

2.1 Theory

For simplicity, but without loss of generality, we consider a generic binary electrolyte gel, consisting of a mixture of solid, fluid, and ionic phases, immersed in an ionic bath at constant electrostatic potential ϕ_0, hydrostatic pressure P_0, and ionic concentration $C_{0+} = C_{0-} = C_0$. Also for simplicity, we shall henceforth take $\phi_0 = 0$. The tissue has a distribution of fixed charges that arise from the tissue proteoglycans and which determine the spatial distribution of the mobile ions and thus (as will be shown) the osmotic pressure. The tissue reference and current configurations are denoted Ω_0 and Ω, respectively, and these configurations occupy volumes V_0 and V, respectively. For $\mathbf{X} \in \Omega_0$, the motion is denoted $\varphi(\mathbf{X})$, the deformation gradient is $\mathbf{F} = \partial\varphi(\mathbf{X})/\partial\mathbf{X}$, and the displacement field is $\mathbf{u}(\mathbf{X}) = \varphi(\mathbf{X}) - \mathbf{X}$. The current configuration is assumed to have a distribution of fixed charge which originates from the ionized glycosaminoglycan (GAG) chains associated with the various proteoglycans (PGs) in the tissue. The charged GAG chains are constrained to follow the tissue motion[3] and do not have any additional freedom of motion (in contrast to the mobile ions which can move freely in the intrafibrillar fluid). At the nanoscale, the GAG charges are highly heterogeneous and are assumed to be modeled by the concentration function $C_{\mathrm{f}}(\mathbf{X}, J)$, where $J = \det\mathbf{F}$ and the units of C_{f} are moles per liter of the current configuration.

In general, the reference configuration Ω_0 will not be stress-free for the solid (elastic) phase and it is therefore convenient to assume the existence of a stress-free reference configuration Ω_0' for this phase with an associated deformation gradient $\mathbf{F}' = \partial\varphi'(\mathbf{X})/\partial\mathbf{X}$. In this case, the right Cauchy-Green deformation tensors $\mathbf{C} = \mathbf{F}^T\mathbf{F}$ and $\mathbf{C}' = \mathbf{F}'^{-T}\mathbf{C}\mathbf{F}'^{-1}$ measure the deformation in the current configuration Ω relative to the reference configuration Ω_0 and stress-free configuration Ω_0', respectively.

It is assumed [31] that the Gibbs free energy of the electrolyte gel can be additively decomposed into constituent components as follows,

$$\mathcal{G}(\phi, \mathbf{C}) = \mathcal{W}_{\mathrm{electrolyte}}(\phi, \mathbf{C}) + \mathcal{W}_{\mathrm{bath}}(\mathbf{C}) + \mathcal{W}_{\mathrm{elastic}}(\mathbf{C}), \tag{1}$$

where the three additive terms on the right hand side of Eq. (1) denote the free energy associated with the electrolyte, bath solution and the elastic solid of the gel, respectively, $\phi : \Omega_0 \to \mathbb{R}$ is the electrostatic potential that arises from the fixed and mobile ionic charges. A suitable form of the Helmholtz free energy density for a binary electrolyte can be derived

[3] In the cornea the PG core protein is attached to the collagen, in cartilage the aggrecan is interwoven around the collagen fibrils.

from the mean-field approximation [32] and results in an electrolyte free energy density $\mathfrak{w}_{\text{electrolyte}}$ (per unit volume of the current configuration) in the form

$$\mathfrak{w}_{\text{electrolyte}}(\mathbf{X},\phi,J)=z_{\text{f}}FC_{\text{f}}(\mathbf{X},J)\phi(\mathbf{X})-2RTC_0\left(\cosh\left(\frac{F}{RT}\phi(\mathbf{X})\right)-1\right)-\frac{\epsilon_w}{2}\left|\nabla\phi(\mathbf{X})\right|^2, \tag{2}$$

where z_{f} is the fixed charge valence value, and R, T, F and ϵ_w are the gas constant, temperature, Faraday constant, and dielectric permittivity of the electrolyte solvent, respectively. The first term on the right hand side measures the free energy of the fixed charge, the second term measures the excess mean concentration of the mobile ions at any point in the electrolyte compared to the bath and, after multiplication by RT, may be interpreted as the osmotic work of introducing excess ions into the neighborhood of the fixed charges C_{f}, and the last term $\frac{\epsilon_w}{2}|\nabla\varphi|^2=\mathbf{E}\cdot\mathbf{D}/2$, where $\mathbf{E}=\nabla\varphi$ is the electric field and $\mathbf{D}=\epsilon_w\mathbf{E}$ is the electric displacement, measures the dielectric free energy. For concreteness, we will take the valence of the fixed charge as $z_{\text{f}}=-1$ (the usual case in connective tissue).

Electrostatic equilibrium requires that $\delta_\phi\mathcal{G}(\phi,\mathbf{C})=\frac{d}{d\varepsilon}\mathcal{G}(\phi_\varepsilon,\mathbf{C})|_{\varepsilon=0}=0$, with $\phi_\varepsilon=\phi+\varepsilon\delta\phi$ and $\varepsilon\in\mathbb{R}$, to hold for all admissible variations $\delta\phi$. In this case,

$$\begin{aligned}\delta_\phi\mathcal{W}_{\text{electrolyte}}&=\delta_\phi\int_\Omega\mathfrak{w}_{\text{electrolyte}}(\phi,J)d\Omega\\&=\int_\Omega\left[-FC_{\text{f}}(\mathbf{X},J)-2FC_0\sinh\left(\frac{F}{RT}\phi(\mathbf{X})\right)+\epsilon_w\nabla^2\phi(\mathbf{X})\right]\delta\phi(\mathbf{X})d\Omega,\end{aligned} \tag{3}$$

and the Euler-Lagrange equation over Ω is

$$\nabla^2\phi(\mathbf{X})=\frac{F}{\epsilon_w}\left(C_{\text{f}}(\mathbf{X},J)+2C_0\sinh\left(\frac{F}{RT}\phi(\mathbf{X})\right)\right). \tag{4}$$

This result may also be displayed in the form of the Poisson-Boltzmann equation

$$-\nabla^2\phi(\mathbf{X})=\frac{F}{\epsilon_w}\left(-C_{\text{f}}(\mathbf{X},J)+\sum_{i=+,-}z_iC_i\big(\phi(\mathbf{X})\big)\right), \tag{5}$$

wherein the mobile ion concentrations C_i (moles per liter) satisfy the Boltzmann distribution given by

$$C_i\big(\phi(\mathbf{X})\big)=C_0\exp\left(-z_i\frac{F}{RT}\phi(\mathbf{X})\right),\quad i=+,- \tag{6}$$

and in which the binary ion valence numbers are $z_+=1$ and $z_-=-1$.

The condition for mechanical equilibrium requires $\delta_{\mathbf{u}}\mathcal{G}(\phi,\mathbf{C}(\mathbf{u}))=\frac{d}{d\varepsilon}\mathcal{G}(\phi,\mathbf{C}(\mathbf{u}_\varepsilon))|_{\varepsilon=0}=0$, with $\mathbf{u}_\varepsilon=\mathbf{u}+\varepsilon\delta\mathbf{u}$, to hold for all admissible variations $\delta\mathbf{u}$. We introduce the pull-back of the electrolyte free energy density $\mathfrak{w}_{\text{electrolyte}}$ to the reference configuration such that $\mathfrak{W}_{\text{electrolyte}}=\varphi^*(\mathfrak{w}_{\text{electrolyte}})=J\mathfrak{w}_{\text{electrolyte}}$ and define the bath free energy density, which measures the potential of the bath hydrostatic pressure P_0 to do work associated with configuration volume change, by

$$\mathfrak{W}_{\text{bath}}(J)=-P_0(J-1). \tag{7}$$

We further define the tissue osmotic pressure to be

$$P_{\text{os}}=-\frac{\partial\mathfrak{W}_{\text{electrolyte}}}{\partial J}, \tag{8}$$

the bath hydrostatic pressure to be

$$P_0 = -\frac{\partial \mathfrak{W}_{\text{bath}}}{\partial J}, \tag{9}$$

and the effective elastic stress to be

$$\mathbf{S}_{\text{elastic}} = 2\frac{\partial \mathfrak{W}_{\text{elastic}}}{\partial \mathbf{C}},$$

where $\mathfrak{W}_{\text{elastic}}(\mathbf{C})$ is the elastic strain energy density. It then follows that

$$\begin{aligned}
\delta_{\mathbf{u}}\mathcal{G}(\phi, \mathbf{C}) &= \delta_{\mathbf{u}} \int_{\Omega_0} \big(\mathfrak{W}_{\text{electrolyte}}(\phi, J) + \mathfrak{W}_{\text{bath}}(J) + \mathfrak{W}_{\text{elastic}}(\mathbf{C})\big) d\Omega_0 \\
&= \int_{\Omega_0} \left(J\left(\frac{\partial \mathfrak{W}_{\text{electrolyte}}}{\partial J} + \frac{\partial \mathfrak{W}_{\text{bath}}}{\partial J} \right) \mathbf{C}^{-1} + \mathbf{S}_{\text{elastic}} \right) : \delta_{\mathbf{u}} \mathbf{E} d\Omega_0 \\
&= \int_{\Omega} \big(-(P_{\text{os}} + P_0)\nabla \cdot \delta\mathbf{u} + \boldsymbol{\sigma}_{\text{elastic}} : \nabla^s \delta\mathbf{u} \big) d\Omega,
\end{aligned} \tag{10}$$

where $\mathbf{E} = \frac{1}{2}(\mathbf{C} - \mathbf{I})$. The effective Cauchy stress $\boldsymbol{\sigma}_{\text{elastic}}$ is obtained from the push-forward of $\mathbf{S}_{\text{elastic}}$ in the standard manner such that $\boldsymbol{\sigma}_{\text{elastic}} = \varphi_*(\mathbf{S}_{\text{elastic}}) = J^{-1}\mathbf{F}\mathbf{S}_{\text{elastic}}\mathbf{F}^T$. By virtue of the symmetry of $\boldsymbol{\sigma}_{\text{elastic}}$, the Euler-Lagrange equation on the current configuration is given by

$$\nabla \cdot \boldsymbol{\sigma}_{\text{elastic}} - \nabla(P_{\text{os}} + P_0) = 0 \quad \text{on } \Omega. \tag{11}$$

It is seen that the tissue fluid pressure may be interpreted, in this equilibrium setting, as the sum of the bath hydrostatic and osmotic pressures.

Finally, for future reference, the osmotic compressibility is defined to be

$$K_{\text{os}}(J) = \frac{\partial^2 \mathfrak{W}_{\text{electrolyte}}}{\partial J^2}. \tag{12}$$

This is a useful formulation because it provides an explicit constitutive equation for the osmotic pressure (8). The variational formulation is also advantageous for finite element implementation. However, it contains two challenges to numerical implementation. The first is the need to solve for both the electrostatic potential and motion, corresponding to the coupled variational equations (5) and (11). This is not convenient and a single-field approach that might make the solution more standard would be desirable. A second challenge concerns the fixed charge C_{f} and electrostatic potential ϕ, both of which will be rapidly varying at the nanoscale. Resolving this spatial variation in the solution is important but direct approaches based on meshing will certainly be infeasible.

2.2 Equivalence with Mixture Theory

Before proceeding, the relationship between the proposed free energy approach and mixture theory is briefly examined in order to demonstrate their broad equivalence. Mixture theory proceeds by providing a suitable description of interactions between the fluid, solid and ionic phases. Mow et al. [6] and Lai et al. [7] proposed a triphasic framework based on a continuum representation of interpenetrating phases and their interactions. The solid, fluid and ions are treated as distinct phases with individual velocity fields. The theory has been generalized to finite deformations and multiple electrolyte systems [8, 33].

Consider a binary electrolyte gel, consisting of a mixture of solid, fluid, and ionic phases, immersed in an ionic bath at constant electrostatic potential ϕ_0, at hydrostatic pressure P_0, and with ionic concentrations $C_{0+} = C_{0-} = C_0$. The triphasic equations for the polyelectrolyte gel, specialized for conditions of thermodynamic equilibrium [7, 14, 33, 34], are

$$\nabla \cdot \boldsymbol{\sigma} - \nabla P = \mathbf{0}, \tag{13}$$

$$\nabla \cdot \mathbf{J}_w = 0, \tag{14}$$

$$\nabla \cdot \mathbf{J}_i = 0, \quad i = +, - \tag{15}$$

$$\nabla \cdot (\epsilon_w \mathbf{E}) = z_f C_f + \sum_{i=+,-} z_i C_i, \tag{16}$$

where $\boldsymbol{\sigma}$ is the effective stress generated by the solid phase and P is the hydrostatic pressure, $\mathbf{J}_w$ and $\mathbf{J}_i$ are the fluid flow and ionic fluxes, respectively, $\mathbf{E}$ is the electric field, ϵ_w is the dielectric permittivity of the electrolyte solvent, C_i and C_f are the concentration of the mobile ions and fixed charges, respectively, z_i and z_f denote the valence number associated with each ionic and charge species. At equilibrium, the fluid flow and ionic fluxes vanish (i.e. $\mathbf{J}_w = \mathbf{J}_i = \mathbf{0}$) in the gel, which implies the gradients of the electrochemical potentials vanish [7, 34] and allows (14) and (15) to be replaced with

$$\nabla \mu_w = \mathbf{0}, \tag{17}$$

$$\nabla \mu_i = \mathbf{0}, \quad i = +, - \tag{18}$$

where the water and ionic electrochemical potentials are given by

$$\mu_w = \mu_w^{\text{ref}} + \big[P - RT(C_+ + C_-)\big]/\rho_w, \tag{19}$$

$$\mu_i = \mu_i^{\text{ref}} + \frac{RT}{M_i} \ln C_i + z_i \frac{F\phi}{M_i}, \quad i = +, - \tag{20}$$

and where ρ_w is the water density, M_i, $i = +, -$, are the atomic weights for anions and cations, respectively, R, T and F are the gas constant, temperature and Faraday constant, respectively, μ_w^{ref} and μ_i^{ref} are the water potential and ionic chemical potentials at the reference state, ϕ is the electrostatic potential. Integrating (17) and (18) between any points in the bath and electrolyte gives the electrolyte fluid pressure and Boltzmann ion distribution

$$P = P_0 + \Pi, \tag{21}$$

$$C_i = C_0 \exp\left(-\frac{z_i F}{RT}(\phi - \phi_0)\right), \tag{22}$$

where

$$\Pi = RT(C_+ + C_- - 2C_0) \tag{23}$$

is the osmotic pressure difference between the electrolyte and bath. Employing (21) in (13) and (22) in (16) allows the governing coupled equations to be expressed as

$$\nabla \cdot \boldsymbol{\sigma} - \nabla\big(\Pi(\phi) + P_0\big) = \mathbf{0}, \tag{24}$$

and

$$-\nabla^2\phi = \frac{F}{\epsilon_w}\left(z_f C_f - 2C_0 \sinh\left(\frac{F\phi}{RT}\right)\right). \tag{25}$$

Equations (24) and (25) may be compared to (11) and (5), respectively. It is simple to show that when the fixed charge concentration is uniform, they are fully equivalent.

The coupled solution of the mixture equations provides the pointwise osmotic pressure Π, which may vary rapidly at the nanoscale. The energy formalism provides a convenient way to homogenize the osmotic pressure leading to an efficient method for non-ideal electrolytes.

3 Unit Cell Approach

3.1 Volume-Averaged Free Energy

We here propose an energy-based homogenization approach based on the use of a unit cell. The spatially varying electrolyte free energy density over the unit cell is replaced by a uniform energy-equivalent free energy density obtained by averaging the Helmholtz free energy density $\mathfrak{w}_{\text{electrolyte}}$ over the volume of a suitably selected unit cell. Further, the electrostatic potential ϕ is analytically approximated over the unit cell by use of Donnan potentials. These approximate potentials identically satisfy the Poisson-Boltzmann equation and finally reduce the problem to solving for the motion within a standard variational framework. The energy homogenization results in a simplified constitutive equation for the tissue osmotic pressure and can also provide a good measure of the tissue compressibility.

A unit cell is introduced containing a representation of the tissue ultrastructure, subject to periodic boundary conditions, and having reference configuration $\omega_0 \subset \Omega_0$ and volume v_0. The deformation gradient, considered uniform over ω_0 by virtue of its small size, maps the unit cell into the current configuration $\omega \subset \Omega$ occupying volume $v = Jv_0$. The fixed charge concentration distribution in the unit cell is $C_f(\mathbf{X}, J)$. Replacing the electrostatic potential ϕ by the Donnan approximation $\widetilde{\phi}$, the unit cell averaged Helmholtz electrolyte free energy density (per unit reference volume) is then given by

$$\overline{\mathfrak{W}}_{\text{electrolyte}}(J) = \frac{1}{v_0}\int_\omega \mathfrak{w}_{\text{electrolyte}}(\widetilde{\phi}, J)\,d\omega, \tag{26}$$

and unit cell osmotic pressure is provided by

$$\overline{P}_{\text{os}} = -\frac{\partial \overline{\mathfrak{W}}_{\text{electrolyte}}}{\partial J}, \tag{27}$$

and the unit cell osmotic compressibility is

$$\overline{K}_{\text{os}} = \frac{\partial^2 \overline{\mathfrak{W}}_{\text{electrolyte}}}{\partial J^2}. \tag{28}$$

Note that because the Donnan potential $\widetilde{\phi}$ is a known explicit function over ω, [20] it does not appear as a dependent variable in $\overline{\mathfrak{W}}_{\text{electrolyte}}$. In general, some effort will be needed to find the analytical form of $\widetilde{\phi}$ which will reflect the tissue ultrastructure over the unit cell. An illustration will be given in Sect. 4.1.

3.2 Special Case of Uniform Fixed Charge

To provide a simple example of a Donnan potential and to illustrate the application of the unit cell free energy density $\overline{\mathfrak{W}}_{\text{electrolyte}}$, consider the special case in which the fixed charge concentration in the unit cell reference configuration is uniform with value C_{f0}. The fixed charge density in the deformed unit cell configuration is then simply obtained from the dilation J as

$$C_f(J) = J^{-1} C_{f0}. \tag{29}$$

For this problem the Donnan electrostatic potential, denoted $\widetilde{\phi}$, is uniform over the unit cell. Substituting (29) into the Poisson-Boltzmann equation (25) allows it to be solved analytically as

$$\widetilde{\phi}(J) = -\frac{RT}{F}\sinh^{-1}\left(\frac{C_f(J)}{2C_0}\right) = -\frac{RT}{F}\sinh^{-1}\left(\frac{C_{f0}}{2JC_0}\right). \tag{30}$$

Substituting (29) and (30) into (2) and using the identity $\cosh(-\sinh^{-1} x) = \sqrt{x^2+1}$ results in

$$\mathfrak{w}_{\text{electrolyte}}(J) = \frac{RTC_{f0}}{J}\sinh^{-1}\left(\frac{C_{f0}}{2JC_0}\right) - 2RTC_0\left(\sqrt{\left(\frac{C_{f0}}{2JC_0}\right)^2 + 1} - 1\right), \tag{31}$$

which is uniform over the current unit cell volume $\upsilon = J\upsilon_0$. Employing (31) in (26) results in

$$\overline{\mathfrak{W}}_{\text{electrolyte}}(J) = \left(\frac{1}{\upsilon_0}\right)(\mathfrak{w}_{\text{electrolyte}})(\upsilon) = J\mathfrak{w}_{\text{electrolyte}}. \tag{32}$$

Employing (32) in (27) yields the unit cell osmotic pressure

$$\overline{P}_{os}(J) = 2RTC_0\left(\sqrt{\left(\frac{C_{f0}}{2JC_0}\right)^2 + 1} - 1\right), \tag{33}$$

and employing (32) in (28) yields the unit cell osmotic compressibility

$$\overline{K}_{os}(J) = \frac{RTC_{f0}}{J^2\sqrt{1 + (\frac{2JC_0}{C_{f0}})^2}}. \tag{34}$$

Equation (33) recovers the classical Donnan osmotic pressure for a binary electrolyte with uniform fixed charge [35] and the osmotic compressibility is convex for all $J \in \mathbb{R}^+$. The osmotic compressibility is central to the volumetric elasticity of the electrolyte, as will be described in Sect. 4.1.

This simple model does manifest swelling behavior. For example, if the electrolyte fixed charge C_{f0} is increased or the bath ionic concentration C_0 is decreased, the osmotic pressure increases by virtue of (33), which in turn increases the electrolyte fluid pressure and the swelling tendency of the electrolyte. In fact, if a sample of such an electrolyte were mechanically unconstrained so that the swelling is not resisted by any external forces, swelling would proceed until the osmotic pressure reduced to zero at $J = \infty$. It is remarked that in gels, the internal elastic fiber system would typically be engaged and resist the swelling pressure [11, 30].

The unit cell method requires two important inputs: modeling the charge concentration function $C_f(\mathbf{X}, J)$ and finding the Donnan potential $\widetilde{\phi}$ over the unit cell. These results will

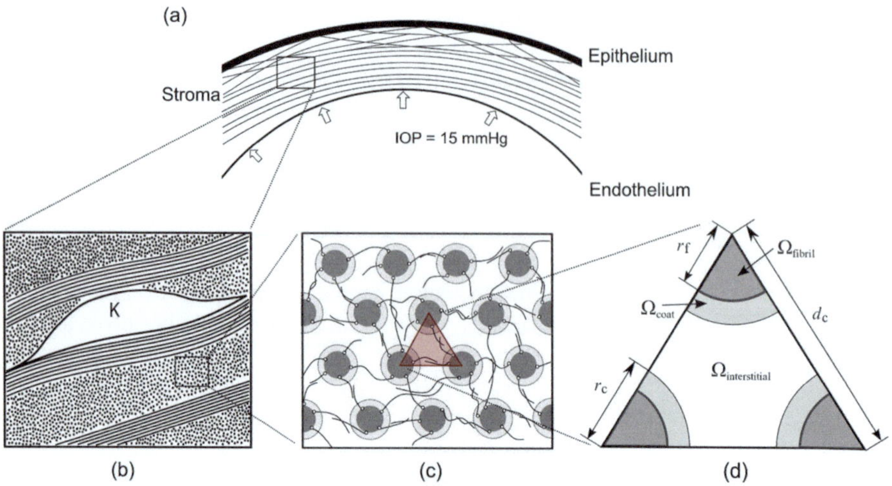

Fig. 1 (**a**) The cornea is a fiber-reinforced electrolyte gel shell that resists the IOP. The collagen lamellae follow the corneal curvature except in the anterior region where the lamellae are interweaving and inclined relative to the corneal surface, and insert into Bowman's layer under the epithelium. (**b**) An illustration of the organization of the corneal stroma showing several lamellae and a keratocyte cell. Collagen fibrils within lamellae are found in parallel arrays following the fibril direction; keratocyte cells are found interspersed between adjacent lamellae. (**c**) Cross-section of an idealized unit cell representing the hexagonal collagen fibril lattice. The coating region around the central cylinder (representing the collagen fibril) and the irregular lines in between are illustrative of the coating GAGs and interfibrillar GAGs, respectively. (**d**) The unit cell domain is prismatic (the end face is shown) and is partitioned into fibril, coating and interstitial regions, denoted Ω_{fibril}, Ω_{coat} and $\Omega_{\text{interstitial}}$, respectively. The interfibrillar distance and fibril radius are $d_c = 53$ nm and $r_f = 15$ nm, respectively, and the PG-coating radius is $r_c = 18$ nm [20]

characterize the electrostatic aspects of the electrolyte gel, as well as the accuracy of the formulation. This is illustrated for a more complex situation in the next section. The unit cell approach, in which the unknown electrostatic potential ϕ is replaced by analytical approximations $\widetilde{\phi}$, allows the Gibbs free energy to be written in the form

$$\mathcal{G}(J, \mathbf{C}) = \overline{\mathcal{W}}_{\text{electrolyte}}(J) + \mathcal{W}_{\text{bath}}(J) + \mathcal{W}_{\text{elastic}}(\mathbf{C}), \tag{35}$$

where $\overline{\mathcal{W}}_{\text{electrolyte}} = \int_{\Omega_0} \overline{\mathfrak{W}}_{\text{electrolyte}}(J) d\Omega_0$. As will be shown, this can be used directly for approximation using the standard finite element method.

4 Application to a Complex Tissue

4.1 Model for the Human Cornea

The corneal stroma is composed of parallel arrays of collagen fibrils aligned with the lamella direction and arranged with psuedo-hexagonal packing [5]. There exists a layer of PG-dense coatings around the fibril surface, which gives rise to distinct charge distributions in nanoscale and can generate significant electrostatic forces to prevent fibril clustering [20]. Keratocyte cells reside between adjacent lamellae and can change volume. The simplest geometric representation of a unit cell within a stromal lamella is that of an equilateral triangular prism with vertices at the centre of three fibrils (Fig. 1c). The unit cell is therefore divided into three regions: fibril, fibril coating and interstitial regions (Fig. 1d). Assuming that collagen fibrils are non-swelling and that keratocyte cells dilate with the tissue dilation [20, 36], the normalized (non-dimensional) charge densities in the fibril coating and interstitial regions are given as:

$$C_{f1}(J) = \left(\frac{\lambda}{J\phi'_k - \phi_f} + \frac{1-\lambda}{\phi_c}\right)\frac{C_{f0}}{2C_0}, \tag{36}$$

$$C_{f2}(J) = \left(\frac{\lambda}{J\phi'_k - \phi_f}\right)\frac{C_{f0}}{2C_0}, \tag{37}$$

where ϕ_k, ϕ_f and ϕ_c denote the volume fractions for keratocytes, collagen fibrils and coatings, respectively, and $\phi'_k = 1 - \phi_k$. C_{f0} is the average charge concentration of the stroma, and λ is the charge partition parameter [20]. The subscripts 1 and 2 denote the coating and interstitial regions, respectively (see Table 1). The current volumes corresponding to these regions are

$$v_1 = \phi_c v_0, \tag{38}$$

$$v_2(J) = \left(J\phi'_k - \phi_f - \phi_c\right)v_0, \tag{39}$$

and the Donnan potentials over both regions are then simply given as

$$\widetilde{\phi}_i(J) = -\frac{RT}{F}\sinh^{-1} C_{fi}. \tag{40}$$

The electrolyte free energy density (per unit volume of the current configuration) is found by specializing (2) with the Donnan potentials $\widetilde{\phi}_i$ so that

$$\mathfrak{w}_i(\widetilde{\phi}, J) = -FC_{fi}\widetilde{\phi}_i - 2RTC_0\left(\cosh\left(\frac{F}{RT}\widetilde{\phi}_i\right) - 1\right), \tag{41}$$

and the unit cell electrolyte free energy density is then given by (26) as

$$\overline{\mathfrak{W}}_{\text{electrolyte}}(J) = \frac{1}{v_0}\sum_{i=1}^{2}\mathfrak{w}_i(\widetilde{\phi}_i, J)v_i(J). \tag{42}$$

The unit cell osmotic pressure follows from (27) and results in

$$\overline{P}_{os}(J) = 2\phi'_k RTC_0\left[\left(\sqrt{C_{f2}^2 + 1} - 1\right) + \frac{\phi_c}{J\phi'_k - \phi_f}C_{f2}\left(\sinh^{-1} C_{f1} - \sinh^{-1} C_{f2}\right)\right]. \tag{43}$$

The first term in this expression describes a Donnan-like osmotic pressure from the GAGs in the interstitial region and the second term describes the contribution from the GAGs in the coating region. The unit cell osmotic compressibility, determined from (28), is

$$\begin{aligned}\overline{K}_{os}(J) = {} & 2\frac{\phi_k'^2}{J\phi'_k - \phi_f}RTC_0\frac{C_{f2}^2}{\sqrt{C_{f2}^2 + 1}} \\ & + 4\frac{\phi_k'^2\phi_c}{(J\phi'_k - \phi_f)^2}RTC_0C_{f2}\left(\sinh^{-1} C_{f1} - \sinh^{-1} C_{f2}\right) \\ & + 2\frac{\phi_k'^2\phi_c}{[J\phi'_k - \phi_f]^2}RTC_0\left(\frac{C_{f2}^2}{\sqrt{C_{f1}^2 + 1}} - \frac{C_{f2}^2}{\sqrt{C_{f1}^2 + 1}}\right).\end{aligned} \tag{44}$$

Table 1 Parameters of the electrolyte model

Parameter	Value	Notes
ϕ_f, volume fraction of the collagen fibrils (%)	24.9	Values derived from [20, 36]
ϕ_k, volume fraction of the keratocytes (%)	11.2	Values derived from [20, 36]
ϕ_c, volume fraction of the PG-dense coating (%)	13.3	Values derived from [20, 36]
C_{f0}, average charge density of the corneal stroma (mM)	38.6	Calibrated value by [20]
λ, the charge partition parameter	0.65	Value estimated by [20]
Q, dimensionless parameter by the ionic pumping effect	0.965, 1.0	Values for the *in vivo* and *ex vivo* corneas
P_0, the hydrostatic pressure in bath solution (mmHg)	15, 0	Values for the *in vivo* and *ex vivo* corneas
C_0, bath concentration (mM)	150	Physiological bath condition
T, temperature (K)	298	

4.2 Swelling Characteristics

It is interesting to compare the unit cell osmotic pressure $\overline{P}_{\mathrm{os}}$ from an ideal electrolyte, as given by (33), with that of the non-ideal (corneal) electrolyte described by (43). One way to do this is to give both unit cells the same total fixed charge, and the result is shown in Fig. 2a. It is seen that the osmotic pressure in the non-ideal case grows exponentially with reducing cell dilation J. At $J = 0.4$, the corneal osmotic pressure is approximately 350 kPa, which exceeds the ideal electrolyte osmotic pressure by a factor of more than eleven. This can be explained by the "volume exclusion" mechanisms that exist in the cornea [15, 20]. As the unit cell shrinks, the "trapped" volumes of collagen fibrils (and keratoctye cells) force the mobile ions into a smaller electrolyte volume, thereby greatly increasing the fixed charge density and osmotic pressure. It may be noted that the coating fixed charge in the non-ideal case is actually not effective at generating osmotic pressure because the coating volume is fixed and invariant with respect to J [37, 38]. However, the volume exclusion effect is much stronger. Fig. 2a provides a comparison of osmotic pressure in the ideal and non-ideal (cornea) electrolyte based on (33) and (43), respectively. Fig. 2b shows the same comparison for osmotic compressibility based on (34) and (44).

The swelling behavior predicted by the model can be illustrated by varying the bath ion concentration C_0 or fixed charge concentration C_{f0}. Fig. 3a shows the predicted osmotic pressure $\overline{P}_{\mathrm{os}}$ versus dilation J for three values of fixed charge concentration. The osmotic pressure increases monotonically with the fixed charge concentration, increasing from 100 to 350 kPa at $J = 0.4$ when C_{f0} increases from 50 % to the normal charge concentration. Figure 3b illustrates the osmotic pressure variation due to changes of bath concentration by plotting $\overline{P}_{\mathrm{os}} - J$ curves for $C_0 = 15, 150$ and 1500 mM. The predicted osmotic pressure decreases monotonically with the increasing bath concentration, which is attributed to the ion-shielding effect due to the abundance of mobile ions.

The tissue has a tendency to swell for all positive values of $\overline{P}_{\mathrm{os}}$; as the value of $\overline{P}_{\mathrm{os}}$ increases, the swelling tendency likewise increases and vice versa. This raises an interesting point in corneal biomechanics. The cornea can easily swell by absorbing water that is transported from the aqueous humor across the leaky endothelial cellular layer. If this occurred

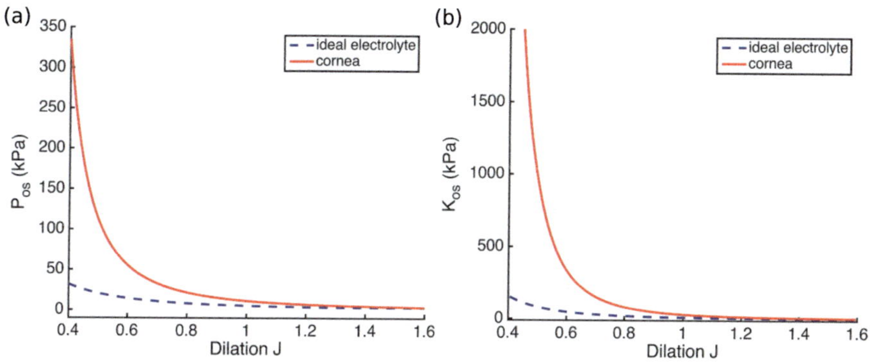

Fig. 2 Comparisons of predicted (**a**) osmotic pressure (P_{os}) and (**b**) osmotic compressibility (K_{os}) for the human cornea and an ideal electrolyte with the same amount of fixed charges

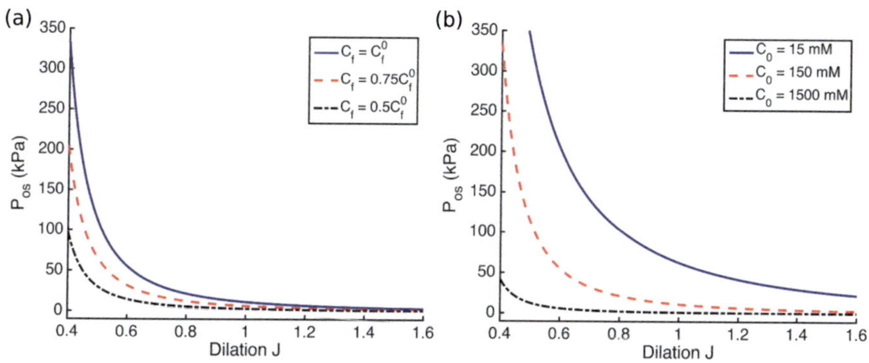

Fig. 3 (**a**) Osmotic pressure of the cornea predicted for three values of fixed concentration C_{f0}, (**b**) Osmotic pressure predicted for three values of bath ionic concentration C_0

the result would be a highly swollen and opaque cornea.[4] Of course, the *in vivo* cornea is stable with regard to hydration and the means by which the *in vivo* cornea achieves volumetric stability is a hypothesized pump-leak mechanism located in the endothelium [21] which facilitates the passive leak of water and active transport of ions. Pumping ions out of the stroma reduces the osmotic pressure and hence the swelling tendency. A model for this mechanism is presented next.

4.3 Active Modulation of Swelling

The corneal endothelium is a 4 μm thin cellular monolayer with a very high density of mitochondria, indicating high metabolic activity, and is located on the posterior surface of the cornea. It is permeable to water, metabolic species, including glucose and lactate ion, and other salt ions. The corneal endothelium is responsible for maintaining the hydration of the cornea through a "pump-leak" mechanism [23]. The "pump" function is provided

[4]The swelling changes the tissue hydration and the quasi-regular lattice of collagen fibrils will be compromised leading to opacity of the tissue. The corneal surface would also be highly distorted and ineffective for accurate light refraction.

by active ion transport across the endothelium [21] which regulates the osmotic pressure of the stroma. The "leak" function is the passive transport of water (solvent) across the endothelium through water channels in response to the stromal osmotic pressure.

In this section we treat the pump function with a macroscopic theory. The goal is to arrive at a simple model for describing how steady-state ion transport across the endothelium affects ionic concentrations and osmotic pressure within the stroma. The analysis is completed in three steps. First, steady-state (but non-equilibrium) conditions are used to establish the jump in electrochemical potential across the endothelial layer as a function of the steady active ion fluxes. Then thermodynamic equilibrium within the stroma is used to establish a modified Boltzmann distribution that governs ionic concentrations throughout the stroma. Finally, a modified free energy provides the stromal osmotic pressure and describes its dependence on the endothelial active ion transport. The resulting model thus provides a macroscropic theory for the corneal "pump-leak" mechanism.

To assess the electrochemical potential jump across the endothelial layer, it is treated as an ideal membrane in which the action of the ion pumps is represented by an independently specified active steady anion flux J_{a-} and cation flux J_{a+}, transporting ions out of the stroma and into the aqueous humor [34]. Because the endothelium is thin compared to the stroma, it is assumed that transport across the endothelium is one-dimensional. A coordinate z is introduced with $z = z_0$ at the endothelium-aqueous humor interface and $z = z^*$ at the endothelium-stroma interface. At steady-state, net ionic fluxes resulting from both active and passive transport will vanish so that

$$J_i = -L_i \frac{d\mu_i}{dz} - J_{ai} = 0, \quad i = +, - \tag{45}$$

must hold across the endothelial layer thickness $z \in [z_0, z^*]$. In (45), L_i is the membrane permeability for ionic species i, μ_i is the electrochemical potential, and $J_{ai} > 0$ is the steady active ionic flux across the endothelium. Integrating (45) across the endothelium results in the jump condition for the electrochemical potential

$$\mu_i^* - \mu_{0i} = -h \frac{J_{ai}}{L_i}, \tag{46}$$

where $h = z^* - z_0$ is the thickness of the endothelial layer, and μ_i^* and μ_{0i} are the ionic electrochemical potentials at $z = z^*$ and $z = z_0$, respectively. Since no active transport is taking place within the stroma, thermodynamic equilibrium requires $\mu_i = \mu_i^*$ where μ_i is the electrochemical potential at any point in the stroma. Using definition (20) for μ_i and μ_{0i}, with ϕ_0 taken to be zero, results in

$$\mu_i - \mu_{0i} = \frac{RT}{M_i} \ln \frac{C_i}{C_0} + z_i \frac{F\phi}{M_i}. \tag{47}$$

Substituting (47) in (46) and rearranging gives

$$C_i = C_0 \exp\left(-\frac{h}{RT} \frac{M_i J_{ai}}{L_i}\right) \exp\left(-z_i \frac{F\phi}{RT}\right). \tag{48}$$

Observing that the first expression in parenthesis on the right hand side of (48) is dimensionless, we define the membrane active transport factor Q_i

$$Q_i = \exp\left(-\frac{h}{RT} \frac{M_i J_{ai}}{L_i}\right). \quad i = +, - \tag{49}$$

Finally, stromal ionic concentration (48) may be written as

$$C_i = Q_i C_0 \exp\left(-z_i \frac{F\phi}{RT}\right), \tag{50}$$

which generalizes the Boltzmann distribution (22). Observe that $0 \leq Q_i \leq 1$ and the value of Q_i reduces with increasing active flux J_{ai}. When no endothelial ionic pumping occurs, $J_{ai} = 0$ and $Q_i = 1$. Equation (50) expresses the idea that when active ionic pumping occurs, $Q_i < 1$ and stromal ionic concentrations are reduced compared to when there is no pumping and $Q_i = 1$. Reduced stromal ionic concentrations produce reduced osmotic (swelling) pressure, and vice versa. Simply stated, active ionic pumping directly reduces the swelling tendency of the stroma. In this macroscopic description, the pump effect is embedded in the membrane active transport factor Q_i.

Based on (50), a modified Poisson-Boltzmann equation (see (5)) for the electrostatic potential in the stroma is introduced as

$$-\nabla^2\phi = \frac{F}{\epsilon_w}\left(-C_{\mathrm{f}} + \sum_{i=+,-} z_i Q_i C_0 \exp\left(-z_i \frac{F\phi}{RT}\right)\right), \tag{51}$$

and the corresponding modified electrostatic free energy density (per unit volume of the current configuration) is

$$\mathfrak{w}_{\mathrm{i}}(\widetilde{\phi}, J) = -F C_{\mathrm{fi}} \widetilde{\phi}_{\mathrm{i}} - RT C_0 \left(\sum_{j=+,-} Q_j \exp\left(-z_j \frac{F}{RT} \widetilde{\phi}_{\mathrm{i}}\right) - 2\right). \tag{52}$$

The Donnan potentials over both the coating and interstitial regions are solved from (51) by setting the right hand side to zero and are given as

$$\widetilde{\phi}_i(J) = \frac{RT}{F} \ln\left(-\frac{C_{\mathrm{fi}}(J)}{Q_-} + \sqrt{\frac{C_{\mathrm{fi}}^2(J)}{Q_-^2} + \frac{Q_+}{Q_-}}\right), \tag{53}$$

where $\widetilde{\phi}_1$ corresponds to the coating region and $\widetilde{\phi}_2$ corresponds to the intercoating region and where the non-dimensional fixed charge concentrations C_{f1} and C_{f2} are given by (36)–(37).

Combining the potentials (53), the Helmholtz free energy densities (52) and (42), the unit cell osmotic pressure then follows from (27) as $\overline{P}_{\mathrm{os}} = -\partial \overline{\mathfrak{W}}_{\mathrm{electrolyte}} / \partial J$, giving

$$\overline{P}_{\mathrm{os}}(J) = 2\phi_{\mathrm{k}}' RT C_0 \left[\left(\sqrt{C_{\mathrm{f2}}^2 + Q} - 1\right) + \frac{\phi_{\mathrm{c}}}{J\phi_{\mathrm{k}}' - \phi_{\mathrm{f}}} C_{\mathrm{f2}} \left(\widetilde{\sinh}^{-1} C_{\mathrm{f1}} - \widetilde{\sinh}^{-1} C_{\mathrm{f2}}\right)\right], \tag{54}$$

in which a modified inverse hyperbolic sine function was introduced to simplify the preceding expression

$$\widetilde{\sinh}^{-1}(x) = \ln\left(x + \sqrt{x^2 + Q}\right), \tag{55}$$

and where $Q = Q_+ Q_-$ is the membrane active transport parameter that describes the combined effect of both the cation and anion active transport.

The osmotic compressibility, defined as $\overline{K}_{\mathrm{os}}(J) = -\partial \overline{P}_{\mathrm{os}} / \partial J$, is the core function in the stroma electrolyte tensor $\mathbb{C}_{\mathrm{stroma}} = 4\partial^2 \mathcal{W}_{\mathrm{stroma}} / \partial \mathbf{C} \partial \mathbf{C}$ required for the finite element tangent operator and is given by

$$\overline{K}_{\mathrm{os}}(J) = 2\frac{\phi_{\mathrm{k}}'^2}{J\phi_{\mathrm{k}}' - \phi_{\mathrm{f}}} RTC_0 \frac{C_{\mathrm{f2}}^2}{\sqrt{C_{\mathrm{f2}}^2 + Q}}$$
$$+ 4\frac{\phi_{\mathrm{k}}'^2\phi_{\mathrm{c}}}{(J\phi_{\mathrm{k}}' - \phi_{\mathrm{f}})^2} RTC_0 C_{\mathrm{f2}}\left(\widetilde{\sinh}^{-1} C_{\mathrm{f1}} - \widetilde{\sinh}^{-1} C_{\mathrm{f2}}\right)$$
$$+ 2\frac{\phi_{\mathrm{k}}'^2\phi_{\mathrm{c}}}{[J\phi_{\mathrm{k}}' - \phi_{\mathrm{f}}]^2} RTC_0\left(\frac{C_{\mathrm{f2}}^2}{\sqrt{C_{\mathrm{f1}}^2 + Q}} - \frac{C_{\mathrm{f2}}^2}{\sqrt{C_{\mathrm{f1}}^2 + Q}}\right). \tag{56}$$

Equations (54)–(56) show that the osmotic pressure and compressibility are dictated by the combined membrane active transport parameter Q. Estimation of the combined membrane active transport parameter Q is challenging. Based on rabbit cornea measurements, a value of $Q = 0.965$ has been proposed [11] and will be used herein; further experimental measurements for the human cornea are needed. It is shown in Sect. 6.2 that the above model can be used to successfully predict *in vivo* corneal swelling in pathological situations. In the *ex vivo* cornea, no active ionic transport occurs since endothelial cells will have no metabolic activity and $Q = 1$. This condition is employed in Sect. 6.1 to model *in vitro* swelling pressure experiments on isolated stroma samples in ionic baths to provide evidence for the accuracy of the proposed model.

5 Volumetric Tangent Operator

We wish to emphasize that since the unit cell electrolyte free energy density (per unit reference volume) is given by

$$\mathfrak{W}(\mathbf{C}) = \overline{\mathfrak{W}}_{\mathrm{electrolyte}}(J) + \mathfrak{W}_{\mathrm{bath}}(J) + \mathfrak{W}_{\mathrm{elastic}}(\mathbf{C}), \tag{57}$$

the variational problem to be solved is similar to the problem of compressible hyperelasticity in which the strain energy is decomposed such that

$$\Psi(\mathbf{C}) = \Psi_{\mathrm{vol}}(J) + \Psi_{\mathrm{iso}}(\mathbf{C}), \tag{58}$$

where $\Psi_{\mathrm{vol}}(J)$ and $\Psi_{\mathrm{iso}}(\mathbf{C})$ are the volumetric and isochoric strain energy densities, respectively [39]. Indeed, it could be suggested that the proposed theory provides a rational approach for the construction of $\Psi_{\mathrm{vol}}(J)$ in the context of polyelectrolyte gels. It should be emphasized that $\overline{\mathfrak{W}}_{\mathrm{electrolyte}}$ incorporates swelling behavior whereas Ψ_{vol} does not. To emphasize this, we briefly summarize the governing variational equation and its linearization, and show how the osmotic energy provides the fundamental constitutive response for the volumetric behavior, with no *ad hoc* constitutive assumptions needed.

Mechanical equilibrium requires $\delta\int_{\Omega_0} \mathfrak{W}\partial\Omega_0 = 0$ which leads to

$$0 = \int_{\Omega_0} (\mathbf{S}_{\mathrm{electrolyte}} + \mathbf{S}_{\mathrm{bath}} + \mathbf{S}_{\mathrm{elastic}}) : \delta\mathbf{E}\, d\Omega_0, \tag{59}$$

with

$$\mathbf{S}_{\mathrm{electrolyte}} = 2\frac{\partial\overline{\mathfrak{W}}_{\mathrm{electrolyte}}(J)}{\partial\mathbf{C}} = -J\overline{P}_{\mathrm{os}}\mathbf{C}^{-1}, \tag{60}$$

$$\mathbf{S}_{\text{bath}} = 2\frac{\partial \mathfrak{W}_{\text{bath}}(J)}{\partial \mathbf{C}} = -JP_0\mathbf{C}^{-1}, \tag{61}$$

and

$$\mathbf{S}_{\text{elastic}} = 2\frac{\partial \mathfrak{W}_{\text{elastic}}}{\partial \mathbf{C}}. \tag{62}$$

Note that the unit cell osmotic pressure $\overline{P}_{\text{os}}(J)$ is given generally by (27) and explicitly for the cornea example by (43).

Let the directional derivative of any functional $\mathcal{F}(\mathbf{u})$ in the direction $\Delta\mathbf{u}$ be denoted $\Delta\mathcal{F} = \frac{d}{d\epsilon}[\mathcal{F}(\mathbf{u} + \epsilon\Delta\mathbf{u})]_{\epsilon=0}$. The linearization of variational equation (59) (needed for a finite element implementation) is expressed by $\mathcal{L}[\delta\mathcal{G}] = \delta\mathcal{G} + \Delta(\delta\mathcal{G})$. The term $\Delta(\delta\mathcal{G})$ can be expressed in standard form in the material description as [39]

$$\begin{aligned}\Delta(\delta\mathcal{G}) = \int_{\Omega_0} \big[&\text{Grad}\,\delta\mathbf{u}(\mathbf{S}_{\text{electrolyte}} + \mathbf{S}_{\text{bath}} + \mathbf{S}_{\text{elastic}}) \cdot \text{Grad}\,\Delta\mathbf{u} \\ &+ \mathbf{F}^T\text{Grad}\,\delta\mathbf{u} : (\mathbb{C}_{\text{electrolyte}} + \mathbb{C}_{\text{bath}} + \mathbb{C}_{\text{elastic}}) : \mathbf{F}^T\text{Grad}\,\Delta\mathbf{u}\big]d\Omega_0,\end{aligned} \tag{63}$$

where the electrolyte modulus tensor is

$$\mathbb{C}_{\text{electrolyte}} = 4\frac{\partial^2 \overline{\mathfrak{W}}_{\text{electrolyte}}(J)}{\partial \mathbf{C}\partial \mathbf{C}} = J(J\overline{K}_{\text{os}} - \overline{P}_{\text{os}})\mathbf{C}^{-1} \otimes \mathbf{C}^{-1} + 2J\overline{P}_{\text{os}}\mathbf{C}^{-1} \odot \mathbf{C}^{-1}, \tag{64}$$

the bath (hydrostatic pressure) modulus tensor is

$$\mathbb{C}_{\text{bath}} = 4\frac{\partial^2 \mathfrak{W}_{\text{bath}}(J)}{\partial \mathbf{C}\partial \mathbf{C}} = -JP_0\mathbf{C}^{-1} \otimes \mathbf{C}^{-1} + 2JP_0\mathbf{C}^{-1} \odot \mathbf{C}^{-1}, \tag{65}$$

and the tissue fiber elasticity is

$$\mathbb{C}_{\text{elastic}} = 4\frac{\partial^2 \mathfrak{W}_{\text{elastic}}(\mathbf{C})}{\partial \mathbf{C}\partial \mathbf{C}}. \tag{66}$$

In the above expressions, $\partial\mathbf{C}^{-1}/\partial\mathbf{C} = -\mathbf{C}^{-1} \odot \mathbf{C}^{-1}$ and $[\mathbf{C}^{-1} \odot \mathbf{C}^{-1}]_{ABCD} = \frac{1}{2}[C^{-1}_{AC}C^{-1}_{BD} + C^{-1}_{AD}C^{-1}_{BC}]$ [39]. If the equilibrium fluid pressure is defined to be

$$P = P_{\text{os}} + P_0, \tag{67}$$

then we can conveniently write

$$\mathbb{C}_{\text{electrolyte}} + \mathbb{C}_{\text{bath}} = J(J\overline{K}_{\text{os}} - P)\mathbf{C}^{-1} \otimes \mathbf{C}^{-1} + 2JP\mathbf{C}^{-1} \odot \mathbf{C}^{-1}. \tag{68}$$

Note again that the unit cell osmotic compressibility $\overline{K}_{\text{os}}(J)$ is given generally by (28) and explicitly for the cornea example by (44).

The spatial form of $\mathbb{C}_{\text{electrolyte}}$ is given by $\mathbb{c}_{\text{electrolyte}} = \varphi_*(\mathbb{C}_{\text{electrolyte}})$, where ϕ_* is the fourth-order push-forward operator based on $\mathbf{F}$. It is found that

$$\mathbb{c}_{\text{electrolyte}} + \mathbb{c}_{\text{bath}} = (JK_{\text{os}} - P)\mathbf{1} \otimes \mathbf{1} + 2P\mathbf{I}. \tag{69}$$

6 Results

6.1 Assessment of Model Accuracy

Before turning to the *in vivo* tissue model, we assess the accuracy of the model described in Sect. 4 using two approaches. In the first we compare the predicted results with those that are obtained by solving the Poisson-Boltzmann equation for the electrostatic potential. Thus we are examining the accuracy of the Donnan potentials. In the second approach we compare the osmotic pressure prediction given by (43) to direct experimental measurements.

To assess the accuracy of the Donnan potential-based approximate solution (40), the Poisson-Boltzmann (PB) equation (25) was solved numerically over the unit cell shown in Fig. 1 using the parameters of Table 1. Three values of the volume dilation $J = 1.0, 0.6$ and 0.5 are considered. The unit cell size reduces with decreasing J [20]. Recall that the collagen fibers and GAG coating regions remain invariant with J. In this case, the interstitial fluid volume is reducing much faster than J, and this "geometric" effect produces a rapidly increasing charge density and osmotic pressure in the interstitial fluid. In order to compare ϕ_{PB} with the piecewise constant Donnan approximation $\widetilde{\phi}$, given by Eq. (40), the profile of the two potentials is plotted along the fibril center-to-center line (edge of the unit cell) in Fig. 4. At $J = 1$, ϕ_{PB} and $\widetilde{\phi}$ are seen to agree well. As J is reduced, the interfibrillar distance decreases and the coating domains approach each other (see Fig. 4d–e). The Donnan approximation $\widetilde{\phi}$ maintains the general profile of ϕ_{PB} but pointwise accuracy deteriorates.

The exact electrostatic potential ϕ_{PB} was then used to compute the electrolyte free energy density (using Eq. (2)) over the unit cell and the result is compared to the approximation in Fig. 5. It is seen that the results computed from these two approaches agree closely for all values of J. While the Donnan potentials may not have pointwise accuracy, the integration of these potentials over the unit cell actually does retain high accuracy (compare area under curves in Fig. 5).

In the next accuracy assessment, predicted unit cell swelling pressure is compared to direct experimental measurement on a sample of isolated corneal stroma immersed in an ionic bath solution (see Fig. 6a) of concentration $C_0 = 150$ mM. The bath hydrostatic pressure is zero and, in this case, the tissue fluid pressure corresponds to the osmotic pressure [15]. The experimental tests employed stroma samples in the form of 7 mm diameter disks having initial (reference) thickness of 0.5 mm and loaded by a porous piston across their thickness (see Fig. 6a) under both confined and unconfined conditions [13, 40]. The piston load was systematically varied and the equilibrium sample thickness recorded. The experiment was simulated using the finite element method based on the above unit cell theory. The detailed description of the stromal elastic energy $\mathfrak{W}_{\text{elastic}}(\mathbf{C})$ is referred to Petsche and Pinsky [26]. The stroma sample was meshed with 2500, 27-node hexahedral elements. In order to model confined [40] and unconfined [13] experimental conditions, edge boundary conditions were taken as zero radial displacement or free, respectively (see Fig. 6a). Sample thickness versus osmotic pressure is shown in Fig. 6b. The results show good agreement between the experimental data and predictions for the full range of experimental measurements in which the sample thickness varied from 0.75 mm (swollen, $J = 1.5$) to 0.2 mm (compressed, $J = 0.4$). In compression, swelling pressures (i.e., osmotic pressure) reach values in excess of 2000 mmHg at $J = 0.2$. Swelling pressure predictions for the confined and unconfined cases are nearly identical. This is expected because lateral expansion of the sample is restricted by the collagen fibrils and, as a consequence, stromal volume change in these tests is predicted to be associated with changes in sample thickness [3].

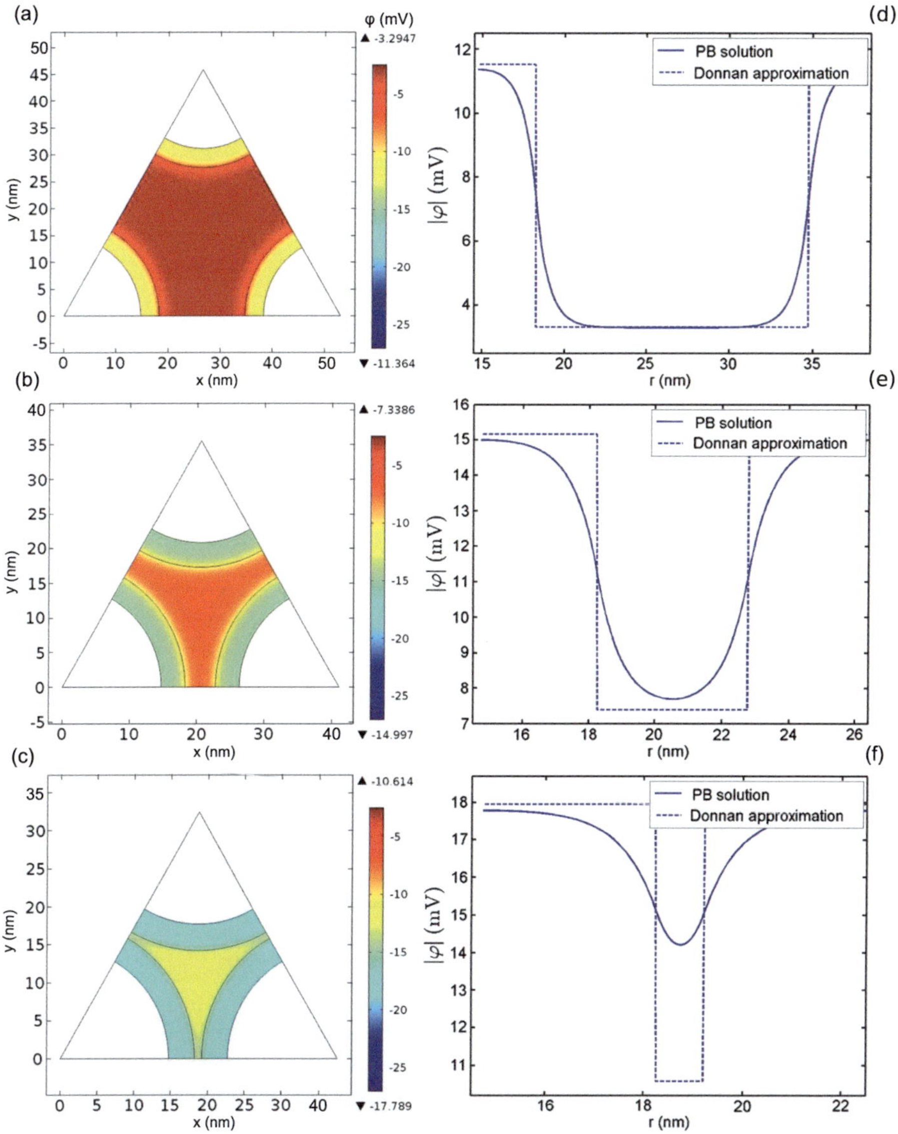

Fig. 4 (**a**)–(**c**): The contour plots of the PB-based electrostatic potential (ϕ_{PB}) within the unit cell with volume dilations $J = 1.0, 0.6$ and 0.5; (**d**)–(**f**) the comparison between the ϕ_{PB} and Donnan-based approximation $\widetilde{\phi}$ along the fibril center-to-center line

6.2 Swelling in the *in Vivo* Cornea

6.2.1 Effect of Active Pumping on Electrolyte Properties

Here we report the effect of active endothelial pumping on the electrolyte properties of the *in vivo* cornea. In the proposed model, the endothelial ionic pumping effect is modeled by the active transport coefficient Q (see Eq. 49). Recall that this term satisfies $0 \leq Q \leq 1$,

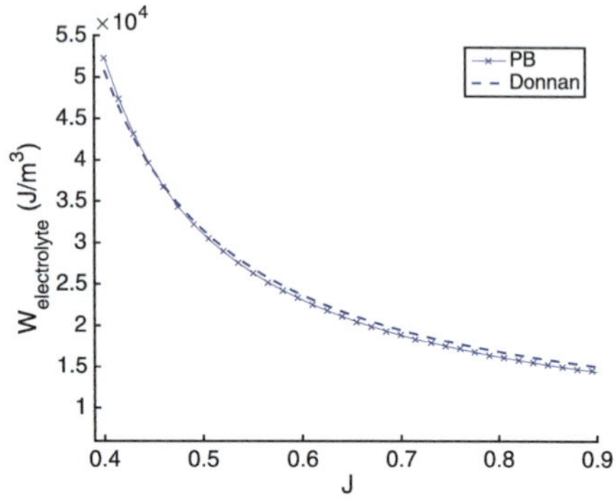

Fig. 5 The effect of the ϕ_{PB} and $\widetilde{\phi}$ on the electrostatic free energy $\mathfrak{W}_{\text{electrolyte}}$. The PB-based and Donnan-based solutions agree well over the entire dilation range

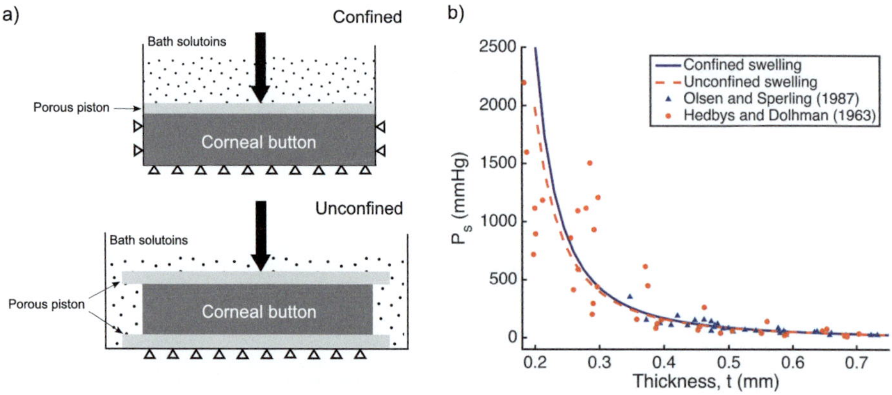

Fig. 6 The predicted confined and unconfined swelling pressure P_s versus corneal thickness t and their comparison with experimental measurements [13, 40] over thickness range 0.2 mm $\leq t \leq$ 0.75 mm. The swelling pressure data have been extracted from the original papers by free graphics software Plot Digitizer, and the hydration data H from [40] have been transformed to thickness t by the observed linear relation $H = 7.09t - 0.44$ [13, 41]

with $Q = 1$ and 0.965 corresponding to no active transport and normal active transport [11], respectively. Figure 7a-c show the computed electrostatic free energy density ($\mathfrak{W}_{\text{electrolyte}}$), osmotic pressure ($P_{\text{os}}$) and osmotic compressibility (K_{os}) as functions of volume dilation J for these two values of Q and also for $Q = 0.930$ which corresponds to hyper-active ionic transport. It can be seen that the profile of electrolyte free energy varies significantly as Q reduces from 1.0 to 0.93. For $Q = 1.0$, which corresponds to the *ex vivo* case without active transport, $\mathfrak{W}_{\text{electrolyte}}$ decreases monotonically with increasing J, and thus gives rise to positive osmotic pressure for all J. For the normal *in vivo* cornea with $Q = 0.965$, $\mathfrak{W}_{\text{electrolyte}}$ has its minimum point at $J = 1$, which corresponds to a zero osmotic pressure at physiological condition [11]. In this case, a reduction in dilation will result in positive stromal osmotic pressure which will produce a tendency to swell. On the other hand, an increase in dilation will result in negative stromal osmotic pressure and produce a tendency to de-swell. Consider the curve for the hyper-active $Q = 0.930$. In this situation, positive or negative deviations in dilation from $J = 1$ will always result in negative osmotic pressure and produce a tendency to de-swell. These conditions illustrate the hydration regulation mechanism of active endothelial ion transport. Increasing active transport shifts the osmotic pressure-dilation curve downwards in Fig. 7b. Predicted values of osmotic compressibility

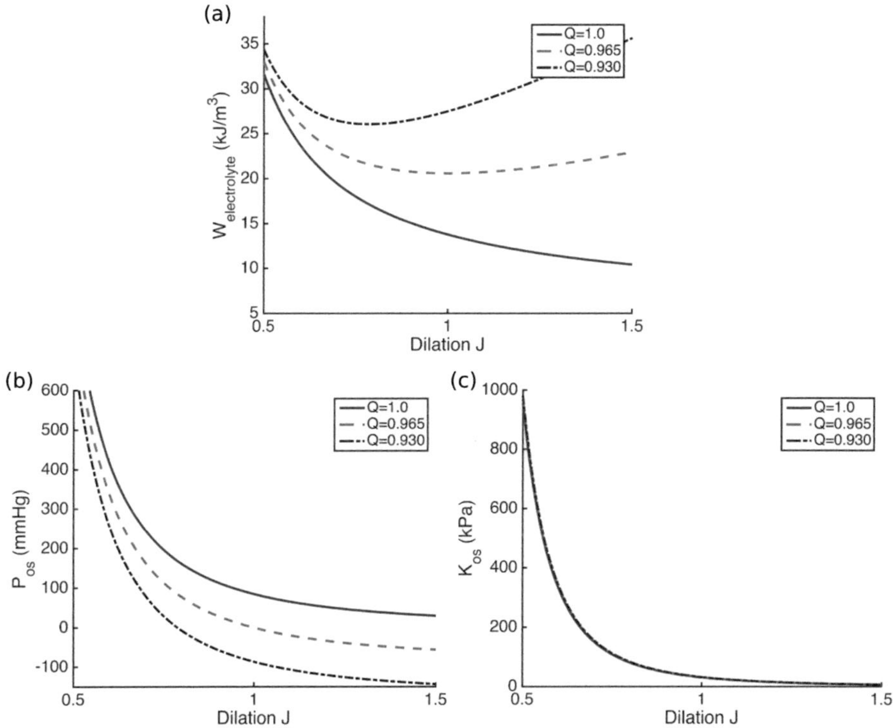

Fig. 7 The computed electrostatic free energy, osmotic pressure and osmotic compressibility versus volume dilation J for three representative values of Q

K_{os} (Eq. (56)) versus dilation J are shown in Fig. 7c for no active transport. Values closely match measurements reported in [13].

6.2.2 Swelling in a Pathological Case

Fuch's corneal dystrophy is a disease in which the cornea swells, causing it to lose transparency and leading to corneal blindness. It is usually characterized by morphological changes in endothelial cells or by an accelerated loss of endothelial cells [42]. In this situation, the cornea will swell due to increasing endothelial permeability, decreasing active ion flux, or both mechanisms simultaneously. These pathological changes can be characterized by the relative changes of the two quantities. For simplicity and without loss of generality, we consider active anion transport only and let L_0 and J_{a0} be the anion membrane permeability and active anion flux rate of the normal cornea, respectively. For any values of L/L_0 and J_a/J_{a0}, their relationship with the active transport coefficient Q (see Eq. (49)) can be expressed by

$$\ln Q = \left(\frac{J_a}{J_{a0}} \cdot \frac{L_0}{L} \right) \ln Q_0, \tag{70}$$

where $Q_0 = 0.965$ is the calibrated value for normal cornea as determined by Cheng et al. [11].

The cornea was modeled geometrically with a central thickness of 520 μm and the anterior and posterior surfaces defined as spherical surfaces with radii of 7.87 and 6.7 mm,

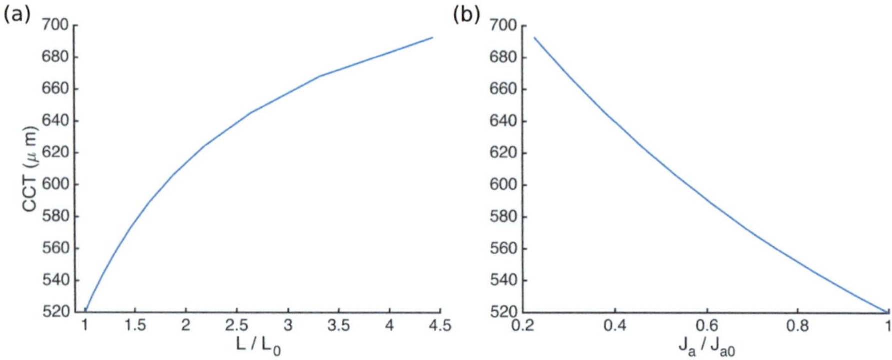

Fig. 8 The predicted swollen CCT in Fuch's dystrophy when (**a**) the endothelial ionic permeability is increased up to a factor of 4.5 ($1 \le L/L_0 \le 4.5$) with normal active ionic flux ($J_a/J_{a0} = 1$) and (**b**) the active ionic flux rate is reduced up to a factor of 0.2 ($0.2 \le J_a/J_{a0} \le 1$) with normal anionic permeability ($L/L_0 = 1$)

respectively. Part of the sclera was also modeled with anterior and posterior surfaces defined with radii of 12.00 and 11.38 mm, respectively. Homogeneous Dirichlet boundary conditions are applied to fix the periphery of the scleral section. The solution proceeds by first solving for the normal cornea with $Q = Q_0 = 0.965$, and then solving a sequence of problems in which Q is systematically varied to simulate changes in endothelial permeability and active anionic flux.

Figure 8a shows the predicted swollen central corneal thickness (CCT) for endothelial ionic permeability increased with $1 \le L/L_0 \le 4.5$ and with normal active ionic flux $J_a/J_{a0} = 1$. Figure 8b shows the predicted swollen CCT for active ionic flux rate reduced up to a factor of 0.2, so that $0.2 \le J_a/J_{a0} \le 1$, with normal anionic permeability $L/L_0 = 1$. In both cases the cornea swells. The predicted maximum swollen CCT of approximately 690 μm lies in the range of clinical observation (500–1000 μm) measured for patients with Fuch's dystrophy [27].

It is interesting to note that beyond the reported range of L/L_0 (e.g., $L/L_0 < 0.2$), the model predicted buckling of the posterior cornea as posterior-oriented swelling produces tissue compression. This is in agreement with clinical measurements [27] which show that the posterior surface presented significant central bulges (toward the anterior chamber) among patients with severe Fuch's dystrophy. As the active pumping effect is weakened and the osmotic pressure is increased (see Fig. 7b), the anterior surface moves outwardly, and the posterior surface moves inwardly towards the aqueous humor (see contour plot of vertical displacement in Fig. 9). It is expected that buckling of the posterior surface occurs when Q increases beyond a critical value, which is predicted to be about $Q_{\text{critical}} \approx 0.992$ in this study.

7 Discussion

In this study, we proposed an energy approach to describe the swelling pressure and volumetric behavior of hydrated tissues in both *ex vivo* and *in vivo* situations. The theory, which is applicable to large strains and swelling, is based on characterization of the tissue electrolyte free energy using the mean-field approximation for ionic interactions and on the nanoscale distribution of fixed charges. The resulting model predicts *ex vivo* swelling

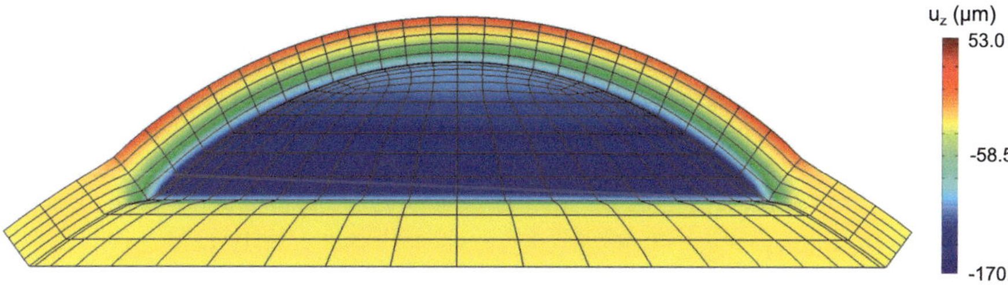

Fig. 9 Fringe plot of the vertical displacement field u_z at $J_a/J_{a0} = 0.22$ for a cornea with Fuch's dystrophy

pressure of the cornea with good accuracy compared to experimental measurements (as described in Sect. 6.1). This provides some confidence in the general framework. Furthermore, the model is shown to be predictive of *in vivo* corneal swelling in pathological cases in which the water leak into the stroma is not balanced by the pumping of ions. Importantly, the formulation does not require the solution of coupled fields, which is in contrast to modeling of hydrated tissues based on classical mixture theory [7, 8]. The proposed energy-based approach is equivalent to mixture theory (see Sect. 2.2) in that it implies satisfaction of the same governing equations but it offers some distinct advantages.

The proposed model provides a simplified constitutive framework for hydrated connective tissues by taking advantage of thermodynamic equilibrium and the Donnan approximation. It has been shown that the electrolyte free energy term can be expressed as a function of the tissue macroscopic volume dilation (J) only. This gives rise to a structural model that only requires solving for the displacement field. The energy approach is also natural for finite element implementation and the current theory is implemented with complexity no greater than for typical hyperelasticity. On the contrary, mixture theories explicitly describe tissue behaviors with multiple fields (e.g., displacement, fluid pressure, ionic concentration etc.). Solving the resulting coupled system is challenging and computationally expensive [43].

The electrolyte free energy (see Eq. (26)) provides a homogenized energy-equivalent value of osmotic pressure over a unit cell (see Eq. (43)). This is useful for consistently embedding nanoscale effects, for example, the distribution of GAG charge into collagen coating and non-coating regions, in a macroscopic theory. By contrast, the triphasic theory produces a pointwise varying expression for the osmotic pressure. The energy approach is also convenient for describing the deviation of the stromal electrolyte from ideal conditions, including electrolyte volume exclusion effects associated with the collagen fibrils and stromal cell (keratocyte) populations.

The proposed theory can be thought of as providing a rational form for the volumetric energy (see Eq. (58))—it can be used to replace the volume energy in a hyperelastic formulation. The proposed approach can accurately predict swelling as demonstrated in Sect. 6 whereas commonly employed hyperelastic models for incompressible or nearly incompressibility behavior have no predictive value for osmotic pressure-driven swelling. The proposed energy functional offers an alternative formulation which is developed from first principles, is rigorously based on physically measurable parameters, and is implemented with relative ease.

The modeling of active endothelial transport and its effect on corneal biomechanics is approached through a macroscopic description, similar to Kedem and Kalchalsky theory [44], in which the membrane osmotic gradient is expressed as a function of the active ionic fluxes and endothelial ionic permeability. This leads to a modified Boltzmann ionic distribution in the stroma and corresponding electrostatic free energy. The effects are contained in the introduced ionic transport parameter Q which describes the combined effect of cation and anion transport. The model suggests that active endothelial ionic transport reduces osmotic pressure and hence swelling pressure and the pump-leak mechanism can be explained as active pumping producing a downward shift of the swelling pressure-hydration curve as discussed in Sect. 6.2.1 and illustrated in Fig. 7. Furthermore, it is worth noting that we have employed the simplest descriptions for the trans-endothelial fluxes without accounting for the potential interactions between solvent and solutes. Recently analysis [45] suggests the coupling effect between water and solutes through membrane transport plays an important role in the modulation of tissue fluid pressure. The incorporation of these effects into the proposed framework would be useful and is currently under development.

The proposed framework can be adapted to model other hydrated tissues such as cartilage and intervertebral disc [7, 17] and synthesized polyelectrolyte gels [30, 46]. For different tissues, the microstructure varies and therefore a different unit cell model is required to characterize the corresponding charge distribution [35]. It should be noted that the Debye length may become comparable to the unit cell size in some cases, for example in cartilage [35]. In this case, the Donnan approximation may not be sufficiently inaccurate. Alternative mathematical techniques such as computing the activity and osmotic coefficient from the unit cell, should be utilized to avoid solving the PB equation numerically and to provide analytical approximations for the osmotic pressure. Moreover, other effect such as the mixing of gel polymers and solvent may be included in the modeling of synthesized hydrogels at low hydration [46].

The current model is limited by the assumption of thermodynamic equilibrium, wherein both fluid and ionic phases have reached equilibrium. In certain situations, understanding the transient behavior of the cornea may become important. For example, the endothelial permeability and relative pumping rates are usually measured by means of a transient swelling experiment [47]. Another practical example is corneal tonometry, in which the measurement of the IOP takes place in the time scale of several seconds. In these cases, the water phase does not reach equilibrium in the short period of time and the displacement-only framework is no longer applicable. However, under Lanir's hypothesis [43], in which the transport of ions is assumed much faster than that of the water, the triphasic theory can still be avoided. Assuming the equilibrium of the ionic phases, the definition of the average osmotic pressure (as given by Eq. (43)) is still valid and a biphasic-swelling theory [43] is then needed to solve for displacement and fluid pressure. This will be an interesting extension of the current framework.

References

1. Fox, A.J.S., Bedi, A., Rodeo, S.A.: The basic science of articular cartilage. Sports Health **1**, 461–468 (2009)
2. Iatridis, J.C., MacLean, J.J., O'Brien, M., Stokes, I.A.: Measurements of proteoglycan and water content distribution in human lumbar intervertebral discs. Spine **32**, 1493–1497 (2007)
3. Elliott, G.F., Hodson, S.A.: Cornea, and the swelling of polyelectrolyte gels of biological interest. Rep. Prog. Phys. **61**, 1325–1365 (1998)
4. Knudson, C.B., Knudson, W.: Cartilage proteoglycans. Semin. Cell Dev. Biol. **12**, 69–78 (2001)
5. Lewis, P.N., Pinali, C., Young, R.D., Meek, K.M., Quantock, A.J., Knupp, C.: Structural interactions between collagen and proteoglycans are elucidated by three-dimensional electron tomography of bovine cornea. Structure **18**, 239–245 (2010)
6. Mow, V.C., Kuei, S.C., Lai, W.M., Armstrong, C.G.: Biphasic creep and stress relaxation of articular cartilage in compression: theory and experiments. J. Biomech. Eng. **102**, 73–84 (1980)
7. Lai, W.M., Hou, J.S., Mow, V.C.: A triphasic theory for swelling and deformation behaviors of articular cartilage. J. Biomech. Eng. **113**, 245–258 (1991)

8. Gu, W.Y., Lai, W.M., Mow, V.C.: A mixture theory for charged-hydrated soft tissues containing multi-electrolytes: passive transport ans swelling behaviors. J. Biomech. Eng. **120**, 169–180 (1998)
9. Weiss, J.A., Makerc, B.N., Govindjeed, S.: Finite element implementation of incompressible, transversely isotropic hyperelasticity. Comput. Methods Appl. Mech. Eng. **135**, 107–128 (1996)
10. Annaidh, A.N., Destrade, M., Gilchrist, M.D., Murphy, J.G.: Deficiencies in numerical models of anisotropic nonlinearly elastic materials. Biomech. Model. Mechanobiol. **12**, 781–791 (2013)
11. Cheng, X., Petsche, S.J., Pinsky, P.M.: A structural model for the *in vivo* human cornea including collagen-swelling interaction. J. R. Soc. Interface **12**, 20150241 (2015)
12. Chahine, N.O., Chen, F.H., Hung, C.T., Ateshian, G.A.: Direct measurement of osmotic pressure of glycosaminoglycan solutions by membrane osmometry at room temperature. Biophys. J. **89**, 1543–1550 (2005)
13. Olsen, T., Sperling, S.: The swelling pressure of the human corneal stroma as determined by a new method. Exp. Eye Res. **44**, 481–490 (1987)
14. Loret, B., Simoes, F.MF.: Effects of the pH on the mechanical behavior of articular cartilage and corneal stroma. Int. J. Solids Struct. **47**, 2201–2214 (2010)
15. Hodson, S.A.: Why the cornea swells. J. Theor. Biol. **33**, 419–427 (1971)
16. Manzano, S., Manzano, R., Doblare, M., Doweidar, M.H.: Altered swelling and ion fluxes in articular cartilage as a biomarker in osteoarthritis and joint immobilization: a computational analysis. J. R. Soc. Interface **12**, 20141090 (2015)
17. Zhu, Q., Jackson, A.R., Gu, W.Y.: Cell viability in intervertebral disc under various nutritional and dynamic loading conditions: 3d finite element analysis. J. Biomech. **45**, 2769–2777 (2012)
18. Kwok, L.S., Klyce, S.D.: Theoretical basis for an anomalous temperature coefficient in swelling pressure of rabbit corneal stroma. Biophys. J. **57**, 657–662 (1990)
19. Muller, L.M., Pels, E., Vrensen, G.FJ.M.: The specific architecture of the anterior stroma accounts for maintenance of corneal curvature. Br. J. Ophthalmol. **85**, 437–443 (2001)
20. Cheng, X., Pinsky, P.M.: Mechanisms of self-organization for the collagen fibril lattice in the human cornea. J. R. Soc. Interface **10**, 20130512 (2013)
21. Bonanno, J.A.: Molecular mechanisms underlying the corneal endothelial pump. Exp. Eye Res. **95**, 2–7 (2012)
22. Waring, G.O., Bourne, W.M., Edelhauser, H.F., Kenyon, K.R.: The corneal endothelium: normal and pathologic structure and function. Ophthalmology **89**, 531–590 (1982)
23. Fischbarg, J., Maurice, D.M.: An update on corneal hydration control. Exp. Eye Res. **78**, 537–541 (2004)
24. Fischbarg, J.: Fluid transport across leaky epithelia: central role of the tight junction and supporting role of aquaporins. Physiol. Rev. **90**, 1271–1290 (2010)
25. Winkler, M., Shoa, G., Xie, Y., Petsche, S.J., Pinsky, P.M., Juhasz, T., Brown, D.J., Jester, J.V.: Three-dimensional distribution of transverse collagen fibers in the anterior human corneal stroma. Investig. Ophthalmol. Vis. Sci. **54**, 7293–7301 (2013)
26. Petsche, S.J., Pinsky, P.M.: The role of 3-D collagen organization in stromal elasticity: a model based on X-ray diffraction data and second harmonic-generated images. Biomech. Model. Mechanobiol. **12**, 1101–1113 (2013)
27. Brunette, I., Sherknies, D., Terry, M.A., Chagnon, M., Bourges, J.L., Meunier, J.: 3-D characterization of the corneal shape in Fuchs dystrophy and pseudophakic keratopathy. Investig. Ophthalmol. Vis. Sci. **52**, 206–214 (2009)
28. Ugryumova, N., Jacobs, J., Bonesi, M., Matcher, S.J.: Novel optical imaging technique to determine the 3-D orientation of collagen fibers in cartilage: variable-incidence angle polarization-sensitive optical coherence tomography. Osteoarthr. Cartil. **17**, 33–42 (2009)
29. Pandolfi, A., Holzapfel, G.A.: Three-dimensional modeling and computational analysis of the human cornea considering distributed collagen fibril orientations. J. Biomech. Eng. **130**, 061006 (2008)
30. Hong, W., Zhao, X., Suo, Z.: Large deformation and electrochemistry of polyelectrolyte gels. J. Mech. Phys. Solids **58**, 558–577 (2010)
31. Katchalsky, A., Michaeli, I.: Polyelectrolyte gels in salt solutions. J. Polym. Sci. **15**, 69–86 (1955)
32. Sharp, K.A., Honig, B.: Calculating total electrostatic energies with the nonlinear Poisson-Boltzmann equation. J. Phys. Chem. **94**, 7684–7692 (1990)
33. Huyghe, J.M., Janssen, J.D.: Quadriphasic theory of swelling incompressible porous media. Int. J. Eng. Sci. **35**, 793–802 (1997)
34. Bryant, M.R., McDonnell, P.J.: A triphasic analysis of corneal swelling and hydration control. Trans. Am. Soc. Mech. Eng. **120**, 370–381 (1998)
35. Buschmann, M.D., Grodzinsky, A.J.: A molecular model of proteoglycan-associated electrostatic forces in cartilage mechanics. J. Biomech. Eng. **117**, 179–192 (1995)
36. Meek, K.M.: The cornea and sclera. In: Fratzl, P. (ed.) Collagen: Structure and Mechanics, pp. 359–396. Springer, Berlin (2008)

37. Fratzl, P., Daxer, A.: Structural transformation of collagen fibrils in corneal stroma during drying. Biophys. J. **64**, 1210–1214 (1993)
38. Twersky, V.: Transparency of pair-correlated, random distributions of small scatterers, with applications to the cornea. J. Opt. Soc. Am. **65**, 524–530 (1975)
39. Holzapfel, G.A.: Nonlinear Solid Mechanics: A Continuum Approach for Engineering. Wiley, New York (2000)
40. Hedbys, B.O., Dohlman, C.H.: A new method for the determination of the swelling pressure of the corneal stroma *in vitro*. Exp. Eye Res. **2**, 122–129 (1963)
41. Hedbys, B.O., Mishima, S.: The thickness-hydration relationship of cornea. Exp. Eye Res. **5**, 221–228 (1966)
42. Elhalis, H., Azizi, B., Jurkunas, U.: Fuchs endothelial corneal dystrophy. Ocul. Surf. **8**, 173–184 (2010)
43. Wilson, W., Donkelaar, C.C.V., Huyghe, J.M.: A comparison between mechano-electrochemical and biphasic swelling theories for soft hydrated tissues. Trans. Am. Soc. Mech. Eng. **127**, 158–165 (2005)
44. Kedem, O., Katchalsky, A.: Thermodynamic analysis of the permeability of biological membranes to non-electrolytes. Biochim. Biophys. Acta **27**, 229–246 (1958)
45. Cheng, X., Pinsky, P.M.: The balance of fluid and osmotic pressures across active biological membranes with application to the corneal endothelium. PLoS ONE **10**, e0145422 (2015)
46. Li, Y., Suo, Z., Vlassak, J.J.: A model of ideal elastomeric gels for polyelectrolyte gels. Soft Matter **10**, 2582–2590 (2014)
47. Bourne, W.M.: Clinical estimation of corneal endothelial pump function. Trans. Am. Ophthalmol. Soc. **96**, 229–242 (1998)

J Elast (2017) 129:171–195
DOI 10.1007/s10659-016-9603-4

Multi-scale Modeling of Vision-Guided Remodeling and Age-Dependent Growth of the Tree Shrew Sclera During Eye Development and Lens-Induced Myopia

Rafael Grytz[1] · **Mustapha El Hamdaoui**[1]

Received: 31 March 2016 / Published online: 24 October 2016

Abstract The sclera uses unknown mechanisms to match the eye's axial length to its optics during development, producing eyes with good focus (emmetropia). A myopic eye is too long for its own optics. We propose a multi-scale computational model to simulate eye development based on the assumption that scleral growth is controlled by genetic factors while scleral remodeling is driven by genetic factors and the eye's refractive error. We define growth as a mechanism that changes the tissue volume and mass while remodeling involves internal micro-deformations that are volume-preserving at the macro-scale. The model was fitted against longitudinal refractive measurements in tree shrews of different ages and exposed to three different visual conditions: (i) normal development; (ii) negative lens wear to induce myopia; and (iii) recovery from myopia by removing the negative lens. The model was able to replicate the age- and vision-dependent response of the tree shrew experiments. Scleral growth ceased at younger age than scleral remodeling. The remodeling rate decreased as the eye emmetropized but increased at any age when a negative lens was put on. The predictive power of the model was investigated by calculating the susceptibility to scleral remodeling and the response to form deprivation myopia in tree shrews. Both predictions were in good agreement with experimental data that were not used to fit the model. We propose the first model that distinguishes scleral growth from remodeling. The good agreement of our results with experimental data supports the notion that scleral growth and scleral remodeling are two independently controlled mechanisms during eye development.

Keywords Myopia · Growth · Remodeling · Emmetropization · Multi-scale modeling · Finite element method

Mathematics Subject Classification 74L15 · 92C10 · 74B20 · 74S05

This work was supported in part by the National Institutes of Health Grant R01 EY05922 and Core Grant P30 EY003039 (Bethesda, Maryland); Eye Sight Foundation of Alabama (Birmingham, Alabama); and Research to Prevent Blindness (New York, New York).

✉ R. Grytz
grytz@uab.edu

[1] Department of Ophthalmology, University of Alabama at Birmingham, Birmingham, AL 35294, USA

1 Introduction

Myopia, also known as nearsightedness, has become an epidemic ocular condition [1] that affects nearly 40 % of the US adult population [2] and increased from 10 % to 90 % in some Asian populations [3, 4]. A myopic eye is characterized by an elongated scleral shell as compared to a normal eye [5]. Due to this axial elongation, the focal plane is in front of the retina causing the blurry vision in a myopic eye. Myopia typically develops during childhood when the eye and its tissues are still growing. Myopia is associated with a higher risk of blinding diseases such as glaucoma, retinal detachment and macular degeneration [6–9]. The purpose of this paper is to gain insight into the mechanisms underlying scleral growth and remodeling (G&R) during normal eye development and myopia development.

For many years, myopia has been the focus of extensive investigations and a substantial effort is being devoted to decipher the underlying mechanisms. During eye development, the sclera is thought to simultaneously grow and remodel. Experimental evidence from animal models increasingly points towards the existence of an active emmetropization mechanism that adapts scleral remodeling to match the eye's axial length to its focal length during eye development, thus, producing eyes with retinal focus [10–13].

Tissue growth involves an increase in tissue mass, which can occur through cell division (hyperplasia), cell enlargement (hypertrophy), and synthesis of extracellular matrix (ECM). Remodeling involves changes in material properties. These material property changes, which often are adaptive, may be due to micro-deformations within the tissue. Most soft tissues such as the sclera are regarded as incompressible or nearly incompressible within the physiological range of deformation due to their high water content. Consequently, micro-deformations that may underly scleral remodeling are thought to lead to volume-conserving deformations at the macro-scale due to the nearly incompressible property of the sclera. Within this paper, we define scleral remodeling as a mechanism that involves the rearrangement of existing material due to micro-deformations that are volume-conserving at the macro-scale, while scleral growth is a mechanism that changes the amount (volume) of scleral tissue.

The scleral ECM exhibits a complex hierarchical structure, which varies across species. We use data obtained from animal experiments in tree shrews. The tree shrew is a well established animal model for myopia and its sclera exhibits a similar hierarchical structure compared to humans [14–18]. The tree shrew as well as the human sclera is mainly composed of type I collagen fibrils at the micro-scale, which form interwoven lamellae at the meso-scale. Inducing myopia in the tree shrew (by either using a negative power lens or a diffuser) has been shown to impact the sclera by: (i) thinning [16]; (ii) reduction in dry weight (3–5 %) [16, 19, 20]; (iii) lowering hyaluronan and sulfated glycosaminoglycan (GAG) levels [20]; (iv) increasing enzymatic degradation [21–24]; (iv) decreasing collagen type I synthesis [25]; and (v) increasing scleral creep rate [26, 27]. The aforementioned findings suggest that axial elongation in myopia is not due to accelerated tissue growth as the dry weight and collagen synthesis are reduced. Instead, myopia is likely due to a remodeling mechanism that involves micro-deformations within the scleral ECM, which alter scleral biomechanics (creep rate). McBrien et al. [28] have suggested that myopia underlies biomechanical scleral weakening, which leads to a creep-like elongation of the sclera at normal intraocular pressure (IOP). Summers Rada et al. [10] suggested that the reduction in aggrecan could make it easier for scleral lamellae to slide across each other during myopia development. In a recent inverse computational study, we have predicted that the collagen fibril crimp increases during myopia development and hypothesized that this may be due to intra- and/or inter-lamellar remodeling deformations [29]. Zhu et al. [30] reported that the eyes of tree-shrews and other species can axially shorten when exposed to a myopic defocus (using a positive power lens or by removing a negative power lens after myopia was induced), which suggests that scleral remodeling may be reversible.

Bryant and McDonnell [31] proposed the first computational model to simulate axial elongation in myopia. This has been the only study that attempted to computationally model myopia development. The model was fitted to a very limited set of experimental data available at that time. Bryant and McDonnell [31] used a growth tensor to model axial elongation based on the assumption that myopia is due to accelerated tissue growth, which is in disagreement with experimental data available now. As a part of our ongoing efforts to elucidate G&R mechanisms in myopia, we propose a new multi-scale computational model to investigate the hypothesis that scleral growth and scleral remodeling are two independently controlled mechanisms. We hypothesize that scleral G&R occur simultaneously and both mechanisms contribute to the axial elongation of the eye during development. We assume that scleral growth is solely determined by genetic factors. In contrast, scleral remodeling is assumed to be controlled by both, genetic factors and visual environment (i.e., vision-guided). Our computational multi-scale model is based on the following main assumptions: (i) the sclera deforms due to external loading and G&R; (ii) scleral G&R ceases with age; (iii) scleral remodeling is modulated by the eye's refractive error; and (iv) scleral remodeling is (initially) reversible.

The model incorporates multiple length scales as the visual stimulus is calculated at the organ-level, which drives scleral remodeling at the lamella-level. Age-dependent scleral growth occurs simultaneously at the tissue-level and ceases with age. Both, scleral G&R lead to scleral elongation at the tissue-level and axial elongation at the organ-level impacting the visual stimulus and closing the vision-guided feedback loop.

This paper is organized as follows: In Sect. 2 we present our G&R formulation. In Sect. 3 we fit our model to new and previously published tree shrew data [32]. In Sect. 4 we investigate the predictive power of our new model by validating it against data that was not used to fit the model [24, 33]. Finally, we discuss the presented model and results in Sect. 5.

2 Computational Modeling of Scleral Growth & Remodeling

2.1 Evolution Equations

Our approach to model scleral G&R is based on the hypothesis that G&R are two independently controlled mechanisms that occur simultaneously at different rates and cease over time. We propose that the scleral remodeling is vision-guided and that the remodeling rate is up- or down-regulated to match the eye's axial length with its focal length during emmetropization. Based on the finding that scleral dry weight is only slightly reduced (3–5 %) [16, 19, 20] after lens induced myopia, we assume that tissue growth is not significantly altered by any visual stimulus.

We define scleral growth as a genetically determined mechanism that increases scleral tissue volume and mass without changing the tissue density. The growth rate is defined at the tissue-level by introducing the following (strictly convex) evolution equation for the so-called growth stretch λ_g

$$\dot{\lambda}_g = \frac{a_g b_g^2}{(t + b_g)^2}, \tag{1}$$

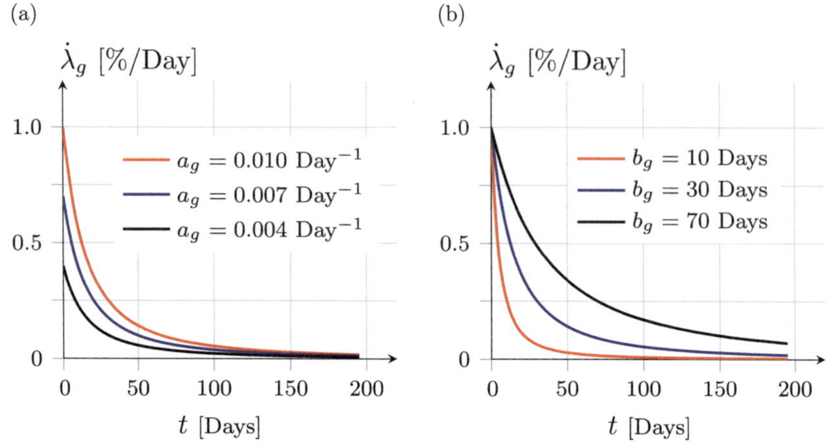

Fig. 1 The growth stretch rate $\dot{\lambda}_g$ as defined in (1). Illustrative plots are shown for different values of a_g and b_g. (**a**) shows the effect of varying the parameter a_g while $b_g = 30$ Days. (**b**) shows the effect of varying the parameter b_g while $a_g = 0.01$ Day^{-1}. The parameter a_g represents the growth rate at $t = 0$, while b_g determines the shape of the growth rate function in (1)

in which, t is time in days, a_g and b_g are parameters to be determined by fitting experimental data. We define $t = 0$ as the day at which the animal opens their eyes for the first time. Consequently, t represents time measured in days of visual experience (DVE). The growth stretch λ_g represents the in-plane elongation of the sclera over time due to tissue growth. a_g is a parameter that is related to the growth rate at $t = 0$ with the unit Days^{-1}. b_g has the same dimension as t and determines how quickly the growth ceases with t. Figure 1 shows the growth stretch rate for different values of a_g and b_g.

We define scleral remodeling as a mechanism that involves internal tissue deformations and restructuring of the ECM at the collagen fibril and/or scleral lamella-level. Similar to our previous work, we model scleral remodeling as a residual stretch between scleral lamellae and the ground substance of the sclera [34]. To mathematically describe the remodeling rate, we propose an independent evolution equation of the remodeling stretch λ_r by the following multiplicative form

$$\dot{\lambda}_r = \dot{\lambda}_r^{\max} \delta_r^E \delta_\alpha, \tag{2}$$

where $\dot{\lambda}_r^{\max}$ is the maximum remodeling rate, $\delta_r^E \in [-1; 1]$ represents the vision-guided stimulus computed at the organ-level, and $\delta_\alpha \in [0; 1]$ represents a factor that accounts for heterogeneous remodeling across the scleral shell. The sclera is assumed to continuously remodel while the eye grows to compensate for environmental changes in refractive error. Based on the refractive state of the eye, scleral remodeling can be upregulated up to its maximum remodeling rate $\dot{\lambda}_r^{\max}$. Consequently, $\dot{\lambda}_r^{\max}$ represents the upper bound of the remodeling rate and is defined as

$$\dot{\lambda}_r^{\max} = \frac{a_r^{\max}\, b_r^{\max\,2}}{(t + b_r^{\max})^2}. \tag{3}$$

The parameter $a_r^{\max}$ represents the maximum remodeling rate at $t = 0$. $b_r^{\max}$ determines the shape of $\dot{\lambda}_r^{\max}$. Similar to the growth rate (1), the maximum remodeling rate is assumed to be genetically determined and to decrease with age. These assumptions are based on the

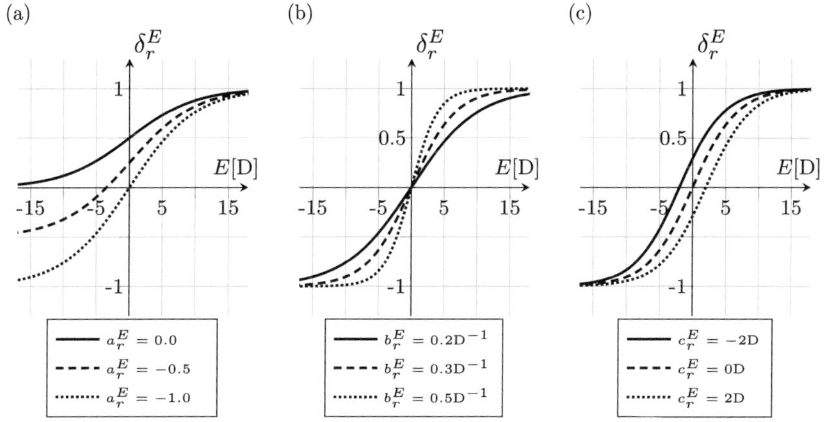

Fig. 2 The visual stimulus function as defined in (4). The function increases with the current refractive error E of the eye. (**a**) shows the effect of varying the parameter a_r^E while $b_r^E = -0.2\ \text{D}^{-1}$ and $c_r^E = 0\ \text{D}$. (**b**) shows the effect of varying the model parameter b_r^E while $a_r^E = -1.0$ and $c_r^E = 0.0\ \text{D}$. (**c**) shows the effect of varying the model parameters c_r^E while $a_r^E = -1.0$ and $b_r^E = 0.3\ \text{D}^{-1}$. The parameter $a_r^E \in [-1; 1]$ represents the lower asymptote. b_r^E represents the slope of the curve at the inflection point, while c_r^E shifts the inflection point along the E-axis

observation that the relative axial elongation rate (treated eye minus control eye) decreases with age in lens treated tree shrews [32].

Scleral remodeling is assumed to be vision-guided, where δ_r^E in (2) represents one potential cue that the eye uses to alter scleral remodeling: the refractive error of the eye. It is unclear what visual cues the retina detects but it is likely that the retina is sensitive to multiple visual stimuli. Note that other visual cues can be incorporated into our model such as the degree of form deprivation, contrast, wave length of light, and light intensity. Additional cues are likely additive terms that would contribute to δ_r^E, i.e., an animal that wears a negative lens that is also a diffuser would result in a higher remodeling stimulus than one of the single cues alone. Based on the findings by Zhu et al. [30] we assume that scleral remodeling can lead to scleral elongation (positive remodeling) and shortening (negative remodeling). In this paper, we use the generalized Richards' functions to define the visual stimulus as a function of the eye's refractive error E

$$\delta_r^E = a_r^E + \frac{1 - a_r^E}{1 + \exp(-b_r^E (E - c_r^E))}. \tag{4}$$

The visual stimulus function δ_r^E is monotonically increasing with E. The impact of the model parameters a_r^E, b_r^E, c_r^E on the vision stimulus function is illustrated in Fig. 2.

Guggenheim and McBrien [35] reported that MMP-2 upregulation was higher at the posterior sclera as compared to equatorial samples in 5 day form deprived tree shrews. Experiments in chicks and tree shrews have shown that if partial diffusers or negative lenses cover only half of the retina, only that half of the eye elongates [36–38]. These findings suggest that scleral remodeling is locally controlled and heterogeneous across the scleral shell. We use the term δ_α in (2) to account for heterogeneous remodeling across the scleral shell. We assume that scleral remodeling is highest at the posterior pole and reduces towards the anterior sclera. The heterogeneity factor δ_α is defined by the following distribution, which

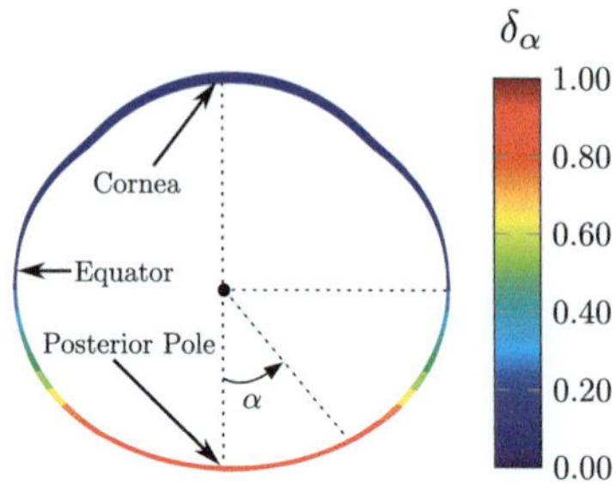

Fig. 3 The heterogeneity factor δ_α as defined in (5) is used to model the heterogeneous remodeling rate across the scleral shell. The heterogeneity factor is $\delta_\alpha = 1$ at the posterior sclera for $|\alpha| < \pi/4$ and it gradually decreases to $\delta_\alpha = 0$ towards the anterior sclera for $\frac{\pi}{4} < |\alpha| < \frac{\pi}{2}$

is illustrated in Fig. 3.

$$\delta_\alpha = \begin{cases} 1 & \text{for } |\alpha| \le \frac{\pi}{4}, \\ 2 - \frac{4}{\pi}|\alpha| & \text{for } \frac{\pi}{4} < |\alpha| < \frac{\pi}{2} \\ 0 & \text{for } |\alpha| \ge \frac{\pi}{2}. \end{cases} \quad \text{with } \alpha \in [-\pi; \pi], \tag{5}$$

The presented G&R model consists of seven unknown parameters: two parameters that define the age-related growth rate $\dot{\lambda}_g$ (a_g, b_g); two parameters that define the age-related maximum remodeling rate $\dot{\lambda}_r^{\max}$ ($a_r^{\max}$, $b_r^{\max}$); and three parameters that define the vision-guided remodeling stimulus δ_r^E (a_r^E, b_r^E, c_r^E). The temporal discretization of the evolution equations (1) and (2) were realized by an explicit update of the growth stretch λ_g and the remodeling stretch λ_r.

2.2 Constitutive Equations

The sclera can deform due to external loads (elastic deformations) and tissue G&R (inelastic deformations). Tissue growth is assumed to change the tissue volume without changing the tissue density. Similar to the work by Bryant and McDonnell [31], we adopt the concept of finite volumetric growth [39] and multiplicatively split the deformation gradient $\mathbf{F}$ into an elastic part $\mathbf{F}_e$ and a growth part $\mathbf{F}_g$

$$\mathbf{F} = \mathbf{F}_e \mathbf{F}_g. \tag{6}$$

We assume that the eye tissues grow isotropically

$$\mathbf{F}_g = \lambda_g \mathbf{I}, \tag{7}$$

where $\mathbf{I}$ is the second order identity tensor and the evolution of the growth stretch λ_g is defined in (1).

The scleral is mostly composed of fibrillar collagen type I, which aggregate into scleral lamellae that are embedded into a soft matrix (ground substance). The ground substance represents all noncollagenous tissue constituents such as elastin, GAGs, proteoglycans, and cells, and is assumed to be isotropic and nearly incompressible. We assume that the total strain energy of the sclera W can be decomposed into the energy contributions of the scleral lamellae W_{col} and ground substance W_{mat}

$$W = W_{\text{col}} + W_{\text{mat}}. \tag{8}$$

The strain energy contribution of the ground substance is approximated by an isochoric Neo-Hookean formulation

$$W_{\text{mat}} = \tfrac{1}{2}\mu\big(J_e^{-2/3}\,\text{tr}(\mathbf{C}_e)\big) + \tfrac{1}{2}\kappa(\ln J_e)^2, \tag{9}$$

which is a function of the elastic right Cauchy-Green tensor $\mathbf{C}_e$ and the elastic Jacobian J_e

$$\mathbf{C}_e = \mathbf{F}_e^{\mathrm{T}}\mathbf{F}_e = \mathbf{F}_g^{-\mathrm{T}}\mathbf{C}\mathbf{F}_g^{-1}, \qquad J_e = \det(\mathbf{F}_e) \quad \text{with } \mathbf{C} = \mathbf{F}^{\mathrm{T}}\mathbf{F}. \tag{10}$$

Two material parameters are used to define W_{mat}: the shear modulus μ and bulk modulus κ.

To account for the hierarchical collagenous structure of the sclera, the strain energy contribution of the scleral lamellae is approximated by a distribution of scleral lamellae that are composed of crimped collagen fibrils [40–42]. At the micro-scale, collagen fibrils crimp and buckle when the eye is unloaded. At the meso-scale, collagen fibrils form scleral lamellae that are strongly interwoven. We use a normalized *von Mises* distribution function $\rho(\phi)$ [42, 43] to define the in-plane distribution of scleral lamellae as a function of an *Eulerian* angle $\phi \in [-\pi/2; \pi/2]$

$$\rho(\phi) = \frac{\exp(b\cos(2\phi))}{I_0(b)\pi}, \tag{11}$$

where b is the concentration parameter and I_0 the modified Bessel function of the first kind and order zero. The three-dimensional orientations of scleral lamellae are defined by the unit vector $\mathbf{e}_0$

$$\mathbf{e}_0(\phi) = \sin(\phi)\mathbf{A}_1 + \cos(\phi)\mathbf{A}_2. \tag{12}$$

$\mathbf{A}_1$ and $\mathbf{A}_2$ are part of an orthonormal basis $\mathbf{A}_i$, where $\mathbf{A}_1$ points in the circumferential and $\mathbf{A}_2$ in the meridional direction of our eye model. Note that for $b = 0$ scleral lamellae are planar isotropic distributed tangential to the scleral shell. We define the strain energy density function of the scleral lamellae as

$$W_{\text{col}} = \int_{-\pi/2}^{\pi/2} \rho(\phi) W_{\text{fib}}\big(\lambda_{\text{lamella}}(\phi)\big) d\phi. \tag{13}$$

Scleral lamellae are composed of crimped collagen fibrils. Consequently, the strain energy in each scleral lamella orientation $\mathbf{e}_0(\phi)$ is represented by our 1-dimensional model for crimped collagen fibrils W_{fib} [40] as a function of the lamella stretch λ_{lamella}.

To account for scleral remodeling, we define the lamella stretch λ_{lamella} in (13) to be the elastic tissue-level stretch λ_e reduced by the remodeling stretch λ_r [34].

$$\lambda_{\text{lamella}}(\phi) = \frac{\lambda_e(\phi)}{\lambda_r} = \frac{(\mathbf{e}_0(\phi)\mathbf{C}_e\mathbf{e}_0(\phi))^{1/2}}{\lambda_r}. \tag{14}$$

While the elastic tissue-level stretch λ_e may differ for different orientations $\mathbf{e}_0(\phi)$, the remodeling stretch λ_r is assumed to be identical for all orientations. The evolution of the remodeling stretch λ_r is defined in (2).

The multi-scale constitutive model consists of six parameters: the shear modulus μ and the bulk modulus κ of the ground substance (tissue-level); the concentration parameter b of scleral lamella orientations (lamella-level); and the collagen fibril crimp angle θ_0, the elastic modulus of collagen fibrils E_{fib}, and the ratio R_0/r_0 between the collagen fibril crimp amplitude and fibril radius (fibril-level). The latter three parameters define the strain energy density of crimped collagen fibrils W_{fib} in (13). We refer to Grytz and Meschke [40] for a detailed derivation of W_{fib}. We assume that the eye tissues are nearly incompressible by setting the bulk modulus to $\kappa = 1000\mu$. We assume that lamellae are planar isotropic distributed ($b = 0$) in the sclera and cornea, and highly aligned in the circumferential direction at the limbus ($b = 10$). The remaining material parameters were obtained from our previous

(a)

μ	=	9.8 kPa
κ	=	9.8 MPa
θ_0	=	27.2°
E_{fib}	=	29.4 MPa
R_0/r_0	=	8.1

(b)

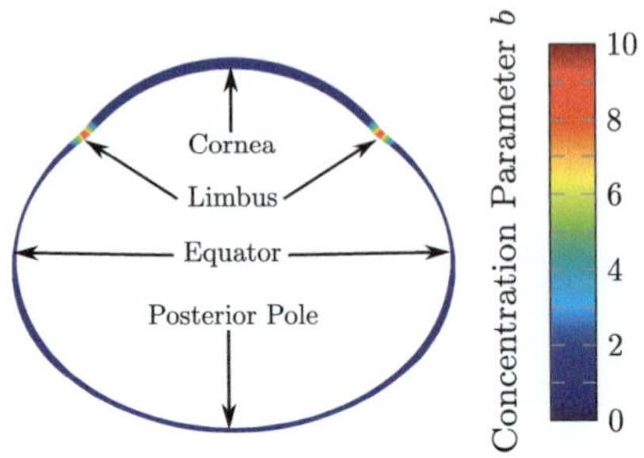

Fig. 4 The elastic model parameters used in this study. (**a**) The tissue and fibril level parameters that were assumed to be constant throughout the sclera and cornea. (**b**) Distribution of the lamella-level concentration parameter b showing high lamellar alignment in circumferential direction at the limbus while the remaining parts of the sclera and cornea were assumed to have a planar isotropic distribution of lamellae tangential to the corneo-scleral shell

inverse fitting of uniaxial tensile test of tree shrew scleral strips [29]. We used the fitted parameters of the youngest normal tree shrew group (24 DVE) from our previous study [29]. The material parameters including the distribution of the concentration factor b across the eye are summarized in Fig. 4. We used the material parameters of the sclera to also model the elastic response of the cornea as no other data was available. Note that the choice of the corneal material parameters plays a minor role in our G&R simulations presented in the following two sections as IOP remained constant in all our simulations. In addition, the corneal parameters play a minor role here as myopia development was simulated by modulating scleral remodeling, which is consistent with experimental observations.

3 Inverse Parameter Identification

3.1 Subjects

Our G&R model was fitted to previously published data from Norton et al. [32] consisting of 6 groups of tree shrews G_{1-6} that were treated with a −5 D lens on one eye at different ages ($n = 5$ per group) and one new group of normal animals (no treatment) G_0 ($n = 7$). One group of Norton et al. [32] was disregarded here as it didn't enrich the cost function used to fit our model. The lens wear started at 11, 16, 24, 35, 48, and 100 DVE in G_{1-6}, respectively. Lens treatment was performed for 11 days in G_{1-5} and for 41 days in the oldest group G_6. The treated eye developed myopia during lens treatment while the refractive development of the control eye remained normal in G_{1-6}. Afterwards, the lens was removed and the animals were allowed to recover from the lens-induced myopia. Only the data of the treated eye of G_{1-6} were used in this study. The time-dependent visual experience of all groups is summarized in Fig. 5. Refractive measurements were made throughout the lens treatment and recovery periods of groups G_{1-6} using a Nidek ARK-700A infrared autorefractor (Marco Ophthalmic, Jacksonville, FL). In addition to refractive measurements, the axial dimensions were obtained for the new normal group G_0 using a Lenstar LS-900 optical biometer (Haag-Streit USA, Mason, OH). Details about the experimental procedures can be found in Norton et al. [32]. The LenStar biometer is known to be calibrated for human eyes [44].

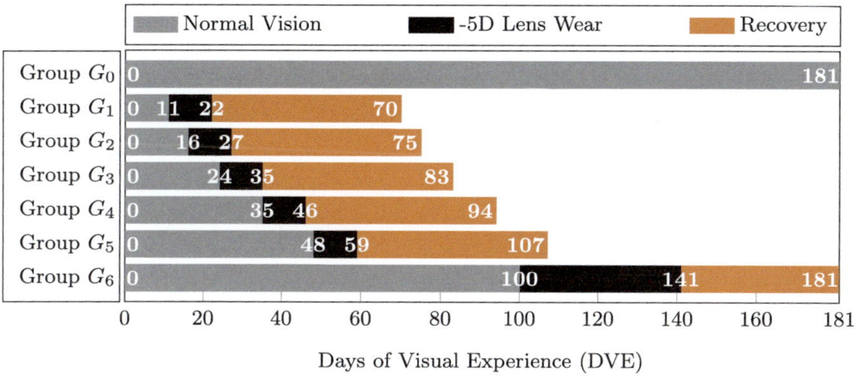

Fig. 5 The visual experience of the animal groups used to fit the G&R model. G_0 represents the normal group without lens treatment. Groups G_{1-6} represent tree shrews that were treated with a −5 D on one eye at different ages. The experimental data of groups G_{1-6} are from Norton et al. [32]. All simulations started at 11 DVE

Gann [45] compared the LenStar biometer with A-scan ultrasonography to measure ocular components in tree shrews showing that LenStar measures are typically smaller compared to A-scan measures of the same eye. Based on the regression reported by Gann [45], the axial length measurements of G_0 were translated into A-scan equivalent values by means of the following equation

$$a_{\text{A-scan}} = 0.70 + 0.93 a_{\text{LenStar}}. \tag{15}$$

The averaged refractive data of each group was used to fit the G&R model. The earliest refractive measurements were performed at 11 DVE in G_0 and G_1, which marks the starting age of all our G&R simulations.

3.2 Reduced Optical Eye Model

Following the idea of Bryant and McDonnell [31], we use the reduced schematic eye model proposed by Norton and McBrien [46] to calculate the refractive error E in our G&R simulations

$$E = \frac{1}{a-n} - \frac{1}{f} - L, \tag{16}$$

where the axial length a, the nodal length n, and the focal length f are illustrated in Fig. 6. L represents the refractive power of a lens when put in front of the eye. $L = 0$ if no lens is used.

To calculate the refractive error in our G&R simulations, two additional assumptions had to be made. First, we assume that the development of the cornea and lens is genetically defined and remains unaffected by the vision-guided mechanism that matches the eye's axial length to the focal length. This assumption is backed up by experimental observation in humans [5] and tree shrews [47]. Based on this assumption, we can consider that the focal length f and nodal length n remain unaltered during lens wear and recovery in the animal experiments G_{1-6}. Second, we assume that the ratio between the nodal and axial length $\tau = n/a$ remains constant over time and unaffected by the emmetropization process. Norton and McBrien [46] experimentally determined this ratio to be $\tau = 0.447$ for tree shrews at

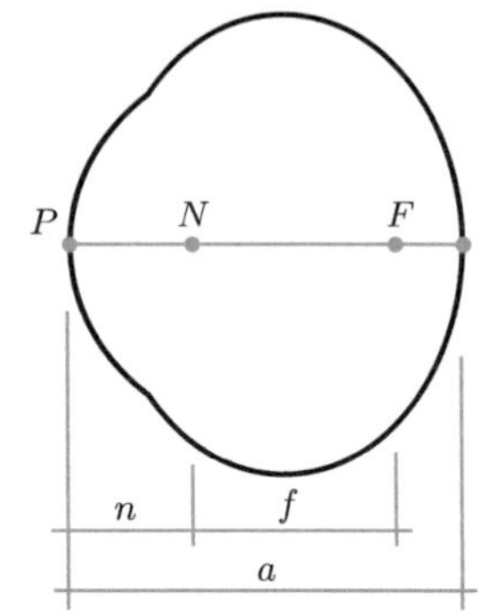

Fig. 6 The reduced schematic eye model as proposed by Norton and McBrien [46]. The points P, N, and F represent the front of the cornea, posterior nodal point, and posterior focal point, respectively. The quantities n, f, and a represent the posterior nodal length, posterior focal length and axial length, respectively. An eye is emmetropic for $a = f + n$, myopic for $a > f + n$, and hyperopic for $a < f + n$

75 DVE. Based on these two assumptions, we can compute the unknown focal length f and nodal length n from the normal group G_0, which includes experimental data of both, the refractive error E and the axial length a. If E and a are given, the nodal and focal length can be simply computed from (16) as

$$n = \tau a, \qquad f = \frac{a}{(1-\tau)^{-1} - aE}, \qquad \text{with } \tau = n/a = 0.447. \tag{17}$$

To calculate n and f at any age, curve fits were performed. The best-fit curves are shown in Fig. 7 and can be written as

$$a = 8.063 - \frac{36.87}{t + 27.14}, \qquad n = 3.604 - \frac{16.48}{t + 27.14}, \qquad f = 4.488 - \frac{26.85}{t + 43.82}, \tag{18}$$

where t represents the time in Days. Our curve-fits differ from [31], who used an exponential fit. The exponential function used by Bryant and McDonnell did not produce good fits for our data, which is mainly caused by the continued axial elongation at older ages in our new normal group G_0. Note that our normal group G_0 includes measurements at much older ages (up to 180 DVE) compared to the data used by Bryant and McDonnell [31] (up to ~ 80 DVE).

3.3 Finite Element Model of the Tree Shrew Eye

The geometry of the tree shrew eye is approximated by means of two concentric ellipses representing the corneo-scleral shells. We estimated the radii of the two ellipses (Fig. 8b) by matching the ellipses to a histological section (Fig. 8a) and the average axial length of our normal group G_0 at 11 DVE (7.095 mm). To model the scleral thickness, we used the average thickness measurements of the youngest normal group (15 DVE) reported by Kang [48]. The measurement locations were translated into the angle β shown in Fig. 8. As we assume axis-symmetry, nasal and temporal thickness measurements at β and $-\beta$ were averaged. Table 1 summarizes the symmetrized thickness measures used to create the FE mesh. We used the LenStar data of our normal group G_0 to define central corneal thickness in our model (231 μm at 11 DVE). Corneal thickness was gradually reduced from the corneal apex to the first finite element of the sclera at the limbus. In contrast to human corneas, the tree shrew cornea thins towards the limbus, which is well appreciated in the histological section in Fig. 8a.

Our constitutive model and G&R formulation was implemented into the open-source code CalculiX [49]. A 2° piece of the axisymmetric tree shrew eye model was discretized

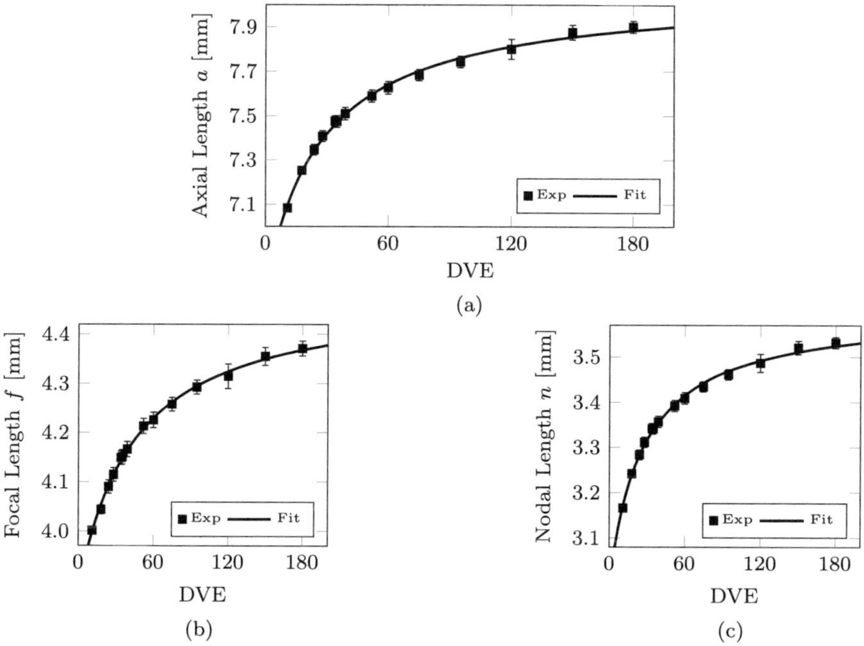

Fig. 7 Curve-fits and experimental data of the normal group G_0 for (**a**) the axial length; (**b**) the focal length; and (**c**) the nodal length. Note that the experimental values of the focal length and the nodal length were computed from the experimentally measured axial length and refractive error by using (17). The reported values in (**a**) were translated into A-scan equivalent axial lengths using (15). *Error bars* represent the standard error of mean

into 170 twenty-node quadratic brick elements with reduced integration (C3D20R) and 10 fifteen-node wedge elements (C3D15) as shown in Fig. 8d. The intraocular surfaces of the corneo-scleral shell were subjected to 15 mmHg IOP. We used our previously published prestressing method to apply IOP and presstress the model without deforming it [50].

The FE simulations of all groups started at 11 DVE and consists of up to four simulations steps: (i) prestressing at 15 mmHg IOP; (ii) normal development; (iii) −5 D lens wear; (iv) recovery from −5 D lens wear. The normal group G_0 underwent the simulation steps (i) and (ii). The group G_1 underwent the steps (i), (iii), and (iv), while the remaining groups $G_{2\text{–}6}$ used all simulation steps (i)–(iv). The simplified model of the eye's optical system in Sect. 3.2 was used to translate the axial length of the FE model into the refractive error. Axial length was defined as the distance between the corneal apex point (P_1) and the intraocular node (P_2) at the posterior pole as shown in Fig. 8d.

The seven G&R model parameters a_g, b_g, $a_r^{\max}$, $b_r^{\max}$, a_r^E, b_r^E and c_r^E were then fitted by minimizing the cost function

$$cost = \sum_{i=0}^{6} \sum_{j=1}^{N_i} \left(E_{\text{FE}}^{i,j} - E_{\text{Exp}}^{i,j}\right)^2, \tag{19}$$

which represents the sum of the squared residuals between the computationally predicted refractive error E_{FE} according to (16) and the experimentally measured refractive error E_{Exp}. $E_{\text{Exp}}^{i,j}$ represents the average experimental refractive errors of group G_i ($i = 0, 1, 2, \ldots 6$)

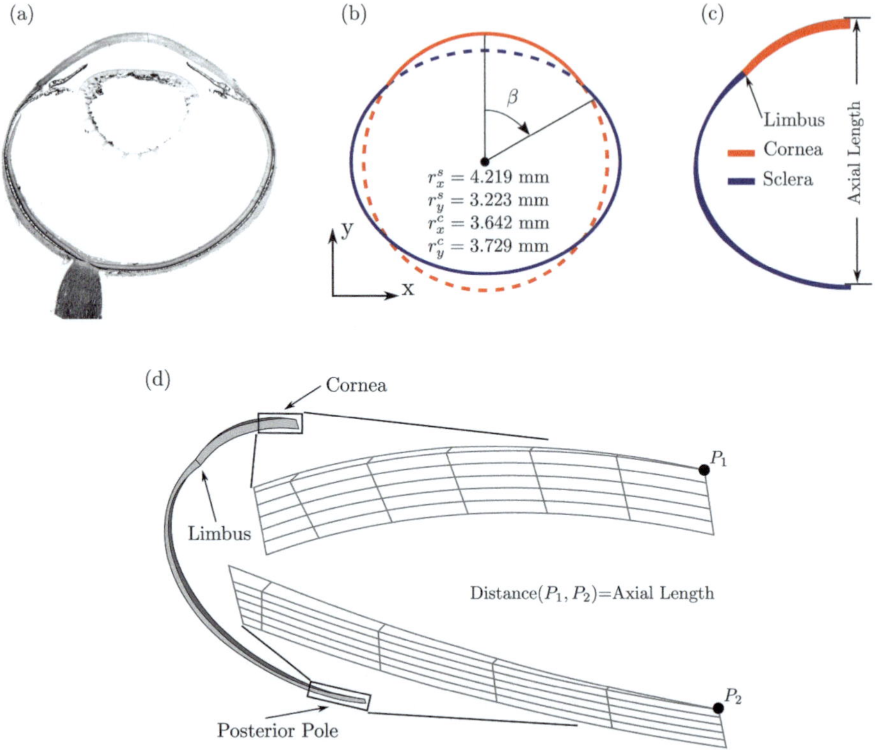

Fig. 8 (**a**) Histological section of a tree shrews eye. (**b**) The *two concentric ellipses* (*dashed* and *solid lines*) used to approximate the tree shrew eye geometry (*solid lines*). The radii in x and y direction of both ellipses are shown. (**c**) A 2D view showing the *scleral thickness* as indicated in Table 1. (**d**) The axisymmetric FE mesh of a tree shrew eye at 11 DVE with zoomed views of the mesh at the corneal apex and posterior pole. Axis-symmetry was enforced by using axisymmetric boundary conditions at the two surfaces in circumferential direction. A constant IOP of 15 mmHg was applied to the interocular surfaces of the mesh. Axial length was calculated as the distance between points P_1 and P_2

Table 1 The corneal and scleral thickness values with corresponding angles β used to generate the FE mesh. The one corneal thickness measure was obtained at the apex ($\beta = 0$) form our normal group G_0 at 11 DVE. The original scleral thickness values were obtained from the 15 DVE normal group reported by Kang [48]. Nasal and temporal scleral thickness values were averaged

β [deg]	0	77.77	84.16	90.55	96.94	103.33	109.72	116.11	122.50
Thickness [μm]	231	90.08	59.20	52.99	66.90	81.95	101.57	140.73	151.15
β [deg]	128.88	135.27	141.66	148.05	154.44	160.83	167.22	173.61	180
Thickness [μm]	144.94	130.11	122.62	114.85	112.00	119.41	110.71	121.98	130.69

at day j of visual experience. N_i represents the total number of refractive measurements of experimental group G_i. As we sum the squared residuals over all groups, the parameters obtained by the inverse fitting represent the best overall fit of our G&R model to all experimental groups.

To ensure that the fitting algorithm converged to the global minimum, we used the genetic algorithm implemented by MATLAB (The MathWorks Inc., version R2014b) for our inverse

Table 2 The upper/lower bounds of the model parameters that were used for the inverse FE analysis

Parameter	a_g	b_g	a_r^{max}	b_r^{max}	a_r^E	b_r^E	c_r^E
Unit	Days^{-1}	Days	Days^{-1}	Days	–	D^{-1}	D
Upper bound	10	10	10	400	2	10	1
Lower bound	10^{-3}	10^{-3}	10^{-4}	0.1	0.1	−10	−1

Table 3 The best fit parameters of the G&R model obtained from the inverse FE analysis

Parameter	a_g	b_g	a_r^{max}	b_r^{max}	a_r^E	b_r^E	c_r^E
Unit	Days^{-1}	Days	Days^{-1}	Days	–	D^{-1}	D
Best Fit	0.00137	10.217	0.01749	69.094	-0.51324	0.3095	0.4988

analysis. The population size was set to 150 and proportional fitness scaling was activated. The seven unknown model parameters were constrained according to Table 2. All other options of the genetic algorithm remained in their default setting.

3.4 Results

The best fit G&R parameters obtained by the inverse FE analysis are shown in Table 3. Figure 9 shows the growth rate (1), normal remodeling rate (2) and vision-guided stimulus function (4) that were obtained by the inverse FE analysis. Compared to the maximum remodeling rate, the growth rate decreased much faster with age impacting the refractive development of the eye only at very young ages (Fig. 9a, b). The vision-guided stimulus function shown in Fig. 9c suggests that scleral remodeling continues at a baseline level when emmetropia is reached. Furthermore, it also supports the notion that scleral remodeling can be reversed by scleral shortening as the stimulus function $\delta_r^E < 0$ for refractive errors $E < -1.66$ D.

The experimental and computationally predicted evolution of the refractive error in G_{0-6} are plotted in Fig. 10. For clarity, we plot the experimental and computationally predicted refractive error of each lens treated group G_{1-6} together with the normal group G_0 in Fig. 11. The evolution of the remodeling rates of each lens treated group G_{1-6} are plotted in Fig. 12 together with the normal remodeling rate of G_0 and the maximum remodeling rate. The G&R model replicated well the experimental data of all groups Figs. 10, 11. In particular, the normal refractive development and the lens treatment periods were well replicated by the model. The recovery periods were less well matched. However, it is important to note that the variation in the experimental data was much higher during the recovery period as can be seen by the error bars in Fig. 11.

Tree shrews are born hyperopic. Consequently, the initial remodeling rate (at 11 DVE) was high in our simulations (close to the maximum remodeling rate) (Fig. 9a) but the normal remodeling rate decreased faster than the maximum remodeling rate as the normal eye emmetropized (Fig. 10). When the −5 D lens was put ON in G_{1-6}, the remodeling rate increased followed by a slow decrease back to the normal remodeling rate as the eye emmetropized to the lens. Due to the age-dependent decrease in the maximum remodeling rate, the remodeling rate slowed with age from group G_1 to G_6 during lens treatment (Fig. 12). When the −5 D lens was taken OFF, the remodeling rate decreased to a negative value in all lens-treated groups (G_{1-6}) suggesting that scleral remodeling was not only slowed but

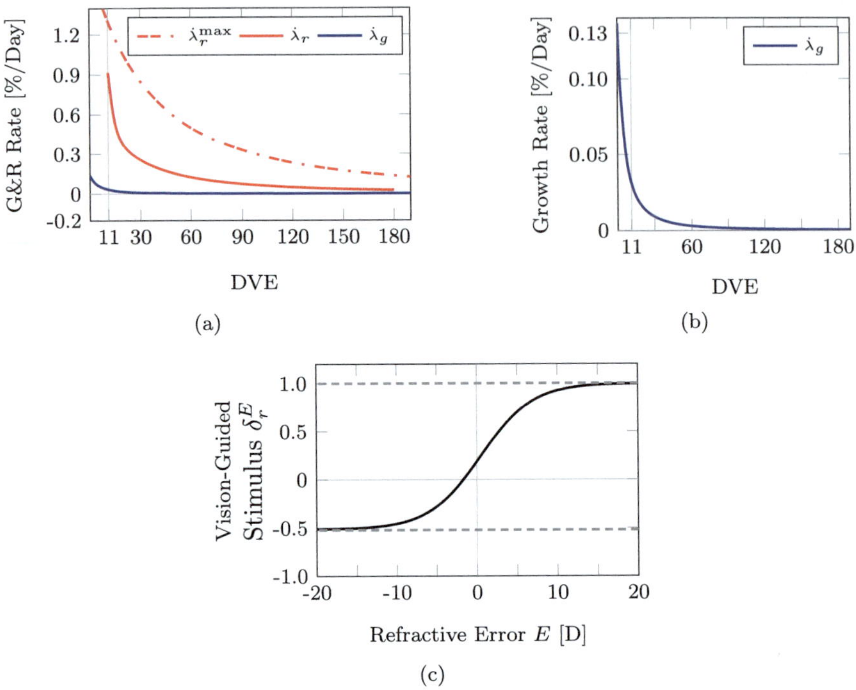

Fig. 9 (**a**) The fitted normal growth rate (1), normal remodeling rate (2), and maximum remodeling rate (3) showing that scleral growth ceased at younger age compared to normal scleral remodeling. (**b**) Magnified view of the fitted normal growth rate suggesting that scleral growth impacts the axial development of the tree shrew eye solely at very young ages. (**c**) The fitted vision-guided stimulus function as defined in (4). At emmetropia ($E = 0$), the visual stimulus is larger than zero ($\delta_r^E = 0.185$) suggesting that scleral remodeling continues at a baseline rate when emmetropia is reached. Scleral shortening ($\delta_r^E < 0$) occurs for $E < -1.66$ D

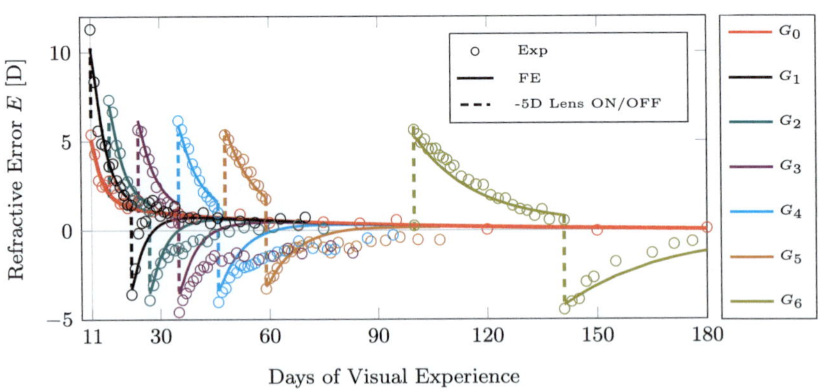

Fig. 10 Plots of the best fit results obtained by the inverse G&R analysis. Shown are the computationally calculated refractive error and the average experimental refractions of groups G_{0-6}. The experimental tree shrew data of groups G_{1-6} was obtained form Norton et al. [32]. The G&R model replicated well the experimental data, in particular, the normal development group (G_0) and lens treatment period of the lens-treated groups (G_{1-6}). The *dashed lines* indicate when the −5 D lens was put ON or OFF in groups G_{1-6} (see Fig. 5 for treatment history)

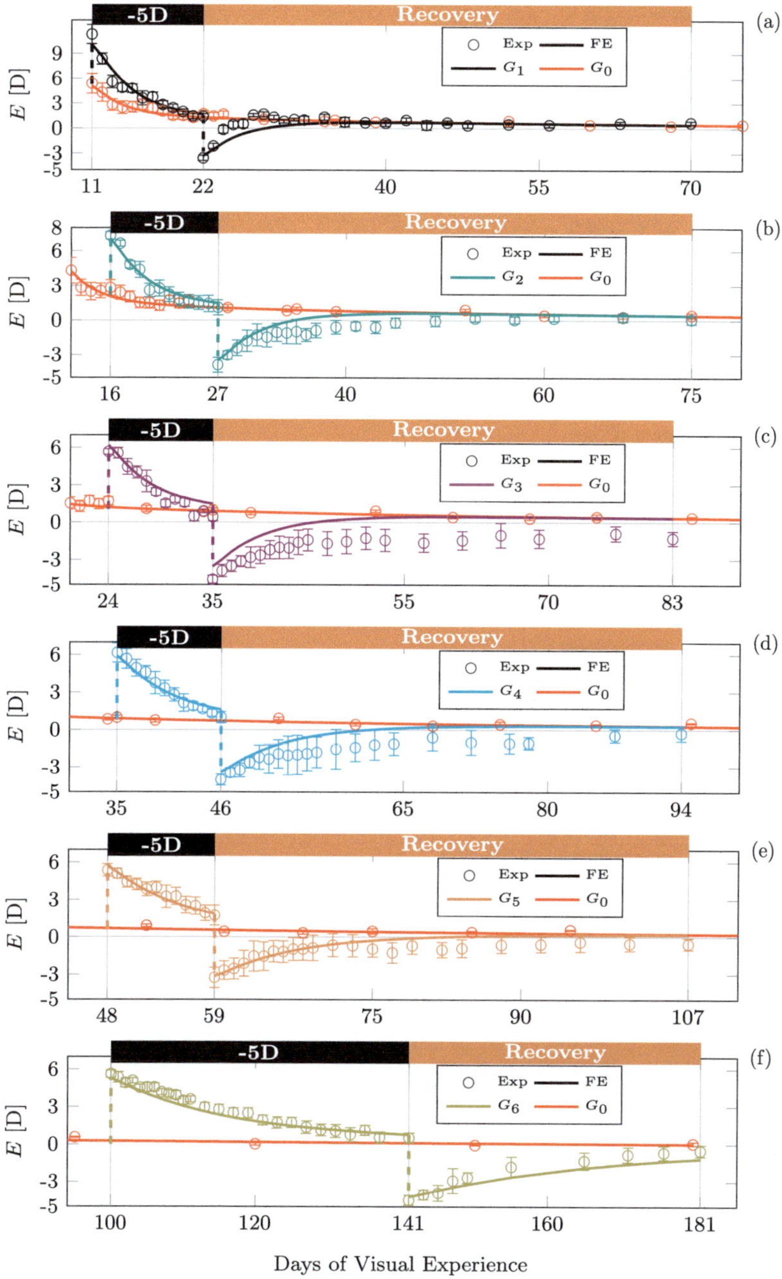

Fig. 11 Magnified plots of the computationally calculated and the experimentally measured refractive errors (average, standard error of mean) of each lens treated group G_{1-6}. The numerical results and experimental data of the normal group G_0 is shown in all plots

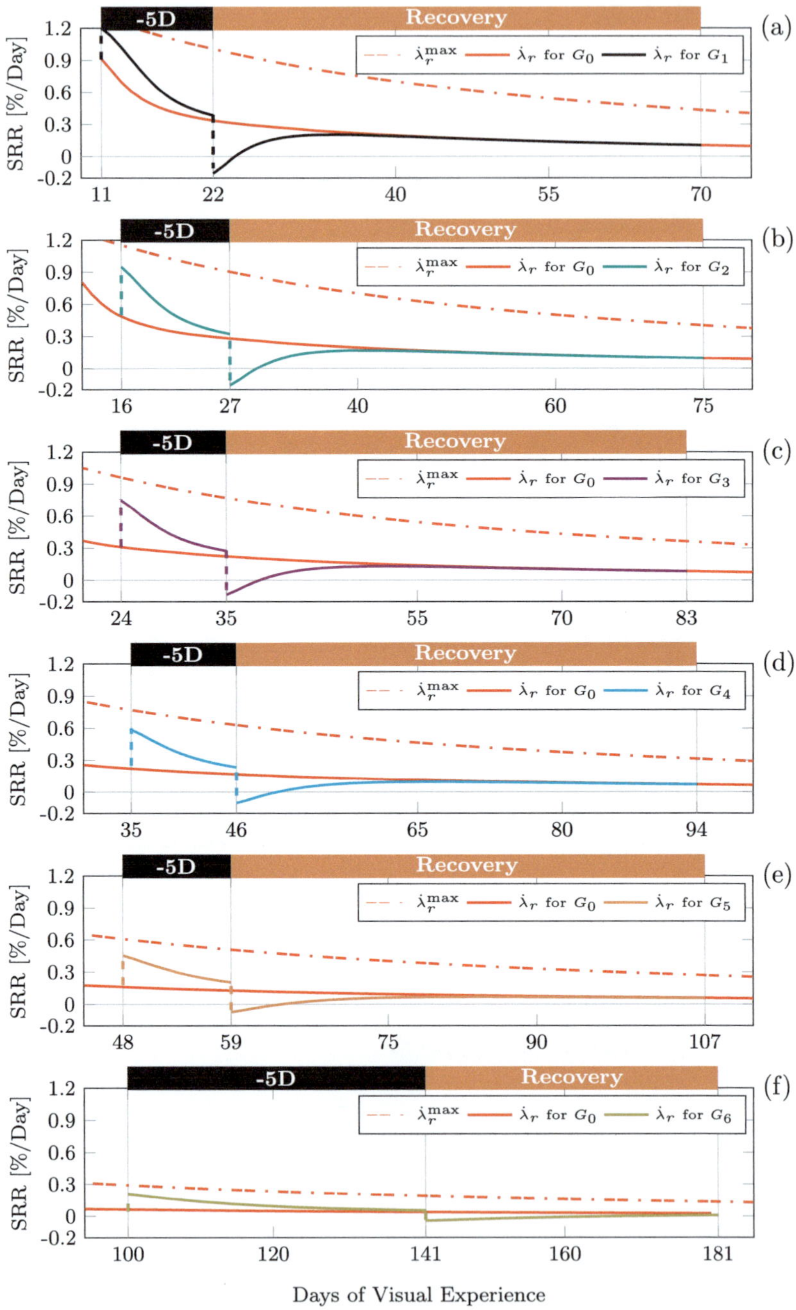

Fig. 12 The scleral remodeling rate (SRR) predicted by the inverse G&R analysis for each lens treated group G_{1-6}. The maximum remodeling rate $\dot{\lambda}_r^{\max}$ and the normal remodeling rate $\dot{\lambda}_r$ for G_0 is shown in all plots

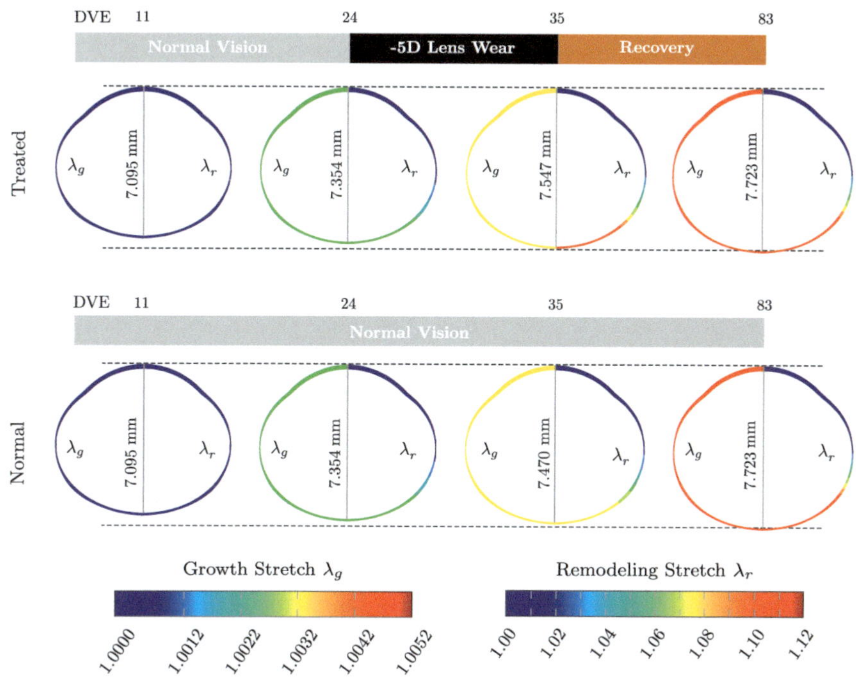

Fig. 13 Evolution of the eye's axial length, the growth stretch λ_g (*contour plot on left half* of the FE model), and the remodeling stretch λ_r (*contour plot on the right half* of the FE model) for the lens treated group G_3 and the normal group G_0 at 11, 24, 35, and 83 DVE. While the growth stretch evolves identically in both models, the remolding stretch increased in G_3 during the lens treatment when compared to normal eye development in G_0. After the recovery period, the axial length and the remodeling stretch of the lens treated FE model evolved back to normal values

temporarily reversed (Fig. 12). The remodeling rate returned back to normal levels as the eye recovered from the lens-induced myopia during the recovery period in G_{1-6}.

Figure 13 shows the evolution of the FE model including the axial length of the lens-treated group G_3 and the normal group G_0 for selected DVEs. The contour plot on the left (right) half of each eye shows the evolution growth (remodeling) stretch across the axisymmetric eye model. Based on our assumption that tissue growth is solely genetically controlled, the growth stretch evolved identically in both models. The normal and treated eyes remodeled in the same fashion until the age of 24 DVE. During lens treatment from 24 to 35 DVE the treated eye elongated at a faster rate due to accelerated scleral remolding induced by the −5 D lens. Due to the lens-induced increase in the remodeling stretch, the treated eye was axially longer at 35 DVE when compared to the normal eye (7.547 mm vs. 7.470 mm). After the lens was removed, the treated eye recovered from myopia by slowing scleral remodeling resulting in a remodeling stretch distribution and axial length that matched the normal eye at 83 DVE.

4 Model Validation

In order to test the predictive power and to validate the proposed G&R model, we compared model predictions to experimental data that was not used to fit the model.

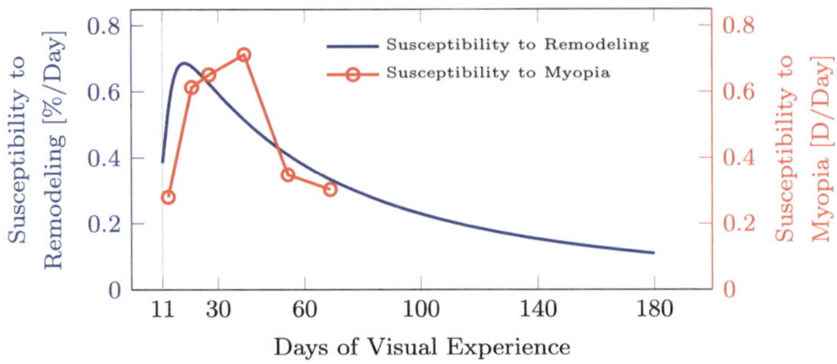

Fig. 14 Susceptibility of tree shrews to scleral remodeling predicted by the G&R model and the susceptibility to myopia experimentally obtained by Siegwart and Norton [33]. The susceptibility to myopia was measured as the difference in refractive error between the treated and control eye after 12 days of monocular form deprivation. The predicted age-dependent susceptibility to scleral remodeling follows the age-dependent trend of the susceptibility to myopia

4.1 Susceptibility to Scleral Remodeling and Myopia

Siegwart and Norton [33] induced myopia in tree shrews for 12 days with a translucent occluder (form deprivation) starting at 7, 15, 21, 33, 48, and 63 DVE ($n = 5$ in each group). In contrast to a negative powered lens, form deprivation induces a constant visual stimulus that leads to progressive axial elongation and progressive myopia. Siegwart and Norton [33] calculated the susceptibility to myopia as the difference in refractive error between the form deprived and control eye after the 12 days of monocular treatment.

Within our G&R model, the scleral remodeling rate $\dot{\lambda}_r$ decreased during normal eye development due to both, the age-dependent reduction of the maximum remodeling rate $\dot{\lambda}_r^{\max}$ and the vision-guided stimulus δ_r^E. Based on our model, a normal eye is susceptible to develop myopia when $\dot{\lambda}_r < \dot{\lambda}_r^{\max}$. We define the susceptibility to scleral remodeling as the difference between the age-dependent maximum remodeling rate $\dot{\lambda}_r^{\max}$ and the vision-guided remodeling rate $\dot{\lambda}_r$ during normal eye development (G_0). Let's assume that form deprivation induces a visual stimulus that up-regulates scleral remodeling, which, in the worst case, can accelerate scleral remodeling up to its maximum rate $\dot{\lambda}_r^{\max}$. Consequently, the susceptibility to scleral remodeling should correlate to the eye's susceptibility to myopia. The numerically predicted susceptibility to scleral remodeling and the experimentally measured susceptibility to myopia [33] are plotted in Fig. 14. The numerical results match well the temporal trend of the experimental data, where both, scleral remodeling and myopia, show an initial increase in susceptibility at juvenile age followed by a decrease in susceptibility at older ages.

4.2 Scleral Remodeling During Form Deprivation

As discussed in the previous subsection, form deprivation provides a constant visual stimulus that induces axial elongation and myopia. Siegwart and Norton [24] form deprived a cohort of tree shrews ($n = 5$) from 24 to 35 DVE followed by 4 days of recovery. We estimated the refractive development of the treated eye by assuming that form deprivation in [24] provided a constant visual stimulus that increased scleral remodeling up to its maximum rate ($\delta_r^E = 1.0$). Figure 15 shows a good agreement between the numerically predicted and experimentally measured refractive development during form deprivation and recovery underlining the predictive power of the G&R model.

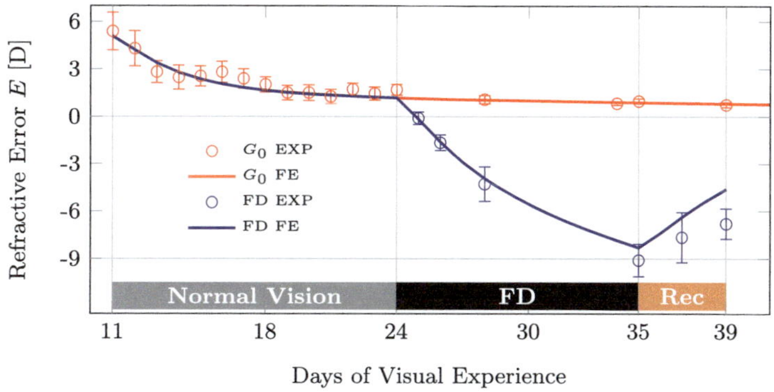

Fig. 15 The computationally predicted (FE) and experimentally measured visual development (EXP) of a tree shrew eye during 11 days of form deprivation (FD) followed by 4 days of recovery. The predicted model replicates well the experimental data. The experimental data of the form deprivation group (average, standard error of mean) were obtained from Siegwart and Norton [24]. The experimental data and numerical prediction of the normal group G_0 are shown for comparison

5 Discussion

We have presented a multi-scale G&R model to simulate refractive development during normal eye development and lens-induced myopia in the tree shrew. The visual stimulus (refractive error) was computed at the organ-level, the age-dependent growth occurred at the tissue-level, and vision-guided remodeling was modeled at the lamella-level. The model was fitted to a wide range of data obtained from tree shrews that experienced normal vision, lens-induced myopia, and recovery from myopia at different ages. The G&R model was able to replicate the age- and vision-dependent response of the tree shrew eye. The model results suggest that scleral growth ceases much faster with age than scleral remodeling. During normal development, scleral remodeling was predicted to occur at a high rate (close to maximum) in very young tree shrews (<15 DVE). The remodeling rate decreased with age and as the eye emmetropized but increased at any age when a negative lens was put on. The fitted vision-guided stimulus function suggests that scleral remodeling does not stop but continues at a baseline rate once emmetropia is reached. This baseline remodeling rate seems to be important to maintain clear vision during childhood as the ocular refractive components are still developing. Recovery from myopia was predicted to occur due to slowed scleral remodeling and a partial recovery of preceding remodeling deformations. These findings support the notion that eyes can detect the sign of blur [51] and can shorten to compensate for myopic defocus [30].

The G&R model was validated against two experiments that were not used to fit the model. The model predicted that tree shrews are most susceptible to lens-induced myopia and scleral remodeling at juvenile age, which is in good agreement with experimental observations by Siegwart and Norton [33]. The model provides a mechanistic explanation for the increased susceptibility at juvenile age. As the tree shrew eye is hyperopic at birth, the visual stimulus is high until the eye is close to emmetropia (~ 20 DVE). Consequently, the susceptibility to myopia is low in very young animals (<15 DVE) as the eye is already remodeling at almost maximum speed. Once the normal tree shrew eye is close to emmetropia (~ 20 DVE), the susceptibility to myopia increases as the vision-guided stimulus has increased potential to accelerate scleral remodeling. At older ages, the susceptibility decreases as the maximum remodeling rate decreases with age, which may be due to an age-related accumulation of collagen crosslinks. The second validation showed that the G&R model can predict the refractive development of a tree shrew eye during form deprivation [24] underlining the predictive power of the model.

It is unclear what visual cues the retina detects. Our model uses the refractive error as a visual stimulus to alter scleral remodeling. Animal experiments and prevalence studies in humans suggest that other visual cues may impact the emmetropization process such as: form deprivation [24]; darkness [52]; near work [53]; contrast sensitivity [54–57]; light intensity [58–62]; chromatic cues [63–67]. Consequently, multiple visual cues may interact and alter scleral remodeling. Additional visual cues can be incorporated into our G&R model by changing the visual stimulus function δ_r^E.

Multiple mechanisms have been suggested to underly collagen remodeling in soft tissues: micro-deformations between fibrils, fibers, or lamellae [10, 29, 68]; collagen turnover [69–75]; or collagen fiber reorientation [76, 77]. We hypothesized that scleral remodeling in myopia is due to micro-deformations between collagen fibers and/or fibrils. We also assumed that these micro-deformations cease with age as predicted by the decrease in maximum scleral remodeling rate with age. Scleral remodeling deformations may be inhibited with age due to age-dependent accumulation of collagen crosslinks. McBrien and Norton [78] have shown that preventing natural collagen crosslinking alone does not alter the refractive development but doubles the axial elongation rate when used together with form deprivation in juvenile tree shrews. A recent study by Wang and Corpuz [79] has shown that artificial crosslinking using genipin slows myopia development in guinea pig. These two experimental observations support the notion that collagen crosslinks modulate the maximum scleral remodeling rate. Different remodeling mechanisms may or may not cease with age as assumed here. For instance, optic nerve head remodeling in glaucoma typically occurs in elderly, where myopic scleral remodeling is ineffective. We and others have suggested that optic nerve head remodeling in glaucoma may be due to altered collagen synthesis and degradation and not due to micro-deformations as proposed for scleral remodeling in myopia [34, 80, 81].

The proposed G&R model is based on many simplifications and assumptions. While its predictive power was investigated, many experimental observation can not be explained yet. Our model makes no difference between a myopic defocus that is induced by a positive powered lens or by removing a negative powered lens after myopia was induced. The experimental data used in this study shows that the tree shrew eye can recover from negative lens-induced myopia at all investigated ages, even in young adult animals [32]. In contrast to the recovery from myopia, the tree shrew eye can only adapt to a plus lens at very young ages [82]. At older ages, at which the recovery from myopia is still working, using a plus lens leads to axial elongation and myopia [83] while our model would predict the opposite. The underlying mechanism that leads to a different response during recovery from myopia and plus lens wear is unclear as both cause a myopic defocus. Additional or alternative visual cues (e.g. contrast), that are currently not accounted for in our model, might modulate scleral remodeling during myopic defocus. Rucker and Wallman [64] suggest that the eye can differentiate between hyperopic and myopic defocus based on changes in luminance and color contrast. Contrast sensitivity was shown to change with the current refractive state of the eye [55], which may underly the different response during plus lens wear and recovery from minus lenses. Our current model did not incorporate the effect of contrast on scleral remodeling, which may be an important visual cue during myopic defocus and may explain why our model was less successful in matching the recovery data.

While the G&R response was modeled using a multi-scale approach, the signaling of the visual defocus from the retina to the sclera was not modeled. The signaling molecules have to pass through the choroid, which may alter the signal [84–87] and contribute to the vision-guided response by choroidal thickening/thinning [88–90]. Furthermore, experimental evidence supports the notion that visual stimulus detection and signaling from the retina to the sclera occur locally [35–38, 91]. In particular, peripheral defocus is thought to promote myopia in humans [92–97]. We imposed heterogeneous scleral remodeling across the scleral shell using the function δ_α. Further advancements of the model are required to incorporate local vision-guided effects of the choroid and sclera.

We have proposed the first multi-scale model that distinguishes scleral growth from remodeling during eye development. The good fit of our model to a wide range of age- and vision-dependent refractive data, and the good agreement of model results to experimental data that were not used to fit the model supports the notion that scleral G&R are independently controlled mechanisms. The multi-scale simulation of scleral G&R provides new insight into the emmetropization process, which cannot be achieved by experiments alone as deformations due to growth cannot be distinguished from remodeling deformations with current experimental techniques.

Acknowledgements This work was supported by the National Institutes of Health Grants R01-EY026588 (RG) and P30-EY003039 (Bethesda, Maryland); Eye Sight Foundation of Alabama (Birmingham, Alabama); and Research to Prevent Blindness (New York, New York). The authors would like to give special thanks to Dr. Thomas T. Norton for sharing his experimental data.

References

1. Dolgin, E.: The myopia boom. Nature **519**(7543), 276–278 (2015)
2. Vitale, S., Sperduto, R.D., Ferris, F.L. III: Increased prevalence of myopia in the United States between 1971–1972 and 1999–2004. Arch. Ophthalmol. **127**(12), 1632–1639 (2009)
3. Lin, L.L.K., Shih, Y.F., Hsiao, C.K., Chen, C.J.: Prevalence of myopia in Taiwanese schoolchildren: 1983 to 2000. Ann. Acad. Med. Singap. **33**(1), 27–33 (2000)
4. Morgan, I., Rose, K.: How genetic is school myopia? Prog. Retin. Eye Res. **24**(1), 1–38 (2005)
5. Heine, L.: Beitrage zur anatomie des myopischen auges. Arch. Augenheilkd. **38**, 277–290 (1899)
6. Saw, S.-M., Gazzard, G., Shih-Yen, E.C., Chua, W.-H.: Myopia and associated pathological complications. Ophthalmic Physiol. Opt. **25**(5), 381–391 (2005)
7. Mitchell, P., Hourihan, F., Sandbach, J., Wang, J.J.: The relationship between glaucoma and myopia: the blue mountains eye study. Ophthalmology **106**(10), 2010–2015 (1999)
8. Xu, L., Wang, Y., Wang, S., Wang, Y., Jonas, J.B.: High myopia and glaucoma susceptibility: the Beijing eye study. Ophthalmology **114**(2), 216–220 (2007)
9. Qiu, M., Wang, S.Y., Singh, K., Lin, S.C.: Association between myopia and glaucoma in the United States populationmyopia and glaucoma in the us population. Investig. Ophthalmol. Vis. Sci. **54**(1), 830 (2013)
10. Summers Rada, J.A., Shelton, S., Norton, T.T.: The sclera and myopia. Exp. Eye Res. **82**(2), 185–200 (2006)
11. McBrien, N.A., Gentle, A.: Role of the sclera in the development and pathological complications of myopia. Prog. Retin. Eye Res. **22**(3), 307–338 (2003)
12. Norton, T.T., Siegwart, J.T.: Animal models of emmetropization: matching axial length to the focal plane. J. Am. Optom. Assoc. **66**(7), 405–414 (1995)
13. Wallman, J., Winawer, J.: Homeostasis of eye growth and the question of myopia. Neuron **43**(4), 447–468 (2004)
14. Walls, G.L.: The Vertebrate Eye and Its Adaptive Radiation. Cranbrook Institute of Science, New York (1942)

15. Torczynski, E.: Normal and abnormal ocular development in man. Prog. Clin. Biol. Res. **82**, 35–51 (1982)
16. Norton, T.T., Rada, J.A.: Reduced extracellular matrix in mammalian sclera with induced myopia. Vis. Res. **35**(9), 1271–1281 (1995)
17. Rada, J.A., Achen, V.R., Perry, C.A., Fox, P.W.: Proteoglycans in the human sclera. Evidence for the presence of aggrecan. Investig. Ophthalmol. Vis. Sci. **38**(9), 1740–1751 (1997)
18. Norton, T.T.: Experimental myopia in tree shrews. Ciba Found. Symp. **155**, 178–194, discussion 194-9 (1990)
19. McBrien, N.A., Lawlor, P., Gentle, A.: Scleral remodeling during the development of and recovery from axial myopia in the tree shrew. Investig. Ophthalmol. Vis. Sci. **41**(12), 3713 (2000)
20. Moring, A.G., Baker, J.R., Norton, T.T.: Modulation of glycosaminoglycan levels in tree shrew sclera during lens-induced myopia development and recovery. Investig. Ophthalmol. Vis. Sci. **48**(7), 2947 (2007)
21. Guo, L., Frost, M.R., He, L., Siegwart, J.T. Jr., Norton, T.T.: Gene expression signatures in tree shrew sclera in response to three myopiagenic conditionsgene expression signatures in myopic sclera. Investig. Ophthalmol. Vis. Sci. **54**(10), 6806 (2013)
22. Gao, H., Frost, M.R., Siegwart, J.T., Norton, T.T.: Patterns of mRNA and protein expression during minus-lens compensation and recovery in tree shrew sclera. Mol. Vis. **17**, 903–919 (2011)
23. Siegwart, J.T., Norton, T.T.: Selective regulation of MMP and timp mRNA levels in tree shrew sclera during minus lens compensation and recovery. Investig. Ophthalmol. Vis. Sci. **46**(10), 3484–3492 (2005)
24. Siegwart, J.T. Jr., Norton, T.T.: The time course of changes in mRNA levels in tree shrew sclera during induced myopia and recovery. Investig. Ophthalmol. Vis. Sci. **43**(7), 2067 (2002)
25. Gentle, A., Liu, Y., Martin, J.E., Conti, G.L., McBrien, N.A.: Collagen gene expression and the altered accumulation of scleral collagen during the development of high myopia. J. Biol. Chem. **278**(19), 16587–16594 (2003)
26. Siegwart, J.T., Norton, T.T.: Regulation of the mechanical properties of tree shrew sclera by the visual environment. Vis. Res. **39**(2), 387–407 (1999)
27. Phillips, J.R., Khalaj, M., McBrien, N.A.: Induced myopia associated with increased scleral creep in chick and tree shrew eyes. Investig. Ophthalmol. Vis. Sci. **41**(8), 2028–2034 (2000)
28. McBrien, N.A., Jobling, A.I., Gentle, A.: Biomechanics of the sclera in myopia: extracellular and cellular factors. Optom. Vis. Sci. **86**(1), E23–E30 (2009)
29. Grytz, R., Siegwart, J.T. Jr.: Changing material properties of the tree shrew sclera during minus lens compensation and recovery scleral material properties in myopia. Investig. Ophthalmol. Vis. Sci. **56**(3), 2065 (2015)
30. Zhu, X., McBrien, N.A., Smith, E.L. III, Troilo, D., Wallman, J.: Eyes in various species can shorten to compensate for myopic defocus. Investig. Ophthalmol. Vis. Sci. **54**(4), 2634 (2013)
31. Bryant, M.R., McDonnell, P.J.: Optical feedback-controlled scleral remodeling as a mechanism for myopic eye growth. J. Theor. Biol. **193**(4), 613–622 (1998)
32. Norton, T.T., Amedo, A.O., Siegwart, J.T. Jr.: The effect of age on compensation for a negative lens and recovery from lens-induced myopia in tree shrews (Tupaia glis belangeri). Vis. Res. **50**(6), 564–576 (2010)
33. Siegwart, J.T. Jr., Norton, T.T.: The susceptible period for deprivation-induced myopia in tree shrew. Vis. Res. **38**(22), 3505–3515 (1998)
34. Grytz, R., Sigal, I.A., Ruberti, J.W., Meschke, G., Downs, J.C.: Lamina cribrosa thickening in early glaucoma predicted by a microstructure motivated growth and remodeling approach. Mech. Mater. **44**, 99–109 (2011)
35. Guggenheim, J.A., McBrien, N.A.: Form-deprivation myopia induces activation of scleral matrix metalloproteinase-2 in tree shrew. Investig. Ophthalmol. Vis. Sci. **37**(7), 1380–1395 (1996)
36. Diether, S., Schaeffel, F.: Local changes in eye growth induced by imposed local refractive error despite active accommodation. Vis. Res. **37**(6), 659–668 (1997)
37. Hodos, W., Kuenzel, W.J.: Retinal-image degradation produces ocular enlargement in chicks. Investig. Ophthalmol. Vis. Sci. **25**(6), 652–659 (1984)
38. Kang, R.N., Norton, T.T.: Alteration of scleral morphology in tree shrews with induced myopia. Investig. Ophthalmol. Vis. Sci. **34**, ARVO Abstract 1209 (1993)
39. Rodriguez, E.K., Hoger, A., McCulloch, A.D.: Stress-dependent finite growth in soft elastic tissues. J. Biomech. **27**(4), 455–467 (1994)
40. Grytz, R., Meschke, G.: Constitutive modeling of crimped collagen fibrils in soft tissues. J. Mech. Behav. Biomed. Mater. **2**(5), 522–533 (2009)
41. Grytz, R., Meschke, G.: A computational remodeling approach to predict the physiological architecture of the collagen fibril network in corneo-scleral shells. Biomech. Model. Mechanobiol. **9**(2), 225–235 (2009)

42. Grytz, R., Fazio, M.A., Girard, M.J.A., Libertiaux, V., Bruno, L., Gardiner, S., Girkin, C.A., Downs, J.C.: Material properties of the posterior human sclera. J. Mech. Behav. Biomed. Mater. **29**, 602–617 (2014)
43. Girard, M.J., Crawford Downs, J., Burgoyne, C.F., Suh, J.F.: Peripapillary and posterior scleral mechanics–part I: development of an anisotropic hyperelastic constitutive model. J. Biomech. Eng. **131**(5), 051011 (2009)
44. Suheimat, M., Verkicharla, P.K., Mallen, E.A.H., Rozema, J.J., Atchison, D.A.: Refractive indices used by the Haag-Streit Lenstar to calculate axial biometric dimensions. Ophthalmic Physiol. Opt. **35**(1), 90–96 (2015)
45. Gann, D.: Comparison of the Lenstar biometer and A-scan ultrasonography to measure ocular components. Master's thesis, University of Alabama at Birmingham (2013)
46. Norton, T.T., McBrien, N.A.: Normal development of refractive state and ocular component dimensions in the tree shrew (tupaia belangeri). Vis. Res. **32**(5), 833–842 (1992)
47. Marsh-Tootle, W.L., Norton, T.T.: Refractive and structural measures of lid-suture myopia in tree shrew. Investig. Ophthalmol. Vis. Sci. **30**(10), 2245–2257 (1989)
48. Kang, R.N.: A light and electronmicroscopic study of tree shrew sclera during normal development, induced myopia, and recovery. Ph.D. thesis, University of Alabama at Birmingham (1994)
49. Dhondt, G.: The Finite Element Method for Three-Dimensional Thermomechanical Applications. Wiley Online Library (2004)
50. Grytz, R., Downs, J.C.: A forward incremental prestressing method with application to inverse parameter estimations and eye-specific simulations of posterior scleral shells. Comput. Methods Biomech. Biomed. Eng. **16**(7), 768–780 (2013)
51. Park, T.W., Winawer, J., Wallman, J.: Further evidence that chick eyes use the sign of blur in spectacle lens compensation. Vis. Res. **43**(14), 1519–1531 (2003)
52. Norton, T.T., Amedo, A.O., Siegwart, J.T. Jr.: Darkness causes myopia in visually experienced tree shrews. Investig. Ophthalmol. Vis. Sci. **47**(11), 4700–4707 (2006)
53. Loman, J., Quinn, G.E., Kamoun, L., Ying, G.-S., Maguire, M.G., Hudesman, D., Stone, R.A.: Darkness and near work: myopia and its progression in third-year law students. Ophthalmology **109**(5), 1032–1038 (2002)
54. Schmid, K.L., Brinkworth, D.R., Wallace, K.M., Hess, R.: The effect of manipulations to target contrast on emmetropization in chick. Vis. Res. **46**(6–7), 1099–1107 (2006)
55. Diether, S., Gekeler, F., Schaeffel, F.: Changes in contrast sensitivity induced by defocus and their possible relations to emmetropization in the chicken. Investig. Ophthalmol. Vis. Sci. **42**(12), 3072–3079 (2001)
56. Ohlendorf, A., Schaeffel, F.: Contrast adaptation induced by defocus—a possible error signal for emmetropization? Vis. Res. **49**(2), 249–256 (2009)
57. Narayanan, R., Metha, A.: Enhanced contrast sensitivity confirms active compensation in blur adaptation. Investig. Ophthalmol. Vis. Sci. **51**(2), 1242–1246 (2010)
58. Ashby, R., Ohlendorf, A., Schaeffel, F.: The effect of ambient illuminance on the development of deprivation myopia in chicks. Investig. Ophthalmol. Vis. Sci. **50**(11), 5348–5354 (2009)
59. Ashby, R.S., Schaeffel, F.: The effect of bright light on lens compensation in chicks. Investig. Ophthalmol. Vis. Sci. **51**(10), 5247–5253 (2010)
60. Siegwart, J.T., Ward, A.H., Norton, T.T.: Moderately elevated fluorescent light levels slow form deprivation and minus lens-induced myopia development in tree shrews. Investig. Ophthalmol. Vis. Sci. **53**, ARVO E-Abstract 3457 (2012)
61. Smith, E.L., Hung, L.-F., Huang, J.: Protective effects of high ambient lighting on the development of form-deprivation myopia in rhesus monkeys. Investig. Ophthalmol. Vis. Sci. **53**(1), 421–428 (2012)
62. Karouta, C., Ashby, R.S.: Correlation between light levels and the development of deprivation myopia. Investig. Ophthalmol. Vis. Sci. **56**(1), 299–309 (2015)
63. Long, Q., Chen, D., Chu, R.: Illumination with monochromatic long-wavelength light promotes myopic shift and ocular elongation in newborn pigmented Guinea pigs. Cutan. Ocul. Toxicol. **28**(4), 176–180 (2009)
64. Rucker, F.J., Wallman, J.: Chicks use changes in luminance and chromatic contrast as indicators of the sign of defocus. J. Vis. **12**(6), 23 (2012)
65. Rucker, F.J.: The role of luminance and chromatic cues in emmetropisation. Ophthalmic Physiol. Opt. **33**(3), 196–214 (2013)
66. Smith, E.L., Hung, L.-F., Arumugam, B., Holden, B.A., Neitz, M., Neitz, J.: Effects of long-wavelength lighting on refractive development in infant rhesus monkeys. Investig. Ophthalmol. Vis. Sci. **56**(11), 6490–6500 (2015)
67. Rucker, F., Britton, S., Spatcher, M., Hanowsky, S.: Blue light protects against temporal frequency sensitive refractive changes the role of blue light in the development of myopia. Investig. Ophthalmol. Vis. Sci. **56**(10), 6121–6131 (2015)

68. Meek, K.M., Tuft, S.J., Huang, Y., Gill, P.S., Hayes, S., Newton, R.H., Bron, A.J.: Changes in collagen orientation and distribution in keratoconus corneas. Investig. Ophthalmol. Vis. Sci. **46**(6), 1948–1956 (2005)
69. Humphrey, J.D., Rajagopal, K.R.: A constrained mixture model for growth and remodeling of soft tissues. Math. Models Methods Appl. Sci. **12**(3), 407–430 (2002)
70. Gleason, R.L., Humphrey, J.D.: A mixture model of arterial growth and remodeling in hypertension: altered muscle tone and tissue turnover. J. Vasc. Res. **41**(4), 352–363 (2004)
71. Machyshyn, I.M., Bovendeerd, P.H.M., Ven, A.A.F., Rongen, P.M.J., Vosse, F.N.: A model for arterial adaptation combining microstructural collagen remodeling and 3D tissue growth. Biomech. Model. Mechanobiol. **9**(6), 671–687 (2010)
72. Ambrosi, D., Ateshian, G.A., Arruda, E.M., Cowin, S.C., Dumais, J., Goriely, A., Holzapfel, G.A., Humphrey, J.D., Kemkemer, R., Kuhl, E., Olberding, J.E., Taber, L.A., Garikipati, K.: Perspectives on biological growth and remodeling. J. Mech. Phys. Solids **59**(4), 863–883 (2011)
73. Kuhl, E.: Growing matter: a review of growth in living systems. J. Mech. Behav. Biomed. Mater. **29**, 529–543 (2014)
74. Cyron, C.J., Humphrey, J.D.: Growth and remodeling of load-bearing biological soft tissues. Meccanica (2016). doi:10.1007/s11012-016-0472-5
75. Cyron, C.J., Humphrey, J.D.: Vascular homeostasis and the concept of mechanobiological stability. Int. J. Eng. Sci. **85**, 203–223 (2014). doi:10.1016/j.ijengsci.2014.08.003
76. Pluijmert, M., Kroon, W., Delhaas, T., Bovendeerd, P.H.M.: Adaptive reorientation of cardiac myofibers: the long-term effect of initial and boundary conditions. Mech. Res. Commun. **42**, 60–67 (2012)
77. Kroon, M.: Modeling of fibroblast-controlled strengthening and remodeling of uniaxially constrained collagen gels. J. Biomech. Eng. **132**(11), 111008 (2010)
78. McBrien, N.A., Norton, T.T.: Prevention of collagen crosslinking increases form-deprivation myopia in tree shrew. Exp. Eye Res. **59**(4), 475–486 (1994)
79. Wang, M., Corpuz, C.C.C.: Effects of scleral cross-linking using genipin on the process of form-deprivation myopia in the Guinea pig: a randomized controlled experimental study. BMC Ophthalmol. **15**(1), 1–7 (2015)
80. Grytz, R., Girkin, C.A., Libertiaux, V., Downs, J.C.: Perspectives on biomechanical growth and remodeling mechanisms in glaucoma. Mech. Res. Commun. **42**, 92–106 (2012)
81. Roberts, M.D., Grau, V., Grimm, J., Reynaud, J., Bellezza, A.J., Burgoyne, C.F., Downs, J.C.: Remodeling of the connective tissue microarchitecture of the lamina cribrosa in early experimental glaucoma. Investig. Ophthalmol. Vis. Sci. **50**(2), 681–690 (2009)
82. Siegwart, J.T., Norton, T.T.: Binocular lens treatment in tree shrews: effect of age and comparison of plus lens wear with recovery from minus lens-induced myopia. Exp. Eye Res. **91**(5), 660–669 (2010)
83. Metlapally, S., McBrien, N.A.: The effect of positive lens defocus on ocular growth and emmetropization in the tree shrew. J. Vis. **8**(3), 1.1–112 (2008)
84. He, L., Frost, M.R., Siegwart, J.T. Jr., Norton, T.T.: Gene expression signatures in tree shrew choroid in response to three myopiagenic conditions. Vis. Res. **102**, 52–63 (2014)
85. Summers Rada, J.A., Palmer, L.: Choroidal regulation of scleral glycosaminoglycan synthesis during recovery from induced myopia. Investig. Ophthalmol. Vis. Sci. **48**(7), 2957–2966 (2007)
86. Li, H., Wu, J., Cui, D., Zeng, J.: Retinal and choroidal expression of BMP-2 in lens-induced myopia and recovery from myopia in guinea pigs. Mol. Med. Rep. **13**(3), 2671–2676 (2016)
87. Summers, J.A.: The choroid as a sclera growth regulator. Exp. Eye Res. **114**, 120–127 (2013)
88. Howlett, M.H.C., McFadden, S.A.: Emmetropization and schematic eye models in developing pigmented Guinea pigs. Vis. Res. **47**(9), 1178–1190 (2007)
89. Summers Rada, J.A., Wiechmann, A.F., Hollaway, L.R., Baggenstoss, B.A., Weigel, P.H.: Increased hyaluronan synthase-2 mRNA expression and hyaluronan accumulation with choroidal thickening: response during recovery from induced myopia. Investig. Ophthalmol. Vis. Sci. **51**(12), 6172–6179 (2010)
90. Troilo, D., Nickla, D.L., Wildsoet, C.F.: Choroidal thickness changes during altered eye growth and refractive state in a primate. Investig. Ophthalmol. Vis. Sci. **41**(6), 1249–1258 (2000)
91. Smith, E.L. 3rd, Hung, L.-F., Huang, J., Blasdel, T.L., Humbird, T.L., Bockhorst, K.H.: Effects of optical defocus on refractive development in monkeys: evidence for local, regionally selective mechanisms. Investig. Ophthalmol. Vis. Sci. **51**(8), 3864–3873 (2010)
92. Backhouse, S., Fox, S., Ibrahim, B., Phillips, J.R.: Peripheral refraction in myopia corrected with spectacles versus contact lenses. Ophthalmic Physiol. Opt. **32**(4), 294–303 (2012)
93. Fedtke, C., Ehrmann, K., Holden, B.A.: A review of peripheral refraction techniques. Optom. Vis. Sci. **86**(5), 429–446 (2009)
94. Benavente-Pérez, A., Nour, A., Troilo, D.: Axial eye growth and refractive error development can be modified by exposing the peripheral retina to relative myopic or hyperopic defocus. Investig. Ophthalmol. Vis. Sci. **55**(10), 6765–6773 (2014)

95. Faria-Ribeiro, M., Queirós, A., Lopes-Ferreira, D., Jorge, J., González-Méijome, J.M.: Peripheral refraction and retinal contour in stable and progressive myopia. Optom. Vis. Sci. **90**(1), 9–15 (2013)
96. Mutti, D.O., Sholtz, R.I., Friedman, N.E., Zadnik, K.: Peripheral refraction and ocular shape in children. Investig. Ophthalmol. Vis. Sci. **41**(5), 1022–1030 (2000)
97. Radhakrishnan, H., Allen, P.M., Calver, R.I., Theagarayan, B., Price, H., Rae, S., Sailoganathan, A., O'Leary, D.J.: Peripheral refractive changes associated with myopia progression. Investig. Ophthalmol. Vis. Sci. **54**(2), 1573–1581 (2013)

J Elast (2017) 129:197–212
DOI 10.1007/s10659-016-9606-1

Bulging Brains

J. Weickenmeier[1] · P. Saez[2] · C.A.M. Butler[3] ·
P.G. Young[4] · A. Goriely[5] · E. Kuhl[1,6]

Received: 23 May 2016 / Published online: 24 October 2016

Abstract Brain swelling is a serious condition associated with an accumulation of fluid inside the brain that can be caused by trauma, stroke, infection, or tumors. It increases the pressure inside the skull and reduces blood and oxygen supply. To relieve the intracranial pressure, neurosurgeons remove part of the skull and allow the swollen brain to bulge outward, a procedure known as decompressive craniectomy. Decompressive craniectomy has been preformed for more than a century; yet, its effects on the swollen brain remain poorly understood. Here we characterize the deformation, strain, and stretch in bulging brains using the nonlinear field theories of mechanics. Our study shows that even small swelling volumes of 28 to 56 ml induce maximum principal strains in excess of 30 %. For radially outward-pointing axons, we observe maximal normal stretches of 1.3 deep inside the bulge and maximal tangential stretches of 1.3 around the craniectomy edge. While the stretch magnitude varies with opening site and swelling region, our study suggests that the locations of maximum stretch are universally shared amongst all bulging brains. Our model has the potential to inform neurosurgeons and rationalize the shape and position of the skull opening,

✉ E. Kuhl
ekuhl@stanford.edu

J. Weickenmeier
weickenmeier@stanford.edu

P. Saez
pablo.saez@upc.edu

A. Goriely
goriely@maths.ox.ac.uk

1 Department of Mechanical Engineering, Stanford University, Stanford, CA 94305, USA

2 Laboratori de Calcul Numeric, Universitat Politècnica de Catalunya BarcelonaTech, 08034 Barcelona, Spain

3 Synopsys/Simpleware, Bradninch Hall, Castle Street, Exeter EX4 3PL, UK

4 College of Engineering, University of Exeter, Exeter, Devon, UK

5 Mathematical Institute, University of Oxford, Oxford OX2 6GG, UK

6 Department of Bioengineering, Stanford University, Stanford, CA 94305, USA

with the ultimate goal to reduce brain damage and improve the structural and functional outcomes of decompressive craniectomy in trauma patients.

Keywords Soft matter · Hyperelasticity · Swelling · Finite element analysis · Neuromechanics · Brain · Craniectomy

Mathematics Subject Classification 74L15 · 92B05 · 92C10 · 92C50

Einen Druck über einen gewissen Grad und über eine gewisse Zeit hinaus hält das Gehirn nicht aus. Darum ist es stets Pflicht, einen Druck auf das ungefährliche Mass von Intensität und Dauer zu reduciren.

Emil Theodor Kocher [1901]

1 Motivation

Under physiological conditions, the mechanical environment of our brain is tightly regulated. The intracranial pressure, for example, lies within a narrow window between 0 and 10 mmHg [18]. An increase in intracranial pressure—most commonly caused by traumatic brain injury, subarachnoid hemorrhage, ischemic stroke or a brain tumor—can be devastating or even fatal: It reduces cerebral perfusion, and limits the supply of metabolites and oxygen [8]. As a method of last resort, neurosurgeons remove part of the skull to allow the swollen brain to bulge outward and facilitate an immediate release of the elevated pressure [27]. This life-saving procedure, known as decompressive craniectomy [35], is typically recommended if the intracranial pressure exceeds 20 mmHg for longer than 30 minutes [25]. While a decompressive craniectomy improves short-term pressure management and survival, its survivors often experience severe long-term disabilities [23]. To date, the precise criteria related to the optimal timing of treatment, the optimal location and size of the skull opening, and the long-term functional outcome remain unclear.

From a mechanical perspective, a decompressive craniectomy is a compromise between maximizing the management of the intracranial pressure and minimizing the deformations induced by the bulging brain [19]. Recent studies have characterized bulge kinematics based on computerized tomography images before and after a decompressive craniectomy using non-linear image registration [44]; yet, little is know about the stress, stretch, and strain inside the brain. While our mechanical intuition tells us that stretch and strain can be reduced by increasing the opening size, a larger opening area is more prone to infection and clinically undesirable [39]. Clinical guidelines suggest opening diameters of about 12 cm [40], but the rationale for this recommendation is rather vague and lacks a clear mechanistic understanding of the bulging process itself.

When aiming to optimize the craniectomy size, it is unclear to which extent the location of the opening influences the stretch and strain profiles across the brain [13]. The most common procedure in clinical practice is a unilateral craniectomy with an opening on either the left or the right lateral skull depending on the side of the swelling [35]. Recent clinical studies have challenged the engineering intuition that a collateral craniectomy with the opening at the site of swelling is less invasive than a contralateral craniectomy with the opening at the opposite, non-injured side [43]. While the opening size for a unilateral craniectomy is anatomically limited, a bilateral craniectomy with a bifrontal opening across both hemispheres provides sufficient anatomic space for large opening sizes [27]. Yet, the precise bulging kinematics for the different types of craniectomy are far from being completely understood.

Mathematical models and computational simulations can provide analytical and numerical insight into the strain, stretch, and stress fields of bulging solids. Using the classic theory of contact mechanics [3, 24], we have recently shown that in the small deformation limit, the bulging problem is conceptually similar to an inverted punch problem [19]. This allows us to solve the bulging problem explicitly for a bulging half-space under plane strain, plane stress, and axisymmetric conditions. The explicit analytical solution for the stress field motivates the introduction of damage drops, drop-shaped zones of high and low shear stress with singularities that scale with the inverse square root of the distance from the opening [45]. Interestingly, the shape of the bulge, the singularities of the stress profile, and the orientation of the drops are generic for all bulging problems and independent of the constitutive model. These characteristic features also agree nicely with computational simulations, both in the linear [12] and in the nonlinear [45] regime. However, it remains unclear how these characteristics evolve in geometries as complex as the human brain. First attempts along these lines have modeled the brain via its convex hull embedded in a rigid skull [16], and shown that shear strains can reach values up to 25 %, even for bulge volumes of only 22 ml [14]. While these numbers clearly highlight the need for a kinematically and constitutively nonlinear formulation, the bulging brain has never been modeled using the nonlinear field theories of mechanics.

Here we introduce a continuum model for bulging brains in the finite deformation setting. We model brain tissue as a swelling, elastically incompressible Mooney-Rivlin solid and illustrate how to translate its mathematical model into a general, nonlinear finite element environment. To demonstrate the features of the bulging problem under finite deformations, we conduct a series of case studies and perform systematic sensitivity analyses with respect to the swelling area, the opening size, and the opening location. We then create a personalized brain model from magnetic resonance images and simulate two different cases of craniectomy, a left unilateral flap and a frontal flap. For both cases, we study three swelling scenarios, swelling in both hemispheres, exclusively in the left hemisphere, and exclusively in the right hemisphere. We report and compare displacements, deformations, radial and tangential stretches, and maximum principal strains.

2 Brain Model

To model brain tissue, we adopt a classical hyperelastic constitutive formulation [22]. We follow the recommendation to approximate brain as an isotropic material since our deformation rates are moderate [47]. To characterize the brain at finite deformations, we introduce the nonlinear deformation map $\boldsymbol{\varphi}$ and its gradient $\boldsymbol{F} = \nabla_X \boldsymbol{\varphi}$ with respect to the coordinates $\boldsymbol{X}$ in the undeformed reference configuration. We allow parts of the brain to swell [29], and decompose the deformation gradient multiplicatively into an elastic part $\boldsymbol{F}^{\mathrm{e}}$ and a swelling part $\boldsymbol{F}^{\mathrm{s}}$,

$$\boldsymbol{F} = \nabla_X \boldsymbol{\varphi} = \boldsymbol{F}^{\mathrm{e}} \cdot \boldsymbol{F}^{\mathrm{s}} \quad \text{with } J = \det(\boldsymbol{F}) = J^{\mathrm{e}} J^{\mathrm{s}}. \tag{1}$$

The Jacobian J denotes the total volume change and $J^{\mathrm{e}} = \det(\boldsymbol{F}^{\mathrm{e}})$ and $J^{\mathrm{s}} = \det(\boldsymbol{F}^{\mathrm{s}})$ denote the volume change associated with the elastic deformation and with swelling. We then make two major kinematic assumptions: We assume that the elastic behavior is incompressible, $J^{\mathrm{e}} = 1$, such that the total volume change is caused exclusively by swelling, $J = J^{\mathrm{s}}$, and that swelling is volumetric, $\boldsymbol{F}^{\mathrm{s}} = (J^{\mathrm{s}})^{1/3}\boldsymbol{I}$, such that the isochoric deformation is purely elastic $\bar{\boldsymbol{F}} = \boldsymbol{F}^{\mathrm{e}}$. These assumptions imply that we can decompose the deformation gradient $\boldsymbol{F}$ into a volumetric contribution purely associated with swelling, $J = J^{\mathrm{s}}$, and an isochoric contribution purely associated with the elastic deformation, $\bar{\boldsymbol{F}} = \boldsymbol{F}^{\mathrm{e}}$,

$$\boldsymbol{F} = \nabla_X \boldsymbol{\varphi} = J^{1/3} \bar{\boldsymbol{F}} \quad \text{with} \quad J = \det(\boldsymbol{F}) \quad \text{and} \quad \bar{\boldsymbol{F}} = J^{-1/3} \boldsymbol{F}. \tag{2}$$

We introduce the left Cauchy-Green deformation tensor $\boldsymbol{b}$ and decompose it into its swelling-induced volumetric contribution in terms of the Jacobian J and its elastic isochoric contribution $\bar{\boldsymbol{b}}$,

$$\boldsymbol{b} = \boldsymbol{F} \cdot \boldsymbol{F}^{\mathrm{t}} = J^{2/3} \bar{\boldsymbol{b}} \quad \text{with} \quad \bar{\boldsymbol{b}} = \bar{\boldsymbol{F}} \cdot \bar{\boldsymbol{F}}^{\mathrm{t}}. \tag{3}$$

To characterize the swelling-induced deformation, we explore three kinematic metrics associated with the Green-Lagrange strain tensor,

$$\boldsymbol{E} = \frac{1}{2}\big[\boldsymbol{F}^{\mathrm{t}} \cdot \boldsymbol{F} - \boldsymbol{I}\big], \tag{4}$$

the maximum principal strain, $\lambda_E^{\max}$, associated with the eigenvalue problem of the Green-Lagrange strain tensor $\boldsymbol{E}$,

$$\boldsymbol{E} \cdot \boldsymbol{n}_E = \lambda_E \boldsymbol{n}_E \quad \text{and} \quad \lambda_E^{\max} = \max\{\lambda_E\}, \tag{5}$$

the normal stretch along the axon, and the tangential stretch perpendicular to the axon. We then introduce the invariants I_1, I_2, and I_3, in terms of the left Cauchy-Green deformation tensor $\boldsymbol{b}$,

$$\begin{aligned} I_1 &= \mathrm{tr}(\boldsymbol{b}) && \text{with} \quad \partial I_1/\partial \boldsymbol{b} = \boldsymbol{I}, \\ I_2 &= \frac{1}{2}\big[\mathrm{tr}^2(\boldsymbol{b}) - \mathrm{tr}\big(\boldsymbol{b}^2\big)\big] && \text{with} \quad \partial I_2/\partial \boldsymbol{b} = I_1 \boldsymbol{I} - \boldsymbol{b}, \\ I_3 &= \det(\boldsymbol{b}) && \text{with} \quad \partial J/\partial \boldsymbol{b} = \frac{1}{2} J \boldsymbol{b}^{-1} \end{aligned} \tag{6}$$

and their elastic, isochoric counterparts $\bar{I}_1$, $\bar{I}_2$, and $\bar{I}_3$, either in terms of the isochoric left Cauchy-Green deformation tensor $\bar{\boldsymbol{b}}$ or in terms of the isochoric principal stretches $\bar{\lambda}_1$, $\bar{\lambda}_2$, and $\bar{\lambda}_3$,

$$\begin{aligned} \bar{I}_1 &= \mathrm{tr}(\bar{\boldsymbol{b}}) &&= J^{-2/3}\, I_1 = \bar{\lambda}_1^2 + \bar{\lambda}_2^2 + \bar{\lambda}_3^2, \\ \bar{I}_2 &= \frac{1}{2}\big[\mathrm{tr}^2(\bar{\boldsymbol{b}}) - \mathrm{tr}\big(\bar{\boldsymbol{b}}^2\big)\big] &&= J^{-4/3}\, I_2 = \bar{\lambda}_1^{-2} + \bar{\lambda}_2^{-2} + \bar{\lambda}_3^{-2}, \\ \bar{I}_3 &= \det(\bar{\boldsymbol{b}}) &&= J^{-6/3}\, I_3 = 1. \end{aligned} \tag{7}$$

Many common constitutive models for brain tissues are special cases of the general Ogden model [33],

$$\bar{\psi} = \sum_{i=1}^{N} \frac{c_i}{\alpha_i}\big[\bar{\lambda}_1^{\alpha_i} + \bar{\lambda}_2^{\alpha_i} + \bar{\lambda}_3^{\alpha_i} - 3\big], \tag{8}$$

parameterized in terms of the Ogden parameters c_i and α_i. For the special case of $N = 2$, with $\alpha_1 = 2$ and $\alpha_2 = -2$, the Ogden model simplifies to the Mooney-Rivlin model [32, 36],

$$\bar{\psi} = \frac{1}{2} c_1\big[\bar{\lambda}_1^2 + \bar{\lambda}_2^2 + \bar{\lambda}_3^2 - 3\big] + \frac{1}{2} c_2\big[\bar{\lambda}_1^{-2} + \bar{\lambda}_2^{-2} + \bar{\lambda}_3^{-2} - 3\big], \tag{9}$$

which we can reformulate in terms of the elastic isochoric invariants $\bar{I}_1$ and $\bar{I}_2$,

$$\bar{\psi} = \frac{1}{2} c_1[\bar{I}_1 - 3] + \frac{1}{2} c_2[\bar{I}_2 - 3]. \tag{10}$$

The Mooney-Rivlin parameters c_1 and c_2 are related to the shear modulus μ as $c_1 + c_2 = \frac{1}{2}\mu$, and their values can be identified through finite deformation experiments [15, 31]. We enforce the elastic incompressibility constraint, $J^e - 1 = 0$, in the form, $J - J^s = 0$, via a Lagrange multiplier p, and add the term $p[J - J^s]$ to the energy functional,

$$\psi = \frac{1}{2} c_1 [\bar{I}_1 - 3] + \frac{1}{2} c_2 [\bar{I}_2 - 3] + p[J - J^s]. \tag{11}$$

To derive the stresses, it proves convenient to reformulate the energy in terms of the overall invariants I_1 and I_2 and the Jacobian J,

$$\psi = \frac{1}{2} c_1 [J^{-2/3} I_1 - 3] + \frac{1}{2} c_2 [J^{-4/3} I_2 - 3] + p[J - J^s]. \tag{12}$$

We can then directly obtain the Kirchhoff stress,

$$\boldsymbol{\tau} = \frac{\partial \psi}{\partial \boldsymbol{F}} \cdot \boldsymbol{F}^t = 2 \frac{\partial \psi}{\partial \boldsymbol{b}} \cdot \boldsymbol{b} = 2 \left[\frac{\partial \psi}{\partial I_1} \frac{\partial I_1}{\partial \boldsymbol{b}} + \frac{\partial \psi}{\partial I_2} \frac{\partial I_2}{\partial \boldsymbol{b}} + \frac{\partial \psi}{\partial J} \frac{\partial J}{\partial \boldsymbol{b}} \right] \cdot \boldsymbol{b} \tag{13}$$

or, with the derivatives of the invariants in Eq. (6),

$$\boldsymbol{\tau} = 2 \left[\frac{\partial \psi}{\partial I_1} + I_1 \frac{\partial \psi}{\partial I_2} \right] \boldsymbol{b} + 2 \frac{\partial \psi}{\partial I_2} \boldsymbol{b}^2 + J \frac{\partial \psi}{\partial J} \boldsymbol{I}. \tag{14}$$

Using the definition of the energy (12), we obtain the following explicit representation of the Kirchhoff stress $\boldsymbol{\tau}$ for a volumetrically swelling, elastically incompressible, Mooney-Rivlin material [17],

$$\boldsymbol{\tau} = [c_1 + \bar{I}_1 c_2] \bar{\boldsymbol{b}} - c_2 \bar{\boldsymbol{b}}^2 - \left[\frac{1}{3} \bar{I}_1 c_1 + \frac{2}{3} \bar{I}_2 c_2 + Jp \right] \boldsymbol{I}. \tag{15}$$

The isochoric contributions to the third term, $\frac{1}{3}\bar{I}_1 c_1 + \frac{2}{3}\bar{I}_2 c_2$, reflect the fact that we have formulated the Mooney-Rivlin model in terms of the isochoric invariants $\bar{I}_1$ and $\bar{I}_2$ and not of the total invariants I_1 and I_2. Even though the elastic behavior is incompressible, the overall behavior is not, and the isochoric invariants $\bar{I}_1$ and $\bar{I}_2$ indirectly depend on the amount of swelling J. Rather than rewriting the energy formulation in Eq. (12), we could have introduced the Kirchhoff stress as $\boldsymbol{\tau} = \partial \Psi / \partial \bar{\boldsymbol{b}} : \mathbb{P} \cdot \boldsymbol{b}$, where $\mathbb{P} = \partial \bar{\boldsymbol{b}} / \partial \boldsymbol{b}$ denotes the spatial fourth order isochoric projection tensor, to obtain the term, $\frac{1}{3}\bar{I}_1 c_1 + \frac{2}{3}\bar{I}_2 c_2$, from the isochoric projection with $\mathbb{P}$ [22].

In our continuum model, we prescribe the amount of swelling J^s pointwise and phenomenologically rather than modeling the swelling process itself [29]. We gradually increase the local tissue volume as $\Delta V = [J^s - 1.0] \cdot 100\ \%$. In our computational model, we represent volumetric swelling via volumetric thermal expansion [1], and only allow selected regions of the cerebral white matter tissue to swell, while all other substructures remain purely elastic with $J^s \doteq 1.0$. We enforce the incompressibility constraint $J - J^s = 0$ by using a hybrid finite element formulation with displacement degrees of freedom for the isochoric part and pressure degrees of freedom for the volumetric part of the deformation.

3 Bulging of a Hemidisk

Our previous analysis of the bulging of a linear elastic half-space through an opening has revealed two interesting features related to damage and stress distributions and relevant to the problem of craniectomy: Large fiber stretches develop deep in the center of the bulge and large shear stresses develop around the opening edge [19, 45]. These previous results

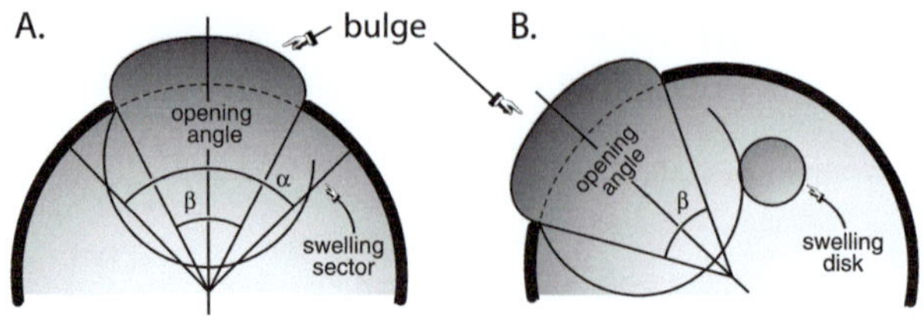

Fig. 1 Bulging of a hemidisk. We allow an elastic body to swell locally, either in a sector of angle α (**A**) or in a disk (**B**). The swelling body bulges out through an opening of angle β

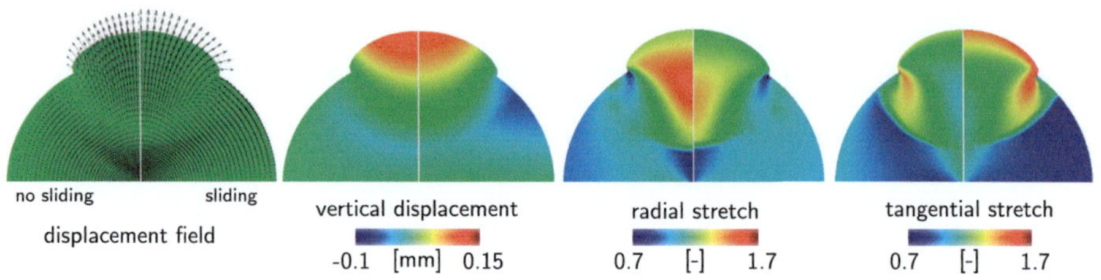

Fig. 2 Bulging of a hemidisk with a swelling sector. Displacement, vertical displacement, and radial and tangential stretches for frictional contact without sliding (*left*) and frictionless contact with sliding (*right*). In frictional contact without sliding, the solid is pushed outward with large displacements along the symmetry axis in the center of the bulge. In frictionless contact with sliding, the solid slides along the boundary and rotates outward around the opening edge

were obtained analytically under the assumption of uniform swelling in a rectangular half-space geometry. In this section, we study the importance of geometric effects in an idealized geometry computationally by systematically varying the location and area of swelling.

3.1 Hemidisk Model

We first consider the bulging problem in a simple two-dimensional geometry. As depicted in Fig. 1, an incompressible isotropic elastic hemidisk is swelling and the deformations are constrained within the hemidisk except in an opening of angle β. We consider two swelling scenarios: the swelling of a sector with an opening α where both α and β are centered about the axis of symmetry as illustrated in Fig. 1A, and the swelling of a disk where the opening β is inclined off the axis of symmetry as illustrated in Fig. 1B. We assume that the axonal direction $\boldsymbol{n}$ is oriented radially outward, with $\boldsymbol{t}$ denoting the tangential direction. For both swelling scenarios, we present the radial or normal stretch $\lambda_n = [\boldsymbol{n} \cdot \boldsymbol{F}^{\mathrm{t}} \cdot \boldsymbol{F} \cdot \boldsymbol{n}]^{1/2}$ and the tangential or shear stretch $\lambda_t = [\boldsymbol{t} \cdot \boldsymbol{F}^{\mathrm{t}} \cdot \boldsymbol{F} \cdot \boldsymbol{t}]^{1/2}$.

3.2 Bulging of a Hemidisk with a Swelling Sector

An important consideration for the stress distribution within the solid is the type of contact. By definition, the boundary of the bulge is traction free. On the base of the hemidisk, we assume no sliding. On the curved part of the contact region, we use two types of boundary conditions: either frictional contact without sliding or frictionless contact with sliding. To visualize both contact conditions side by side, we only show half of the hemidisk for each contact condition, frictional contact on the left and frictionless contact on the right.

Figure 2 illustrates the impact of the contact condition for a swelling sector of $\alpha = 80°$ and an opening angle of $\beta = 60°$ at a swelling magnitude of $J = 1.2$. For frictional contact without sliding shown on the left, the boundary nodes are fixed. Upon swelling, the solid is

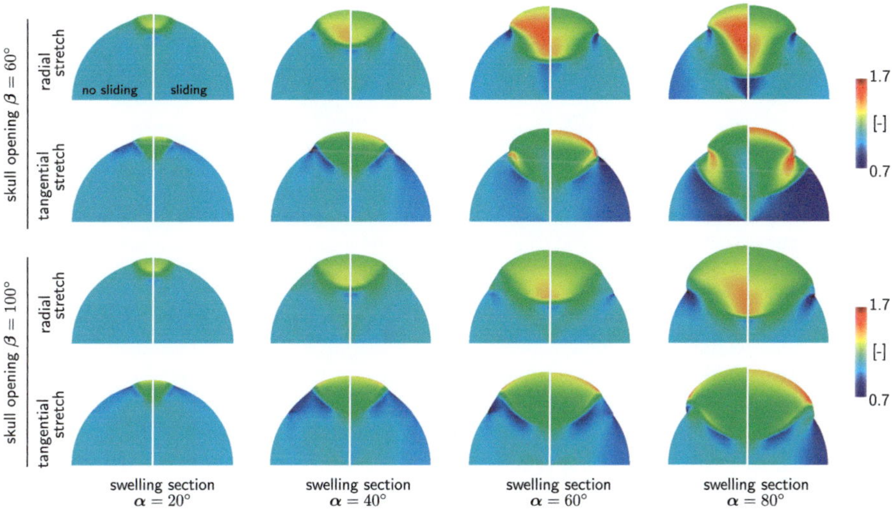

Fig. 3 Bulging of a hemidisk with a swelling sector. Radial and tangential stretches for varying opening angles of β (*rows*) and for varying swelling sector angles α (*columns*) for frictional contact without sliding (*left*) and frictionless contact with sliding (*right*). Radial stretches take maximum values of 1.7 in regions deep inside the bulge; tangential stretches take maximum values of 1.7 in regions localized around the craniectomy edge

pushed outward with large displacements along the symmetry axis in the center of the bulge. For frictionless contact with sliding, the boundary nodes are allowed to slide freely along the contact region. Upon swelling, the solid slides along the boundary and rotates outward around the opening edge.

Figure 3 illustrates a sensitivity analysis with respect to the opening angle β, the swelling sector angle α, and the contact condition for a swelling magnitude of $J = 1.2$. As the angle α of the swelling sector increases with $\alpha = 20°, 40°, 60°, 80°$, from left to right, the relative swelling area increases as $\Delta A/A = \alpha/\pi[J - 1]$. Figure 3 reveals a number of expected and new features: (i) As expected, an increase in skull opening reduces the maximal deformation and with it the maximal stretch; (ii) The radial stretch is maximal in a zone deep inside the bulge and increases rapidly as the swelling increases; (iii) The tangential stretch is maximal at the opening edge in a zone that takes the form of a drop; (iv) The maximal radial stretch is markedly higher in the case of frictional contact without sliding than in the frictionless contact case; (v) The maximal tangential stretch is markedly higher in the case of frictionless contact with sliding than in the frictional contact case. These features and trends appear to be shared broadly by all bulging cases.

3.3 Bulging of a Hemidisk with a Swelling Disk

In the case of tumor-induced swelling, it is likely that the swelling region takes a spherical rather than a sector shape. To explore the effects of a swelling disk and analyze the sensitivity of the swelling location with respect to the location of the skull opening, we study five cases with varying swelling locations for a swelling of $J = 1.3$. For all five cases, we model the contact region as frictionless with sliding.

Figure 4 illustrates the radial and tangential stretches for the bulging hemidisk with a varying position of the swelling disk. In all five cases, swelling is a local event. Maximal

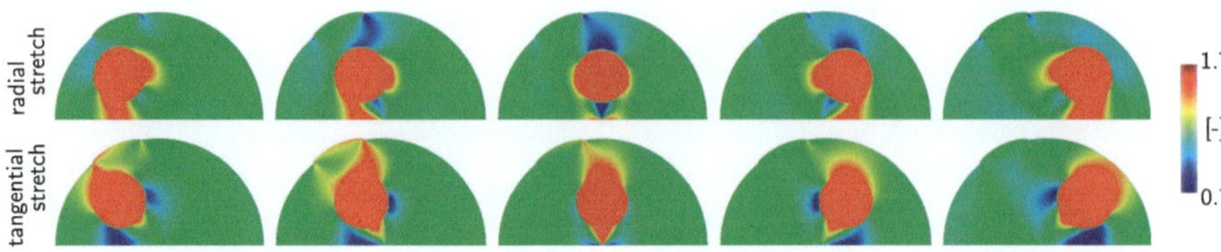

Fig. 4 Bulging of a hemidisk with a swelling disk. Radial and tangential stretches (*rows*) for five different swelling locations (*columns*). Radial and tangential stretches take maximal and mimimal values around the swelling disk, while large regions of the hemidisk are unaffected by the local swelling

and minimal stretches are localized close to the swelling disk. Except for the swelling region itself, the overall stretch profile is rather insensitive to the location of swelling.

4 Bulging of a Personalized Brain

To simulate the effects of swelling in an anatomically detailed brain geometry, we create a personalized human head model from magnetic resonance images and simulate six different scenarios: a decompressive craniectomy with either unilateral flap or frontal flap subjected to both left and right, exclusively left, and exclusively right hemispherical swelling of the white matter tissue.

4.1 Personalized Brain Model

Figure 5 shows representative sagittal, coronal, and transverse slices of an adult female head that form the basis of our anatomic model. The brain has a total volume of 1,108 cm^3, a surface area of 1,673 cm^2, and an average cortical thickness of 0.252 mm. Our magnetic resonance image set contains a total of 190 slices in the sagittal plane at a spacing of 0.9 mm. Each slice has a matrix representation of 256×256 pixels with an in-plane resolution of 0.9×0.9 mm [38]. From the magnetic resonance images, we create a personalized high-resolution anatomic model of the brain using the ScanIP software environment of Simpleware [48]. This semi-automatic software iteratively produces an anatomically detailed and geometrically accurate three-dimensional reconstruction of all relevant substructures including the cerebral gray and white matter, the cerebrospinal fluid, the cerebellum, the skin, and the skull [9]. From these substructures, we create a finite element model with 1,275,808 linear tetrahedral elements and 241,845 nodes with the help of Simpleware [48]. Our model represents the skull with 43,614 elements and 14,591 nodes, the skin with 33,821 elements and 11,250 nodes, the cerebral gray matter with 666,570 elements and 99,124 nodes, the cerebral white matter with 338,346 elements and 53,719 nodes, the cerebellum with 10,208 elements and 3,466 nodes, and the remaining cerebrospinal fluid, the veins, the meninges, part of the brain stem, the sinuses, the falx, and the ventricles with 185,249 elements and 59,595 nodes [46]. For simplicity, our current model does not include an explicit representation of the falx and the tentorium although recent studies suggest that the falx could play an important role in distributing deformations more evenly across the brain [21]. We import our head model into the finite element software package Abaqus, in which we prescribe the constitutive models as well as the boundary, contact, and loading conditions [1].

For the constitutive model, we adapt a Mooney-Rivlin model with gray matter parameters $c_1 = 0.28$ kPa and $c_2 = 333$ kPa [31]. We assume that the cerebellum is as stiff as the gray

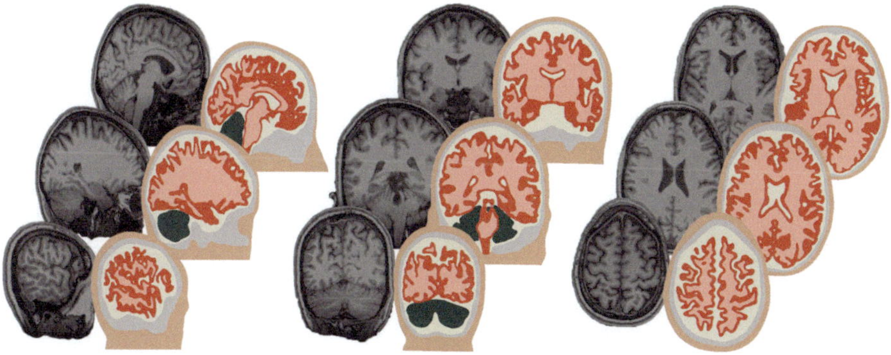

Fig. 5 Personalized decompressive craniectomy model. Magnetic resonance images (*left*) and computational model (*right*). Anatomically detailed and geometrically accurate three-dimensional reconstructions of the individual substructures including the gray matter (*red*), the white matter (*pink*), the cerebellum (*green*), the skin (*brown*), the skull (*gray*), and the remaining cerebrospinal fluid, the veins, the meninges, part of the brain stem, the sinuses, the falx, and the ventricles (*beige*); shown for selected sagittal, coronal, and transverse slices

Table 1 Material parameters of the Mooney-Rivlin model in different regions of the brain

Substructure	Parameter c_1 [kPa]	Parameter c_2 [kPa]
Cerebral gray matter	0.28	333.0
Cerebral white matter	0.56	666.0
Cerebellum	0.28	333.0
Cerebrospinal fluid	0.03	33.3

matter tissue, and that the white matter tissue is twice as stiff [4, 5, 26, 28]. For simplicity, we model the cerebrospinal fluid as an ultrasoft solid with a stiffness ten times lower than the gray matter stiffness. We assume that all soft tissues are incompressible and enforce the incompressibility constraint using hybrid linear tetrahedral C3D4H elements [1]. Table 1 summarizes our material parameters for the individual substructures of the brain.

Figure 6 illustrates our boundary and loading conditions across the brain. For the boundary conditions, we use a combination of fixed and sliding contact. To limit the motion of the inferior soft tissue regions, we fix the lower boundary of the cerebrospinal fluid shown in red [14]. To allow the brain to slide freely along the skull, we apply frictionless contact at the upper interface between the cerebrospinal fluid and the skull, shown in purple [46]. This implies that the shared nodes between the cerebrospinal fluid and the skull are duplicated to allow for relative sliding between them. To prohibit separation and penetration, tensile and compressive forces can be transmitted between these pairs of nodes. We enforce the contact constraint by allowing for a relative motion between the cerebrospinal fluid and the skull while penalizing separation and penetration at the fluid-skull interface. To reduce the computational time, we ignore the skin layer, model the skull as a rigid body, and assume a tight contact between gray and white matter, the cerebellum, and the cerebrospinal fluid [14]. For the loading conditions, we simulate brain swelling by prescribing a local volumetric expansion in a predefined white matter region. We gradually increase the amount of swelling from $J^s = 1.0$ to $J^s = 1.1$ to model a volumetric expansion of 10 % in selected regions of the white matter tissue. We solve the resulting non-linear system of equations using Abaqus/Standard based on an implicit Newton-Raphson iteration scheme. We adopt a

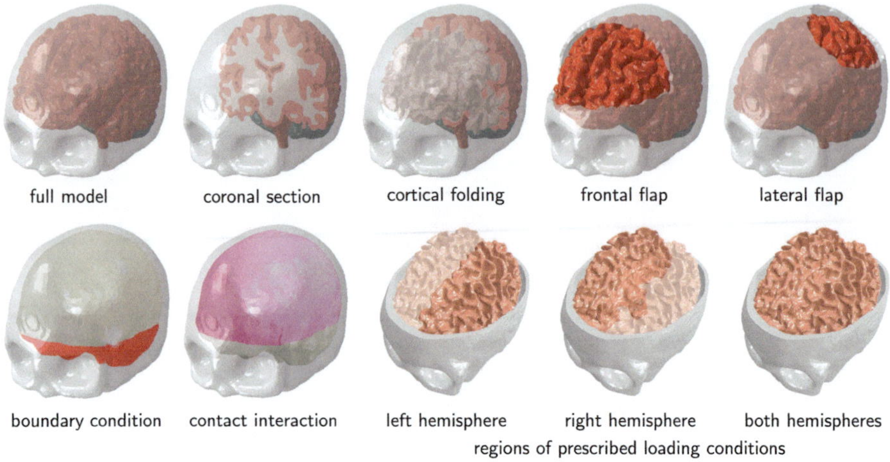

Fig. 6 Personalized decompressive craniectomy model. Boundary conditions and loading conditions. *Top row*: Full model discretized with 1,275,808 linear tetrahedral elements and 241,845 nodes; representative coronal section; anatomic details with cortical folds; frontal flap with 4,279 skull elements removed; lateral flap with 2,494 elements removed. *Bottom row*: Boundary conditions with lower red region fixed relative to the skull and upper purple region allowed to slide along the skull; swelling of left, right, and both white matter hemispheres

quasi-static analysis with an automatic time incrementation, which would terminate if there was no convergence for a minimal pseudo-time increment specified to 10^{-5} [1].

4.2 Bulging of a Personalized Brain with Swelling White Matter Tissue

We simulate three different cases of swelling, in both hemispheres, exclusively in the left hemisphere, and exclusively in the right hemisphere. To release the swelling-induced pressure, we simulate two different decompressive craniectomies, a frontal flap with 4,279 skull elements removed and a unilateral flap with 2,494 elements removed. For all cases, we quantify and compare the mechanical response in terms of the overall deformation, the maximum principal strain, the radial and tangental stretch, and the midline shift. The midline shift is a common clinical indicator to characterize the degree of subcortical swelling and axonal damage.

Figure 7 illustrates the displacement and the superposed deformation in transverse sections facing downward and in sagittal sections facing left for unilateral and frontal flaps with both left and right, only left, and only right hemispherical swelling. The surgical area available for a frontal flap is about twice as large as the area for a unilateral flap. Consequently, for the same amount of swelling, the displacements of the frontal flap are significantly smaller than for the unilateral flap. This finding is in agreement with our intuition and with our idealized hemidisk simulation in Fig. 3, for which larger opening angles generate smaller radial and tangential stretches. The superposed deformation in transverse and sagittal sections in Fig. 7 highlights the relative motion of different regions of the brain as the brain bulges outward. Swelling naturally causes a shift of all intracranial tissues. The shift of the midline, which is clearly visible in this sequence of images, is a key clinical indicator for the degree of trauma.

Figure 8 illustrates the radial and tangential stretches in transverse sections facing downward and in sagittal sections facing left for unilateral and frontal flaps with both left and

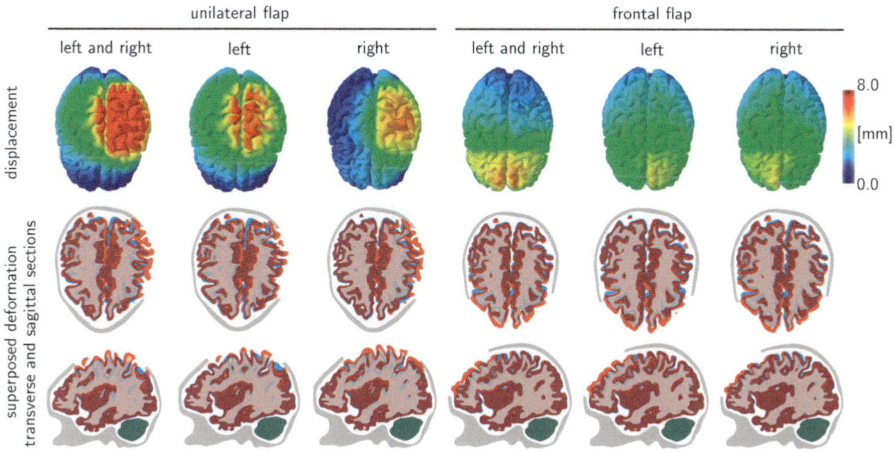

Fig. 7 Decompressive craniectomy. Displacement and superposed deformation in transverse section facing downward and sagittal section facing left for unilateral and frontal flaps with left and right, left, and right hemispherical swelling. Swelling causes a shift of intracranial tissues, a key indicator of the trauma's severity in clinical practice. The midline shift of the cortical and subcortical layers highlights the immediate release of tissue strain upon removal of the unilateral and frontal flaps

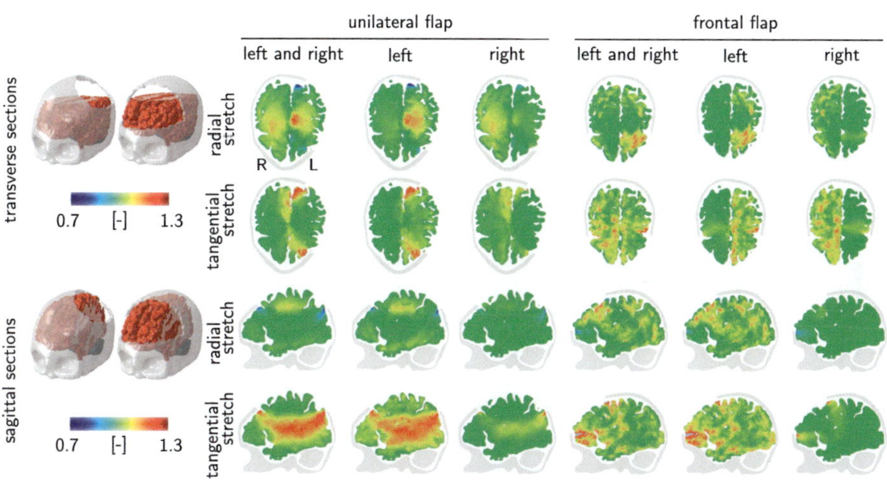

Fig. 8 Decompressive craniectomy. Radial and tangential stretches in transverse section facing downward and sagittal section facing left for unilateral and frontal flaps with left and right, left, and right hemispherical swelling. Swelling causes maximum radial stretches of up to 1.3 deep inside the bulge, minimum radial stretches of 0.7 around the opening, and maximum tangential stretches of up to 1.3 around the opening

right, only left, and only right hemispherical swelling. If we assume that axons are primarily oriented outward, we can associate the radial stretch with the axonal stretch and the tangential stretch with the axonal shear. For a swelling of 10 %, the radial stretch takes maximal values of up to 1.3 deep inside the bulge and minimal values of 0.7 around the edge of the opening. The tangential stretch takes maximal values of up to 1.3 in a ring around the open-

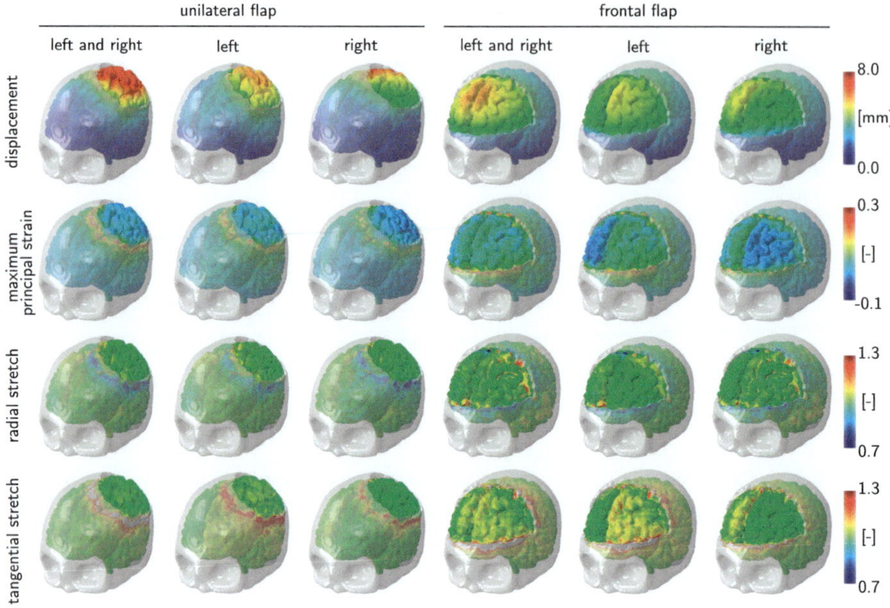

Fig. 9 Decompressive craniectomy. Displacement, maximum principal strain, radial stretch, and tangential stretch for unilateral and frontal flaps with left and right, left, and right hemispherical swelling. Swelling causes maximum principal strains of up to 30 % localized around the opening, maximum radial stretches of up to 1.3 deep inside the bulge, minimum radial stretches of 0.7 around the opening, and maximum tangential stretches of up to 1.3 around the opening

ing. These three regions might be associated with potential zones of herniation and axonal failure, either by tension or compression, or by shear.

Figure 9 illustrates the displacement, maximum principal strain, radial stretch, and tangential stretch for unilateral and frontal flaps with both left and right, only left, and only right hemispherical swelling. The displacement field confirms that the deformation is larger for the unilateral flap than for the frontal flap. Naturally, the displacements are largest in the center of the bulge, which explains the large radial strains in the bulge region. In agreement with Fig. 8, the radial stretch takes maximum values of 1.3 deep inside the bulge and minimum values of 0.7 around the opening. Similarly, the tangential stretch takes maximum values of 1.3 in a ring around the opening.

5 Discussion

Decompressive craniectomy is an invasive neurosurgical procedure to release elevated pressures in a swollen brain. Although the technique is highly controversial, it is often performed as a method of last resort; yet, little is known about how the opening of the skull affects the strain and stress fields inside the brain. Here we introduce a computational model to explore the effects of decompressive craniectomy in idealized and personalized geometries. Through a systematic analysis of different sets of simulations, we identify several common features and trends that could help make the overall procedure less invasive.

In all cases, a unified stretch pattern with three extreme stretch regions emerges: a tensile zone deep inside the bulge, a highly localized compressive zone around the opening, and a shear zone around the opening. This suggests that regions deep inside the bulge are most vulnerable to damage by axonal stretch, while regions near the craniectomy edge are most vulnerable to damage by axonal shear and herniation. These findings are in agreement with our analytical prediction [19]. Axonal stretch has been studied quite extensively in single axon experiments in vitro [41], but axonal shear has been given little attention as a potential failure mechanism. Only a few studies distinguish between tension/compression- and shear-type damage [11]. Our study shows that the tangential stretch, a surrogate measure for the axonal shear, can take values as high as 1.3 for swelling volumes as small as 28 ml, corresponding to only 5 % of the overall white matter volume. These numbers agree well with a recent simulation that predicted shear strains of the order of 25 % for swelling volumes of 22 ml [14]. In view of the long and slender ultrastructure of an axon [42], it seems fairly reasonable to assume that it could be highly vulnerable to shear-type loading. We have recently shown that shear stresses in bulging solids are highly localized at the bulging edge in regions that we have termed damage drops [45]. The drop-shaped regions in the tangential stretch profiles of Fig. 3 agree remarkably well in shape, location, and orientation with our previous analytical predictions [19].

The recent awareness to shear loading is in line with a current trend in mild traumatic brain injury: For a long time, scientists have thought that linear accelerations are the major origin of brain damage in traumatic brain injury, and that damage would be mainly a result of axonal stretch [37]. Stretch-based injury criteria suggests that there is a 50 % chance of brain tissue damage at strain levels of 18 % in vitro [2] and a 50 % chance of mild traumatic brain injury at strain levels of 19 % in vivo [49]. We now know that rotational accelerations could play an equally important role in traumatic brain injury [20]. With more information becoming available, we might soon recognize that brain damage results not only from stretch [17], but also from shear [11], and that the critical damage thresholds need to be considered for each mechanism individually or for both mechanisms in combination [47].

Our simulations suggest that a frontal craniectomy, which provides anatomic space for a larger opening, creates significantly lower displacements, strains, and stretches in comparison to a unilateral craniectomy [27]. Typically, a craniectomy is performed to release swelling volumes on the order of 50–150 ml [12]. A recent study reported swelling volumes of 27–127 ml [44]. If we assume that gray and white matter are approximately of the same volume, and that both hemispheres are of equal size, for a total brain volume of $1,108\ \text{cm}^3$, our swelling of 10 % corresponds to a swelling volume of 56 ml for swelling of both hemispheres and to 28 ml for unilateral swelling. While our swelling volumes are clearly on the lower end and would probably not be sufficient to require surgery in clinical practice, we already observe significant local strains that could exceed the functional and morphological damage thresholds of 18 % and 21 % reported in the literature [2]. However, we would like to note that these thresholds were found by dynamically stretching brain tissue. During a decompressive craniectomy, the stretch rates might be lower, and the injury thresholds could be different. Additional experimental studies would be needed to accurately characterize the relevant strain safety limits.

Our study only presents a first prototype analysis of strains and stretches inside the brain in response to intracranial swelling and decompressive craniectomy. To gain insight, we have made several simplifying assumptions: First, on the time scale of interest, on the order of hours, days, and weeks, we have modeled brain tissue as nonlinearly elastic keeping in mind that on shorter times scales, viscous effects might play an important role [10]. Second, we have modeled brain tissue using a Mooney-Rivlin model although we have recently shown that a higher order Ogden model [33] might be better suited to model the nonlinear behavior of the brain [6, 31]. Since our analysis is mainly based on kinematic quantities—swelling, deformation, stretch, and strain—and we do not discuss the final stress distribution in the bulging brain, we believe this is a reasonable first assumption, although a four- or six-parameter Ogden model would more accurately capture the characteristic shear stiffening of brain tissue [34]. Third, while recent experiments suggest that the elastic response of brain tissues may reasonably well be approximated as isotropic [47], the damage response could very well be anisotropic with different failure mechanisms and different damage thresholds associated with axonal tension and axonal shear [7]. Fourth, for simplicity, we have assumed that all axons point radially outward. A more realistic model would take into account the discrete axonal orientation at each individual point of the brain [30]. Conceptually, our analysis itself would remain the same; yet, the post-processing to calculate the normal and shear stretches would use the true axonal direction $\boldsymbol{n}$ from diffusion tensor images rather than the simplified assumption that $\boldsymbol{n}$ points radially outward.

6 Conclusion

Taken together, our study of bulging brains illustrates how swelling-induced deformations propagate across the brain when opening the skull. It underlines the notion that a decompressive craniectomy is a highly invasive surgical procedure that releases an elevated intracranial pressure at the expense of inducing local zones of extreme strain and stretch. Mathematical models and computational simulations can help identify regions of extreme tissue kinematics. This approach could guide neurosurgeons to optimize the shape and position of the craniectomy with the goal to avoid placing the craniectomy edge near functionally important regions of the brain.

Acknowledgements We thank Allan L. Reiss and his group for providing the MRI scans. This work was supported by the Timoshenko Scholar Award to Alain Goriely and by the Humboldt Research Award and the National Institutes of Health grant U01 HL119578 to Ellen Kuhl.

References

1. Abaqus 6.14. Analysis user's manual. SIMULIA. Dassault Systèmes (2014)
2. Bain, A.C., Meaney, D.F.: Tissue-level thresholds for axonal damage in an experimental model of central nervous system white matter injury. J. Biomech. Eng. **122**, 615–622 (2000)
3. Barber, J.R.: The solution of elasticity problems for the half-space by the method of Green and Collins. Appl. Sci. Res. **40**, 135–157 (1983)
4. Brauna, J., Guob, J., Lutzkendorf, R., Stadler, J., Papazoglou, S., Hirsch, S., Sack, I., Bernarding, J.: High-resolution mechanical imaging of the human brain by three-dimensional multifrequency magnetic resonance elastography at 7T. NeuroImage **90**, 308–314 (2014)
5. Budday, S., Nay, R., de Rooij, R., Steinmann, P., Wyrobek, T., Ovaert, O.C., Kuhl, E.: Mechanical properties of gray and white matter brain tissue by indentation. J. Mech. Behav. Biomed. Mater. **46**, 318–330 (2015)
6. Budday, S., Sommer, G., Birkl, C., Langkammer, C., Haybäck, J., Kohnert, J., Bauer, M., Paulsen, F., Steinmann, P., Kuhl, E., Holzapfel, G.: Mechanical characterization of human brain tissue. Acta Biomater. (2017), accepted for publication
7. Cloots, R.J.H., van Dommelen, J.A.W., Nyberg, T., Kleiven, S., Geers, M.G.D.: Micromechanics of diffuse axonal injury: influence of axonal orientation and anisotropy. Biomech. Model. Mechanobiol. **10**, 413–422 (2011)
8. Cooper, D.J., Rosenfeld, J.V., Murray, L., Arabi, Y.M., Davies, A.R., D'Urso, P., Kossmann, T., Ponsford, J., Seppelt, I., Reilly, P., Wolfe, R.: Decompressive craniectomy in diffuse traumatic brain injury. N. Engl. J. Med. **364**, 1493–1502 (2011)
9. Cotton, R.T., Pearce, C.W., Young, P.G., Kota, N., Leung, A.C., Bagchi, A., Qidwai, S.M.: Development of a geometrically accurate and adaptable finite element head model for impact simulation: the naval research laboratory-simpleware head model. Comput. Methods Biomech. Biomed. Eng. **19**, 101–113 (2016)

10. de Rooij, R., Kuhl, E.: Constitutive modeling of brain tissue: current perspectives. Appl. Mech. Rev. **68**, 010801 (2016)
11. ElSayed, T., Mota, A., Fraternali, F., Ortiz, M.: Biomechanics of traumatic brain injury. Comput. Methods Appl. Mech. Eng. **197**, 4692–4701 (2008)
12. Fletcher, T.L., Kolias, A.G., Hutchinson, P.J.A., Sutcliffe, M.P.F.: Development of a finite element model of decompressive craniectomy. PLoS ONE **9**, e102131 (2014)
13. Fletcher, T.L., Kolias, A.G., Hutchinson, P.J.A., Sutcliffe, M.P.F.: An improved method for assessing brain deformation after decompressive craniectomy. PLoS ONE **9**, e110408 (2014)
14. Fletcher, T.L., Kolias, A.G., Adams, H., Hutchinson, P.J.A., Sutcliffe, M.P.F.: Modelling of brain deformation after decompressive craniectomy. Ann. Biomed. Eng. (2016). doi:10.1007/s10439-016-1667-7
15. Franceschini, G., Bigoni, D., Regitnig, P., Holzapfel, G.A.: Brain tissue deforms similarly to filled elastomers and follows consolidation theory. J. Mech. Phys. Solids **54**, 2592–2620 (2006)
16. Gao, C.P., Ang, B.T.: Biomechanical modeling of decompressive craniectomy in traumatic brain injury. Acta Neurochir., Suppl. **102**, 279–282 (2008)
17. Goriely, A., Budday, S., Kuhl, E.: Neuromechanics: from neurons to brain. Adv. Appl. Mech. **48**, 79–139 (2015)
18. Goriely, A., Geers, M.G.D., Holzapfel, G.A., Jayamohan, J., Jerusalem, A., Sivaloganathan, S., Squier, W., van Dommelen, J.A.W., Waters, S., Kuhl, E.: Mechanics of the brain: perspectives, challenges, and opportunities. Biomech. Model. Mechanobiol. **14**, 931–965 (2015)
19. Goriely, A., Weickenmeier, J., Kuhl, E.: Stress singularities in swelling soft solids. Phys. Rev. Lett. **117**, 138001 (2016)
20. Hernandez, F., Wu, L.C., Yip, M.C., Laksari, K., Hoffmann, A.R., Lopez, J.R., Grant, G.A., Kleiven, S., Camarillo, D.B.: Six degree-of-freedom measurements of human mild traumatic brain injury. Ann. Biomed. Eng. **43**, 1918–1934 (2015)
21. Hernandez, F., Giordano, C., Kleiven, S., Camarillo, D.B.: Coronal head rotation, falx cerebri displacement, and corpus callosum strain are related and implicated in sport-related MTBI. J. Neurotrauma **33**, A34–A35 (2016)
22. Holzapfel, G.A.: Nonlinear Solid Mechanics: A Continuum Approach for Engineering. John Wiley & Sons, New York (2000)
23. Hutchinson, P.J., Corteen, E., Czosnyka, M., Mendelow, A.D., Menon, D.K., Mitchell, P., Murray, G., Pickard, J.D., Rickels, E., Sahuquillo, J., Servadei, F., Teasdale, G.M., Timofeev, I., Unterberg, A., Kirkpatrick, P.J.: Decompressive craniectomy in traumatic brain injury: the randomized multicenter RESCUEicp study. Acta Neurochir., Suppl. **96**, 17–20 (2006)
24. Johnson, K.L.: Contact Mechanics. Cambridge University Press, Cambridge (1987)
25. Jones, H.R., Burns, T.M., Aminoff, M.J., Pomeroy, S.L.: Netter Collection of Medical Illustrations: Nervous System: Part I, Brain, vol. 7. Saunders Elsevier, Philadelphia (2013)
26. Kaster, T., Sack, I., Samani, A.: Measurement of the hyperelastic properties of ex vivo brain tissue slices. J. Biomech. **44**, 1158–1163 (2011)
27. Kolias, A.G., Kirkpatrick, P.J., Hutchinson, P.: Decompressive craniectomy: past, present and future. Nat. Rev. Neurol. **9**, 405–415 (2013)
28. Kruse, S.A., Rose, G.H., Glaser, K.J., Manduca, A., Felmlee, J.P., Jack, C.R., Ehman, R.L.: Magnetic resonance elastography of the brain. NeuroImage **39**, 231–237 (2008)
29. Lang, G., Stewart, P.S., Vella, D., Waters, S.L., Goriely, A.: Is the Donnan effect sufficient to explain swelling in brain tissue slices? J. R. Soc. Interface **11**, 20140123 (2014)
30. Li, X., van Holst, H., Kleiven, S.: Decompressive craniectomy causes significant strain increase in axonal fiber tracts. J. Clin. Neurosci. **20**, 509–513 (2013)
31. Mihai, L.A., Chin, L.K., Janmey, P.A., Goriely, A.: A hyperelastic constitutive model for compression stiffening applicable to brain and fat tissues. J. R. Soc. Interface **12**, 20150486 (2015)
32. Mooney, M.: A theory of large elastic deformation. J. Appl. Phys. **11**, 582–592 (1940)
33. Ogden, R.W.: Large deformation isotropic elasticity – on the correlation of theory and experiment for incompressible rubberlike solids. Proc. R. Soc. Lond. A **326**, 565–584 (1972)
34. Pogoda, K., Chin, L.K., Georges, P.C., Byfield, F.R.J., Bucki, R., Kim, R., Weaver, M., Wells, R.G., Marcinkiewicz, C., Janmey, P.A.: Compression stiffening of brain and its effect on mechanosensing by glioma cells. New J. Phys. **16**, 075002 (2014)
35. Quinn, T.M., Taylor, J.J., Magarik, J.A., Vought, E., Kindy, M.S., Ellegala, D.B.: Decompressive craniectomy: technical note. Acta Neurol. Scand. **123**, 239–244 (2011)
36. Rivlin, R.S.: Large elastic deformations of isotropic materials. IV. Further developments of the general theory. Philos. Trans. R. Soc. Lond. **241**, 379–397 (1948)
37. Rowson, S., Duma, S.M., Beckwirth, J.G., Chu, J.J., Greenwald, R.M., Crisco, J.J., Brolinson, P.G., Duhaime, A.C., McAllister, T.W., Maerlender, A.C.: Rotational head kinematics in football impacts: an injury risk function for concussion. Ann. Biomed. Eng. **40**, 1–13 (2012)

38. Saggar, M., Quintin, E.M., Kienitz, E., Bott, N.T., Sun, Z., Hong, W.C., Chien, Y., Liu, N., Dougherty, R.F., Royalty, A., Hawthorne, G., Reiss, A.L.: Pictionary-based fMRI paradigm to study the neural correlates of spontaneous improvisation and figural creativity. Sci. Rep. **5**, 10894 (2015)
39. Stiver, S.I.: Complications of decompressive craniectomy for traumatic brain injury. Neurosurg. Focus **26**, E7 (2009)
40. Tagliaferri, F., Zani, G., Iaccarino, C., Ferro, S., Ridolfi, L., Basaglia, N., Hutchinson, P., Servadei, F.: Decompressive craniotomies, facts and fiction: a retrospective analysis of 526 cases. Acta Neurochir. **154**, 916–919 (2012)
41. Tang-Schomer, M.D., Patel, A.R., Baas, P.W., Smith, D.H.: Mechanical breaking of microtubules in axons during dynamic stretch injury underlies delayed elasticity, microtubule disassembly, and axon degeneration. FASEB J. **24**, 1401–1410 (2010)
42. van den Bedem, H., Kuhl, E.: Tau-ism: the Yin and Yang of microtubule sliding, detachment, and rupture. Biophys. J. **109**, 2215–2217 (2015)
43. von Holst, H., Li, X.: Decompressive craniectomy (DC) at the non-injured side of the brain has the potential to improve patient outcome as measured with computational simulation. Acta Neurochir. **156**, 1961–1967 (2014)
44. von Holst, H., Li, X., Kleiven, S.: Increased strain levels and water content in brain tissue after decompressive craniectomy. Acta Neurochir. **154**, 1583–1593 (2012)
45. Weickenmeier, J., Kuhl, E., Goriely, A.: The mechanics of decompressive craniectomy: bulging in idealized geometries. J. Mech. Phys. Solids **96**, 572–590 (2016)
46. Weickenmeier, J., Butler, C., Young, P.G., Goriely, A., Kuhl, E.: The mechanics of decompressive craniectomy: personalized simulations. Comput. Methods Appl. Mech. Eng. (2017). doi:10.1016/j.cma.2016.08.011
47. Wright, R.M., Ramesh, K.T.: An axonal strain injury criterion for traumatic brain injury. Biomech. Model. Mechanobiol. **11**, 245–260 (2011)
48. Young, P.G., Beresford-West, T.B.H., Coward, S.R.L., Notarberardino, B., Walker, B., Abdul-Aziz, A.: An efficient approach to converting 3D image data into highly accurate computational models. Philos. Trans. R. Soc. Lond. A **366**, 3155–3173 (2008)
49. Zhang, L., Yang, K.H., King, A.I.: A proposed injury threshold for mild traumatic brain injury. J. Biomech. Eng. **126**, 226–236 (2004)

J Elast (2017) 129:213–238
DOI 10.1007/s10659-017-9622-9

Mimicking Cortex Convolutions Through the Wrinkling of Growing Soft Bilayers

Martine Ben Amar[1,2] **· Adrien Bordner**[1,2]

Received: 12 May 2016 / Published online: 27 January 2017

Abstract Brain convolutions are a specificity of mammals. Varying in intensity according to the animal species, it is measured by an index called the gyrification index, ratio between the effective surface of the cortex compared to its apparent surface. Its value is close to 1 for rodents (smooth brain), 2.6 for newborns and 5 for dolphins. For humans, any significant deviation is a signature of a pathology occurring in fetal life, which can be detected now by magnetic resonance imaging (MRI). We propose a simple model of growth for a bilayer made of the grey and white matter, the grey matter being in cortical position. We analytically solved the neo-Hookean approximation in the short and large wavelength limits. When the upper layer is softer than the bottom layer, the selection mechanism is shown to be dominated by the physical properties of the upper layer. When the anisotropy favors the growth tangentially as for the human brain, it decreases the threshold value for gyri formation. The gyrification index is predicted by a post-buckling analysis and compared with experimental data. We also discuss some pathologies in the model framework.

Keywords Brain · Gyrification index · Buckling · Post-buckling · Finite Elasticity · Neo-Hookean · Surface tension

Mathematics Subject Classification 35A15 · 35J50 · 74B20 · 92C05

1 Introduction

This paper intends to clarify the onset of buckling of a bilayer during growth. Indeed, if the one layer linear mode of instability is well understood since the pioneering work of Biot [11] concerning soft hyperelastic materials, the theoretical prediction of the wavelength remains

✉ M. Ben Amar
benamar@lps.ens.fr

1 Laboratoire de Physique Statistique, Ecole Normale Supérieure, PSL Research University, Sorbonne Universités UPMC Univ Paris 06, CNRS, 24 rue Lhomond, 75005 Paris, France

2 Institut Universitaire de Cancérologie, Faculté de médecine, Université Pierre et Marie Curie-Paris 6, 91 Bd de l'Hôpital, 75013 Paris, France

matter of debates, especially for bilayers. It is also true that the wavelength selection for one layer is not obvious: the linear analysis predicts a threshold but at vanishing wavelength or infinite wavenumber, a result which contradicts experimental observations. Then, the wavelength selection at threshold is a matter of a not trivial singular perturbation. In the case of a bilayer, the multiplicity of independent parameters of the top layer 1 versus the bottom layer 2 may give some hope for a simpler determination of the wavelength. Differential growth between both layers, the ratio μ of shear moduli (defined by $\mu = \mu_1/\mu_2$) or the two different thicknesses represent new degrees of freedom that can modify the wavelength selection, compared to the one layer model. In fact, from previous works by Biot [10], achieved several years before [11], the wrinkling of a bilayer, made of a thin film on top of an infinite soft substrate, under external compression, can be simply explained by the Hookean elasticity of the stiff thin plate modified by a local contribution due to the substrate which reminds of the Winkler foundation model. According to Biot's result [10], it gives an instability with a critical pressure threshold which varies in $\mu^{-2/3}$, a selected wavelength λ at threshold in $\mu^{1/3}$ which is large since the dimensionless ratio μ is a large dimensionless number in this limit. This situation of an infinite soft substrate falls in the domain of long wavelengths while the Biot's singularity of a neo-Hookean monolayer concerns the short-wavelength limit.

Morphogenesis and embryogenesis reveal a lot of examples of bilayer wrinkling due to growth such as our fingerprints [4, 19, 37], intestine villi [8, 29], esophagus [39] and brain convolutions [5, 28, 56]. Occurring in fetal life, at a rather precise stage post fertilization, these instabilities are common to most mammals, but not only, and concern differential growth of bilayers, with a bottom layer, in most cases, stiffer than the upper layer.

The aim of this paper is then to clarify this problem of wavelength selection with a special focus on brain convolutions. Our analysis consists of a minimal set of mechanical parameters to describe the growth of a planar bilayer: both layers of different thicknesses are represented by the neo-Hookean elasticity with different shear moduli, the growth in each layer is treated with the multiplicative gradient decomposition formulation [48]. The upper layer is free, the bottom layer infinite or attached to a stiff substrate. Our purely analytical treatment is limited to harmonic patterns derived close to the stability limit, in contrast to more elaborate numerical investigations where detailed phase-diagrams are derived in the nonlinear regime of parameters [60]. The chosen formulation [48] treats growth in the same way as initially pre-strained materials which, next, are remodeled under physical constraints, such as the minimization of the elastic energy and the boundary conditions. Most of these simulations are based on this analogy. Comparison can be achieved with our results if the growth and the pre-strain per unit volume correspond. The advantage of numerical investigations [13, 34, 60] is the possibility to detect non harmonic patterns, which is impossible here.

The two limiting cases, short and long wavelengths, are examined for completeness, although the brain is mostly concerned by the short wavelength limit. The gyrification index is analytically predicted. The paper is organized as follows: Sect. 2 is devoted to a short overview of the brain in fetal life. Sect. 3 presents the biomechanical model and the exact results for the threshold determination for a growing bilayer. Sect. 4 focusses on the short wavelength determination and by a nonlinear analysis, Sect. 5 shows that for the physical parameters concerning the brain, the bifurcation is supercritical allowing the determination of the gyrification index that we compare to experiments [1, 68]. In Sect. 6, the long wave-limit is recovered and is compared to the Biot's result [10] in the case of growth. Finally in Sect. 7, we conclude that to some extent, the bilayer model of soft nonlinear elasticity with growth explains the onset of gyri formation of the brain.

2 The Brain Convolutions, the Gyrification Index and Associated Pathologies

We can distinguish mammals in two groups according to the wrinkling level of the brain. Smooth brains are called lissencephalic (and concern mostly rodents) whereas folded brain are gyrencephalic. The human brain is formed of two hemispheres, highly convoluted, but less than the lama, the dolphin, etc. At macroscopic scales, the hemispheres are composed of two main layers: the *grey matter* and *white matter*. The *grey matter* forms the cortex and is located below the skull and the meninges. It contains the neuronal bodies and the glial cells (which are cells of support). The axons and dendrites (very long extension of the neural cells) are radially oriented, then inducing anisotropy in the brain tissue. The *white matter* is mainly composed of glial cells and a high number of myelinated axons which transmit long distance signals between the cortex and the other brain regions.

In the human fetal life, the brain originates from a smooth and homogeneous structure named telencephalon which appears at months 2–3 of gestation. During months 3–4, the neuron precursors proliferate in a thin zone at the basis of telencephalon (the subventricular zone). Then it follows a radial migration of these cells resulting in the formation of the cortical layer (*grey matter*). The timing and the direction of this neuronal migration is crucial, and any perturbation of these processes leads to major disorders and folding abnormalities in the newborn brain [59]. The migration is under the high control of radial glial cells, forming a radial scaffold along which the neurons migrate. Although oriented radially, these cells also deviate at the level of the cortex, and contribute to the tangential growth of the cortex as shown recently [47]. The meninges are thought to play here a crucial role by secreting and organizing the pial basement membrane which makes tight contact with the cortical layer. The radial glial cells are anchored to this membrane and any defect of the meninges disrupts the scaffold leading to malformations [50]. Once neuronal migration ends at month 4, the growth of the brain really begins, together with the surface buckling and the first cerebral indentations (sulci): this is called gyrification. The gyrification coincides with the period of highest tissue growth [61].

Recently, mechanical measurements have been performed on brain tissues and have shown that these two layers have different mechanical properties induced by the various micro-structures. Experimental results showed that the *white matter* is stiffer than the *gray matter* for adults and fetuses [5, 33, 64]. In Table 1, recent data measurements are reported for mammals, showing a noticeable dispersion on experimental values which are strongly dependent on the technique, the data analysis, the regional variation but also on the age of the brain and the strain intensity. Some consensus appears for a stiffness ratio μ of order 0.7 [55, 64] except perhaps for rodents. Rodent cerebellum seems to be in opposition with $\mu = 1.5$, larger than one [22]. However, if one must be cautious about cerebellum and cerebrum data comparisons, it is important to recall that most of rodents do not have gyri. For human brain, this consensus is based on experimental biomechanical measurements [5, 33, 64] and on analytical works on wrinkling [55, 64]. However, during the final revision of this work, an extensive experimental study concerning mechanical tests in static and dynamics [14] announces opposite conclusions concerning the stiffness of the cortex and of the white matter. Achieved on human cadavers, with an average age of 66 years after autopsy, these authors study the human brain elasticity and viscoelasticity *in vitro* in various location such as the corpus callosum, the corona radiata but not immediately below the cortex, that is in the sub-plate. For various loading conditions: tension, compression and shear, they conclude that the cortex always appears stiffer than the white matter, in opposition with other results [33]. Clearly the question of the comparison of the stiffness of both layers: the cortex and

Table 1 Brain stiffness

	Bovine[a] cerebrum	Bovine[b] cerebrum	Pig[a] cerebrum	Pig[a] cerebrum	Rat[c] cerebellum
Grey matter (kPa)	0.689	1.5	2.23	1.195	0.454
White matter (kPa)	1.440	2	3.083	1.787	0.29
Ratio μ	0.48	0.75	0.72	0.64	1.565
Reference	[12]	[12]	[58]	[35]	[22]

[a]Indentation technique (low strain).

[b]Indention at large strains.

[c]Indentation and scanning force microscopy.

the white matter, so important for the modeling, is matter of debate. In any case, we must keep in mind that the gyrification process we examine here concerns embryonic stages so fetuses and not cadavers of old human beings. As underlined in [64], the stiffness of the brain, whatever the anatomic regions, increases with aging and it seems that it increases more for the cortex [64]. As also mentioned in [64], when experimental data are lacking or debatable, a theoretical treatment may help to clarify the situation.

These properties motivate our two-layer model and explain why we will focus our study on the shear moduli ratio μ. Many models have been proposed to explain the brain folding [42, 52]. It first has been suggested that the considerable growth of the telencephalon, under the external constraint of the limited volume by the cranium, is responsible for gyrification ("crumpled sheet model") [18]. However, clinical and animal experiments have shown that the cranium constraint is a minor factor for buckling [3, 26]. Another model [42] treats the cortex as a one smectic layer subject to axons pulling from the white matter. According to the last biomechanics review [28], two hypotheses emerge and are in competition, differential growth between both layers and mechanical forces such as axonal tension. Although it remains controversial, this last hypothesis disagrees with dissection experiments [64]. Knowing that the cortical surface area increases faster than the brain volume has inspired numerical models [54, 56]. It is shown that the tangential expansion of an outer layer, bounded to a more slowly growing central core, induces buckling. In addition, cortical development studies reveal unequal cell migration mechanisms for radial and tangential expansion inducing anisotropy in the cortical layer growth [47]. Then, we can conclude that the human cortex growth of healthy beings is anisotropic and this anisotropy will be introduced in the growth representation in our two-layer model.

Many pathologies are now related to cerebral folding. Recent observations link autistic disorders to abnormal deepness of the cerebral sulci [15], affecting the social communication functions. Moreover, the abnormal regulation of the brain growth results in an early overgrowth followed by an abnormally slow-down during development [21]. This increase concerns the cortical area rather than the cortical thickness [44]. It also exists rare generalized abnormalities of gyrification, for instance the lissencephaly which consists in a low number of sulci and, on contrary, polymicrogyria where the gyri are too many and too small. Physiopathology of these diseases is mainly unknown, but neuronal migration and the failure of cell proliferation inducing abnormal growth could be involved [23, 43]. It is also hypothesized that the thickened meninges potentially influence folding and could lead to polymicrogyria [51].

Our understanding of the growth dynamics of the brain in fetal life will increase in the next future at the same time as its connection with pathologies [25, 38]. However, the simple buckling of a double elastic layer of soft tissues is itself rather poorly understood and many works accept wavelength relationships out of their domain of application. Some analytical results such as the wavelength varying like a power law of the ratio of both stiffnesses [10] are very simple, but not always used in the range of physical parameters where they are valid. In addition, even if there is a difference between mechanical properties or growth dynamics for the *white matter* and *grey matter*, we cannot expect large differences in physical properties, which mostly prohibit the use of limiting models in nonlinear elasticity. Focusing only on the brain morphogenesis, the main ingredients seem to be the ratio between the shear modulus of the two layers μ (grey matter versus white matter shear modulus) but remaining of order one, a growth anisotropy which favors tangential growth in the cortex and a differential growth between both layers. In addition the cortex is much thinner that the *white matter* so we will pay more attention of the case of infinite substrate. We introduce these ideas in the following section.

3 The Model

We consider two growing layers of soft materials, joined together, having different thicknesses. Such ideal situation is known to generate buckling, folding, creases [31, 57, 60] and even other more complex patterns [2, 8, 60], especially in three dimensions, as soon as the growth differs in the two layers. Complex patterns are out of reach of the linear analysis presented here and they require sophisticated numerical treatments. However, most of them are issued of the linear regime presented here and fall in the nonlinear regime of the first bifurcation. The first bifurcation, which is analyzed here, only allows to explain the existence of the wrinkling in function of the physical parameters and the pattern length scale. Since both layers are strongly connected, axial growth in the vertical direction induces a configuration of lowest elastic energy. This geometry differs from simulations based on pre-stretched substrate [60] in the horizontal direction which may lead to different conclusions, although, at first sight, both problems are similar. To limit the number of independent parameters, the growth is assumed homogeneous in both samples, anisotropic in the upper layer, the materials are considered as hyperelastic and neo-Hookean [30, 45] and the horizontal dimensions infinite in both directions. The upper layer, called 1 in the following, has a thickness H, while, the thickness of the substrate 2 is either infinite or chosen as unity (see Fig. 1). The shear modulus of the substrate is also chosen as unity and we call μ the ratio between the shear modulus of the upper slab compared to the substrate one. These hypotheses can appear rather restrictive to fully represent the biological complexity of growth during morphogenesis or embryogenesis, however a growing number of independent parameters obscures the wrinkling results, changing the threshold value for existence without always changing the physical reality [27]. Here we mostly focus on the ratio between both shear moduli which is often smaller than one in biological systems [8, 19, 56] contrary to material sciences [36, 65, 66]. Capital letters (X, Y) represent the cartesian coordinates of a point in the reference configuration, assumed free of stresses at initial time, and, small letters (x, y) are devoted to the current configuration, at time t. Both samples are assumed incompressible.

3.1 The Growth Model

Assuming the classical multiplicative decomposition [48], growth is represented by a tensor $\mathbf{G}$, the geometric deformation gradient tensor by $\mathbf{F}$ such as $\mathbf{F} = \mathbf{F_e}.\mathbf{G}$, where $\mathbf{F_e}$ [30, 45]

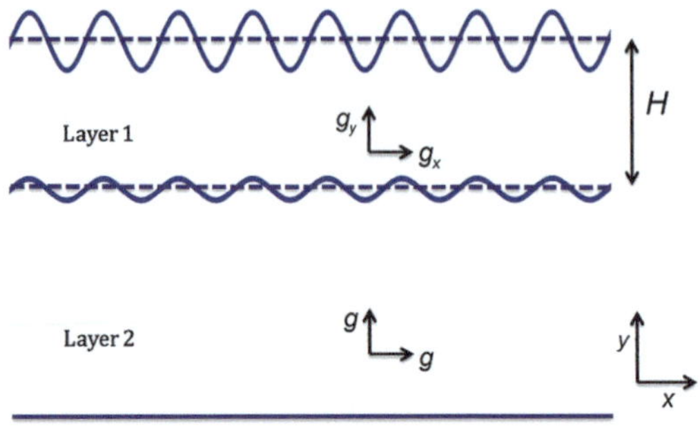

Fig. 1 The bilayer model under buckling

is the elastic deformation gradient tensor. Once the hypothesis that the characteristic mechanical time is much shorter compared to the typical time scale of growth processes, one can neglect the dynamics and determine the current configuration at time t by a variational principle of minimization of the elastic energies.

The growth tensor will be chosen anisotropic in the upper layer with an eigenvalue g_x along the x direction, tangent to the interface, larger than the one g_y along the perpendicular direction (see Fig. 1). This will reflect the structure of the grey matter and its tendency to proliferate more in the direction parallel to the cranium. This growth anisotropy is not known experimentally but may have some consequences on final patterns at large strains, as shown for villi [8]. For the white matter, there is no reason to introduce such difference and we will choose an isotropic growth modeling, to limit the number of parameters. The growth of our bilayer is then characterized by three eigenvalues. The growth tensors $\mathbf{G}_i$ and elastic tensors $\mathbf{F}_{e_i}$ are then:

$$\mathbf{G}_1 = \begin{bmatrix} g_x & 0 \\ 0 & g_y \end{bmatrix}, \qquad \mathbf{G}_2 = \begin{bmatrix} g & 0 \\ 0 & g \end{bmatrix},$$
$$\mathbf{F}_{e_1} = \begin{bmatrix} g_x^{-1} x_{,X} & g_y^{-1} x_{,Y} \\ g_x^{-1} y_{,X} & g_y^{-1} y_{,Y} \end{bmatrix}, \qquad \mathbf{F}_{e_2} = \begin{bmatrix} g^{-1} x_{,X} & g^{-1} x_{,Y} \\ g^{-1} y_{,X} & g^{-1} y_{,Y} \end{bmatrix}; \tag{1}$$

Where the notation with a comma in subscript means partial derivative. Hereafter we use the index 1 and 2 when it is necessary to distinguish the layers. We drop the index otherwise. An index with a coma indicates a partial derivative. In the following, we will assume that fiber-like structures such as axons or vessels, present in the brain, do not contribute to an anisotropic mechanical behavior. Indeed, Budday *et al.* [14] have extensively analyzed this question experimentally in various parts of the white matter, especially the *corpus callosum*, the most fibrous area of the human brain and they conclude that, from the biomechanical viewpoint, the white matter can be represented as the cortex by an isotropic constitutive law. In addition we assume that the brain is incompressible which means that any change of volume is due to cell proliferation and not to elastic stresses. Then we will adopt the constraint of incompressibility and the relative volume variation J is given by: $J = \det\mathbf{F} = \det\mathbf{G}$ so $J_1 = g_x g_y$ and $J_2 = g^2$. These relations are automatically satisfied by the definition of stream-functions [6], as shown in the next section. Before going further, let us compare our model with the pre-strain method used in numerical simulations [60]. Considering a sample of initial length L_s for the substrate and L_f for the upper layer, the process consists in extending first the substrate from L_s up to L_f, then in gluing both samples up to a final state which, in absence of buckling, leads to a final length L for the bilayer. It gives the following elastic deformation gradients which can be compared to our uniaxial growth model along the Y direction:

$$\mathbf{F}_{ef} = \begin{bmatrix} L/L_f & 0 \\ 0 & L_f/L \end{bmatrix}; \qquad \mathbf{F}_{e1} = \begin{bmatrix} 1/g_x & 0 \\ 0 & g_x \end{bmatrix};$$
$$\mathbf{F}_{es} = \begin{bmatrix} L/L_s & 0 \\ 0 & L_s/L \end{bmatrix}; \qquad \mathbf{F}_{e2} = \begin{bmatrix} 1/g & 0 \\ 0 & g \end{bmatrix}. \tag{2}$$

In this simplified configuration, the unknown L value is derived from the bilayer elastic energy minimization and in the limit of thin substrate gives:

$$L \sim L_s + \mu H \frac{L_f^4 - L_s^4}{4L_f^2 L_s}, \tag{3}$$

while the compressive strain, used as a control parameter is $\epsilon = (L_f - L)/L_f = 1 - 1/g_x$. At this stage, it is then possible to compare both approaches.

3.2 The Definition of Non Linear Stream Functions

One way to impose the constraint of incompressibility consists in introducing a Lagrange parameter called the pressure. Another way consists in minimizing the elastic energy in the sub-set of deformations which automatically satisfy the incompressibility constraint and preserve the area in two dimensions. The construction of such deformations proposed by Carroll [17], in elasticity, is based on the mixed coordinate systems where one coordinate is chosen in the reference frame and the other in the current configuration [6]. The advantage of such representation is to eliminate the pressure which, otherwise, remains an unknown function coupled to the deformations and to perform the variation with respect to a unique function called the stream function by analogy with the Eulerian hydrodynamic formulation for Darcy and Stokes flows in two dimensions. Notice that this method can be applied to any imposed preserving area mapping between the initial and current configuration, once it remains only dependent on a unique coordinate so it can be extended to the radial geometry [63]. We then introduce two stream functions Φ and Ψ [6] which are defined by the following implicit relationships involving a mixed coordinate representation:

$$X_1 = \frac{1}{J_1} \frac{\partial \Phi(x_1, Y_1)}{\partial Y_1}; \qquad y_1 = \frac{\partial \Phi(x_1, Y_1)}{\partial x_1};$$
$$X_2 = \frac{1}{J_2} \frac{\partial \Psi(x_2, Y_2)}{\partial Y_2}; \qquad y_2 = \frac{\partial \Psi(x_2, Y_2)}{\partial x_2}. \tag{4}$$

In the following, notations for partial derivatives with respect to coordinates (x_i, Y_j) also will be defined by a subscript with a coma followed by x_i and/or Y_j. The components of the geometric deformation gradient tensor read:

$$\begin{cases} \frac{\partial x}{\partial X}\Big|_Y = J\Phi_{,xY}^{-1}, & \frac{\partial y}{\partial Y}\Big|_X = (\Phi_{,xY}^2 - \Phi_{,xx}\Phi_{,YY})\Phi_{,xY}^{-1}, \\ \frac{\partial x}{\partial Y}\Big|_X = -\Phi_{,YY}\Phi_{,xY}^{-1}, & \frac{\partial y}{\partial X}\Big|_Y = \Phi_{,xx}\Phi_{,xY}^{-1}. \end{cases} \tag{5}$$

We can now proceed to the variational calculation.

3.3 Variational Calculation

The mechanical energy of the bilayer is the sum of the elastic energies of both layers E_i and the interfacial energy at the frontier between them, E_{c2}, and on top of the upper layer E_{c1}. We choose the simplest hyperelastic elastic model, the neo-Hookean one [30, 45], for both layers which reads:

$$E_1 = \frac{1}{2} J_1 \mu \int_0^H dX dY \operatorname{Tr}(\mathbf{F_e}.\mathbf{F_e}^{\mathrm{T}} - 2\mathbf{I}) \quad \text{and}$$
$$E_2 = \frac{1}{2} J_2 \int_{-H_2}^0 dX dY \operatorname{Tr}(\mathbf{F_e}.\mathbf{F_e}^{\mathrm{T}} - 2\mathbf{I}). \tag{6}$$

Eventually, H_2 can be considered as infinite, and

$$E_{c_i} = d_i \int_{Y=Int} dx \sqrt{\left(\frac{\partial x}{\partial X}\right)^2 + \left(\frac{\partial y}{\partial X}\right)^2}. \tag{7}$$

As before, $\mu = \mu_1/\mu_2$ and d_i represents the capillary parameter of the upper layer or of the intermediate interface between both layers. Replacing the elastic tensor components by the partial derivatives of the stream function for the upper layer reads:

$$E_1 = \frac{1}{2} \mu g_y^2 \int_0^H dx dY \Phi_{,xY}^{-1} \left(1 + \Phi_{,xx}^2 + g_y^{-4} \left[\Phi_{,YY}^2 + \left(\Phi_{,xY}^2 - \Phi_{,xx}\Phi_{,YY}\right)^2\right] - 2 g_y^{-2} \Phi_{,xY}^2\right), \tag{8}$$

and a similar relationship for E_2 involving Ψ. A trivial configuration can be found where the growth is forced in the Y direction which also satisfies the constraints on the stresses at the boundaries. This basic state corresponds to a thickening of both layers and the stream functions can be decomposed into two parts according to:

$$\Phi(x, Y) = x Y g_x g_y + \epsilon \phi(x, Y) \quad \text{and} \quad \Psi(x, Y) = x Y g^2 + \epsilon \psi(x, Y). \tag{9}$$

The first term represents a purely axial growth. In the following, the elastic energies will be expanded up to fourth order in power of ϵ: $E_i = E_0 + \epsilon^2 E_i^{(2)} + \epsilon^4 E_i^{(4)} + o(\epsilon^4)$. The derivation of the Euler-Lagrange equation and the bifurcation analysis are achieved at second order of the energy expansion but the post-buckling treatment requires the expansion up to fourth order, see Sect. 5. Once ϕ and ψ are expanded in Fourier modes $F(Y)\sin(kx)$, these modes are uncoupled at second order but also at fourth order. Indeed a solution of non-linear equations in Fourier modes generates secular terms and a better approximation consists in the elimination of mode coupling [9]. This simplifies the analysis and gives for the upper layer:

$$E_1^{(2)} = \frac{\mu}{4 J_1^2 p} \int dY \left(J_1^2 k^4 p^2 F_1^2 + k^2 (3 J_1^2 + p^2) F_1'^2 + 2 J_1^2 k^2 F_1 F_1'' + F_1''^2\right), \tag{10}$$

$$E_{c1}^{(2)} = \frac{1}{2} d_1 k^3 \pi F_1(H)^2, \tag{11}$$

where F' (respectively F'') means the first (respectively the second) derivative with respect to Y and the anisotropy parameter p is equal to $p = g_y/g_x$. Differential tangential expansion leads to $p < 1$.

$$E_1^{(4)} = \frac{\mu k^2}{16pJ_1^4}\int dY\big[k^2p^2F_1'^2\big(J_1^2k^2F_1^2+3F_1'^2\big)+F_1''^2\big(3J_1^2k^2F_1^2+F_1'^2\big)\big], \tag{12}$$

$$E_{c1}^{(4)} = -\frac{3}{32}d_1k^7\pi F_1(H)^4, \tag{13}$$

with similar relationships for $E_2^{(2)}$ and $E_2^{(4)}$ (with $\mu = p = 1$ and J_1 changed to J_2) for the bottom layer. The Euler-Lagrange equation is a fourth-order linear differential equation:

$$\frac{d^4F(Y)}{dY^4} - k^2\big(J^2+p^2\big)\frac{d^2F(Y)}{dY^2} + J^2p^2k^4F(Y) = 0, \tag{14}$$

with the boundary conditions explicated in Appendix A.

3.4 The Profile Function at Linear Order

When both layers grow, each solution of Eq. (14) can be written as superposition of hyperbolic functions, according to:

$$F_1(Y) = \frac{\epsilon}{1+A_1}\big(\cosh(J_1kY) + A_1\sinh(J_1kY) + A_2\cosh(kpY) + A_3\sinh(kpY)\big), \tag{15}$$

$$F_2(Y) = \frac{\epsilon}{1+A_1}\big(B_1\cosh(J_2kY) + B_2\sinh(J_2kY) + B_3\cosh(kY) + B_4\sinh(kY)\big). \tag{16}$$

Our choice for $F_2(Y)$ is valid for a finite thickness of the bottom layer and the boundary conditions must be prescribed for this layer too, at bottom. For a finite bilayer, we choose a condition of no-sliding of the bottom layer on a rigid support. When the second layer is infinite, one needs to change F_2 which becomes:

$$F_2(Y) = \frac{\epsilon}{1+A_1}\big(C_1\exp(J_2kY) + C_2\exp(kY)\big), \tag{17}$$

where $Y < 0$ and the interface is chosen at $Y = 0$. When the second layer does not grow but serves only as a support, Eq. (16) is modified and F_2 is changed into

$$F_2(Y) = \frac{\epsilon}{1+A_1}\big(D_1\cosh(kY)(1+D_2Y) + D_3\sinh(kY)(1+D_4Y)\big). \tag{18}$$

If the second layer does not grow and is also infinite, F_2 reads:

$$F_2(Y) = \frac{\epsilon}{1+A_1}E_1\exp(kY)(1+E_2Y). \tag{19}$$

In Eq. (15) up to Eq. (19), ϵ is a parameter of expansion for weakly nonlinear analysis which is evaluated in the post-buckling analysis (see Sect. 5). Since the amplitude of the perturbation is ϵ, one arbitrary coefficient has been fixed to 1 in F_1. However, it remains 7 coefficients to fix for 8 equations so k cannot be chosen arbitrarily. The k determination is a challenging problem of selection. However the ability of such models to describe the buckling is given only once k is theoretically determined. Despite the simplicity of the model, explicit analytical solution cannot be found (see Appendix B), but two opposite cases can be examined: (a) the short wavelength limit equivalent to infinite k ($\lambda = 2\pi/k$) and (b) the long wavelength limit and vanishing wavenumber k. Both cases have been partially studied in the past but here, we aim to clarify all possible scenarios.

4 The Dispersion Relation at Short-Wavelengths

The short-wavelength range concerns spatial oscillatory modes with a wavelength smaller than the lateral dimensions of the sample and all lengths such as the two thicknesses of the layer. It means that kH is a priori a large dimensionless parameter. The dispersion relation results from the calculation of the coefficients defined in Eqs. (15), (16), (17), (18), (19) and introduced in the boundary conditions (see Appendix A). After an easy but tedious calculation, all coefficients are determined and the solvability condition, or equivalently the dispersion relation, can be written as the sum of four exponentials with rather complicated polynomial coefficients of the physical quantities introduced in the model. It is possible to solve numerically this dispersion relation, however we can go further in the analysis by noticing that, if the upper layer grows, ($J_1 > 1$), two exponentials have the largest positive exponents and so will dominate the dispersion relation in the limit of infinite wavenumber k. As shown in [6], these exponentials control the selection mechanism. Once restricted to dominant quantities, the approximated dispersion relation becomes:

$$\mathcal{D} = \mathcal{D}_a - \mathcal{D}_b e^{-2Hkp}. \tag{20}$$

As shown in [6], for one growing layer, the key of the analysis is the existence of two polynomials that we introduce at first:

$$\begin{cases} \mathbf{P}_0(J) = -1 - J - 3J^2 + J^3, \\ \mathbf{Q}_0(J) = 1 - J + 3J^2 + J^3, \end{cases} \tag{21}$$

which allows to write simply $\mathcal{D}_a$ and $\mathcal{D}_b$ for the one layer case. For a bilayer, these terms are more complicated and read:

$$\mathcal{D}_a = \big(\mathbf{P}_0(\tilde{J}_1) - d_1 k \tilde{J}_1(\tilde{J}_1 + 1)\big)\mathcal{P}(\mu, \tilde{J}_1, J_2, d_2), \tag{22}$$

where $\tilde{J}_1 = J_1/p$ and $\mathcal{P}$ is a polynomial of degree two in μ. In the same way, we derive $\mathcal{D}_b$:

$$\mathcal{D}_b = \left(\frac{1 + \tilde{J}_1}{1 - \tilde{J}_1}\right)^2 \mathbf{Q}_0(\tilde{J}_1)\mathcal{Q}(\mu, \tilde{J}_1, J_2). \tag{23}$$

The new polynomials coming from the bottom layer are then, for $\mathcal{D}_a$:

$$\begin{aligned} \mathcal{P} = {} & \mu^2 J_2^2\big(\mathbf{P}_0(\tilde{J}_1) + d_2 k \tilde{J}_1(1 + \tilde{J}_1)\big) - \mu \tilde{J}_1 J_2\big(\mathbf{P}_1(\tilde{J}_1, J_2) - d_2 k \tilde{J}_1(1 + J_2)\big) \\ & + p\tilde{J}_1^2 \mathbf{P}_0(J_2), \end{aligned} \tag{24}$$

where

$$\mathbf{P}_1(\tilde{J}_1, J_2) = 1 - J_2^2 + p + p\tilde{J}_1^2(J_2 - 1) + J_2(2 + p) + \tilde{J}_1\big(1 + J_2^2 + 2p\big). \tag{25}$$

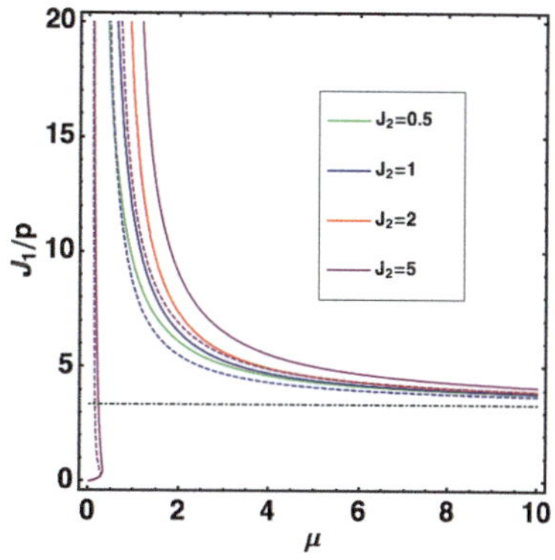

Fig. 2 Secondary bifurcation parameter J_1/p as a function of μ, ratio between both shear moduli of the upper layer versus the substrate. *Continuous colored lines* correspond to $p=1$, *colored dashed lines* $p=0.5$. As expected the growth anisotropy lowers the required growth rate. The *dot-dashed line in black* correspond to $J_c \sim 3.38$

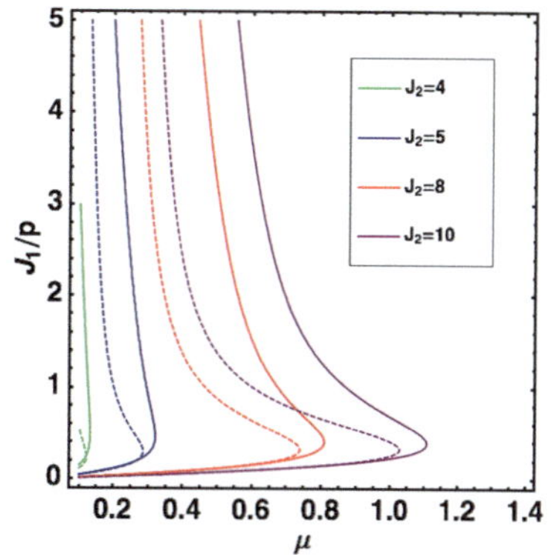

Fig. 3 Secondary bifurcation parameter J_1/p as a function of μ, ratio between both shear moduli of the upper layer versus the substrate. The substrate is stiffer than the upper layer when $\mu < 1$. The growth rate in the substrate is above the critical parameter $J_c \sim 3.38$. *Continuous colored lines* correspond to $p=1$, *colored dashed line* $p=0.5$. As expected the growth anisotropy lowers the required growth rate

and for $\mathcal{D}_b$:

$$\mathcal{Q}(\mu, \tilde{J}_1, J_2) = \mu^2 J_2^2 \mathbf{Q}_0(\tilde{J}_1) + \mu \tilde{J}_1 J_2 \mathbf{Q}_1 + p \tilde{J}_1^2 \mathbf{P}_0(J_2). \tag{26}$$

For $\mathcal{D}_b$, we can neglect the surface tension. Equation (23) is not valid if $J_1 = p = 1$ (which corresponds to a non growing cortex) but remains valid if $J_2 = 1$ (which corresponds to the case where the white matter does not grow). $\mathbf{Q}_1$ is explicited in Appendix B.

The cancellation of the dispersion relation $\mathcal{D}$, in the short wavelength approximation, is satisfied by imposing either $P_0(\tilde{J}_1) = 0$ or $\mathcal{P}(\mu, \tilde{J}_1, J_2, d_2) = 0$. It then exits two possibilities giving respectively the first and second bifurcation for J_1 at fixed J_2. The root of $\mathbf{P}_0$, for $\tilde{J}_1 = J_c$, gives the first buckling threshold and reads:

$$J_c = \frac{1}{3}\left(3 + (54 + 6\sqrt{33})^{1/3} + (54 - 6\sqrt{33})^{1/3}\right) \sim 3.383. \tag{27}$$

However, in the case of a bilayer, a second threshold appears giving by the cancellation of $\mathcal{P}(\mu, \tilde{J}_1, J_2, d_2) = 0$. This secondary instability is plotted in Figs. 2 and 3. In Fig. 3, a wrinkling can occur for $\tilde{J}_1$ smaller than J_c if $J_2 > J_c$ and μ smaller or of order one. If $\mu > 1$, $\tilde{J}_1$ must be bigger than J_c as shown by Fig. 2, independently of the value of J_2. However, since we restrict our discussion to $\mathcal{D}_a$, at this stage, the found wavelength is exactly zero in both cases which is the Biot's singularity. It is necessary to go beyond this threshold to reach a finite wavelength by considering both exponentials in Eq. (20). It turns out that no solution has been found for the one layer model in absence of capillarity [6]. To eliminate the singularity, one may consider a competitive energy in the variational elasticity problem such as the capillarity [6] or perhaps the coupling of two layers (and the existence of a new parameter μ) which may induce the cancellation of Eq. (20). Indeed, known solutions have been found in material sciences with a soft substrate leading to a selection mechanism by the ratio μ without capillarity [36, 65]. So there is perhaps hope to solve the Biot's singularity without surface tension. Despite the simplicity of the model, the dispersion relation remains difficult for the analysis and we begin by considering extreme cases: very soft or stiff substrate. The reader, only interested in the biomechanics of the human brain, can go directly to Sect. 4.2.1 and then to Sect. 5 which concern the range of μ parameter smaller than one, so the human brain bilayer. For completeness, the full analysis is given here for arbitrary parameters which may concern other biological cases or brain pathologies.

4.1 The Case of a Very Soft/Very Stiff Substrate

4.1.1 *Very Soft Substrate*

This limit allows to reduce the dispersion relation Eq. (22) to an expansion in μ^2 and μ only, giving:

$$\mathbf{P}_0(\tilde{J}_1)\left(\mathbf{P}_0(\tilde{J}_1) - \frac{\tilde{J}_1}{\mu J_2}\mathbf{P}_1(\tilde{J}_1, J_2)\right)e^{2kHp} \sim \left(\frac{1+\tilde{J}_1}{1-\tilde{J}_1}\right)^2 \mathbf{Q}_0(\tilde{J}_1)^2. \tag{28}$$

The critical threshold is then given by $\tilde{J}_1 \sim J_c$, and the wave number is simply:

$$kH = \frac{1}{p}\operatorname{Log}\frac{(1+J_c)}{|J_c-1|}\frac{\mathbf{Q}_0(J_c)}{\mathbf{P}_0'(J_c)(\tilde{J}-J_c)}. \tag{29}$$

When μ is large, $(\tilde{J} - J_c)$ can be evaluated from the cancellation of the left-hand-side of Eq. (28), provided that $\tilde{\mathbf{P}}_1(J_c, J_2)$ is positive. This is clearly the case when p is varied between $0 < p < 1$, the wavenumber k and the buckling wavelength λ can be derived according to:

$$\begin{aligned} kH &= \frac{1}{p}\operatorname{Log}\left(\frac{2\mu J_2}{J_c}\frac{1+J_c}{|J_c-1|}\frac{(3J_c^2+1)}{\mathbf{P}_1(J_c, J_2)}\right) \quad \text{and} \\ \lambda &= (2\pi p H)/\operatorname{Log}\left(\frac{2\mu J_2}{J_c}\frac{1+J_c}{|J_c-1|}\frac{(3J_c^2+1)}{\mathbf{P}_1(J_c, J_2)}\right). \end{aligned} \tag{30}$$

For very soft substrates, surface tension is not required to select the wavenumber. The wavelength remains proportional to H and its weakness is due to the large value of μ. However, the dependence in the μ values being logarithmic is weak and $kH \sim 1/p$.

4.1.2 *Very Stiff Substrate*

The same kind of analysis can be achieved for very stiff substrate which means $\mu \to 0$. Expanding the dispersion relation in this limit, remaining at zero or first order in μ we have

$$\mathbf{P}_0(\tilde{J}_1)\left(\mathbf{P}_0(J_2) - \frac{\mu J_2}{p\tilde{J}_1}\mathbf{P}_1(\tilde{J}_1, J_2)\right)e^{2Hkp} \sim \left(\frac{1+\tilde{J}_1}{1-\tilde{J}_1}\right)^2 \mathbf{Q}_0(\tilde{J}_1)\mathbf{P}_0(J_2). \tag{31}$$

For this case, it seems that two bifurcations occur, one controlled by the upper layer and $\tilde{J}_1 \sim J_c$, the other controlled by the bottom layer and $J_2 \sim J_c$. In fact, in absence of capillarity in the upper layer ($d_1 = 0$), only the bifurcation with $J_2 \sim J_c$ seems possible if $P_1(\tilde{J}_1, J_c)$ remains positive in this neighborhood. However also the left-hand side vanishes for the same value of $J_2 = J_c$. So the solvability condition requires in this case to expand to second order in μ Eq. (24) once $J_2 = J_c + \mu \delta J_2$ where

$$\delta J_2 = \frac{J_2 \mathbf{P}_1}{p \tilde{J}_1 \mathbf{P}'_0(J_2)},$$

and $J_2 \sim J_c$. A solution can be found if the following quantity is positive:

$$\mathcal{R} = \mathbf{P}_0(\tilde{J}_1)\left(\frac{J_2^2}{\delta J_2}\mathbf{P}_0(\tilde{J}_1) - \tilde{J}_1 \frac{\partial (J_2 \mathbf{P}_1)}{\partial J_2} + \frac{p \tilde{J}_1}{2} \delta J_2 \frac{d^2 \mathbf{P}_0(J_2)}{d J_2^2}\right), \tag{32}$$

where $J_2 = J_c$. The interest of such analysis comes from the fact that we can expect lower value for J_1 if growth occurs in the sub-layer. If not, the folding is dominated by the first bifurcation, where $\tilde{J}_1 \sim J_c$. Considering $1 < \tilde{J}_1 < J_c$, the term in parenthesis must be negative for a selection mechanism based on the growth in the substrate. A numerical exploration of this term as $\tilde{J}_1$ varies between 1 and J_c gives effectively negative values so one can expect solutions without surface tension due to the substrate. We conclude that for a stiff substrate, growing with J_2 larger than J_c, the wrinkling instability still occurs but finds its origin from growth in the substrate and the wavenumber is given by:

$$kH = \frac{1}{2p} \operatorname{Log}\left(\frac{2\tilde{J}_1^2}{\mu \mathcal{R}}\left(\frac{1+J_c}{J_c - 1}\right)^2 (3J_c^2 + 1)\mathbf{P}'_0(J_c)\right). \tag{33}$$

Notice that our limit corresponds to the green curve in Fig. 3 for very small μ. For intermediate values of μ only a numerical investigation is possible and it is the aim of the next paragraph.

4.2 Intermediate μ Values

In the short-wavelength number limit, bifurcations occur when $\mathcal{D}_a$ vanishes. According to Eq. (22), this gives two possibilities since $\mathcal{D}_a$ can be decomposed into two factors. We will call first bifurcation the cancellation which occurs when the first factor vanishes, which is also the bifurcation of a single layer [6].

4.2.1 The Primary Bifurcation

The first factor in $\mathcal{D}_a$ does not depend on the bottom layer parameters and on μ. This bifurcation is then the classical one and can only be observed if the associated wavelength has a finite non vanishing value. It is why the capillarity is considered in the model to eliminate the Biot's singularity and to determine the wavenumber k or the spatial wavelength λ of the nascent folds. If not, the k value is infinite or $\lambda = 0$ as explained by Biot in [11]. To solve this paradox, also called the Biot's singularity, as shown in [6], a weak surface tension coefficient is introduced, but one can observe that only the surface tension on top of the upper layer is concerned. To find the k number, we expand the solvability condition Eq. (38) in the vicinity of the root of $\mathcal{D}_a$, given by Eq. (27), and define the key parameter R_1:

$$R_1 = \frac{\mathcal{D}_b}{\mathcal{D}'_a} = \frac{\mathcal{Q}_0(J_c)\mathcal{Q}(\mu, J_c, J_2)}{\mathbf{P}'_0(J_c)\mathcal{P}(\mu, J_c, J_2, 0)} \sim \frac{2(3J_c^2 + 1)\mathcal{Q}(\mu, J_c, J_2)}{(3J_c^2 - 6J_c - 1)\mathcal{P}(\mu, J_c, J_2, 0)}, \tag{34}$$

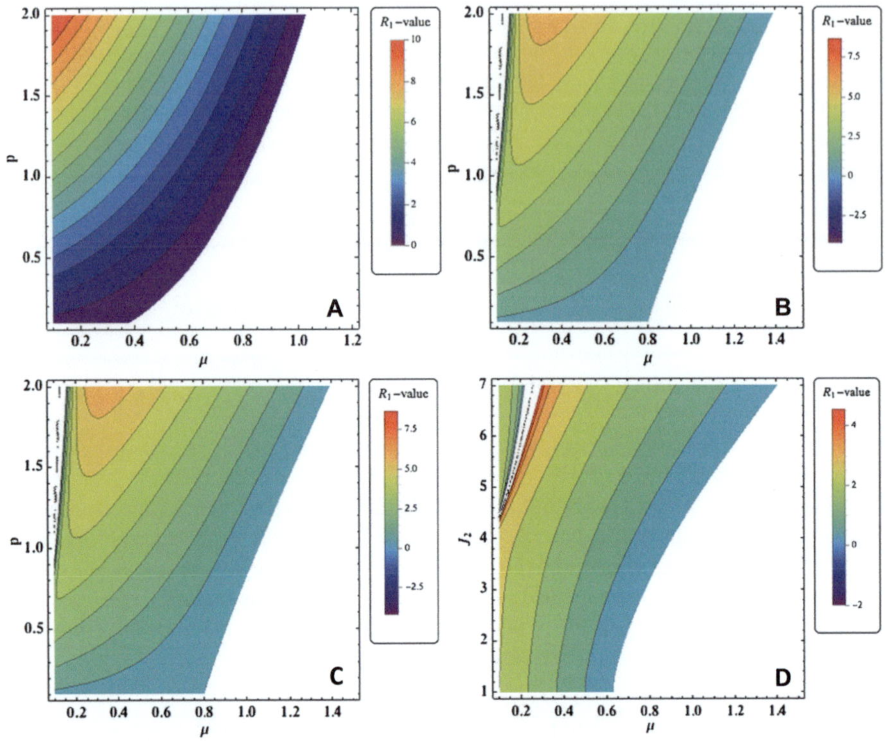

Fig. 4 Phase-diagram for the parameter R_1 (Eq. (34)) for the primary bifurcation (vicinity of $J_1 \sim J_c$) as μ and p vary, for $J_2 = 1$, (no growth in the substrate, **A**) and $J_2 = 4$ (the growth in the substrate layer exceeds the threshold of instability of a single layer, **B**). The same parameter as a function of the growth in the substrate for isotropic growth ($p = 1$, **C**), and, anisotropic growth ($p = 0.5$, **D**). In each figure, notice that the domains were $R_1 > 0$ is in color, with a color bar legend which indicates the magnitude of $R_1 > 0$. *White area* corresponds to domains where the bifurcation disappears and concerns μ values larger than one, so soft substrate

where $\mathcal{D}'_a$ is the derivative of $\mathcal{D}_a$ with respect to $\tilde{J}_1 = J_c$. Only if R_1 is positive, a wavy pattern exists with a given wavelength at threshold. If not, there is no such bifurcation for $\tilde{J}_1 \sim J_c$ and we must consider the secondary bifurcation corresponding to the cancellation of $\mathcal{D}_a$ given by $\mathcal{P}$. The numerical value of R_1 can be calculated. In Fig. 4, a phase-diagram gives R_1 when μ and p vary for not growing ($J_2 = 1$) and growing ($J_2 = 4$) substrate (see Figs. 4A and 4B) while in Figs. 4C and 4D, μ and J_2 vary with $p = 1$ and $p = 0.5$. Not surprisingly, the domain of folding patterns increases when J_2 increases, when the stiffness of the substrate increases and the parameter p decreases. In this case, the wavenumber is:

$$kH = \frac{1}{2p} \operatorname{Log} \frac{R_1}{(J - J_c)} \quad \text{with } (J - J_c) \sim k d_1 \frac{J_c(J_c + 1)}{3J_c^2 - 6J_c - 1}, \tag{35}$$

which leads to:

$$kH = \frac{1}{2p} \operatorname{Log} \frac{2pR_1(3J_c^2 - 6J_c - 1)}{d_1 J_c (J_c + 1)}. \tag{36}$$

4.2.2 The Secondary Bifurcation

We consider now the possibility to observe another bifurcation by the cancellation of $\mathcal{P}(\mu, \tilde{J}_1, J_2, d_2)$. The interest to study such secondary bifurcation appears clearly from the diagrams displayed in Fig. 4 which gives solutions only for rigid substrate $\mu < 1$ corresponding to the analysis of Sect. 4.1.2. Folding seems to not exist in the white domains where μ is larger than unity. It may also happen that the cancellation of $\mathcal{P}$ occurs before the cancellation of $\mathcal{P}_0$, then corresponding to lower growth rates. From Figs. 2 and 3, it seems possible to reach new bifurcation parameter J_1 at a lower value than before for a growing substrate stiffer than the upper layer ($\mu < 1$). In Fig. 2, a new group of solutions for $\mu > 2$ (soft substrate) and for $\tilde{J}_1 = J_1/p \sim 5$ appear in the domain where there is no primary bifurcation. This limiting value is poorly dependent on the growth inside the substrate. According to our previous paragraph, one also needs to consider the sign of the new parameter R_2 given by:

$$R_2 = \frac{\mathcal{D}_b}{\mathcal{D}'_a} = \frac{\mathcal{Q}_0(J_c)\mathcal{Q}(\mu, J_c, J_2)}{\mathbf{P}_0(J_c)\mathcal{P}'(\mu, J_c, J_2, 0)}, \tag{37}$$

calculated in the vicinity of the root of $\mathcal{P}(\mu, \tilde{J}_1, J_2, 0)$. However the main difficulty arises from the fact that, contrary to the primary bifurcation, the new critical growth rate J_t, (the real root of $\mathcal{P}(\mu, \tilde{J}_1, J_2, 0)$), is not a simple function but depends on all the other parameters such J_2, μ, p. It is why a numerical analysis is performed. The accuracy is not perfect as shown in Fig. 5 but gives a fast overview of the results. $\mathcal{P}$ is a polynomial of degree three in J_1 so has either one real eigenvalue or three, as shown by Figs. 2 and 3. Focusing on small μ values, according Fig. 3, for J_2 value higher than the critical threshold J_c, it exists two possible positive solutions for J_1/p: one which is small, eventually smaller than one, the other which is higher. However these solutions do not give a positive R_2 value so are rejected in this limit. For larger values of μ, Fig. 5A ($p = 1$) and Fig. 5B ($p = 0.5$) show the unique eigenvalue $J_1 = J_t$ when the substrate becomes softer than the upper layer. Such solution corresponds obviously to Fig. 2. Figs. 5C and 5D confirm that $R_2 > 0$ quite in the same domain of existence. The wavenumber can be calculated but in this case the selection is done by the surface tension d_2 between the two layers and not by the upper layer. To conclude, the bilayer folding results mainly from the upper layer if $\mu < 1$ and from the bottom layer if $\mu > 1$. If $\mu < 1$, (as it is believed for the brain), the analysis is simple: only the first bifurcation and the cortex layer play a role in the analysis. Instability occurs for $J_1 \sim p3.38$ and the wavelength scales like $\lambda \sim 4\pi p H$.

4.3 Comparison with Pre-Strain Models

In a recent work, a phase-diagram has been established with $3D$ simulations in plain-strain elasticity [60]. Based on a pre-stretching process of the substrate, a similarity with our treatment is obvious if one takes into account only Eq. (2). However, differences occur at the level of boundary conditions. Differential growth of bilayers modifies the mechanical stresses by a factor which takes into account the volumetric growth which is different in both layers so the boundary conditions at the interface are not identical (see Appendix A). Also in [60], capillarity is not introduced but a delamination energy is taken into account. Boundary conditions are crucial for threshold determination and post-buckling results in the bifurcation theory. In addition, our analysis is done in two dimensions while the Wang and Zhao work involves a three-dimensional analysis. In fact the last point is not essential in plain-strain elasticity for uniaxial buckling, it requires simply a renormalization of coefficients and pg_x must be

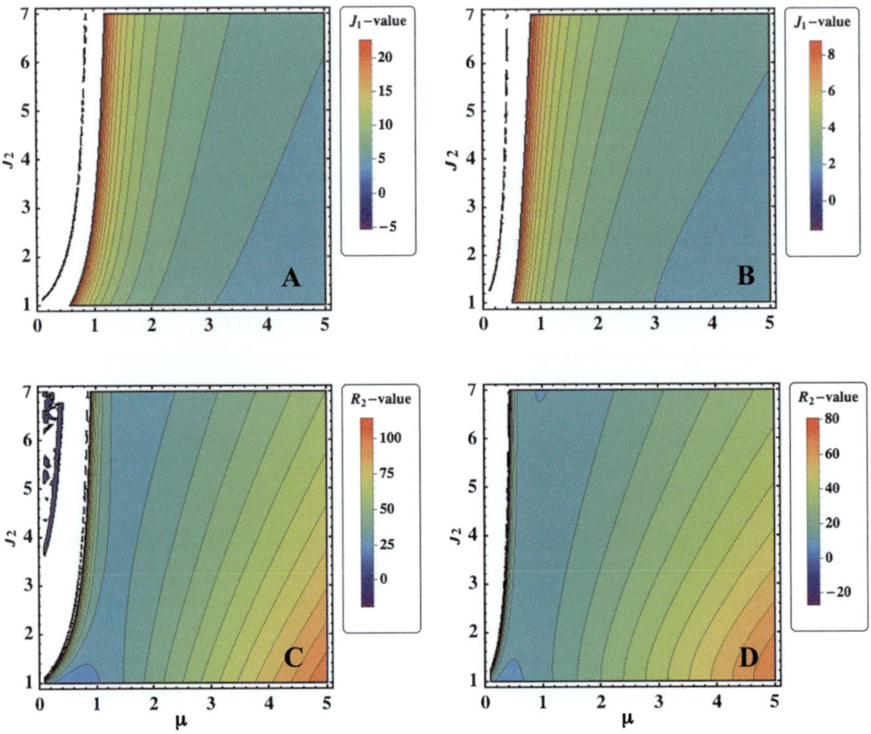

Fig. 5 Phase-diagrams for the secondary bifurcation (Sect. 4.2.2) as a function of the shear modulus ratio μ and the growth on the substrat J_2. **A** and **B** represent the threshold value for the growth in the upper layer J_1 as μ and J_2 vary. J_1 is the root of the polynomial Q of Eq. (26). Notice the high values of J_1 for small μ which decreases when μ increases. This graph corresponds to Fig. 2. As shown in **B** ($p = 0.5$), the results are very similar to **A** but with a lower value for J_1 (notice the change of scales in the bar legend). In **C** and **D** are presented the values of R_2, when it is positive, a necessary condition for folding. Despite some numerical inaccuracies, the phase diagrams **A** and **C** coincide with **B** and **D**

replaced by $(pg_x)^{2/3}$ on each horizontal axis. Considering $\mu = \mu_f/\mu_s$ values smaller than one, our analysis indicates that the critical parameter is $J_1 \sim 3.383$ which gives in a $3D$ space, a growth $g_1 \sim 1.501$. If we transpose this growth value, the critical pre-stretch value is then $\epsilon_c \sim 1 - 1/g_1 \sim 0.333$ while numerics indicates 0.35. In absence of surface tension, singular patterns such as creases are observed. However, this limit $\epsilon_c \sim 0.33$ remains a line of separation as μ increases as can be observed in Figure (4) of [60] and in Fig. 2. So the two models recover quite analogous results in the short wavelength limits except that in our case, J_1 and J_2 can vary independently which is not the case in the pre-strain model. For infinite μ values, the results differ by a numerical factor which depends probably on the boundary conditions. The net advantage of numerical simulations is to provide informations on changes in patterns as the mismatch conditions increase so in the fully nonlinear regime.

4.4 Two Layers of Arbitrary Thicknesses

In the previous section, we have considered the dispersion relation for the wrinkling of a bilayer with an infinite substrate, an hypothesis which simplifies the analytical treatment. However, one can wonder if the ratio of thicknesses is not also a pertinent parameter for the wavelength selection. Let us come back to the more realistic case when the two layers have comparable thicknesses. The dispersion relation is then:

$$\mathcal{D} \sim \mathcal{D}_a - \mathcal{D}_b e^{-2Hkp} - \mathcal{D}_c e^{-2k}, \tag{38}$$

where $\mathcal{D}_a$ and $\mathcal{D}_b$ has the same meaning as before in Eq. (20) and

$$\mathcal{D}_c = \left(\frac{1+J_2}{1-J_2}\right)^2 \left(\mathbf{P}_0(\tilde{J}_1) - d_1 k \tilde{J}_1(\tilde{J}_1+1)\right) \times \left\{\mu^2 J_2^2 \mathbf{P}_0(\tilde{J}_1) - \mu \tilde{J}_1 J_2 \mathbf{P}_2(\mu, J_1, J_2) + J_1^2 p \mathbf{Q}_0(J_2)\right\}. \tag{39}$$

In the short wavelength limit that we have considered previously, we can discard e^{-2k} if Hp is smaller than one so if the thickness of the upper layer multiplied by the anisotropy coefficient is smaller than the thickness of the bottom layer. In this case, the result of the previous section (for an infinite substrate) is valid. For a thin substrate, the analysis is modified although the same strategy has to be maintained. Because for the brain, the first case prevails, the cortex thickness being much smaller, we do not make the analysis for a thin substrate.

5 Post-Buckling and the Gyrification Index

We now examine the buckling and post-buckling by considering the energy expansion at fourth order in the amplitude ϵ of the profile function F_1. Since experimentally μ is smaller than one for the brain (see Table 1), we focus on the primary bifurcation. We expect, according to our previous analysis, that the buckling threshold is given by the volumetric growth in the upper layer, with a rate value slightly larger than pJ_c. ϵ gives the amplitude of the buckling which must be related to the gap: $\Delta = \tilde{J}_1 - J_c$. Averaging Eqs. (10) and (12) per unit length, we derive to leading order in the Δ expansion:

$$\mathcal{E} = \frac{\mu k^3 \epsilon^2}{4\tilde{J}_1} \frac{(\tilde{J}_1 - 1)^2 e^{2HJ_1k}(1+\tilde{J}_1)}{(1+\tilde{J}_1^2)^2} \left(-\mathbf{P}_0'(\tilde{J}_1)\Delta + \epsilon^2 P_2(\tilde{J}_1) k^4 e^{2HkJ_1}\right), \tag{40}$$

$$P_2 = \frac{(\tilde{J}_1 - 1)^2(963 + 1256\tilde{J}_1 + 3265\tilde{J}_1^2)}{16p^2(\tilde{J}_1+1)^2(\tilde{J}_1+3)(3\tilde{J}_1+1)(1+\tilde{J}_1)^2}. \tag{41}$$

Minimization of $\mathcal{E}$ with respect to ϵ gives the amplitude as the profile function $F_1(Y)$:

$$\epsilon = \frac{e^{-HkJ_1}}{k^2}\sqrt{\Delta}\sqrt{\frac{\mathbf{P}_0'(\tilde{J}_1)}{2P_2(\tilde{J}_1)}} \sim 1.3544 \frac{e^{-HkJ_1}}{k^2}\sqrt{p(J_1 - pJ_c)}. \tag{42}$$

P_2 being positive, we can conclude that the bifurcation is supercritical, the amplitude of the folds varies as the square-root of the gap from threshold Δ. From Eqs. (4) and (42), an evaluation of the fold amplitude can be achieved since, on top of the layer, y_1 reads:

$$y_1 = J_1 H + \epsilon e^{kJ_1H} k \cos(kx) \sim J_1 H + 1.354 \frac{\sqrt{p(J_1 - pJ_c)}}{k} \cos(kx). \tag{43}$$

The quantity measured usually in brain embryogenesis is the gyrification index, ratio between the undulated surface compared to the supposed averaged surface without folds.

From our model, this quantity that we call G_I can be calculated, noticing that $y' \sim \epsilon e^{HJ_1k}k^2\sin(kx) \sim 1.3544\sqrt{p(J_1 - pJ_c)}\sin(kx)$,

$$G_I = \frac{k}{2\pi}\int_{-\pi/k}^{\pi/k}\sqrt{1+y'^2}dx = \frac{1}{2\pi}\int_{-\pi}^{\pi}\sqrt{1+m\sin(u)^2}du, \tag{44}$$

where $m = 1.3544^2 p(J_1 - pJ_c)$. So the theory predicts an index which is independent of k and depends only on the growth of the upper layer if the wrinkling results from the primary bifurcation. In fact it depends mainly on the growth anisotropy coefficient p and the volumetric growth J_1 at time t. However, our calculation results from an expansion in ϵ of the elastic energy and so has a limited range of validity. In addition, the post-buckling analysis only gives the amplitude of the buckling but maintains the harmonic shape of the folds, which is not observed in simulations [60]. To test the range of validity of our results, we need to compare them to human brain data concerning fetus and newborns [1, 68] and first to calibrate J_1. The initial cortex mass of the initial configuration has to be estimated and J_1 will be the ratio between the current cortex mass to the initial one. Since we do not have a separate information for the cortex and the white matter, as a first approximation, we will take the total mass of the brain compared to its initial value. From the experimental data [1, 68], we estimate the constitution of the initial configuration at 23 weeks of ontogeny with a brain mass of 50 g, and we notice that the detectable folding events occur at 28 weeks with a brain mass of 114 g, which gives us threshold for folding about $J_1 \sim 114/50 \sim 2.28$ and an anisotropy coefficient of order $p \sim 0.7$. These informations are sufficient to derive the gyrification index according to Eq. (44) which does not take into account other physical quantities. We plot G_I and compare it to the experimental measurements in Fig. 6. One can notice the very good agreement and, surprisingly, the model works even in a range of parameters where our formula looses its validity. Indeed our analysis includes the first nonlinearities in $(J_1 - pJ_c)$ and the agreement extends up to J_1 values of order 10. For the gyrification index, the post-buckling analysis, in this case, seems to be a valuable tool allowing to reach an important quantity for the experimentalists. In addition, there is no adjustable parameter. The only required quantities to plot the theoretical G_I curve consist of the evaluation of J_1 and the anisotropy parameter p. From the data [1, 68], the initial mass is deduced and the ratio between the brain mass at the onset of buckling and the initial mass gives us J_1, which once compared to the theory furnishes the anisotropy parameter $p \sim 0.7$. When the gyri are small and the gyrification index close to one, it is difficult to get an accurate adjustment of parameters with the formula Eq. (44). As all bifurcations in physics, the threshold is always difficult to reach experimentally. In Fig. 7, the gyrification index is displayed as a function of J_1 for different growth anisotropy coefficient, p varying from 0.5 up to 1.5. One can notice the evolution, as the growth proceeds, varying the p values. In Fig. 8, a comparison with measurements made with MRI [38], close to the gyri initiation, is performed, G_I being smaller than 1.6 while the brain mass is smaller than 200 mL. The two data sets (blue and red points) correspond respectively to pre-term babies and fetuses after a similar period of gestation, for comparison. The data sets cover more the gyri initiation where the data are lacking in Fig. 6. We choose the best fit both for the anisotropy coefficient p and the initial brain volume. For fetuses, the initial volume is estimated to be 40 mL (so close to the estimation of 50 g extracted from [1, 68]). As for the anisotropy coefficient p, it is stronger $p \sim 0.3$ which indicates a stronger proliferation rate of the neurons along the interface between the grey and white matter in this early period of brain development. For pre-terms, the initial volume is about 27 mL and the anisotropy index is of order 0.55. Then, the anisotropy coefficient is intermediate between the one of fetuses and the value extracted

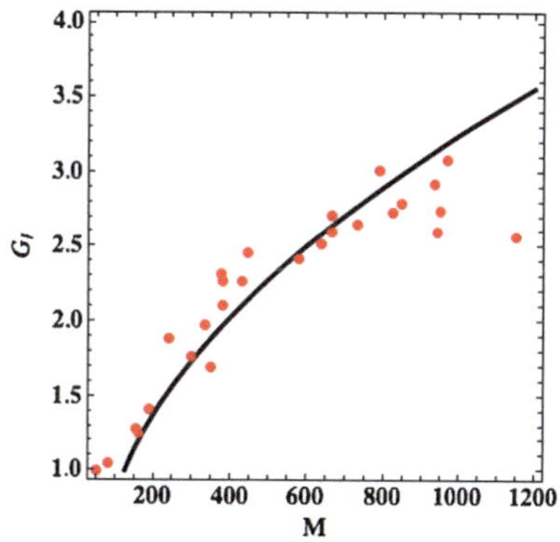

Fig. 6 The theoretical gyrification index (*smooth line*) and experimental data on human brains from [1, 68]

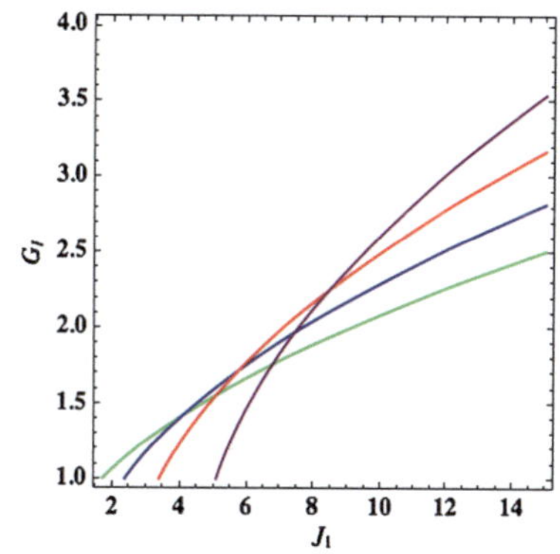

Fig. 7 Evolution of the theoretical gyrification index with the anisotropy coefficient $p = g_y/g_x$. In *green*, $p = 0.5$, in *blue* $p = 0.7$, in *red*, $p = 1$, in *purple* $p = 1.5$ (in this case, we assume the proliferation more important in the y direction)

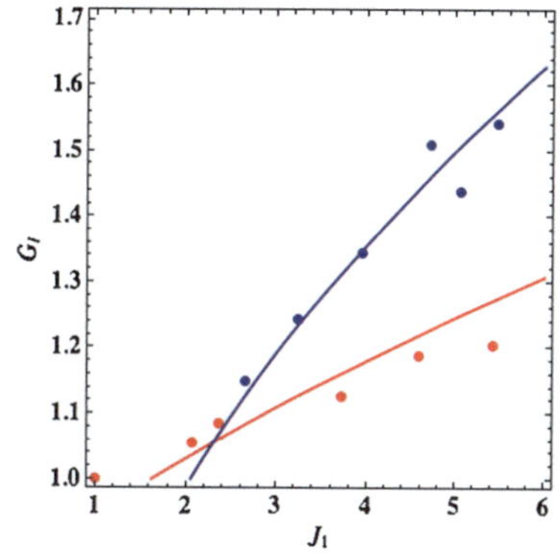

Fig. 8 Comparison of the gyrification index of fetuses obtained *in utero* (in *red*) and from pre-term babies (in *blue*) after the same period of gestation [38]. Comparison is made with the model and estimation of parameters is achieved via fitting

from the data of [1, 68] which cover more the last period of brain development. Here again, it seems that the early period of gyrification presents more growth anisotropy.

6 Long Wavelength Approximation

The aim of this section consists in recovering the classical results of material science for the wavelength selection when $\lambda \to \infty$ and is much larger than the thickness of the sample

[10, 36, 65]. This limit has been considered by Biot [10] and has been revisited in other contexts [16, 32, 53] but not in the framework of growth. Except in [36] where the substrate is assumed neo-Hookean, usually linear elasticity is used to describe the buckling of a very thin layer put on a very soft substrate. To treat this limit, which does not concern the brain folding, we consider our dispersion relation that we expand up to $(kH)^4$ and keep only the leading order terms proportional to μ:

$$\mu C_4(kH)^4 - C_2(kH)^2\mu + C_1(kH) = 0, \tag{45}$$

with

$$\begin{cases} C_4 = -(1/6)J_2 p^7(\tilde{J}_1^6 - 3\tilde{J}_1^4 - 5\tilde{J}_1^2 - 1) \sim \frac{4}{3}J_2 p^7, \\ C_2 = J_2 p^5(3\tilde{J}_1^4 - 2\tilde{J}_1^2 - 1) \sim 8J_2 p^5 \delta J, \\ C_1 = (1+J_2)p^4(1 + 3\tilde{J}_1^2 + \tilde{J}_1 p - \tilde{J}_1^3 p) \sim 4(1+J_2)p^4, \end{cases} \tag{46}$$

where $\delta = \tilde{J}_c - 1$. Indeed, for consistency, δ must be small and the reduced equation is then

$$p^3(kH)^3 - 6p\delta g(kH) + \frac{3(1+J_2)}{\mu J_2} = 0. \tag{47}$$

It is worth noticing and contrary to material sciences, all coefficients depend on the growth problem, of J_1 and J_2. The imposed pressure in Biot's model is replaced by the volumetric growth $\delta g = J_1 - 1$ and the growth increase is weak in this limit. However, as in [10], a threshold must be reached for the buckling and we can derive easily its value and the corresponding wavelength:

$$\delta g_c = \frac{1}{2}\left(\frac{3(1+J_2)}{2J_2\mu}\right)^{2/3} \quad \text{and} \quad \lambda = 2\pi p \mu^{1/3} H \left(\frac{2J_2}{3(1+J_2)}\right)^{1/3}. \tag{48}$$

The classical results in power 1/3 and 2/3 is easily recovered in this limit $\mu \to \infty$.

7 Discussion and Conclusion

In this paper, we have examined the buckling instability of a bilayer in differential growth with the purpose to explain the brain gyrification occurring in fetal life. Our study covers the whole range of growth parameters in both layers (J_1 and J_2) and stiffness ratio values μ, some of them are known for the brain via previous anatomy studies and biomechanical measurements [12]. In addition, MRI imaging provides new data concerning the G_I values [25, 38] *in utero*. Of course the complex brain geometry, even at 20 weeks of development, is very far from a double layer. It presents observable structures, not taken into account in a planar bilayer. By a careful analysis of data from [1, 68] concerning gyri measurements, we identify the brain growth with gyri beginning at 20/22 weeks and sulci apparition at 27 weeks. Sulci appearance is mentioned earlier in MRI studies [24, 67]. The stiffness ratio has been measured by indentation measurements giving $\mu \sim 0.5$ for bovines and pigs but it presents an important regional variability [12]. If it was accepted that for the human brain, the μ value was smaller than one, a recent and careful study on cadavers post autopsy arrives to the opposite conclusion [14]. Measurements concerning fetuses, required to explain the gyri formation occurring during pregnancy, are not available. It is why our study covers the whole range of parameters for the μ values but our results, compared to the data [1, 25, 38,

68] concerning the G_I values during pregnancy favors the hypothesis of a cortex less stiffer than the white matter.

As mentioned previously, the gyri appearance at 20/22 weeks from Fig. 6 and from the data of [1, 68] reveals a bifurcation occurring for a growth rate $J_1 \sim 2.28$ which allows to identify the growth anisotropy to be $p = 0.7$. Then knowing the initial thickness value H which is about 0.1 mm at 11 weeks [46], we deduce the cortex thickness at 27 weeks: $J_1 H \sim 0.34$ mm and at birth $J_1 H \sim 20H \sim 2$ mm. Comparison with MRI studies *in vitro* are satisfactory since estimation from Figure (4) of [49] gives approximatively 0.5 mm at 27 weeks, in reasonable agreement with our zero-order approximation of purely axial growth. However, very recent measurements, also achieved with MRI [40, 41] indicate that the average thickness of newborn cortex is 1.99 ± 0.1 mm for the left hemisphere, 2.03 ± 0.12 for the right hemisphere, values which will increase by 42%, in the first year, much less in the second year. It must be emphasized that the cortex thickness of humans increases with aging. For example an average value of 2.59 mm is estimated for humans of fifty years old [62]. However, our study concerns only embryonic development and our naive estimation of 2 mm at birth is in perfect agreement with the last measurements [40, 41]. Knowing that the size of the brain is approximatively two centimeters, at 27 weeks, we can assume the substrate infinite.

Let us now discuss the prediction of the buckling threshold according to the μ value. In the range smaller than one, we have shown that the instability originates mainly from the growth in the cortex layer. We predict the folding threshold occurring for $J_{1c} = p3.38 \sim 2.3$ and eventually a secondary instability occurring for $J_2 \sim 10.9$, which is completely unrealistic at 23 days of ontogeny. So the buckling process at short wavelength is for sure due to the cortex layer. The experimental data, given in [1, 68], concern the whole brain growth and not separately the cortex and the white matter. Data concerning size increases, layer per layer, are only known for the pre-gyrification period, unfortunately [49]. The good agreement we have obtained for the growing cortex thickness indicates that the growth mismatch between both layers of the brain is not really significant, since we take, for the cortex, these data normalized by the initial size. Remember that J_1 is a relative growth parameter so a ratio between the current mass versus the initial mass. Clearly, the growth anisotropy p and the difference between stiffnesses are much more crucial for the comparisons between the experimental and predicted gyrification index, shown in Figs. 6, 8.

The post-buckling treatment, at leading order of approximation, allows to theoretically evaluate the gyrification index by a simple relationship given by Eq. (44) and a rather good agreement is reached with experimental results for human brain [1, 25, 38, 68]. Here again, we must emphasize that if $\mu < 1$, G_I which evaluates the distorted surface versus the supposed smooth one only depends on the growth characteristics of the cortex. If $\mu > 1$, Eq. (44) is no more valid. In our comparison, there is no adjustable parameter since the parameter p as the initial size of the cortex are extracted from the data.

One can wonder how to explain pathologies such as lissencephaly or microgyria. Lissencephaly (corresponding to a lower G_I) value can be explained by the fact that some hypothesizes assumed in the model are not checked. For example the cortical growth looses its anisotropy, rejecting the threshold J_1 to larger values. The contrast of stiffnesses may be modified becoming larger than one. Microgyria is explained by a pathology of the meninges which consists in an increase of their thickness. Then, it can correspond to a buckling of the meninges instead of the cortex. For healthy brain, the meninges are stiffer than the cortex: they contain fibroblasts and fibers. But in pathological situation, perhaps, the observed microgyria correspond to the buckling of one of the layers of the meninges. If the meninges are thinner than the cortex at the same age, then the wavelength is expected to be smaller, giving tiny wrinkles. More informations on these pathologies are required, from the biological and physical sides, to refine the model.

To conclude, we show the complexity to find the upset of wrinkling in the short wavelength limit for bilayers with similar physical parameters. The limit where $\mu < 1$ (case which gives a better agreement with experimental data for the gyrification index) is easier since the buckling answer only originates from the upper layer and the wavelength must scale with the thickness of this layer. Also the threshold for the onset of buckling can be predicted without difficulty, being approximatively pJ_c. It is generally believed that the thickness of the cortex does not grow during fetal life indicating perhaps that the gyrification originates from the secondary instability. In this case, only the substrate grows and the cortex keeps its thickness constant. Numerical threshold values derived for the brain with our model shows that both thresholds are well separated and the secondary instability has no physical meaning. Our main difficulty remains the Biot's singularity and the evaluation of the microscopic length of regularization for precise comparison with experiments. In our model, the wavelength is found by a sophisticated balance between a surface energy or capillarity and the elastic energy of buckling. In the case of the brain, it is possible that the meninges, which envelop the cortex play this role as shown in Appendix C and can be considered as a surface tension. However their constitutions are complex, made of three different layers. Hopefully, the prediction of the gyrification index does not require the precise value of the wavelength. Up to an unknown constant, our estimation for the wavelength is $2\pi pH$ so approximatively 8 mm for a cortex of thickness 2 mm. To be more quantitative is difficult due to the brain geometry and to obvious structural inhomogeneities.

Finally, the growth bilayer model requires a threshold for buckling which can explain why the brain of rodent escapes to this instability. The detailed analysis of pathologies inducing modification of the cortex buckling can also be explained by modifications of the growth process. Anomalous thickening of the meninges for example may favor the buckling of this layer compared to the cortex giving a weakening of the amplitude of the gyri and a shorter wavelength, then corresponding to polymicrogyria.

Acknowledgements This research was supported in part by Institut Universitaire de France and by the National Science Foundation under Grant No. PHY11-25915. It was motivated by enlightening discussions on folding with Ellen Kuhl and Alain Goriely in Oberwolfach session: "Mathematics of Differential Growth, Morphogenesis, and Pattern Selection" and with Oksana Manyuhina during the KITP session: "Geometry, elasticity, fluctuations and order in soft matter". MBA wants to thank the organizers: Professor Ogden and Holzapfel for their kind invitation to participate to this special issue.

Appendix A: Boundary Conditions

The boundary conditions concern the equality of the displacement at the interface between the two layers (2 conditions) and the mechanical equilibrium of both interfaces for $Y = 0$ and $Y = H$ which finally gives 6 conditions. Let us focus on the interfacial conditions in both models: the growth and the "pre-strain" model [60], where both layers remain glued together. First we have the equality of displacements at the interface in both layers for both models, then equality of nominal stresses in the normal direction: S_{12} and S_{22}. In the case of growing layers, it has been shown [7] that the nominal stress is given by:

$$\mathbf{S} = G^{-1}\left(\frac{\partial E}{\partial F_e} - pJ\mu F_e^{-1}\right), \tag{49}$$

which gives, for neo-Hookean materials and for the upper layer, in a bidimensional approach:

$$\mathbf{S} = \mu g_x \left(F_e - p F_e^{-1}\right), \tag{50}$$

p being the Lagrange multiplier which ensures the constraint of incompressibility. Even if we adjust correctly the pre-strain in order to represent the growth so F_e in each layer, the boundary conditions concerning the mechanical equilibrium of the interface is different. There is no pre-factor corresponding to the growth of the sample in thc pre-strain model. These pre-factors are important for bilayers with differential growth and play a role for the threshold and the wavelength selection. Let us now come back to the boundary conditions written with the stream-functions. For $Y = 0$, $x_1 = x_2$ and $y_1 = y_2$ are transformed into:

$$F_1(Y) = F_2(Y) \quad \text{and} \quad J_2 F_1'(Y) = J_1 F_2'(Y) \quad \text{for } Y = 0. \tag{51}$$

Once $\mathcal{M}$ is defined as $\mathcal{M} = \mu/p \cdot (J_2/J_1)^2$, cancellation of the shear stress (see Chapter 5 of [45]) gives:

$$\mathcal{M}\left(\frac{d^2 F_1}{dY^2} + J_1^2 k^2 F_1(Y)\right) = \left(\frac{d^2 F_2(Y)}{dY^2} + J_2^2 k^2 F_2(Y)\right), \tag{52}$$

while the normal stress reads:

$$\begin{aligned} &\mathcal{M}\left(\frac{d^3 F_1(Y)}{dY^3} - k^2\left(p^2 + 2J_1^2\right)\frac{dF_1(Y)}{dY}\right) - J_2^2 d_2 k^4 F_1(Y) \\ &\quad = \frac{d^3 F_2(Y)}{dY^3} - \left(1 + 2J_2^2\right)k^2 \frac{dF_2(Y)}{dY}. \end{aligned} \tag{53}$$

On top on the bilayer, for $Y = H$, Eqs. (52) and (53) simplify to give:

$$\frac{d^2 F_1}{dY^2} + J_1^2 k^2 F_1(Y) = 0, \tag{54}$$

$$\frac{d^3 F_1(Y)}{dY^3} - k^2\left(p^2 + 2J_1^2\right)\frac{dF_1(Y)}{dY} - d_1 k^4 F_1(Y) = 0. \tag{55}$$

Appendix B: The Polynomials of the Dispersion Relations

In Sect. 4, we have introduced the polynomial $\mathbf{P}_1(\tilde{J}_1, J_2)$:

$$\mathbf{P}_1(\tilde{J}_1, J_2) = 1 - J_2^2 + p + p\tilde{J}_1^2(J_2 - 1) + J_2(2 + p) + \tilde{J}_1\left(1 + J_2^2 + 2p\right). \tag{56}$$

Similar polynomials enter the dispersion relation, Eqs. (22) and (38), that we do not explicit such as: $\mathbf{Q}_1$ in Eq. (26) and $\mathbf{P}_2$ in Eq. (39). It reads:

$$\mathbf{Q}_1(\tilde{J}_1, J_2) = 1 - J_2^2 - p - \tilde{J}_1\left(1 + J_2^2 - 2p\right) - J_2(-2 + p) + p\tilde{J}_1^2(1 - J_2), \tag{57}$$

$$\mathbf{P}_2(\tilde{J}_1, J_2) = 1 - J_2^2 + \tilde{J}_1\left(1 + J_2^2 - 2p\right) + J_2(-2 + p) - p + pJ_1^2(1 + J_2). \tag{58}$$

Appendix C: Skin Effect and Meninges

Let us consider the case of a very thin layer on top of an infinite substrate. Only the substrate grows: $J_1 = p = 1$ and $J_2 > 1$. The dispersion relation reads at first order in H:

$$\mathcal{D} = \mathbf{P}_0(J_2) - d_0 k J_2(1 + J_2), \tag{59}$$

where $d_0 = 4H\mu$. Comparing with Eq. (22), we deduce that, in a first approximation, this system behaves like a unique layer and the very thin layer only contributes to a surface tension d_0. The meninges and more specifically the pial basement membrane, one layer among the three layers of the meninges, can be considered as a surface tension.

References

1. Armstrong, E., Scleicher, A., Omran, H., Curtis, M., Zilles, K.: The ontogeny of human gyrification. Cereb. Cortex **5**, 56–63 (1995)
2. Audoly, B., Boudaoud, A.: A simple buckling model for studying the emergence of herringbone patterns. J. Mech. Phys. Solids **56**(7), 2422–2443 (2008)
3. Barron, D.H.: An experimental analysis of some factors involved in the development of the fissure pattern of the cerebral cortex. J. Exp. Zool. **113**(3), 553–581 (1950)
4. Basan, M., Joanny, J.F., Prost, J., Risler, M.: Undulation instability of epithelial tissues. Phys. Rev. Lett. **106**, 158101 (2011)
5. Bayly, P.V., Taber, L.A., Kroenke, C.D.: Mechanical forces in cerebral cortical folding: a review of measurements and models. J. Mech. Behav. Biomed. Mater. **29**, 568–581 (2014)
6. Ben Amar, M., Ciarletta, P.: Swelling instability of surface-attached gels as a model of tissue growth under geometric constraints. J. Mech. Phys. Solids **58**, 935–954 (2010)
7. Ben Amar, M., Goriely, A.: Growth and instability in elastic tissues. J. Mech. Phys. Solids **53**, 2284–2319 (2005)
8. Ben Amar, M., Jia, F.: Anisotropic growth shapes intestinal tissues during embryogenesis. Proc. Natl. Acad. Sci. USA **110**, 10525–10530 (2013)
9. Bender, C.M., Orzag, C.A.: Advanced Mathematical Methods for Scientists and Engineers. McGraw-Hill International Book Company
10. Biot, M.A.: Folding instability of a layered viscoelastic medium under compression. Proc. R. Soc. A **242**(1231), 444–454 (1953)
11. Biot, M.A.: Surface instability of rubber in compression. Appl. Sci. Res. **12**, 168–182 (1963)
12. Budday, S., Nay, R., de Rooij, R., Steinmann, P., Wyrobek, T., Ovaert, T.C., Kuhl, E.: Mechanical properties of gray and white matter brain tissue by indentation. J. Mech. Behav. Biomed. Mater. **46**, 318–330 (2015)
13. Budday, S., Kuhl, E., Hutchinson, J.W.: Period-doubling and period-tripling in growing bilayered systems. Philos. Mag. **95**(28–30), 3208–3224 (2015)
14. Budday, S., Sommer, G., Birkl, C., Langkammer, C., Haybaeck, J., Kohnert, J., Bauer, M., Paulsen, F., Steinmann, D., Kuhl, E., Holzapfel, G.A.: Acta Mater. (2017, preprint)
15. Brun, L., Auzias, G., Viellard, M., Villeneuve, N., Girard, N., Poinso, F., Da Fonseca, D., Deruelle, C.: Localized misfolding within Broca's area as a distinctive feature of autistic disorder. Biol. Psychiatry: Cogn. Neurosci. Neuroimaging **1**, 160–168 (2015)
16. Cao, Y.P., Jia, F., Zhao, Y., Feng, X.Q., Yu, S.W.: Buckling and post-buckling of a stiff film resting on an elastic graded substrate. Int. J. Solids Struct. **49**, 1656–1664 (2012)
17. Carroll, M.M.: A representation theorem for volume-preserving trasformations. Int. J. Non-Linear Mech. **39**, 219–224 (2004)
18. Clark, W., &Le Gros, E.: In: Johnson, J. (ed.) Deformation Patterns in the Cerebral Cortex. Oxford University Press, London (1945)
19. Ciarletta, P., Ben Amar, M.: Papillary networks in the dermal-epidermal junction of the skin: a biomechanical model. Mech. Res. Commun. **42**, 68–76 (2011)
20. Ciarletta, P., Ben Amar, M.: Growth instabilities and folding in tubular organs: a variational method in nonlinear elasticity. Int. J. Non-Linear Mech. **47**, 248–257 (2012)
21. Courchesne, E., Karns, C.M., Davis, H.R., Ziccardi, R., Carper, R.A., Tigue, Z.D., Chisum, H.J., Moses, P., Pierce, K., Lord, C., et al.: Unusual brain growth patterns in early life in patients with autistic disorder an MRI study. Neurology **57**(2), 245–254 (2001)

22. Christ, A.F., Franze, K., Gautier, H., Moshayedi, P., Fawcett, J., Franklin, R.J.M., Karadottir, R.T., Guck, J.: Mechanical difference between white and gray matter in the rat cerebellum measured by scanning force microscopy. J. Biomech. **43**(15), 2986–2992 (2010)
23. Dobyns, W.B., Truwit, C.L.: Lissencephaly and other malformations of cortical development: 1995 update. Neuropediatrics **26**(3), 132–147 (1995)
24. Garel, C., Chantrel, E., Brisse, H., Elmaleh, M., Luton, D., Oury, J.F., Sebag, G., Hassan, M.: Fetal cerebral cortex: normal gestational landmarks identified using prenatal MR imaging. Am. J. Neuroradiol. **22**(1), 184–189 (2001)
25. Germanaud, D., Lefèvre, L., Fischer, C., Bintner, M., Curie, A., des Portes, V., Eliez, S., Elmaleh-Bergès, M., Lamblin, D., Passemard, S., Operto, G., Schaer, M., Verloes, A., Toro, R., Mangin, F.G., Hertz-Pannier, L.: Simplified gyral pattern in severe developmental microcephalies? New insights from allometric modeling for spatial and spectral analysis of gyrification. NeuroImage **102**, 317–333 (2014)
26. Goldman, P.S., Galkin, T.W.: Prenatal removal of frontal association cortex in the fetal rhesus monkey: anatomical and functional consequences in postnatal life. Brain Res. **152**(3), 451–485 (1978)
27. Goriely, A., Destrade, M., Ben Amar, M.: Instabilities in elastomers and soft tissues. Q. J. Mech. Appl. Math. **59**, 615–630 (2006)
28. Goriely, A., Geers, M., Holzapfel, G.A., Jayamohan, J., Jérusalem, A., Sivaloganathan, S., Squier, W., van Dommelen, J.A.W., Waters, S., Kuhl, E.: Mechanics of the brain: perspectives, challenges, and opportunities. Biomech. Model. Mechanobiol. **14**(5), 931–965 (2015)
29. Hannezo, E., Prost, J., Joanny, J.F.: Instabilities of monolayerd epithelia: shape and structure of villi and crypts. Phys. Rev. Lett. **107**, 078104 (2011)
30. Holzapfel, G.A.: In: Wiley, J. (ed.) Nonlinear Solid Mechanics: A Continuum Approach for Engineering (2000)
31. Hong, W., Liu, Z., Suo, Z.: Formation of creases on the surfaces of elastomers and gels. Appl. Phys. Lett. **95**, 111901 (2009)
32. Jia, F., Cao, Y.P., Liu, T.S., Jiang, Y., Feng, X.Q., Yu, S.W.: Wrinkling of a bilayer resting on a soft substrate under in-plane compression. Philos. Mag. **92**(12), 1554–1568 (2012)
33. Jin, X., Zhu, F., Mao, H., Shen, M., Yang, K.H.: Comprehensive experimental study on material properties of human brain tissue. J. Biomech. **46**(16), 2795–2801 (2013)
34. Jina, L., Takeic, A., Hutchinson, J.W.: Mechanics of wrinkle/ridge transitions in thin film/substrate systems. J. Mech. Phys. Solids **81**, 22–40 (2015)
35. Kaster, T., Sack, I., Samani, A.: Measurement of the hyperelastic properties of ex vivo brain tissue slices. J. Biomech. **44**, 1158–1163 (2011)
36. Khang, D.-H., Rogers, J., Lee, H.H.: Mechanical buckling: mechanics, metrology, and stretchable electronics. Adv. Funct. Mater. **18**, 1–11 (2008)
37. Kucken, M., Newell, A.C.: Fingerprint formation. J. Theor. Biol. **235**, 71–83 (2005)
38. Lefévre, J., Germanaud, D., Dubois, J., Rousseau, F., de Macedo Santos, I., Angleys, H., Mangin, J.F., Hüppi, P.S., Girard, N., De Guio, F.: Are developmental trajectories of cortical folding comparable between cross-sectional datasets of fetuses and preterm newborns? Cereb. Cortex **26**(7), 3023–3035 (2016)
39. Li, B., Cao, Y.P., Feng, X.Q.: Growth and surface folding of esophageal mucosa: a biomechanical model. J. Biomech. **44**, 182–188 (2011)
40. Li, G., Nie, J., Wang, L., Shi, F., Gilmore, J.H., Lin, W., Shen, D.: Measuring the dynamic longitudinal cortex development in infants by reconstruction of temporally consistent cortical surfaces. NeuroImage **90**, 266–279 (2014)
41. Li, G., Lin, W., Gilmore, J.H., Shen, D.: Spatial patterns, longitudinal development, and hemispheric asymmetries of cortical thickness in infants from birth to 2 years of age. J. Neurosci. **35**(24), 9150–9162 (2015)
42. Manyuhina, O.V., Mayett, D., Schwarz, J.M.: Elastic instabilities in a layered cerebral cortex: a revised axonal tension model for cortex folding. New J. Phys. **16**(12), 123058
43. Moon, H.M., Wynshaw-Boris, A.: Cytoskeleton in action: lissencephaly, a neuronal migration disorder. Wiley Interdiscip. Rev.: Dev. Biol. **2**(2), 229–245 (2013)
44. Ohta, H., Nordahl, C.W., Iosif, A.M., Lee, A., Rogers, S., Amaral, D.G.: Increased surface area, but not cortical thickness, in a subset of young boys with autism spectrum disorder. In: Autism Research. Wiley, New York (2015)
45. Ogden, R.W.: Non-linear Elastic Deformations. Dover, New York (1997)
46. Rapalska, K., Wozniak, K.: Differentiation of the cortical plate in early human foetuses. Folia Morphol. **61**(4), 233–236 (2002)
47. Reillo, I., de Juan Romero, C., García-Cabezas, M.A., Borrell, V.: A role for intermediate radial glia in the tangential expansion of the mammalian cerebral cortex. Cereb. Cortex **21**(7), 1674–1694 (2011)
48. Rodriguez, A.K., Hoger, A., McCulloch, A.: Stress-dependent finite growth in soft elastic tissue. J. Biomech. **27**, 455–467 (1994)

49. Scott, J.A., Habas, P.A., Kim, K., Rajagopalan, V., Hamzelou, K.S., Corbett-Detig, J.M., Barkovich, A.J., Glenn, O.A., Studholme, C.: Growth trajectories of the human fetal brain tissues estimated from 3D reconstructed in utero MRI. Int. J. Dev. Neurosci. **29**, 529–536 (2011)
50. Siegenthaler, J.A., Pleasure, S.J.: We have got you 'covered': how the meninges control brain development. Curr. Opin. Genet. Dev. **21**(3), 249–255 (2011)
51. Squier, W., Jansen, A.: Polymicrogyria: pathology, fetal origins and mechanisms. Acta Neuropathol. Commun. **2**(80), 1–16 (2014)
52. Striedter, G.F., Srinivasan, S.: Knowing when to fold them. Science **349**(624), 31–32 (2015)
53. Sultan, E., Boudaoud, A.: The buckling of a swollen thin gel layer bound to a compliant substrate. J. Appl. Mech. **75**, 051002 (2008)
54. Tallinen, T., Chung, J.J.Y., Biggins, J.S., Mahadevan, L.: Gyrification from constrained cortical expansion. Proc. Natl. Acad. Sci. USA **111**(35), 12667–12672 (2014)
55. Tallinen, T., Chung, J.J.Y., Rousseau, F., Girard, N., Lefèver, J., Mahadevan, L.: On the growth and form of cortical convolutions. Nat. Phys. **12**, 588–593 (2016)
56. Toro, R., Burnod, Y.: A morphogenetic model for the development of cortical convolutions. Cereb. Cortex **15**, 1900–1913 (2005)
57. Trujillo, V., Kim, J., Hayward, R.C.: Creasing instability of surface-attached hydrogels. Soft Matter **4**, 564–569 (2008)
58. van Dommelen, J.A.W., van der Sande, T.P.J., Hrapko, M., Peters, G.W.M.: Mechanical properties of brain tissue by indentation: interregional variation. J. Mech. Behav. Biomed. Mater. **3**, 158–166 (2010)
59. Volpe, J.J.: Neurology of the Newborn. Elsevier, Amsterdam (2008)
60. Wang, Q., Zhao, X.: A three-dimensional phase diagram of growth-induced surface instabilities. Sci. Rep. **5**, 8887 (2015)
61. Welker, W., Jones, E.G.: In: Peters, A. (ed.) Cerebral Cortex: Comparative Structure and Evolution of Cerebral Cortex, Part II. Springer, Berlin (1990)
62. Winkler, A.M., Kochunovc, P., Blangerod, J., Almasyd, L., Zillese, K., Foxc, P.T., Duggiralad, R., Glahna, D.C.: Cortical thickness or grey matter volume? The importance of selecting the phenotype for imaging genetics studies. NeuroImage **53**(3), 1135–1146 (2010)
63. Wu, M., Ben Amar, M.: Modelling fibers in growing discs of soft tissues. Math. Mech. Solids **39**(2), 219–224 (2014)
64. Xu, G., Knutsen, A.K., Dikranian, K., Kroenke, C.D., Bayly, P.V., Taber, L.A.: Axons pull on the brain, but tension does not drive cortical folding. J. Biomech. Eng. **32**(7), 071013 (2010)
65. Yin, J., Yagüe, J.L., Eggenspieler, D., Gleason, K.K., Boyce, M.C.: Surface micropatterning: deterministic order in surface micro-topologies through sequencial wrinkling. Adv. Mater. (2012). doi:10.1002/adma.20120937
66. Zang, J., Zhao, X., Cao, Y., Hutchinson, J.W.: Localized ridge wrinkling of stiff films on compliant substrates. J. Mech. Phys. Solids **60**, 1265–1279 (2012)
67. Zhang, Z., Hou, Z., Lin, X., Teng, G., Meng, H., Zang, F., Fang, F., Liu, S.: Development of the fetal cerebral cortex in the second trimester: assessment with 7T postmortem MR imaging. Am. J. Neuroradiol. **34**(7), 1462–1467 (2013)
68. Zilles, K., Armstrong, E., Scleicher, A., Kretschmann, H.: The human pattern in gyrification in the cerebral cortex. Anat. Embryol. **170**, 173–179 (1988)

J Elast (2017) 129:239–256
DOI 10.1007/s10659-016-9604-3

Axonal Buckling Following Stretch Injury

Georgina E. Lang[1] · **Sarah L. Waters**[1] · **Dominic Vella**[1] ·
Alain Goriely[1]

Received: 9 May 2016 / Published online: 30 May 2017

Abstract Diffuse brain injury is caused by rapid rotation of the head, and causes strain injury to tissue throughout the brain. Following strain injury, axons exhibit delayed recovery, showing regional buckling behavior immediately after stretch and returning to their original appearance over an extended period of time. This axonal buckling is hypothesized to occur as a result of localized stretching within the axon: Rapid strain causes mechanical damage to microtubules, increasing the effective length of axons. This damage is repaired gradually returning the axon to its initial length.

Here, we test the hypothesis that localized stretching is a possible explanation for the regional buckling behavior. An elongated region of axon is modeled as an Euler beam on an elastic foundation, where the foundation represents the surrounding brain tissue, which consists of glial cells and extracellular matrix. After stretch the elastic foundation returns immediately to its pre-stretch length, while the axon is initially elongated and returns to its original length over a longer period of time. The model exhibits solutions similar to those observed experimentally in post-stretch axons, with undulations that have a similar wavelength and amplitude.

Keywords Axon · Stretch injury · Buckling · Elastic foundation

Mathematics Subject Classification 92C10 · 92C20 · 74K10

1 Introduction

When axons are damaged by stretch injury due to tissue elongation, an interesting sequence of events occurs: Regions of the axon exhibit buckling immediately after injury, then the axon returns to its original appearance over a longer period of time. Review of the experimental literature suggests that this behavior is caused by localized stretching of regions of the axon, due to mechanical damage caused to the cytoskeleton within the axon. In this

✉ A. Goriely
goriely@maths.ox.ac.uk

[1] Mathematical Institute, University of Oxford, Oxford, United Kingdom

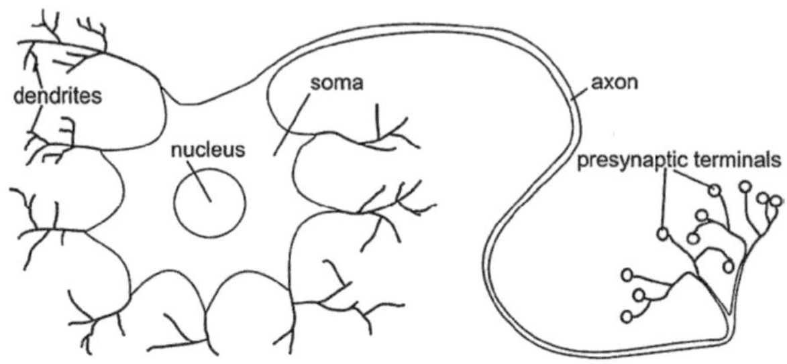

Fig. 1 Schematic illustration showing the structure of a typical neuron (picture courtesy of Kristian Franze from [17] copyright)

work we test whether localized stretching is a feasible mechanism to explain the observed behavior. To answer this question, we model an axon and the surrounding tissue as a beam on an elastic foundation.

1.1 Axon Mechanics

The brain is the 'control centre' for the body and is a highly complex organ made from soft tissue. During everyday activity the brain is protected from damage by the skull and brain tissue is flexible enough to withstand the low forces that are applied from normal movement or light impact, such as running and jumping. However, extreme forces or impacts can lead to brain damage.

Brain tissue is very soft [20] and composed primarily of two cell types: neurons, which carry out information processing and signalling, and glial cells which are thought to carry out several support roles including providing electrical insulation to neurons, regulating extra-cellular ionic conditions, and offering structural support [31]. As illustrated in Fig. 1, a typical neuron has a cell body, several processes called dendrites (which receive information from other neurons), and one axon (which transmits information to other neurons through connections with their dendrites). Since axons may be several centimetres long, they require both mechanical stability to prevent them from breaking under small forces, and efficient transport mechanisms to enable transport of nutrients and organelles within the cell. The cytoskeleton of a neuron extends through the cell body, axon and dendrites, offering structural support and facilitating intracellular transport [36]. Three protein polymers make up the cytoskeleton; microtubules, neurofilaments, and microfilaments. Microtubules are the stiffest component of the cytoskeleton [35], and maintain axonal shape by resisting compression [15]. They also have a role in intracellular transport, as motor proteins can 'walk' along the microtubules to transport components (such as mitochondria and vesicles) from one part of the cell to another [10]. Neurofilaments are the most abundant cytoskeletal element in adult vertebrate neurons, and the core of an axon is made up of microtubules and neurofilaments, connected by crosslinking proteins [12, 22]. Interestingly, some invertebrates such as Drosophila do not have neurofilaments [37] and neurofilaments are relatively rare in embryonic neurons (Yuan *et al.*, 2009). Microfilaments, or actin filaments, are arranged in a layer just under the cell membrane and also exist throughout the cytoskeleton [18]. They are involved in axon growth at the growth cone, and also enable intracellular transport by acting as tracks for motor proteins [5].

Viewed as a mechanical structure, the entire axon can be modeled as a viscoelastic fluid that actively generates forces, and can elongate and shorten as the result of stretching [7, 34]. Equivalently they can be modeled as a morphoelastic rod [29] as discussed in [36]. However, in the first instance, when trying to understand the role of elastic forces in the buckling of axons, we can neglect internal viscous re-organization and growth. In that case, the initial

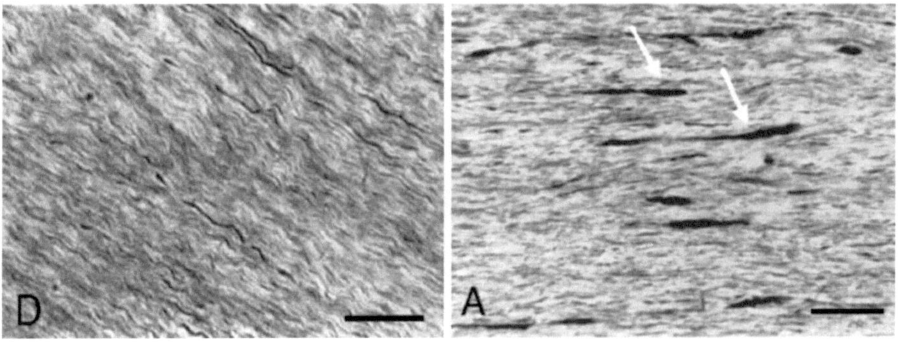

(A) Healthy tissue (B) Damaged tissue

Fig. 2 Comparison of healthy and stretch damaged brain tissue, taken from an optic nerve section. *White arrows* show swollen axons. Scale bar 40 μm. (Reproduced with permission from [6])

response of an axon can be understood as an elastic rod on a foundation. Interestingly, finite-element modeling based on the interactions of microtubules also show buckling instability [43]. On longer time scales, there is an internal re-organization. This remodeling process can be modeled as a change in the effective length for the elastic problem.

1.2 Axonal Injury

Traumatic brain injury is a general term describing damage or injury caused to the brain by an external mechanical force. There are two types of traumatic brain injury: focal and diffuse. Focal injury is caused by a direct blow to the head, and results in damage of brain tissue close to the site of impact. Symptoms can include accumulations of blood within the brain tissue (haematoma), bleeding into the subarachnoid space (subarachnoid haemorrhage) and bruising (contusions) [24, 45]. In contrast, diffuse brain injury is caused when the brain undergoes rapid acceleration, for example in a car crash [40]. When the resulting force is applied to the brain the white matter can deform, resulting in axons being stretched and damaged [21].

The strain rate is important: If axons are stretched over a period of several minutes they are able to withstand strains of over 100 % and return to their original configuration with no evidence of damage [44]; indeed low strain rates can stimulate axon growth [9, 42]. However, axons are damaged when rapidly stretched over a typical timescale of milliseconds. Accidents that lead to diffuse brain injury, such as car crashes, falls and assaults, all involve large strain rates being applied to the brain leading to this *dynamic stretch injury*.

Following dynamic stretch injury axons may become broken and swollen [6]. This diffuse brain injury may be identified in tissue by regions of broken and swollen axons throughout the white matter (Fig. 2). In *primary axotomy*, axons break immediately when strain is applied. However, in vitro experiments have shown that axons have good tolerance to rapid strain, and may not undergo primary axotomy even when strains of up to 65 % are applied sufficiently slowly [41].

1.3 Secondary Axonal Damage

The majority of axons become disconnected as a result of *secondary axotomy*, in which a sequence of mechanical and chemical changes result in disconnection some time after the

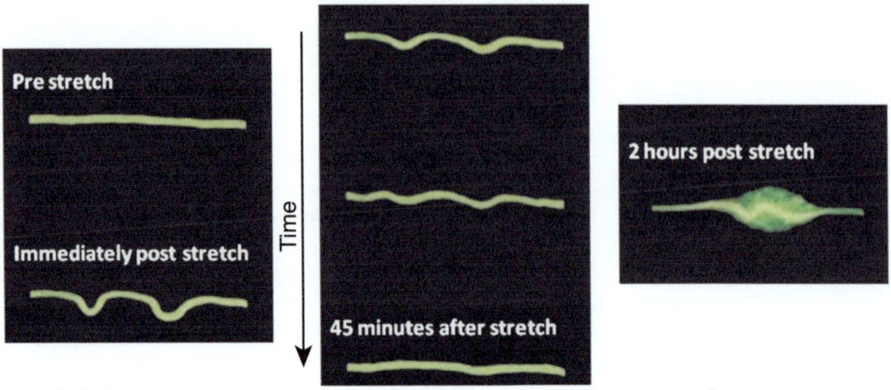

Fig. 3 Demonstration of the delayed elastic behavior of an axon after dynamic stretch injury, illustrated by fluorescence microscopy. Adapted from [40] (picture courtesy of Doug Smith)

strain is applied. After this stretch, unbroken axons show regions of delayed recovery: they exhibit buckling behavior immediately after the stretch is released then recover to a normal appearance within about 45 minutes (Fig. 3). As shown experimentally by Smith *et al.* [41], after a longer period of time (from 1 hour onwards) the axons can become swollen at the sites that originally showed buckling behavior, and eventually disconnect [40, 41].

Axons are brittle under rapid loading but elastic under slow loading [23]. The model of 'sliding filaments' of Chetta *et al.* [12] proposed that when a force is applied, crosslinks between microtubules break so that the microtubules can slide over each other. The crosslinks then reform once the material is at rest. As there is no damage to individual microtubules, transport mechanisms within the cell are not disrupted.

When exposed to dynamic stretch, it is thought that the microtubules connections are mechanically damaged [44]. Once the stretch is released, microtubules fail to realign properly, creating blocks that prevent the axons returning to their pre-stretch length, and resulting in undulations. After trauma, as the result of the forces acting on the system, microtubules then buckle and slide back together, allowing the shape of the axons to appear to return to normal. However since microtubules are critical to transport within a cell, transport mechanisms with buckled microtubules [39] are disrupted causing long-term neurological problems.

Experiments in which single axons are stretched have showed that buckled regions will often exhibit swelling some time afterwards, subsequently causing secondary axotomy [40, 41]. The swellings appear to contain a central core of neurofilaments, which can be straight or undulated, surrounded by less dense accumulation of neurofilaments. The mechanism of axonal swelling and potential subsequent axotomy is a complicated process. It is thought to involve mechanical factors due to mechanical damage to cell membranes and microtubules, and chemical factors triggered by ionic changes in damaged regions [11].

The chemical changes that lead to axon swelling and secondary axotomy are very complex; several interacting processes are responsible for the changes that take place [16]. An influx of calcium is believed to be the chemical trigger leading to secondary axotomy: Mechanical damage to the cell membrane [3] and channel proteins [48] allows abnormal amounts of calcium to enter the cell. Experiments have shown abnormally high calcium concentrations within axons following traumatic injury [48]. Furthermore, high levels of calcium within the cell stimulate the release of calcium-dependent proteases [28], which break down microtubules and damage neurofilaments in the cytoskeleton. Damage to the

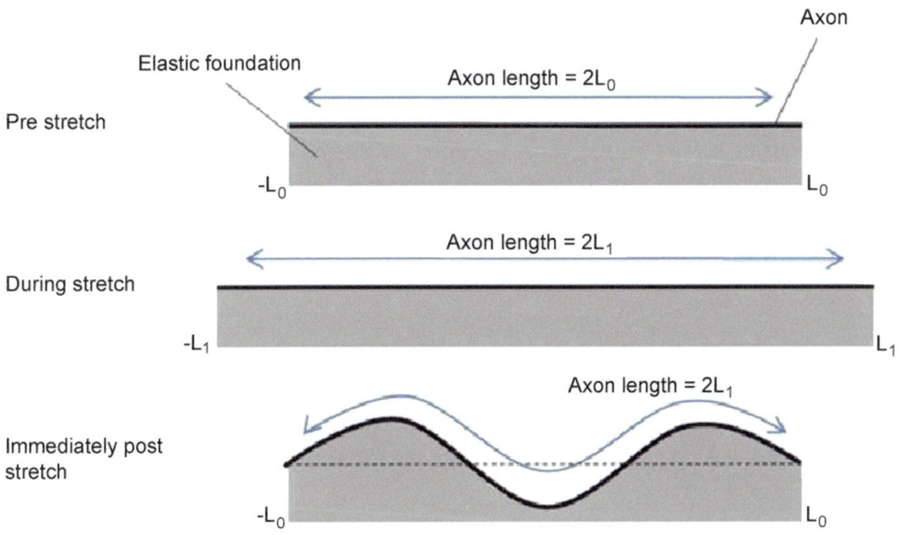

Fig. 4 Diagram illustrating the model: an Euler beam on an elastic foundation

cytoskeleton disrupts axonal transport, leading to accumulation of transport proteins and swelling [11]. High calcium concentrations can also damage mitochondria. The cell membrane of damaged mitochondria releases cytochrome-c [3], a protein that triggers apoptosis within the cell. After secondary axotomy has occurred the disconnected region of the axon is dissolved by Wallerian degradation and the neuron will either die or the axon will regenerate: axons within the peripheral nervous system may regenerate whilst those in the central nervous system may not [47]. An understanding of the internal damage of axons seems to be crucial to address problems related to tau-pathologies and various forms of dementia [1, 46].

2 Problem Set up

2.1 Model Description

The aim of this work is to test the hypothesis that localized internal stretching of axons is responsible for the buckling behavior observed following stretch injury. The elongated region of an axon is modeled as a planar Euler beam on an elastic foundation representing the brain tissue. A Winkler foundation is used to model the elastic foundation, which supposes that the force applied to the beam by the foundation is proportional to the transverse displacement of the beam.

The hypothesis of localized stretching is tested by assuming that after dynamic stretch the elastic foundation immediately returns to its original length, whilst the beam returns to its original length over an extended period of time. Immediately after stretch the elongated beam is longer than the foundation and therefore subject to compressive forces. A diagram of the model is shown in Fig. 4. The balance between the resistance of the beam to bending and the elastic response of the foundation will determine the wavelength of the buckling.

2.2 Assumptions

In this simplified model, an axon is modeled as a one-dimensional beam, which may only be displaced transversely. It is assumed that the beam undergoes only small displacements and uniform elongation. We further assume that the foundation is homogeneous along the beam, and that contact is never broken between the beam and foundation. Since we assume that the elongation is confined within a given region of typical size $2L_0$, there is no displacement outside the domain and we use clamped boundary conditions.

2.3 Governing Equations

The transverse displacement of the beam is described by $w(x)$ on the domain $-L_0 < x < L_0$. The solution will be given by the function $w(x)$ that minimizes the energy within the beam, subject to the constraint that the length of the beam, $2L(t)$, is a prescribed function of time. Mathematically this is equivalent to minimizing the function

$$\text{Energy} = \overbrace{\frac{B}{2}\int_{-L_0}^{L_0} \frac{w''^2}{(1+w'^2)^3}\,\mathrm{d}x}^{\text{bending energy}} + \overbrace{\frac{k}{2}\int_{-L_0}^{L_0} w^2\,\mathrm{d}x}^{\text{foundation energy}}, \tag{1}$$

subject to the constraint

$$2L(t) = \int_{-L_0}^{L_0} \left(1+w'^2\right)^{\frac{1}{2}}\,\mathrm{d}x, \tag{2}$$

where $'$ denotes differentiation with respect to x. Here, B (units Nm^2) is the bending stiffness of the beam and k (units Nm^{-2}) is the elasticity of the foundation. We will discuss estimates for these parameters in Sect. 3. Introducing a Lagrange multiplier, P, gives the energy functional

$$J\big[w(x)\big] = \int_{-L_0}^{L_0} \mathcal{F}\big[w, w', w''\big]\,\mathrm{d}x, \tag{3}$$

where

$$\mathcal{F}\big[w, w', w''\big] = \frac{B}{2}\frac{w''^2}{1+w'^2} + \frac{k}{2}w^2 - P\left(\left(1+w'^2\right)^{\frac{1}{2}} - \frac{L}{L_0}\right). \tag{4}$$

The integrand is linearized by the assumption that the beam undergoes only small displacements, and the Euler-Lagrange equation for the problem reads

$$\frac{\partial \mathcal{F}}{\partial w} - \frac{\mathrm{d}}{\mathrm{d}x}\frac{\partial \mathcal{F}}{\partial w'} + \frac{\mathrm{d}^2}{\mathrm{d}x^2}\frac{\partial \mathcal{F}}{\partial w''} = 0. \tag{5}$$

Substituting $\mathcal{F}$ into the Euler-Lagrange equation gives:

$$B\frac{\mathrm{d}^4 w}{\mathrm{d}x^4} + P\frac{\mathrm{d}^2 w}{\mathrm{d}x^2} + kw = 0. \tag{6}$$

This equation can also be obtained as a reduction of the Kirchhoff equation to small planar transverse motion [30] or from a general description of Kirchhoff rods that can grow while attached to a support [26, 29]. The force P, which was introduced as a Lagrange multiplier, can equivalently be thought of as the compressive force acting along the beam.

We have not considered any dynamical effect so far. Indeed, we are primarily interested in the quasi-static equilibrium states of the rod but not the dynamics associated to the buckling to these modes. However, we are interested in the evolution of the rod after a remodeling time (5 to 45 minutes) much longer than the typical viscous time scales (of the order of seconds or smaller). Denoting the extended length immediately after stretch by L_1, and allowing the length of the beam to decay over a characteristic time scale α, the length of the beam at any time is prescribed by,

$$L(t) = L_0 + (L_1 - L_0)e^{-t/\alpha}. \tag{7}$$

This particular form for the relaxation can be understood by recalling that the axon is a viscoelastic material, for which the typical relaxation scale is given as a ratio of internal viscosity to Young's modulus. When the axon is stretched, the elastic elements are stretched. Over time the stress in these elements relaxes and they return to their rest length. The rate at which this process occurs is determined by the ratio of the viscosity over the Young's Modulus of the axon.

To close these equations, a clamped boundary condition is imposed at both ends of the domain,

$$w(\pm L_0) = 0, \qquad \frac{dw}{dx}(\pm L_0) = 0. \tag{8}$$

Equations (6) to (8), together with the length constraint (2) are the full equations for the system.

2.4 Non-dimensionalization

Non-dimensionalizing lengths with L_0 such that $w = L_0 w^*$ and $x = L_0 x^*$ and time with α so that $t = \alpha t^*$ gives the final system (dropping the stars),

$$\frac{\mathrm{d}^4 w}{\mathrm{d}x^4} + \lambda \frac{\mathrm{d}^2 w}{\mathrm{d}x^2} + \beta w = 0, \tag{9}$$

where

$$\lambda = \frac{P L_0^2}{B}, \qquad \beta = \frac{k L_0^4}{B}, \tag{10}$$

are dimensionless parameters. Linearizing equation (2) gives,

$$2\big(\hat{L}(t) - 1\big) = \int_{-1}^{1} \frac{1}{2}\left(\frac{\mathrm{d}w}{\mathrm{d}x}\right)^2 \mathrm{d}x, \tag{11}$$

where $\hat{L}(t) = L(t)/L_0$, $\hat{L}_1 = L_1/L_0$, and $\hat{L}(t)$ is the prescribed function of time,

$$\hat{L}(t) = 1 + (\hat{L}_1 - 1)e^{-t}. \tag{12}$$

The dimensionless clamped boundary conditions are

$$w(\pm 1) = 0, \qquad \frac{\mathrm{d}w}{\mathrm{d}x}(\pm 1) = 0. \tag{13}$$

Equations (9) to (13) comprise the full system, and constitute a well defined problem. The only parameters in the system are the dimensionless parameters β and λ: here β is a control parameter that can be interpreted as the length of the beam L_0 in terms of the characteristic buckling wavelength $\sim (\beta/k)^{1/2}$; conversely λ is an unknown specified by the constraint (11).

3 Parameter Analysis

As with many physiological problems, obtaining precise parameter values can be a significant challenge. The properties of brain tissue vary with species, region, age, and even gender [17]; furthermore, experimental results are often inconsistent with one another due to different methodologies, subjects, and experimental conditions [14, 20].

3.1 Axon Dimensions

The length and radius of an individual axon is variable, and depends upon the species and age of the subject, as well as the location of the axon in the nervous system. The radius of an axon can range from 0.5 mm for the giant squid axon [13] down to fractions of microns. Here, we are primarily interested in understanding the behavior observed experimentally by Smith *et al.* [41], for which a culture of human cells was used. Smith *et al.* measured the diameter of axons in their culture to be in the range of $0.5\ \mu\text{m} \le 2r_0 \le 1.5\ \mu\text{m}$; we therefore take $r_0 = 0.5\ \mu\text{m}$ as a typical axon radius. This value is consistent with values reported for other mammals: axon diameters ranging from 0.1–2.5 μm have been reported for monkey optic nerve neurons [38], 1–2.5 μm for myelinated axons in the spinal cord of cats, cows and rats [27], and up to 4 μm for porcine optic nerves [4].

The characteristic length of the region in which buckling occurs can be obtained from images of post-stretch axons published by Tang-Schomer *et al.* [44] and Smith *et al.* [41]. In each case, it is observed that the buckled region has length in the range 10 μm to 20 μm and so we shall consider the foundation length to lie within this range. Immediately after stretch, the overall length of axons has been reported to increase by up to 8 %, however this does not reflect the localized length increase within buckled regions since stretching is heterogeneous along the axon. Buckled regions may undergo a length increase of up to 60 % [41], and therefore the elongated axon length L_1 is taken to be 60 % greater than L_0.

It is consistently reported that axons return to the pre-stretch length over a period of approximately 45 minutes, and that the greatest rate of return to original length occurs within the first five minutes following stretch injury [40, 44]. Therefore α, the timescale for the recovery to original length, is taken as $\alpha = 300$ s.

3.2 Evaluating Bending Stiffness of an Axon and Elasticity of the Foundation

The key dimensionless parameter in the problem β, given in (10), depends on the axon bending stiffness, B, the foundation length, $2L_0$, and the elasticity of the foundation, k. We therefore turn to the estimation of these parameters.

3.2.1 Bending Stiffness

We consider two different ways to estimate the bending stiffness of an axon, either directly from the radius (modelling the axon as a homogeneous elastic rod) or by taking into account its internal structure.

For rods made of an isotropic, homogeneous, and linearly elastic material, the bending stiffness B is the product of Young's modulus E and the second moment of area I, i.e., $B = EI$. The second moment of area for a cylindrical beam with radius r_0 is $I = \pi r_0^4/4$. Therefore in order to obtain the second moment of area, only the radius of a typical axon is required. Using $r_0 = 0.5$ μm gives $I = 4.9 \times 10^{-26}$ m^4. Bernal *et al.* [7] experimentally measured the Young's modulus by of an individual axon to be 12 kPa. This was done by measuring the extensional response of axons, essentially measuring the product $\pi E r_0^2$ directly. Independent work obtained a value of 9 kPa by nano-indentation [35]. These results are of the same order of magnitude, and so we take $E \approx 10$ kPa, giving $B = 4.9 \times 10^{-22}$ Nm^2. In Appendix B, we compute an upper bound for the bending stiffness of an axon by considering the constituent parts of the cytoskeleton. This strict upper bound assumes internal rigid structures and leads to an estimate of $B \approx 6 \times 10^{-19}$ Nm^2, three order of magnitude larger than the above estimate.

3.2.2 Foundation Stiffness

It is difficult to measure the elasticity of the glial matter and extracellular matrix surrounding an axon [32]. Given an estimate for the elastic modulus of the material, G, the stiffness of the surrounding foundation may then be estimated assuming that the surrounding medium is homogeneous, incompressible and isotropic: in particular, the elasticity of the foundation is given by [8] $k = 4\pi G/\log(L_{\text{buck}}/r_0)$, where L_{buck} is the characteristic length scale for buckling and r_0 is the rod radius. The problem with this estimate is that it depends on a quantity L_{buck} that is itself the main unknown. It is therefore not suitable for modeling. Instead, we combine experimental measurements and theoretical estimates for the growth velocity of an axon under tensional force to obtain an estimate for β.

3.3 Experimental Determination of β

The main observation of O'Toole *et al.* [33] is that when an axon of length $L(t)$ is pulled with force F_0, the force profile along the axon is given by

$$F(x) = F_0 \frac{\cosh(x/a)}{\cosh(L(t)/a)}, \tag{14}$$

where a is a typical length scale that can be measured from the length profile as a function of time and estimated experimentally to be $a \approx 3.4 \times 10^{-5}$ m. The problem of a pulled axon on an elastic substrate can be modeled as a growing elastic structure with an adhesive force attaching it to the substrate [29]. If the adhesive force is modeled as a constant multiple of the displacement of the axon, and it is assumed that the foundation has an isotropic response to displacement, the constant of proportionality is k. The axon stretches according to Hooke's Law, so that the strain acting on the axon is proportional to the stress experienced. Then the force profile across the axon is given by (14) with

$$a = \sqrt{\frac{EA}{k}}, \tag{15}$$

where $A = \pi r_0^2$ is the cross-sectional area of the axon. Using the above of a to eliminate E/k from β, we find

$$\beta = \frac{kL_0^4}{B} = \frac{4L_0^4}{r_0^2 a^2}. \tag{16}$$

With the measurement of a from [33], we find that β is dependent only upon the characteristic length scale L_0 and axon radius r_0. Indeed, using $a = 3.4 \times 10^{-5}$ m and $r_0 = 0.5 \times 10^{-6}$ m, gives $\beta = 1.4 \times 10^{22} L_0^4$. This value for β shall be used when solving the model. Note that this estimate relies on the underlying assumption that the typical relaxation times for the internal stress relaxation in the axons and the relaxation time of the substrate attachment are similar.

4 Solution

Since the equation for the deflection is linear, the solution of the system (9)–(13) can easily be obtained. General solutions to (9) have the form $w \sim e^{ikx}$ where k satisfies the dispersion relation

$$k^4 - \lambda k^2 + \beta = 0. \tag{17}$$

Depending on the value of $\lambda^2 - 4\beta$, there are three different forms for the general solution. If $\lambda^2 - 4\beta = 0$ then there are two real repeated roots for k, while if $\lambda^2 - 4\beta < 0$ then the four roots of k take the form $k = \pm r \pm i\rho$. With clamped boundary conditions there are no non-trivial solutions to the problem for either of these cases, as shown in Appendix A. Thus only the case where $\lambda^2 - 4\beta > 0$ (corresponding to a buckled axon) is considered. Solutions take the form

$$w = A\cos(k^+x) + B\cos(k^-x) + C\sin(k^+x) + D\sin(k^-x), \tag{18}$$

where

$$k^- = \sqrt{\frac{2\lambda - \sqrt{\lambda^2 - 4\beta}}{2}}, \qquad k^+ = \sqrt{\frac{2\lambda + \sqrt{\lambda^2 - 4\beta}}{2}}. \tag{19}$$

The boundary conditions on $w(\pm 1) = 0$ can be applied, and the resulting equations added and subtracted from each other to give

$$A\cos(k^+) + B\cos(k^-) = 0, \tag{20}$$

$$C\sin(k^+) + D\sin(k^-) = 0, \tag{21}$$

and similarly the conditions $w'(\pm 1) = 0$ can be manipulated to give another pair of equations:

$$Ak^+\sin(k^+) + Bk^-\sin(k^-) = 0, \tag{22}$$

$$Ck^+\cos(k^+) + Dk^-\cos(k^-) = 0. \tag{23}$$

Solving these equations yields two possible solutions for $w(x)$. Either

$$w(x) = A_0\left(\frac{\cos(k^+x)}{\cos(k^+)} - \frac{\cos(k^-x)}{\cos(k^-)}\right), \tag{24}$$

if the dispersion relation $k^+\tan k^+ = k^-\tan k^-$ is satisfied, or

$$w(x) = B_0\left(\frac{\sin(k^+x)}{\sin(k^+)} - \frac{\sin(k^-x)}{\sin(k^-)}\right), \tag{25}$$

if $k^+\cot k^+ = k^-\cot k^-$.

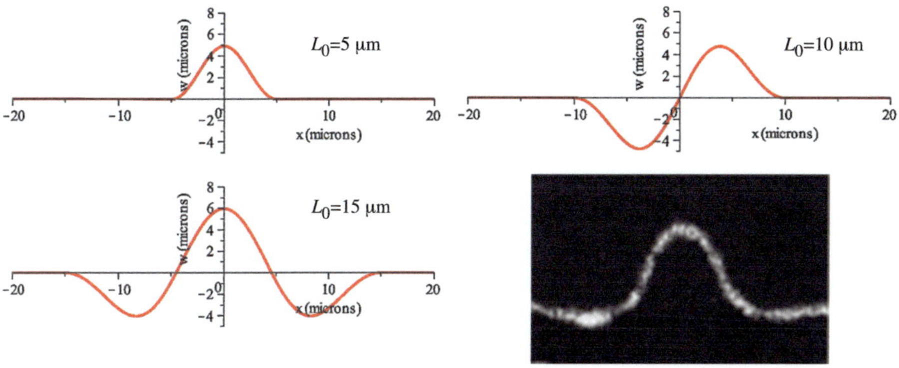

Fig. 5 Solutions to the model at time $t = 0$, for elongated regions of different lengths and Image of a buckled region of axon, shortly following stretch injury (picture courtesy of Doug Smith [44]). Width of image is 15 μm

Since k^+ and k^- are functions of λ, λ can be found from the solution to the dispersion relation. Note that for any β, each dispersion relation exhibits infinitely many solutions for λ. However, the solution that minimizes the energy of the system is the smallest value of λ satisfying one or other of the dispersion relations, subject to the constraint $\lambda^2 - 4\beta > 0$. Thus the solution takes the form (24) or (25) depending upon which dispersion relation permits the smaller value of λ.

A numerical solver is used to solve each of the dispersion relations and obtain values k^+, k^-, and λ. Once these have been found, the amplitude of the solution at any time is found by substituting the solution into the length constraint, given in (11). The system can then be redimensionalized to obtain the dimensional solution.

5 Results

5.1 Behavior Immediately Following Stretch

Through our rescaling, the value of β is dependent on the fourth power of the unknown length, L_0; as a result, the shape of the beam is sensitively dependent on the value L_0. Figure 5 shows the results of our model for foundation lengths from 10 μm $\leq 2L_0 \leq$ 30 μm. As the length of the beam increases, the number of peaks and troughs increases. Physically, a larger value of β means that deforming the foundation has become more 'expensive' in comparison to bending the beam and so, to minimize the energy of the system while accommodating the imposed compression, the beam must adopt a more convoluted, bent shape.

Comparing results of these simulations with published experimental results demonstrates qualitatively similar behavior. Tang-Schomer *et al.* [44] show images of a 15 μm section of axon shortly following stretch injury. We note that the observed shape is qualitatively very similar to that predicted by our model (see Fig. 5c), though we emphasize that the parameters here are not precisely known.

5.2 Evolution of the Axon After Injury

Following injury, an axon will return to its original appearance over a period of approximately 45 minutes. This can easily be modeled in our system by introducing the time-

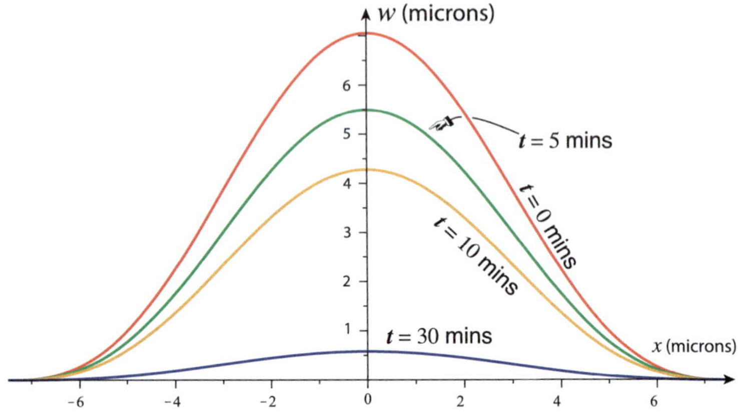

Fig. 6 Results of the model for $L_0 = 7.5$ μm, showing evolution of the solution over time

dependent axon length (12). Figure 6 demonstrates that our model also displays this behavior: the effect of reducing the axon length over time is to reduce the amplitude of the displacement. Indeed, returning to Fig. 3, and considering a localized region close to one of the 'buckles', it can be seen that the behavior of the experiment is qualitatively similar to that predicted by the model.

6 Discussion

Our results support the idea that axon behavior after stretch is a consequence of buckling and localized elongation: initially, the elongated regions buckle due to elongation. On a longer time scale, the internal remodeling of the axon shortens the length and the amplitude of the buckle decreases, as observed experimentally [44]. It is therefore plausible that mechanical damage to microtubules causes localized lengthening of the axon. However, the model includes numerous simplifications. It is unlikely that a region of axon increases uniformly in length, and therefore more realistic behavior may be observed by considering heterogeneous elongation within the stretched region.

The elasticity of the foundation proved very difficult to evaluate, and it was assumed that the foundation was elastic, isotropic and homogeneous. Since the foundation surrounding an axon is made from glial cells and extracellular matrix, in reality these may also be affected by stretch injury so that the buckling may be caused by a combination of localized stretching of the axon and heterogeneous elastic response of the surrounding tissue. Furthermore, the bending stiffness of the axon is likely to be affected by stretch injury — the effects of a heterogeneous bending stiffness should also be considered. It would be a simple modelling exercise to consider a hierarchy of increasingly sophisticated models for the response of the foundation and the axon by including such effects as the nonlinear response of the foundation, a more accurate geometric model of the axon as well as heterogeneity, anisotropy and viscoelasticity. However, we do not expect any different qualitative behavior to be observed with such complications. The real bottleneck in our study has been the estimation of the biologically relevant parameters; as such the simple model presented here is a useful first step to motivate further experimental work. Recent numerical work on the buckling and torsional behavior of bundles of microtubules is particularly relevant here as it will allow us to bridge between cellular and sub-cellular levels [25, 43]. Despite the various simplifications and unknown parameters, the model compares favorably with experimental results and we conclude that theoretical results support the hypothesis that localized stretching is a plausible mechanism to explain the behavior of axons following stretch injury.

Acknowledgements The research leading to these results has received funding from the European Research Council under the European Union's Horizon 2020 Programme/ERC Grant Agreement no. 637334 (DV). The authors are grateful to Professors Doug Smith and Kristian Franze for sharing the images reproduced in this paper.

Appendix A: Solutions to the Model for $\lambda^2 \leq 4\beta$

This appendix shows that there are no non-trivial solutions to the system of (9)–(13) unless $\lambda^2 > 4\beta$.

A.1 $\lambda^2 = 4\beta$

Initially, we consider the case $\lambda^2 = 4\beta$. Note that since β is positive constant, this implies that λ is non-zero and so the general solution to (9) is

$$w = A\cos(kx) + Bx\cos(kx) + C\sin(kx) + Dx\sin(kx), \tag{26}$$

where $k = \lambda/2$. By assumption, the beam is clamped at the boundaries, so that the appropriate boundary conditions are those of (13). Applying the boundary condition $w(\pm 1) = 0$ and adding and subtracting the equations obtained gives:

$$A\cos(k) + D\sin(k) = 0, \tag{27}$$

$$B\cos(k) + C\sin(k) = 0, \tag{28}$$

while proceeding similarly with the boundary condition $w'(\pm 1) = 0$ we find another pair of equations:

$$(D - Ak)\sin(k) + Dk\cos k = 0, \tag{29}$$

$$(Ck + B)\cos(k) - Bk\sin k = 0. \tag{30}$$

Solving (27) and (29) gives that either $A = D = 0$ or the dispersion relation $\sin k\cos k - k = 0$ is satisfied. Graphically it can be shown that the only solution to this dispersion relation is $k = 0$, which is inconsistent with $k = \lambda/2 = \sqrt{\beta} > 0$. We therefore conclude that $A = D = 0$. Similarly, solving (28) and (30) gives that either $B = C = 0$ or $\sin k\cos k + k = 0$; the only solution to this equation is again $k = 0$ and so $B = C = 0$. As a result, no non-trivial solution exists that satisfies the clamped boundary conditions with $\lambda^2 = 4\beta$.

A.2 $\lambda^2 < 4\beta$

In the case $\lambda^2 < 4\beta$ the general solution to (9) is given by:

$$w = A\cos(rx)\sinh(\rho x) + B\cos(rx)\cosh(\rho x)$$

$$+ C\sin(rx)\sinh(\rho x) + D\sin(rx)\sinh(\rho x), \tag{31}$$

where $k = \pm r \pm i\rho$ are the four solutions of the auxiliary equation $k^4 - \lambda k^2 + \beta = 0$. Applying the boundary conditions at $w(\pm 1)$ and adding and subtracting the results gives:

$$A\cos r \sinh\rho + D\sin r \cosh\rho = 0, \tag{32}$$

$$B\cos r \cosh\rho + C\sin r \sinh\rho = 0. \tag{33}$$

Similarly imposing $w'(\pm 1) = 0$ and manipulating we find another pair of equations:

$$0 = A(-r\sin r \sinh\rho + \rho\cos r \cosh\rho) + D(r\cos r \cosh\rho + \rho\sin r \sinh\rho), \tag{34}$$

$$0 = B(-r\sin r \cosh\rho + \rho\cos r \sinh\rho) + C(r\cos r \sin\rho + \rho\sin r \cosh\rho). \tag{35}$$

Solving (32) and (34) we obtain that either $A = D$ or the dispersion relationship $\cos r \sin r/r = \sinh\rho \cosh\rho/\rho$ holds. Graphically it can be shown that the only solution to the dispersion relation is $\rho = r = 0$, contradicting our assumption that the root of the auxiliary equation is complex. Similarly solving (33) and (35) gives that either $B = C$ or r and ρ satisfy $\cos r \sin r/r = -\sinh\rho\cosh\rho/\rho$, which also has no non-trivial solutions. We therefore conclude that no non-trivial solutions exist that satisfy the clamped boundary conditions when $\lambda^2 < 4\beta$.

Appendix B: Second Moment of Area for a Circular Arrangement of Microtubules

An upper bound, for the bending stiffness of an axon can be estimated by considering the constituent parts within the axon. Microtubules are the stiffest component of the cytoskeleton [35], and so we consider a structure consisting only of microtubules, whose Young's modulus and radius have been extensively characterized [2, 19]. It is assumed that microtubules are arranged in a circular pattern located close to the perimeter of the axon, as shown in Fig. 7(a). This assumption of arranging the microtubules around the outermost boundary of the axon and neglecting sliding between axons gives an upper bound for stiffness. If E_m is the Young's modulus of microtubules and I_m is the second moment of area for the arrangement of microtubules, then $B = E_m I_m$. For an arrangement of N microtubules, each with radius r_m, and arranged at a radius R from the centre of an axon, the second moment of area can be shown to be $\pi r_m^2 R^2 N/2$.

The second moment of area for an arrangement of N microtubules each with radius r_m and positioned in a circular arrangement at a distance R from the centre of an axon, as illustrated in Fig. 7(b) is given by $\pi r_m^2 R^2 N/2$. First consider the moment of area I of a single microtubule, with centre located at distance a_j from the vertical axis. The second moment of area is given by,

$$\begin{aligned}
I(a_j) &= \int_D z^2 \,dz\,dy = \int_{-r_m}^{r_m} \int_{a-\sqrt{r_m^2-y^2}}^{a+\sqrt{r_m^2-y^2}} z^2 \,dz\,dy, \\
&= \frac{1}{3}\int_{-r_m}^{r_m} \left(a + \sqrt{r_m^2 - y^2}\right)^3 - \left(a - \sqrt{r_m^2 - y^2}\right)^3 dy, \\
&= \frac{1}{3}\int_{-r_m}^{r_m} 8\sqrt{r_m^2 - y^2}^{\,3} + 6\left(a^2 - (r_m^2 - y^2)\right)\left(\sqrt{r_m^2 - y^2}\right) dy.
\end{aligned}$$

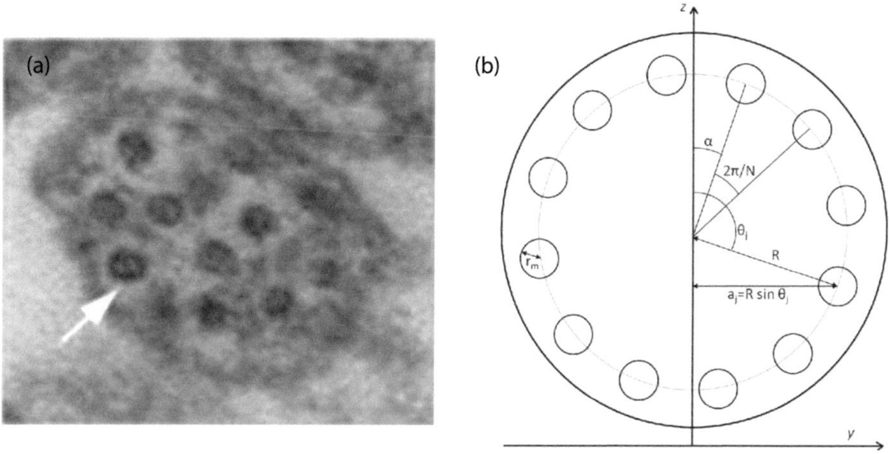

Fig. 7 (**a**) Cross section of an axon, showing the microtubules (*arrow*). Axon radius = 0.5 μm. Image taken from [44] with kind permission from Doug Smith. (**b**) Diagram illustrating the arrangement of $N = 12$ microtubules which are used to calculate the second moment of area of an axon

Parameterizing y by $r_m \sin\theta$ for $\theta \in [-\pi/2, \pi/2]$. Then the integral simplifies to,

$$\begin{aligned} I(a_j) &= \tfrac{1}{3}\int_{-\frac{\pi}{2}}^{\frac{\pi}{2}} 8r_m^4\cos^4\theta + 6\big(a^2 - r_m^2\cos^2\theta\big)\,\mathrm{d}\theta, \\ &= \frac{2}{3}\int_{-\frac{\pi}{2}}^{\frac{\pi}{2}} 2r_m^4\left(\frac{3}{8} + \frac{1}{2}\cos 2\theta + \frac{1}{8}\cos 4\theta\right) + 3a_j^2 r_m^2(1+\cos\theta)\,\mathrm{d}\theta, \\ &= \pi r_m^2\left(\frac{r_m^2}{4} + a_j^2\right). \end{aligned}$$

Now considering a circular arrangement of N microtubules, evenly spaced at radius R from the centre of the axon. The distance from the vertical axis of each microtubules is given by $a_j = R\sin\theta_j$ where $\theta_j = \alpha + \frac{2\pi(j-1)}{N}$. The second moment of area of the collection is given by:

$$\begin{aligned} I &= \sum_{j=i}^{N} I(a_j) = \sum_{j=i}^{N} \pi r_m^2\left(\frac{r_m^2}{4} + (R\sin\theta_j)^2\right), \\ &= \frac{N\pi r_m^4}{4} + \pi r_m^2 R^2 \sum_{j=i}^{N} \sin^2\left(\alpha + (j-1)\frac{2\pi}{N}\right). \end{aligned}$$

Finally it remains to evaluate the sum, it can be shown that for all $N > 2$ and all α the sum in the equation above is $N/2$

$$\sum_{j=i}^{N} \sin^2\left(\alpha + (j-1)\frac{2\pi}{N}\right) = \sum_{j=0}^{N-1}\left(\frac{1}{2i}\right)^2 \left(e^{(\alpha+\frac{2\pi j}{N})i} - e^{-(\alpha+\frac{2\pi j}{N})i}\right)$$

$$= \frac{N}{2} - \frac{1}{4}\left(e^{2\alpha i} \sum_{j=0}^{N-1} e^{\frac{4\pi i}{N} j} + e^{-2\alpha i} \sum_{j=0}^{N-1} e^{-\frac{4\pi i}{N} j} \right)$$

$$= \frac{N}{2} - \frac{1}{4}\left(e^{2\alpha i} \left(\frac{1 - e^{4\pi i}}{1 - e^{4\pi i/N}} \right) + e^{-2\alpha i} \left(\frac{1 - e^{-4\pi i}}{1 - e^{-4\pi i/N}} \right) \right).$$

The numerators $1 - e^{\pm 4\pi i}$ are always equal to zero, and the denominator $1 - e^{\pm 4\pi i/N}$ is non zero unless $N = 1$ or 2. Thus for $N > 2$,

$$I = \frac{N\pi r_m^4}{4} + \pi r_m^2 R^2 \frac{N}{2} = \frac{\pi r_m^2 N}{2}\left(\frac{r_m^2}{2} + R^2 \right) \approx \frac{\pi r_m^2 R^2 N}{2},$$

where an approximation is made since the radius of a microtubule $r_m \ll R$ — the radius at which the microtubules are arranged within the axon R.

The typical diameter of a microtubule $r_m \approx 25$ nm [2] and, as previously, we take the radius of an axon $r_0 \approx 0.5$ μm. It is assumed that the microtubules are arranged close to the edge of the axon so that $R = 0.4$ μm, and we assume that there are $N = 12$ microtubules in a cross section. This number varies between axons but, as the bending stiffness varies only linearly with N, variations in N will not have a great effect. Taking $E_m \approx 1.2$ GPa [19] we have,

$$B = E_m I_m = E_m \times \frac{\pi r_m^2 R^2 N}{2} \approx 6 \times 10^{-19}\ \text{Nm}^2. \tag{36}$$

The cytoskeleton also consists of neurofilaments and microfilaments. Their overall contribution to the bending stiffness of axons are not fully understood.

References

1. Ahmadzadeh, H., Smith, D.H., Shenoy, V.B.: Mechanical effects of dynamic binding between tau proteins on microtubules during axonal injury. Biophys. J. **109**(11), 2328–2337 (2015)
2. Alberts, B.: Molecular Biology of the Cell: Reference Edition. Molecular Biology of the Cell, vol. 1. Garland, New York (2008)
3. Andriessen, T.M.J.C., Jacobs, B., Vos, P.E.: Clinical characteristics and pathophysiological mechanisms of focal and diffuse traumatic brain injury. J. Cell. Mol. Med. **14**(10), 2381–2392 (2010)
4. Assaf, Y., Blumenfeld-Katzir, T., Yovel, Y., Basser, P.J.: Axcaliber: a method for measuring axon diameter distribution from diffusion MRI. Magn. Reson. Med. **59**(6), 1347–1354 (2008)
5. Baas, P.W., Karabay, A.: Axonal Transport and the Neuronal Cytoskeleton. Wiley, New York (2001)
6. Bain, A.C., Meaney, D.F.: Tissue-level thresholds for axonal damage in an experimental model of central nervous system white matter injury. J. Biomech. Eng. **122**(6), 615–622 (2000)
7. Bernal, R., Pullarkat, P.A., Melo, F.: Mechanical properties of axons. Phys. Rev. Lett. **99**(1), 018301 (2007)
8. Brangwynne, C.P., MacKintosh, F.C., Kumar, S., Geisse, N.A., Talbot, J., Mahadevan, L., Parker, K.K., Ingber, D.E., Weitz, D.A.: Microtubules can bear enhanced compressive loads in living cells because of lateral reinforcement. J. Cell Biol. **173**(5), 733–741 (2006)
9. Bray, D.: Axonal growth in response to experimentally applied mechanical tension. Dev. Biol. **102**(2), 379–389 (1984)
10. Bressloff, P.C., Newby, J.M.: Stochastic models of intracellular transport. Rev. Mod. Phys. **85**(1), 135 (2013)
11. Büki, A., Povlishock, J.T.: All roads lead to disconnection? Traumatic axonal injury revisited. Acta Neurochir. **148**, 181–194 (2006)
12. Chetta, J., Kye, C., Shah, S.B.: Cytoskeletal dynamics in response to tensile loading of mammalian axons. Cytoskeleton **67**(10), 650–665 (2010)
13. Cole, K.S., Curtis, H.J.: Electric impedance of the squid giant axon during activity. J. Gen. Physiol. **22**(5), 649–670 (1939)

14. de Rooij, R., Kuhl, E.: Constitutive modeling of brain tissue: current perspectives. Appl. Mech. Rev. **68**(1), 010801 (2016)
15. Dennerll, T.J., Joshi, H.C., Steel, V.L., Buxbaum, R.E., Heidemann, S.R.: Tension and compression in the cytoskeleton of pc-12 neurites. ii: Quantitative measurements. J. Cell Biol. **107**(2), 665–674 (1988)
16. Farkas, O., Povlishock, J.T.: Cellular and subcellular change evoked by diffuse traumatic brain injury: a complex web of change extending far beyond focal damage. In: Weber, J.T., Maas, A.I.R. (eds.) Neurotrauma: New Insights Into Pathology and Treatment. Progress in Brain Research, vol. 161, pp. 43–59. Elsevier, Amsterdam (2007)
17. Franze, K., Guck, J.: The biophysics of neuronal growth. Rep. Prog. Phys. **73**(9), 094601 (2010)
18. Ganguly, A., Tang, Y., Wang, L., Ladt, K., Loi, J., Dargent, B., Leterrier, C., Roy, S.: A dynamic formin-dependent deep f-actin network in axons. J. Cell Biol. **210**(3), 401–417 (2015)
19. Gittes, F., Mickey, B., Nettleton, J., Howard, J.: Flexural rigidity of microtubules and actin filaments measured from thermal fluctuations in shape. J. Cell Biol. **120**(4), 923–934 (1993)
20. Goriely, A., Budday, S., Kuhl, E.: Neuromechanics: from Neurons to Brain. Adv. Appl. Mech. **48**, 79–139 (2015)
21. Goriely, A., Geers, M.G.D., Holzapfel, G.A., Jayamohan, J., Jérusalem, A., Sivaloganathan, S., Squier, W., van Dommelen, J.A.W., Waters, S.L., Kuhl, E.: Mechanics of the brain: perspectives, challenges, and opportunities. Biomech. Model. Mechanobiol. **14**, 931 (2015)
22. Hirokawa, N.: Cross-linker system between neurofilaments, microtubules and membranous organelles in frog axons revealed by the quick-freeze, deep-etching method. J. Cell Biol. **94**(1), 129–142 (1982)
23. Holland, M.A., Miller, K.E., Kuhl, E.: Emerging brain morphologies from axonal elongation. Ann. Biomed. Eng. **43**(7), 1640–1653 (2015)
24. Lang, G.E., Vella, D., Waters, S.L., Goriely, A.: Propagation of damage in brain tissue: coupling the mechanics of edema and oxygen delivery. Biomech. Model. Mechanobiol. **14**, 1197–1216 (2015)
25. Lazarus, C., Soheilypour, M., Mofrad, M.R.: Torsional behavior of axonal microtubule bundles. Biophys. J. **109**(2), 231–239 (2015)
26. Lessinnes, Th., Moulton, D.E., Goriely, A.: Morphoelastic rods. Part II: Growing birods. J. Mech. Phys. Solids (2015). doi:10.1016/j.jmps.2015.07.008
27. Matthews, M.A.: An electron microscopic study of the relationship between axon diameter and the initiation of myelin production in the peripheral nervous system. Anat. Rec. **161**(3), 337–351 (1968)
28. Maxwell, W.L., Povlishock, J.T., Graham, D.L.: A mechanistic analysis of nondisruptive axonal injury: a review. J. Neurotrauma **14**, 419–440 (2007)
29. Moulton, D.E., Lessinnes, T., Goriely, A.: Morphoelastic rods. Part I: A single growing elastic rod. J. Mech. Phys. Solids **61**(2), 398–427 (2012)
30. Neukirch, S., Frelat, J., Goriely, A., Maurini, C.: Vibrations of post-buckled rods: the singular inextensible limit. J. Sound Vib. **331**(3), 704–720 (2012)
31. Nolte, J.: The Human Brain: An Introduction to Its Functional Anatomy, 4th edn. Mosby, St. Louis (1998)
32. O'Keeffe, S.G., Moulton, D.E., Waters, S.L., Goriely, A.: Growth-induced axial buckling of a slender elastic filament embedded in an isotropic elastic matrix. Int. J. Non-Linear Mech. **56**, 94–104 (2013)
33. O'Toole, M., Lamoureux, P., Miller, K.E.: A physical model of axonal elongation: force, viscosity, and adhesions govern the mode of outgrowth. Biophys. J. **94**(7), 2610–2620 (2008)
34. O'Toole, M., Lamoureux, P., Miller, K.E.: Measurement of subcellular force generation in neurons. Biophys. J. **108**(5), 1027–1037 (2015)
35. Ouyang, H., Nauman, E., Shi, R.: Contribution of cytoskeletal elements to the mechanical property of axons. In: Micro/Nano Symposium (UGIM), 2010 18th Biennial University/Government/Industry, pp. 1–5 (2010)
36. Recho, P., Jerusalem, A., Goriely, A.: Growth, collapse, and stalling in a mechanical model for neurite motility. Phys. Rev. E **93**(3), 032410 (2016)
37. Rolls, M.M.: Neuronal polarity in drosophila: sorting out axons and dendrites. Dev. Neurobiol. **71**(6), 419–429 (2011)
38. Sanchez, R.M., Dunkelberger, G.R., Quigley, H.A.: The number and diameter distribution of axons in the monkey optic nerve. Investig. Ophthalmol. Vis. Sci. **27**(9), 1342–1350 (1986)
39. Schaedel, L., John, K., Gaillard, J., Nachury, M.V., Blanchoin, L., Théry, M.: Microtubules self-repair in response to mechanical stress. Nat. Mater. **14**(11), 1156–1163 (2015)
40. Smith, D.H., Meaney, D.F.: Axonal damage in traumatic brain injury. Neuroscientist **6**(6), 483–495 (2000)
41. Smith, D.H., Wolf, J.A., Lusardi, T.A., Lee, V.M.-Y., Meaney, D.F.: High tolerance and delayed elastic response of cultured axons to dynamic stretch injury. J. Neurosci. **19**(11), 4263–4269 (1999)
42. Smith, D.H., Wolf, J.A., Meaney, D.F.: A new strategy to produce sustained growth of central nervous system axons: continuous mechanical tension. Tissue Eng. **7**(2), 131–139 (2001)

43. Soheilypour, M., Peyro, M., Peter, S.J., Mofrad, M.R.: Buckling behavior of individual and bundled microtubules. Biophys. J. **108**(7), 1718–1726 (2015)
44. Tang-Schomer, M.D., Patel, A.R., Baas, P.W., Smith, D.H.: Mechanical breaking of microtubules in axons during dynamic stretch injury underlies delayed elasticity, microtubule disassembly, and axon degeneration. FASEB J. **24**(5), 1401–1410 (2010)
45. Unterberg, A.W., Stover, J., Kress, B., Kiening, K.L.: Edema and brain trauma. Neuroscience **129**(4), 1019–1027 (2004). Brain Water Homeostasis
46. van den Bedem, H., Kuhl, E.: Tau-ism: the yin and yang of microtubule sliding, detachment, and rupture. Biophys. J. **109**(11), 2215–2217 (2015)
47. Vargas, M.E., Barres, B.A.: Why is Wallerian degeneration in the CNS so slow? Annu. Rev. Neurosci. **30**(1), 153–179 (2007)
48. Wolf, J.A., Stys, P.K., Lusardi, T., Meaney, D., Smith, D.H.: Traumatic axonal injury induces calcium influx modulated by tetrodotoxin-sensitive sodium channels. J. Neurosci. **21**(6), 1923–1930 (2001)

J Elast (2017) 129:257–281
DOI 10.1007/s10659-017-9631-8

A Comparison of Phenomenologic Growth Laws for Myocardial Hypertrophy

Colleen M. Witzenburg[1] · **Jeffrey W. Holmes**[1,2,3]

Received: 3 June 2016 / Published online: 1 March 2017

Abstract The heart grows in response to changes in hemodynamic loading during normal development and in response to valve disease, hypertension, and other pathologies. In general, a left ventricle subjected to increased afterload (pressure overloading) exhibits concentric growth characterized by thickening of individual myocytes and the heart wall, while one experiencing increased preload (volume overloading) exhibits eccentric growth characterized by lengthening of myocytes and dilation of the cavity. Predictive models of cardiac growth could be important tools in evaluating treatments, guiding clinical decision making, and designing novel therapies for a range of diseases. Thus, in the past 20 years there has been considerable effort to simulate growth within the left ventricle. While a number of published equations or systems of equations (often termed "growth laws") can capture some aspects of experimentally observed growth patterns, no direct comparisons of the various published models have been performed. Here we examine eight of these laws and compare them in a simple test-bed in which we imposed stretches measured during *in vivo* pressure and volume overload. Laws were compared based on their ability to predict experimentally measured patterns of growth in the myocardial fiber and radial directions as well as the ratio of fiber-to-radial growth. Three of the eight laws were able to reproduce most key aspects of growth following both pressure and volume overload. Although these three growth laws utilized different approaches to predict hypertrophy, they all employed multiple inputs that were weakly correlated during *in vivo* overload and therefore provided independent information about mechanics.

Keywords Overload · Remodeling · Adaptation · Cardiac mechanics · Ventricle · Computer model

---

Electronic supplementary material The online version of this article (doi:10.1007/s10659-017-9631-8) contains supplementary material, which is available to authorized users.

✉ J.W. Holmes
holmes@virginia.edu

1 Department of Biomedical Engineering, University of Virginia, Charlottesville, VA, USA

2 Department of Medicine, University of Virginia, Charlottesville, VA, USA

3 Robert M. Berne Cardiovascular Research Center, University of Virginia, Charlottesville, VA, USA

Mathematics Subject Classification 74D10 · 74E99 · 74L15

1 Introduction

Hypertrophy of the adult left ventricle (LV) is frequently described in terms of two prototypical patterns: concentric and eccentric hypertrophy. Experimentally, pressure overloading of the LV leads to concentric hypertrophy (thickening of individual myocytes and the heart wall), whereas volume overloading of the LV leads to eccentric hypertrophy (lengthening of myocytes and dilation of the cavity) [1, 2]. While some conditions such as hypertension are clearly associated with pressure overloading and concentric hypertrophy and others such as mitral regurgitation are clearly associated with volume overloading and eccentric hypertrophy, many involve some combination of the two. Myocardial infarction, congenital heart defects, some heart valve diseases, pregnancy, and athletic training all induce hypertrophy falling along a spectrum ranging from concentric to eccentric growth. Thus predicting growth and remodeling in these settings requires models that capture responses across this entire spectrum. Since hypertrophy is often progressive, the most pertinent clinical questions surrounding disorders tend to be prognostic: even simply anticipating the rate of progression can be essential when planning treatment. Thus, though the underlying mechanisms responsible for hypertrophy remain elusive [3, 4], phenomenologic models have the potential to provide insight into the development of hypertrophy and guide clinical decision making, in addition to aiding in the design of device-based therapies to slow or reverse remodeling.

Rodriguez et al. [5] presented a general theory for volumetric growth. Together with a set of equations (usually termed a "growth law") that specify how rapidly a region of the heart will grow in any given situation, this framework can predict how cardiac dimensions, residual stresses, and the response to repeated cyclic loading will evolve over time. Because changes in mechanical loading of the heart clearly induce growth and remodeling, many growth laws utilize measures of local mechanics to drive predicted growth. However, there has been considerable debate concerning which specific growth stimuli are actually sensed by the cells in the heart, or (perhaps more practically) which metrics provide the best basis for formulating growth laws. Grossman et al. [6] found that estimated wall stresses in patients with pressure overload were not different from those in normal subjects, and proposed that peak systolic stress may act as a stimulus for concentric hypertrophy, eventually normalizing the initial increases in wall stress due to elevated pressures. Emery and Omens [7] concluded that end-diastolic fiber strain was normalized during volume overload, making it a likely candidate stimulus for eccentric hypertrophy. Holmes [8] investigated potential stimuli during volume overload more systematically and concluded that volume amplitude or related metrics such as ejection strain could provide a basis for distinguishing between pressure and volume overload. Due to the uncertainty surrounding both the underlying biology and the predictive value of the many available potential stimuli, selecting or formulating a growth law is challenging.

A number of growth laws have been proposed and demonstrated to fit aspects of the growth response following pressure and/or volume overloading. It is difficult to compare published laws from various studies, however, since each study uses a different left ventricular geometry, passive constitutive relation, active model of contraction, and set of circulation conditions. Here, we compared eight different growth laws directly. We employed published reports to identify the changes in stretch that occur *in vivo* during acute pressure and volume overloading. Then, we applied these stretches in a simple test-bed and predicted growth using each of the eight growth laws. Finally, we compared features of the growth predicted by each law to reported measurements of myocyte dimensions following pressure or volume overload to assess the ability of each law to capture experimentally observed trends.

2 Methods

2.1 Growth Framework

We performed a direct comparison of several published growth laws (Table 1). All of these growth laws assumed constant density of added or subtracted myocardium (so that growth manifests as a change in volume) and utilized the theoretical framework of Rodriguez et al. [5]. In this framework, material is added or removed through a growth deformation, F_g, changing the local stress-free configuration. The overall deformation is given by

$$F_{eg} = F_e F_g, \tag{1}$$

where the growth deformation, F_g, maps the original reference state to an intermediary unstressed (and potentially incompatible) grown state. The elastic deformation, F_e, then maps the intermediary state into the intact, loaded state. Only the elastic deformation contributes to stress.

As outlined below, we prescribed diastolic and systolic circumferential and longitudinal stretches typical of normal hearts or of hearts subjected to hemodynamic overload, then used the various growth laws to predict the resulting growth. In the context of Eq. (1), this was implemented by prescribing F_{eg} values, computing F_g for a single growth step using one of the growth laws, computing the new diastolic and systolic elastic deformations ($F_e = F_{eg} F_g^{-1}$), and iterating until steady-state was reached or a pre-specified growth limit was exceeded. In order to visually compare the responses of the various growth laws detailed below, Figure S1 plots the growth rate of each law as elastic fiber or crossfiber stretch is varied over the range explored in this study.

2.2 Growth Laws

2.2.1 *A Model for Stress-Induced Growth in the Developing Heart: L&T*

Lin and Taber [9] considered how embryonic growth is affected by mechanical loading. While they acknowledged that pre-natal growth is dominated by hyperplasia and post-natal growth by hypertrophy, they argued that the basic patterns of growth in which the ventricle increases in size in response to increases in pressure and volume are similar. They created a stress-based growth law in which they approximated the LV as a cylindrical thick-walled pressure vessel, simulated passive inflation during diastole, and prescribed growth based on end-diastolic stresses. The fiber direction (f) was assumed to be circumferential and the cross-fiber direction (c) to be parallel to the cylinder axis, with both axes perpendicular to the radial direction (r). Their growth law took the form:

$$F_{g,j}^{i+1} = \frac{F_{g,j}^{i}}{[1 + F_{g,j}^{i} * -\sum_k D_{jk}(\sigma_k - \sigma_{set,k})]}, \tag{2}$$

where $(k, j) = (f, c, r)$, D is a matrix of growth rate coefficients determined by fitting data from embryonic chick left ventricles in both normal and pressure-overloaded states, σ is the Cauchy stress at end diastole, and σ_{set} is the end-diastolic Cauchy stress at growth equilibrium. The superscript i indicates the growth time step. Lin and Taber originally chose $\sigma_{set} = 0$ for all three directions; as explained below, we chose new setpoints to achieve zero growth in all directions under baseline loading conditions.

Table 1 Growth laws compared and modified input set-points

Law	Geometry	Material description		Input parameters	Original set-point	Modified set-point
		Passive	Active			
L&T	Thick walled cylinder	compressible isotropic Fung-type exponential	none	end-diastolic stress in the fiber, cross-fiber, and radial directions	$\sigma_{set,f} = 0$ mmHg $\sigma_{set,c} = 0$ mmHg $\sigma_{set,r} = 0$ mmHg	$\sigma_{set,f} = 1.92$ mmHg $\sigma_{set,c} = 1.92$ mmHg $\sigma_{set,r} = -1.38$ mmHg
LT2	Thick walled cylinder	linear elastic	linear elastic	diastolic and systolic stress in the fiber direction	$\sigma_{set,a} = 30$ kPa $\sigma_{set,p} = 3$ kPa	$\sigma_{set,a} = 1170$ kPa $\sigma_{set,p} = 88$ kPa
KFR	Truncated ellipsoid	slightly compressible anisotropic exponential	none	end-diastolic fiber strain	$s_{hom} = 0.13$	$s_{hom} = 0.44$
KUR	Truncated ellipsoid	slightly compressible anisotropic exponential	none	end-diastolic fiber strain	$s_{hom} = 0.13$	$s_{hom} = 0.44$
GEG	Truncated ellipsoid	compressible isotropic neo-Hookean	none	end-diastolic fiber strain	$F_{g,f,max} = 1.50$ $\lambda^{crit} = 1.01$	$F_{g,f,max} = 1.50$ $\lambda^{crit} = 1.44$
GCG	Truncated ellipsoid	compressible isotropic neo-Hookean	none	first invariant of the Mandel stress at end systole	$F_{g,c,max} = 3.00$ $p^{crit} = 0.012$ Mpa	$F_{g,c,max} = 3.00$ $p^{crit} = 0.053$ Mpa
ART	Chamber	compressible anisotropic exponential	Hill-type model	maximum and minimum sarcomere lengths	$L_{s,max,adapt} = 2.30$ μm $L_{s,min,adapt} = 1.75$ μm	$L_{s,max,adapt} = 2.07$ μm $L_{s,min,adapt} = 1.61$ μm
KOM	Finite element model of canine left ventricle	slightly compressible transversely isotropic	Hill-type model	maximum fiber strain and the minimum maximum principal cross-section strain	$E_{f,set} =$ not given $E_{cross,set} =$ not given	$E_{f,set} = 0.537$ $E_{cross,set} = 0.124$

2.2.2 *Biomechanical Growth Laws for Muscle Tissue: LT2*

Taber [10] also considered muscle development more generally, creating growth laws for skeletal muscle, heart muscle, and arteries. In this case the full range of post-natal growth, from birth to maturity, was simulated. For heart growth, the LV was again approximated as a cylindrical thick-walled pressure vessel. Both passive and active stresses were considered, and growth was determined based on stress in the circumferential direction. Thus, in L&T all directions were considered (circumferential, radial, and longitudinal), but only at end-diastole, whereas in LT2 stress both at end-diastole and end-systole was considered, but only in the circumferential direction. Here again, the fiber direction (f) was assumed to be circumferential and the cross-fiber direction (c) to be parallel to the cylinder axis, with both axes perpendicular to the radial direction (r). The growth law took the form:

$$F_{g,r}^{i+1} = F_{g,r}^{i} * \left(\frac{1}{\overline{T}_r} * \frac{1}{(\sigma_{\theta a0})_m} * \left[\sigma_{\theta a} - (\sigma_{\theta a0})_m\right] + 1\right), \tag{3}$$

$$F_{g,f}^{i+1} = F_{g,f}^{i} * \left(\frac{1}{\overline{T}_\theta} * \frac{1}{(\sigma_{\theta p0})_m} * \left[\sigma_{\theta p} - (\sigma_{\theta p0})_m\right] + 1\right), \tag{4}$$

$$F_{g,c}^{i+1} = 1 \tag{5}$$

where $\overline{T}_r$ and $\overline{T}_\theta$are normalized time constants and $(\sigma_{\theta a0})_m$ and $(\sigma_{\theta p0})_m$ are the active and passive fiber stresses at growth equilibrium. These equilibrium stresses were adjusted to achieve zero growth in all directions under baseline conditions.

2.2.3 *Computational Modeling of Volumetric Soft Tissue Growth: Application to the Cardiac Left Ventricle: KFR and KUR*

Like Lin and Taber, Kroon et al. [11] also simulated passive inflation during diastole, but used a finite-element truncated ellipsoidal model of the LV to simulate the response to volume overload rather than pressure overload, employing growth laws that depended on strain rather than stress. In the “fixed” reference configuration growth law (denoted KFR), the stress-free reference configuration remained constant throughout the simulation,

$$F_{g,j}^{i+1} = \left[\beta(\sqrt{2 * E_{ff} + 1} - 1 - s_{hom}) + 1\right]^{\frac{1}{3}} F_{g,j}^{i}. \tag{6}$$

By contrast, their “updated” reference configuration growth law considered the possibility that turnover of tissue components might allow growth-induced residual stresses to relax over time. In this growth law (denoted KUR),

$$F_{g,j}^{i+1} = \left[\beta(\sqrt{2 * E_{ff} + 1} - 1 - s_{hom}) + 1\right]^{\frac{1}{3}}. \tag{7}$$

In both laws, β is the product of the growth rate constant and the time interval, s_{hom} is the homeostatic end-diastolic myofiber strain, and E_{ff} is the elastic fiber strain at end diastole for growth time step i. In both cases growth was assumed to occur isotropically, thus $j = (f, c, r)$.

2.2.4 *A Multiscale Model for Eccentric and Concentric Cardiac Growth Through Sarcomerogenesis: GEG and GCG*

Göktepe et al. [12] used a finite-element model to simulate end diastole and end systole and proposed two different growth laws: one to simulate growth following volume overload (eccentric hypertrophy) and one to simulate growth following pressure overload (concentric hypertrophy). For volume overload they utilized a strain-based growth law (here denoted GEG),

$$F_{g,f}^{i+1} = \frac{1}{\tau}\left[\frac{F_{g,f,max} - F_{g,f}^{i}}{F_{g,f,max} - 1}\right]^{\gamma} * \left[F_{e,f}^{i} - \lambda^{crit}\right] + F_{g,f}^{i} \quad \text{and} \tag{8}$$

$$F_{g,c}^{i+1} = F_{g,r}^{i+1} = 1, \tag{9}$$

where $F_{g,f,max}$ specifies the maximum fiber growth allowed and λ^{crit} is the homeostatic elastic fiber stretch. Note that growth is not permitted in the cross-fiber or radial directions. For pressure overload they utilized a stress-based growth law (denoted GCG),

$$F_{g,c}^{i+1} = \frac{1}{\tau}\left[\frac{F_{g,c,max} - F_{g,c}^{i}}{F_{g,c,max} - 1}\right]^{\gamma} * \left[\mathrm{tr}(M^{e}) - p^{crit}\right] + F_{g,c}^{i} \quad \text{and} \tag{10}$$

$$F_{g,f}^{i+1} = F_{g,r}^{i+1} = 1, \tag{11}$$

where $F_{g,c,max}$ specifies the maximum cross-fiber growth allowed, M^e is the Mandel stress (Cauchy stress weighted by the Jacobian) associated with the elastic deformation at end systole, and p^{crit} is the homeostatic value for the Mandel stress. Note that, in this case, growth is not permitted in the fiber or radial directions. In both laws, τ controls how much growth occurs per growth time step i and γ controls the shape of the growth-limiting function.

2.2.5 Adaptation to Mechanical Load Determines Shape and Properties of Heart and Circulation: the CircAdapt Model: ART

Arts et al. [13] combined an active model of sarcomere mechanics, analytic expressions relating fiber stretch and stress to cavity pressures and volumes, and a circuit model of the circulation to simulate the full cardiac cycle before and during pressure overload. They proposed a set of growth laws in which the ratio of maximum to minimum fiber stretch determined fiber growth, while a more complicated function of maximum and minimum fiber stretch determined radial growth:

$$F_{g,f}^{i+1} = \frac{\max(F_{e,f}^{i})/\min(F_{e,f}^{i})}{L_{s,max,adapt}/L_{s,min,adapt}} F_{g,f}^{i}, \tag{12}$$

$$F_{g,r}^{i+1} = F_{g,r}^{i}\sqrt{\left(\frac{1+a}{1+ae^{b(L_{s,max,adapt}-L_{s,0}^{i}*\max(F_{e,f}^{i}))}}\right) * \left(\frac{L_{s,max,adapt}/L_{s,min,adapt}}{\max(F_{e,f}^{i})/\min(F_{e,f}^{i})}\right)}, \tag{13}$$

$$L_{s,0}^{i+1} = L_{s,0}^{i}\left(\frac{L_{s,max,adapt}/L_{s,min,adapt}}{\max(F_{e,f}^{i})/\min(F_{e,f}^{i})}\right), \quad \text{and} \tag{14}$$

$$F_{g,c}^{i+1} = F_{g,r}^{i+1}, \tag{15}$$

where $L_{s,max,adapt}$ is the homeostatic set point for sarcomere length at the beginning of ejection, $L_{s,min,adapt}$ is the homeostatic set point for sarcomere length at the end of ejection, and a and b control the sensitivity to stretch. The value specified for the initial reference sarcomere length, $L_{s,0}^{1}$, did not affect the long-term behavior of this law, because $L_{s,0}^{i}$ changes with each growth time step i.

2.2.6 A Single Strain-Based Growth Law Predicts Concentric and Eccentric Cardiac Growth During Pressure and Volume Overload: KOM

Finally, Kerckhoffs et al. [14] proposed a single law consisting of a pair of sigmoidal strain-based equations to model the response to both pressure and volume overload. Similar to Arts et al. [13], they included a Hill-type model to simulate myocyte contraction and coupled their LV model to a model of the circulatory system to determine loading conditions for the ventricles. They did not allow the circulation parameters to change with growth, however, instead prescribing cardiac output. The stimuli for myocyte growth,

$$s_l = \max(E_{ff}) - E_{f,set} \quad \text{and} \tag{16}$$

$$s_t = \min\left(\frac{E_{cc}+E_{rr}}{2} + \sqrt{\left(\frac{E_{cc}-E_{rr}}{2}\right)^2 + E_{cr}^2}\right) - E_{cross,set}, \tag{17}$$

are dependent on the elastic fiber strain, E_{ff}, cross-fiber strain, E_{cc}, radial strain E_{rr}, and cross-fiber radial shear, E_{cr}.$E_{f,set}$ is the strain set-point at which no axial fiber growth occurs and $E_{cross,set}$ is the strain set-point at which no cross-fiber or radial growth occurs. Growth in the fiber direction is determined by

$$F_{g,f}^{i+1} = \begin{cases} F_{g,f}^{i} * [\frac{1}{1+e^{d_l*(F_{g,f}^{i}-F_{f,50})}} * \frac{a_l}{1+e^{-b_l*(s_l-s_{l,50})}} + 1] & s_l \geq 0 \\ F_{g,f}^{i} * [\frac{-a_l}{1+e^{b_l*(s_l+s_{l,50})}} + 1] & s_l < 0. \end{cases} \tag{18}$$

Growth is the same in the radial and cross-fiber directions and is determined by

$$F_{g,c}^{i+1} = F_{g,r}^{i+1} = \begin{cases} F_{g,c}^{i} * [\frac{1}{1+e^{d_t*(F_{g,c}^{i}-F_{c,50})}} * \frac{a_t}{1+e^{-b_t*(s_t-s_{t,50})}} + 1] & s_t \geq 0 \\ F_{g,c}^{i} * [\frac{-a_t}{1+e^{b_t*(s_t+s_{t,50})}} + 1] & s_t < 0. \end{cases} \tag{19}$$

Equations (15) and (16) prescribe sigmoidal functions such that maximum and minimum growth allowed within a single step is determined by the parameters a_l and a_t and there is a small nearly quiescent (low-growth) region near the setpoints specified by $s_{l,50}$, $s_{t,50}$, b_l, and b_t. When the stimulus is positive, i.e., $s_l \geq 0$ or $s_t \geq 0$, a limiting term is included to avoid cases of runaway growth.

2.3 Inputs to the Growth Laws: Control and Overloaded Stretches and Stresses

The LV has a complex fiber architecture in which myocardial fiber orientation changes as a function of both depth within the myocardial wall and distance from the base to the apex [15]. At the midwall, the fibers are aligned with the circumferential direction of the LV and the in-plane cross-fiber direction is aligned with the longitudinal axis of the LV. Measurements of circumferential and longitudinal strains during active contraction [16–19] and during passive inflation [20, 21] indicate that they are roughly equal. During the cardiac cycle, peak stretches in both the circumferential and longitudinal directions occur at end diastole, associated with significant radial thinning [8, 22–24]. Peak radial thickening occurs near end systole, when circumferential and longitudinal stretches are low. Thus, to provide a simple test-bed for comparing different growth laws, we simulated cyclic equibiaxial stretch of a thin slab of myocardium (Fig. 1) between prescribed end-diastolic and end-systolic stretches. Since the growth laws we tested were not sensitive to strain rate, we specified linear stretching/shortening between these extremes. We simulated changes in regional mechanics typical of pressure overload and volume overload by adjusting end-diastolic and

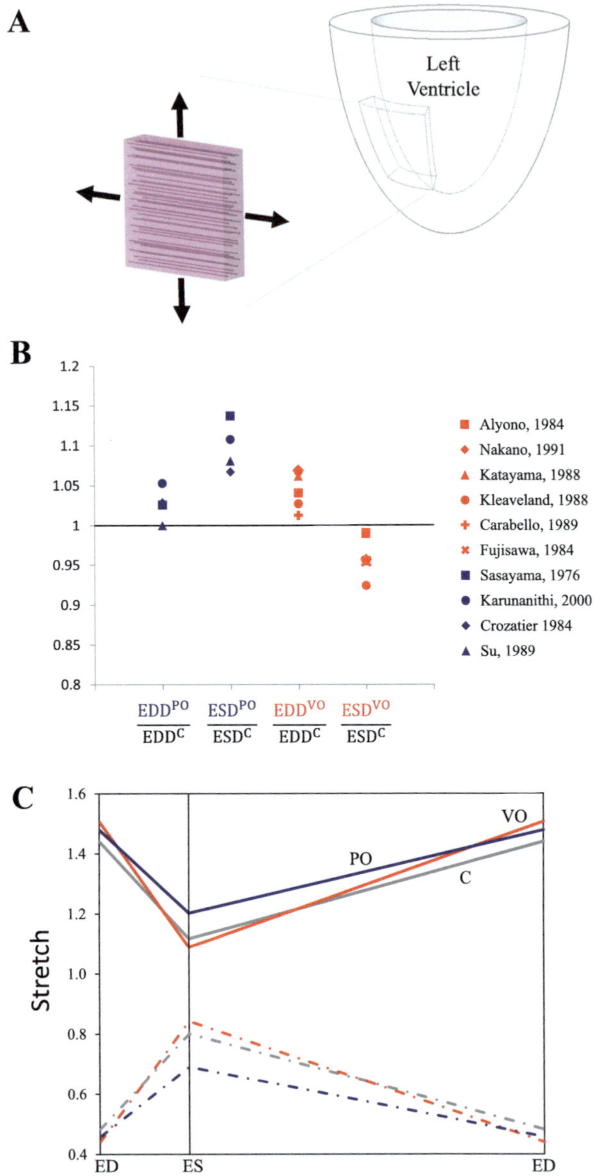

Fig. 1 (**A**) Schematic of mid-wall slab of left ventricular equatorial subjected to prescribed stretches in our simulations. (**B**) End-systolic and end-diastolic dimension ratios relative to control during acute pressure [34–37] and volume overload [28–33] studies. To avoid potentially confounding effects of growth normalizing the inciting stimulus, only studies that recorded dimensions prior to appreciable change in ventricular mass were considered. (**C**) Average stretch vs. cardiac cycle time for control (C), volume overload (VO), and pressure overload (PO) in reference to the stress-free configuration. ED signifies end diastole and ES signifies end systole. Equibiaxial loading was assumed, thus the fiber (*solid lines*) and cross-fiber (*dashed lines*) directions experienced the same stretch profile and radial (*dash-dot lines*) thinning was prescribed to maintain constant tissue volume

end-systolic stretches to match reported *in vivo* values, and observed the predicted growth responses.

The majority of the growth laws compared here predict growth based on elastic stretches or strains computed relative to a stress-free state. This stress-free state is a convenient reference when formulating models, but never occurs *in vivo*. We utilized end-diastolic pressure (EDP) and segment length data from Fomovsky et al. [25] to estimate end-diastolic circumferential stretch relative to the unloaded state in normal dogs. Fomovsky and colleagues recorded data during inferior vena cava occlusions, which produced a series of beats with gradually decreasing EDP. We plotted end-diastolic segment length as a function of ln(EDP) and used a linear fit to estimate the intercept of this plot, corresponding to circumferential segment length at an EDP of 1 mmHg. We treated this segment length as an estimate of the unloaded length, and retained only the estimates with a coefficient of variation $< 5.0\%$. The average estimated end-diastolic circumferential stretch relative to the unloaded state in these 6 animals was 1.44 ± 0.10 (mean $\pm$ SEM, see Figure S2). However, unloaded hearts are not stress-free [26], and stretch ratios computed from unloaded segment lengths would not account for residual strains. Luckily, Rodriguez et al. [27] compared unloaded and stress-free sarcomere lengths in rats and found that while the transmural distribution of sarcomere lengths differed in these two states, the sarcomere length at the midwall was the same.

Next, we reviewed *in vivo* data to determine how stretch changes following the creation of experimental pressure and volume overload in dogs. To avoid potentially confounding effects of growth normalizing the inciting stimulus, only studies that recorded end-diastolic and end-systolic dimensions after the imposition of hemodynamic overload but prior to appreciable change in left ventricular mass were considered. Changes in left ventricular diameter or volume measured using angiography or sonomicrometry were used to estimate stretches during overload; where volumes were provided instead of dimensions, stretch ratios were estimated using volume ratio raised to the one-third power. Six studies induced experimental volume overload and reported changes in volumes within four days; two created aorto-caval fistulas [28, 29] and four induced mitral regurgitation [30–33]. In these studies, volume overload increased end-diastolic volume or diameter (end-diastolic stretch ratios of 1.05 ± 0.02 relative to control, mean $\pm$ SD, $p = 0.02$, Fig. 1B) and decreased end-systolic volume or diameter (end-systolic stretch ratios of 0.96 ± 0.02 relative to control, mean $\pm$ SD, p $= 0.04$). Four studies induced experimental pressure overload using ascending [34–36] or descending [37] aortic constriction and reported changes in left ventricular dimensions immediately. In these studies, pressure overload increased end-systolic volume or diameter (end-systolic stretch ratios of 1.10 ± 0.03 relative to control, mean $\pm$ SD, $p = 0.01$, Fig. 1B), while increases in end-diastolic volume or diameter were not significant (end-diastolic stretch ratios of 1.03 ± 0.02 relative to control, mean $\pm$ SD, $p = 0.09$).

The *in vivo* experimental stretch ratios were expressed relative to end diastole, rather than the unloaded configuration. Thus, for overload studies we computed end-diastolic stretches relative to the unloaded configuration by combining our estimate of control end-diastolic stretch relative to the unloaded state (λ_{ED}^{C}) with measurements of control (EDD^{C}), and overload (EDD^{O}) end-diastolic dimensions:

$$\lambda_{ED}^{O} = \lambda_{ED}^{C} * \frac{EDD^{O}}{EDD^{C}}. \tag{20}$$

Similarly, we computed end-systolic stretches for both control and overload states using the end-diastolic stretch relative to unloaded state and measurements of control and overload dimensions:

$$\lambda_{ES}^{C} = \lambda_{ED}^{C} * \frac{ESD^{C}}{EDD^{C}} \quad \text{or} \quad \lambda_{ES}^{O} = \lambda_{ED}^{O} * \frac{ESD^{O}}{EDD^{O}}. \tag{21}$$

The resulting stretch profiles used to simulate pressure and volume overload are shown in Fig. 1C. Control fiber and cross-fiber end-diastolic and end-systolic stretches were set to 1.44 and 1.12 at baseline, 1.48 and 1.20 to simulate pressure overload, and 1.51 and 1.09 to simulate volume overload. Myocardial tissue was assumed incompressible within each loading cycle. Thus, radial end-diastolic and end-systolic stretches were set to 0.48 and 0.80, 0.46 and 0.69, and 0.44 and 0.84, for control, pressure overload, and volume overload conditions, respectively. Three of the growth laws (L&T, LT2 and GCG) utilized stresses (rather than stretches or strains) as inputs; in these cases the stresses were computed from the prescribed stretches using the constitutive equation provided in the original reference. Finally, to provide the most useful comparison of the predictions of these different growth laws during simulated pressure and volume overload, we adjusted the set-points (Table 1) for each growth law such that repeated cycling between baseline end-diastolic and end-systolic stretches induced zero growth.

2.4 Comparison Data: Changes in LV and Myocyte Dimensions Following Overload

Most *in vivo* studies of hypertrophy track left ventricular wall thickness in one or more locations as well as changes in cavity diameter, radius, or circumference. Changes in wall thickness should be analogous to the growth stretches $F_{g,r}$ predicted by the models considered here. Similarly, changes in the left ventricular circumference should be roughly equivalent to $F_{g,f}$ at the midwall, although not necessarily in other layers of the heart wall; note that no commonly reported global dimension is comparable to $F_{g,c}$ predicted by the models. Experimentally, the ratio of LV radius-to-thickness or myocyte length-to-width is frequently used to quantify shape change following growth. Thus, for simulations of both pressure and volume overload for each growth law, the growth ratio was calculated as fiber growth stretch, $F_{g,f}$, divided by radial growth stretch, $F_{g,r}$.

Only some of the acute *in vivo* pressure and volume overload studies used above to quantify changes in stretch also tracked subsequent growth, and those few studies measured growth at very different timepoints. Thus, in order to better determine the levels of expected growth following pressure and volume overload, we included additional studies that quantified LV dimensions. The ratio of LV radius-to-wall thickness decreased during experimental pressure overload and increased during volume overload in dogs (Fig. 2, left column), with considerable variation among the published studies [30–34, 38–41]. Similar trends but even wider ranges were reported in rats (Fig. 2, second column) [2, 7, 42–63]. In order to avoid the confounding effects of failure and decompensation, only studies that reported dimensional changes within 2 to 12 weeks of overload were considered.

There has been considerable evidence showing that changes in myocyte dimensions and shape parallel changes in ventricular dimensions and shape. Myocyte dimensional changes may provide a more relevant local measure of growth for comparison with growth laws that are applied locally, on an element-by-element basis. Thus, we also reviewed studies in which myocyte length and cross-sectional area changes were measured following *in vivo* inducement of pressure or volume overload. Reported myocyte length-to-width ratios increased following volume overload and decreased following pressure overload (Fig. 2, right columns) in both dogs [64, 65] and rats [66–73]. Compared to LV radius-to-wall thickness ratios, myocyte length-to-width ratios changed less in response to overload; the reason for this difference is not clear, particularly given that the rat ventricular and myocyte dimensions were measured at similar times following the onset of overload (mean ± SEM indicated in Fig. 2). We pooled the dog and rat myocyte dimension measurements for subsequent comparison to the predictions of the various growth models. Tables S1 and S2 provide a comprehensive list of the data plotted in Fig. 2 with associated references.

3 Results

Figure 3A shows the fiber, cross-fiber, and radial growth levels (F_g) predicted by the eight different growth laws in response to prescribed biaxial stretches typical of pressure overload. The range of experimentally reported changes in myocyte width (dashed light green

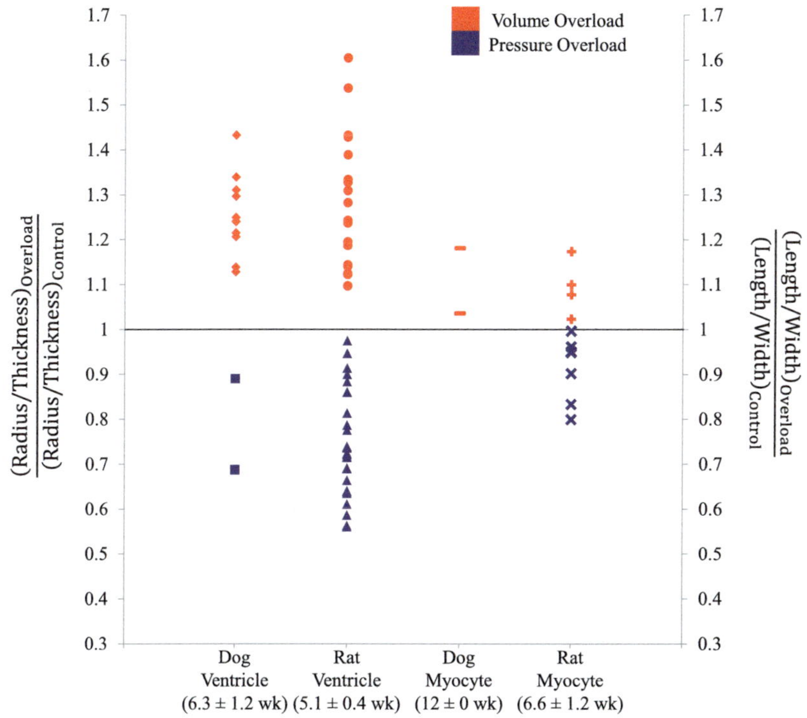

Fig. 2 *Left columns* show reported changes in LV radius-to-wall thickness ratio for *in vivo* studies following the creation of volume and pressure overload in dogs [30–34, 38–41] and rats [2, 7, 42–63], displayed as normalized radius-to wall thickness ratios (*left vertical axis*). *Right columns* show reported changes in myocyte shape for *in vivo* studies following the creation of volume and pressure overload conditions in dogs [64, 65] and rats [66–73], displayed as normalized myocyte length-to-width ratios (*right vertical axis*). The mean ± SEM of the time from onset of overload to dimension measurement is shown below the column labels for each group of studies

lines) and length (dashed dark green lines) following 2 to 12 weeks of pressure overload are shown for comparison. Pressure overload is typically associated with ventricular wall thickening and myocyte width increase *in vivo*. Both KOM and L&T predicted steady-state radial growth within the range of experimentally observed values, while simultaneously predicting realistic levels of steady-state fiber growth. LT2 produced runaway growth in the radial direction. Since LT2 did not reach a steady-state, we halted the pressure overload simulation after two growth time steps, when the predicted fiber-to-radial growth ratio first surpassed the experimental mean; at this growth step, both predicted radial growth and fiber growth were within the range of experimentally observed values. ART also produced runaway growth in the radial and cross-fiber directions; if halted after two growth time steps, when the predicted fiber-to-radial growth ratio first surpassed the experimental mean, predicted radial growth was within the range of experimentally observed values, but the model predicted atrophy in the fiber direction that was inconsistent with experiments. Göktepe et al. [12] created separate laws designed specifically to simulate concentric growth following pressure overload (GCG) and eccentric growth following volume overload (GEG). Similarly to ART, GCG produced maximal growth (limited by the law to $F_{g,r} = 3.0$) in response to prescribed biaxial stretches typical of pressure overload; when halted after growth step 4, when the growth ratio first surpassed the experimental mean, GCG predicted realistic radial

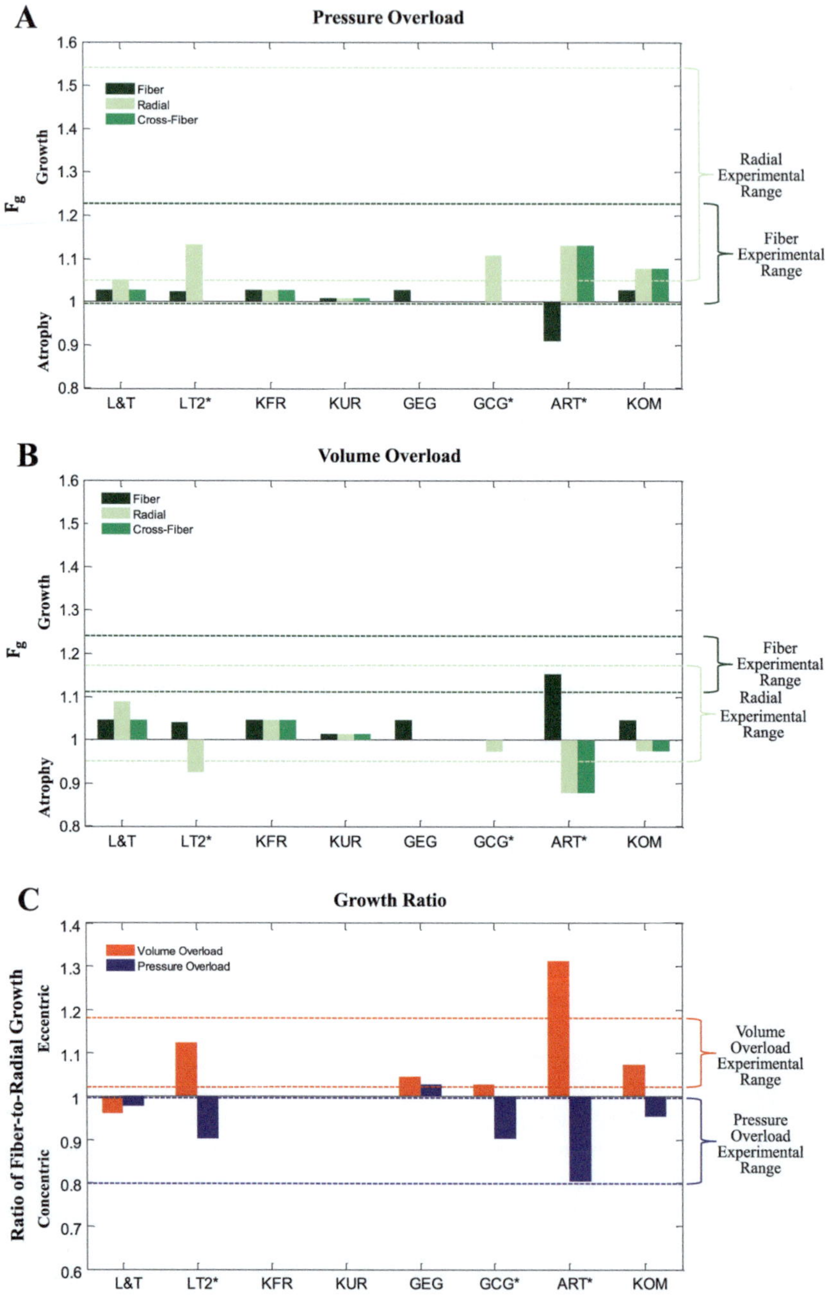

Fig. 3 The fiber, radial, and cross-fiber growth levels (F_g) predicted following pressure (**A**) and volume overload (**B**) by the various growth laws; *green dashed lines* indicate the experimental ranges in myocyte length (*dark green*) and width (*light green*) observed experimentally. (**C**) The fiber-to-radial growth ratio predicted for both pressure and volume overload; *red and blue dashed lines* indicate the experimental ranges for the normalized ratio of myocyte length to width. Three growth laws (GCG, ART, and LT2) predicted runaway growth or atrophy ($F_g > 20$ or $F_g < 0.05$), so simulations were stopped after the time step when the ratio of fiber-to-radial growth first surpassed the value observed experimentally following pressure overload

growth, but by design predicted no fiber or cross-fiber growth. KFR and KUR predicted low levels of isotropic growth that fell outside the experimental range for relative myocyte thickness. By design, GEG was constrained such that there was no growth in the radial or cross-fiber directions and therefore did not produce radial growth in our simulations.

Figure 3B shows the fiber, cross-fiber, and radial growth levels (F_g) predicted by the eight different growth laws in response to prescribed biaxial stretches typical of volume overload. The experimental studies compiled for comparison consistently showed growth in the fiber direction (increased cavity radius and myocyte length), but much more variable changes in ventricular radius and myocyte width. Similarly, most of the growth laws tested (L&T, KFR, GEG, KOM) predicted modest growth in the fiber direction ($\sim 5\%$), with variable responses in the cross-fiber and radial directions ranging from 9% growth (L&T) to 2% atrophy (KOM). ART produced runaway growth, and when halted after two growth time steps (same time point as the pressure overload simulation) yielded fiber growth well within the experimental range but more radial atrophy than observed experimentally. When stopped after 2 growth time steps (same time point as the pressure overload simulation), LT2 predicted more modest fiber growth and substantial radial atrophy. Finally, when stopped after 4 growth time steps (same time point as the pressure overload simulation), GCG produced radial growth within experimental ranges but, by design, no fiber growth.

Figure 3C shows the ratio of fiber growth to radial growth produced by each law in response to prescribed biaxial stretches typical of pressure or volume overload. The ranges of normalized experimental myocyte length-to-width ratios from Fig. 2 are indicated by the blue and red dotted lines, respectively. Consistent with the changes in fiber and radial growth reviewed above, only KOM predicted concentric growth (fiber:radial growth ratio <1) in response to prescribed stretches consistent with experimental pressure overload, predicted eccentric growth (fiber:radial growth ratio >1) in response to prescribed stretches consistent with volume overload, and achieved steady-state. As noted above, the fiber-to-radial growth ratio for the LT2, ART, and GCG simulations using stretches typical of pressure overload fell within the experimental range because this was used as a stopping criterion. At the same growth step, LT2 predicted fiber-to-radial growth ratios near the middle of the experimental range in response to stretch inputs typical of volume overload, while GCG predicted much lower ratios and ART much higher ratios. L&T predicted a steady-state growth ratio within the experimental range for pressure overload, but not volume overload, while GEG captured the trend for volume overload but not pressure overload. Finally, by design KFR and KUR produced isotropic growth with no change in fiber:radial growth ratios. Figure S3 shows the number of growth time steps required for each simulation.

In addition to simulating responses to mean experimentally measured stretches during pressure overload and volume overload, we explored the behavior of each growth law in response to a wide range of end-diastolic and end-systolic stretches. We plotted the predicted ratio of fiber:radial growth for each combination as a contour plot (Fig. 4), and included the experimental stretch ranges (white boxes, Fig. 4A–E) and average myocyte length:width ratios (colored boxes, Fig. 4F) for comparison. Bands of constant predicted growth ratio run diagonally for KOM (Fig. 4A), ART (Fig. 4B), and LT2 (Fig. 4C), reflecting the fact that these laws predict growth based on a combination of end-diastolic and end-systolic stretches or stresses. By contrast, L&T and GEG respond directly to diastolic mechanics (vertical color bands in Figs. 4C–D), while GCG responds to systolic stretch (horizontal color bands in Fig. 4E). Because the *in vivo* stretches computed from experimental pressure and volume overload data differed more in systole than in diastole, only laws that incorporated systolic information were able to match observed trends in fiber:radial growth for both cases.

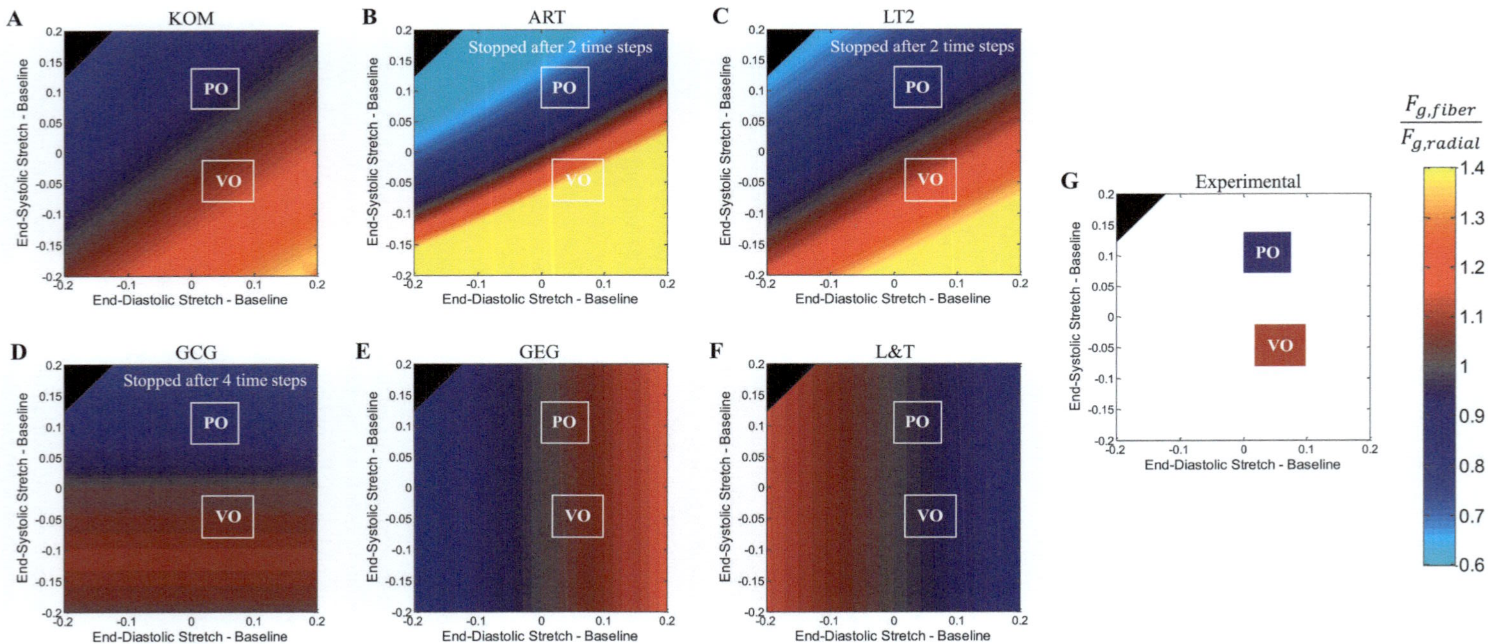

Fig. 4 Predicted fiber-to-radial growth ratios (*color*) given perturbations in end-diastolic (x-axis) and end-systolic stretch (y-axis) for (**A**) KOM, (**B**) ART, (**C**) LT2, (**D**) GCG, (**E**) GEG, and (**F**) L&T. Note that the simulations for ART, LT2, and GCG were stopped after growth time steps 2, 2, and 4, respectively. *White boxes* (**A–F**) indicate the experimental ranges of end-systolic and end-diastolic stretches immediately following pressure and volume overload from Fig. 1B. (**G**) Average myocyte length-to-width ratios following pressure overload and volume overload from Fig. 2 are plotted using the same color scheme as the other panels

4 Discussion

Computational models that correctly predict growth and remodeling of the heart in response to exercise, disease, and various drugs and devices could be very useful in designing new therapies for a range of pathologic conditions. Several groups have now formulated and published equations (often termed "growth laws") that can correctly predict cardiac growth in certain situations. However, to our knowledge no comprehensive comparison of these published growth laws is available to guide researchers and engineers interested in employing models of heart growth in their work. Accordingly, in this study we conducted a systematic comparison of eight published growth laws originally developed to model various experiments that induce growth and remodeling of the heart by surgically altering hemodynamic loading. All of the growth laws tested here were phenomenologic in nature; in other words, they attempted to predict growth from one or more measures of regional mechanics, without representing the complex underlying biology by which those mechanical inputs are transduced in myocytes. They were developed based on the experimental observation that suddenly changing heart mechanics triggers growth [2, 30–34, 38–42], and that the speed or extent of growth appears to correlate with the magnitude of the alteration in mechanics [32, 74, 75].

The growth laws tested here all utilized either myocardial stretches or stresses that were easily computed from stretches and the constitutive parameters provided. Furthermore, they were all originally intended to simulate the response of the LV to experimental pressure overload (an increase in systolic cavity pressure usually induced by constricting the aorta to increase outflow resistance) or volume overload (an increase in cardiac output due to creation of an arteriovenous fistula or mitral regurgitation). Therefore, in order to standardize comparisons in this study, we surveyed the literature to establish a range for the changes in stretch induced by each of these surgical procedures, and simulated the growth response of a section of the LV midwall to these prescribed changes in stretch. We then compared growth predicted by each model to the range of actual growth reported from experimental pressure and volume overload studies.

Most of the published laws reproduced some features of the experiment they were originally designed to simulate—for example, predicting radial growth during pressure overload or growth in the fiber direction during volume overload. However, only three of the laws were able to capture experimentally observed growth patterns in multiple directions at once (Table 2), and only one of these laws produced growth that returned stretches to homeostatic values in our biaxial simulation. As discussed below, these simulations revealed two principles that may prove useful in future cardiac growth models. First, only laws that employed multiple inputs that are poorly coupled during hemodynamic perturbations (such as fiber and crossfiber strain, or end-diastolic and end-systolic stress) were able to capture most of the features of both pressure-overload and volume-overload responses. Second, only laws that used inputs from multiple directions (fiber, cross-fiber, etc.) to drive growth in multiple directions could return to homeostasis following simulated overload in the absence of coupling to a circulatory model.

4.1 Capturing Both Pressure-Overload and Volume-Overload Responses

In many clinically important situations such as myocardial infarction, the changes in mechanics induced in the heart involve some combination of the prototypical changes induced by experimental pressure and volume overload. In such cases a growth model that can correctly predict both the individual pressure and volume overload responses will likely be

Table 2 Summary of growth law predictions in response to pressure and volume overload

Experimental observations		L&T	LT2*	KFR	KUR	GEG	GCG*	ART*	KOM
PO	Radial direction	√	√			●	√	√	√
	Fiber/radial	√	*	●	●	●	*	*	√
	Fiber direction	√	√	√	√	√	√		√
VO	Fiber direction						●	√	
	Fiber/radial	●	√	●	●	√	√	√	√
	Radial direction	√		√	√	√	√		√

√ Predicted value was inside experimental range

* Simulation was halted when predicted value was within experimental range

● Predicted value did not follow experimental trend

required. Of the models tested here, both the KOM and LT2 growth equations were explicitly developed by the authors to capture both overload responses. Therefore, is it not surprising that they captured the key features of both pressure (radial growth stretch >1 and fiber:radial growth ratio <1) and volume (fiber growth stretch and fiber:radial growth ratio both >1) overload in our simulations. However, the ART growth law—originally applied by its authors solely to model pressure overload—also reproduced the key trends. Upon further examination, comparison of these three models with the others offers some potential insight into how to design phenomenologic growth equations to capture multiple responses.

One feature shared by KOM, ART, and LT2 is that all models employed a system of equations involving more than one input. KOM used stretch in two directions (fiber stretch and principal stretch in the plane perpendicular to the fibers) as inputs, while ART and LT2 utilized fiber stretch or stress at two different times during the cardiac cycle (maximum, usually at end diastole, and minimum, usually at end systole). Importantly, each law employed two measures that were weakly correlated with one another in experimental studies of pressure and volume overload (Fig. 5A, D, and G), so that both inputs provided independent information about heart mechanics. By contrast, L&T employed a system of three growth equations involving three different components of the stress tensor, but because all stress components were computed at end-diastole, they were strongly coupled to one another through the constitutive law and did not provide independent information (Fig. 5J). Interestingly, the ranges of experimental data we gathered for this study (Fig. 1B) suggest that it should be possible to capture both pressure and volume overload responses with a single input related to end-systolic fiber stretch: experimentally, pressure and volume overload induced similar changes in diastolic stretch compared to control, but very different changes in systolic stretch (increased in pressure overload, decreased in volume overload). However, most single-input growth laws compared here utilized end-diastolic stretch or stress and therefore predicted similar growth in response to prescribed stretches representing pressure or volume overload (Fig. 4E and F). One law (GCG) did utilize end-systolic stress as its only input, but assumed that growth occurred only in the radial direction, which prevented it from matching experimentally measured volume-overload responses.

4.2 Stopping or Limiting Growth

We had to halt simulations for some of the growth laws compared here after just a few growth time steps in order to prevent runaway growth, while others included explicit limits on the

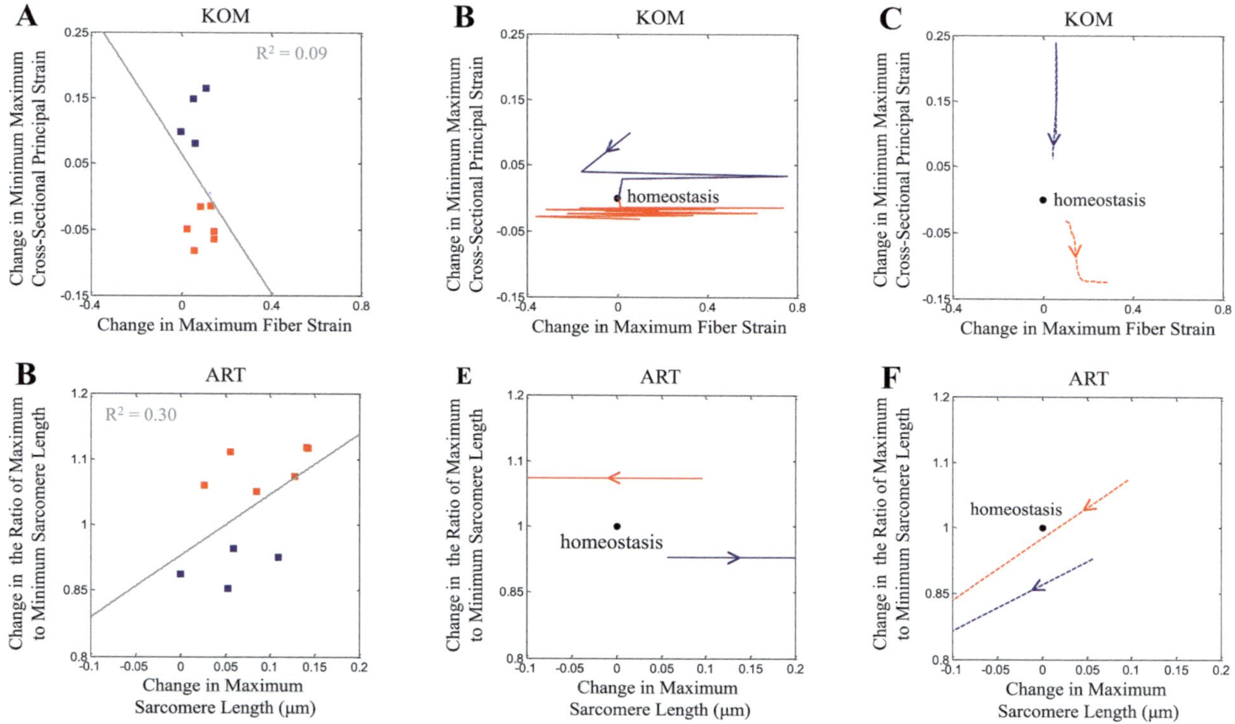

Fig. 5 The correlation of growth stimuli for (**A**) KOM, (**D**) ART, (**G**) LT2, and (**J**) L&T following acute pressure (*blue squares*) and volume overload (*red squares*). Stimulus values were determined from the experimental studies shown in Fig. 2B. Laws utilizing multiple inputs that did not covary strongly across the different experiments were more successful at predicting experimentally measured growth. The evolution of the growth stimuli when overall stretch was prescribed in the fiber and cross-fiber directions for (**B**) KOM, (**E**) ART, (**H**) LT2, and (**K**) L&T. *Arrows* indicate how the stimuli change with increasing growth for both the pressure (*blue*) and volume (*red*) overload simulations. The evolution of the growth stimuli when force was prescribed in the fiber and cross-fiber directions for (**C**) KOM, (**F**) ART, (**I**) LT2, and (**L**) L&T. Again, *arrows* indicate how the stimuli change with increasing growth for both the pressure (*blue*) and volume (*red*) overload simulations

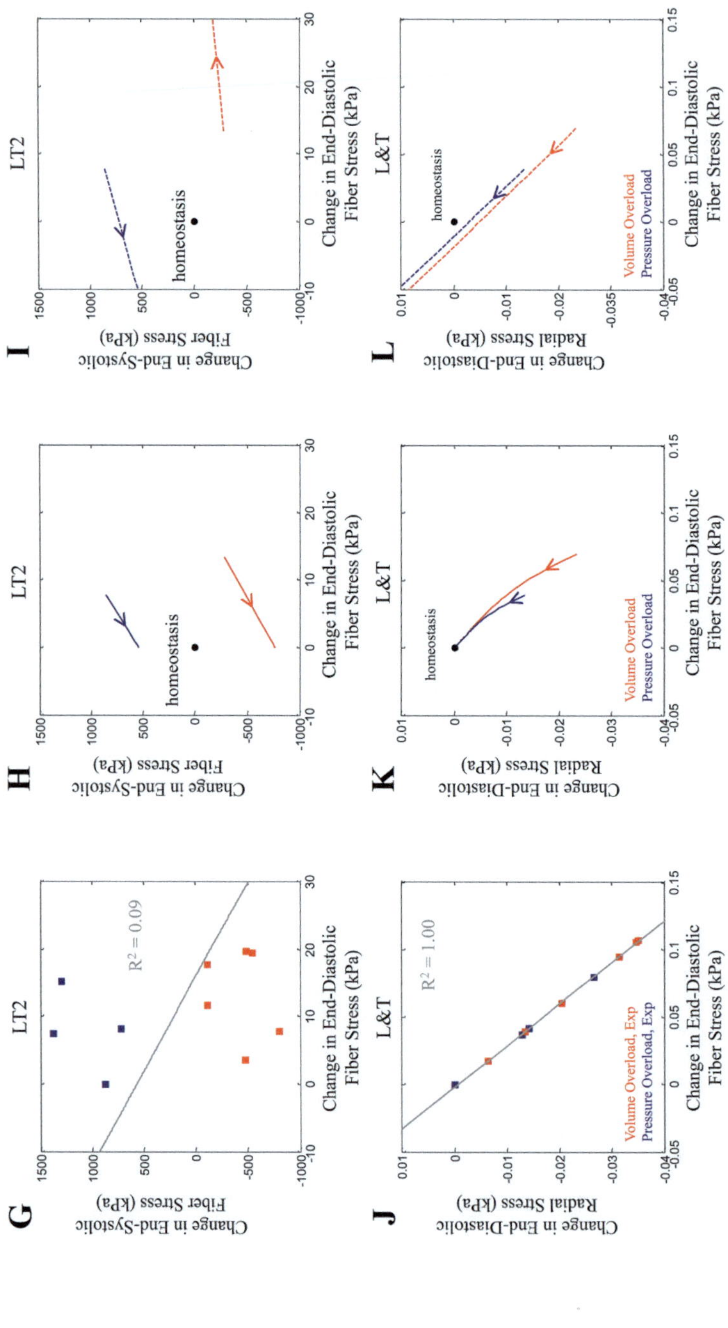

Fig. 5 (*Continued*)

maximum growth allowed. These are two of several approaches that have been employed to prevent growth models from producing experimentally unrealistic magnitudes of growth. One possibility for achieving a stable grown state is if growth normalizes the value of a critical input, returning it to its homeostatic value and thereby preventing additional growth. In the heart, the best-known example of this approach is the (conceptual) model proposed by Grossman et al. [6], where systolic wall stress drives radial growth, and the increase in wall thickness eventually returns wall stress to its homeostatic value.

Our analysis suggests that it is remarkably difficult to devise a growth law that achieves homeostasis following multiple different hemodynamic perturbations (Fig. 5). In Fig. 5B, E, H, and K, we plotted the evolution of the growth stimuli for four of the laws considered here to better visualize how they change as growth proceeds. Only KOM (Fig. 5B) and L&T (Fig. 5K) returned to homeostasis in our prescribed-stretch simulations; these two laws used inputs from two different directions to drive growth in those directions, providing enough degrees of freedom to satisfy two homeostatic conditions simultaneously. For example, in the KOM pressure-overload simulation (blue solid line in Fig. 5B), the increase in prescribed maximum fiber stretch produced growth in the fiber direction, which reduced elastic stretch and returned fiber strain to its homeostatic value; at the same time, elevated cross-fiber strain triggered cross-fiber growth, which reduced cross-fiber strain back to its homeostatic value. (Note that the jagged path is due to oscillation along the sigmoidal stretch-growth curves shown in Figure S1 and could be reduced by re-parameterizing this law for the specific stretch range used here). By contrast, the LT2 law employs two inputs from the same direction, end-diastolic and end-systolic fiber stress. These inputs were nicely decoupled during overload (Fig. 5G), allowing the law to predict different growth trends in response to stretch inputs typical of pressure-overload vs. volume-overload. However, model-predicted growth in the fiber direction (Eq. (4)) reduced diastolic and systolic elastic stretches and stresses proportionally (solid blue line in Fig. 5H), while growth in the radial direction (Eq. (3)) did not affect either input. Thus, under prescribed-stretch boundary conditions, this law was unable to return both stresses to their homeostatic values following perturbation and exhibited runaway growth.

This last example also emphasizes the role of boundary conditions. Many of the original papers describing the growth laws discussed here prescribed diastolic and/or systolic LV pressures typical of control and overload states, then studied how growth evolved in the setting of those applied pressures. In order to represent these boundary conditions in our biaxial simulation, we performed additional simulations in which we cycled between prescribed systolic and diastolic forces rather than between prescribed stretches. Under prescribed-force boundary conditions (Fig. 5I), the LT2 law was again unable to reach homeostasis, but for a different reason: overload altered ED and ES fiber stresses by different fractions of their control values, while subsequent radial growth altered both stresses proportionally. Thus, radial growth could never restore both stresses to their homeostatic values simultaneously. For example, in the pressure-overload simulation, radial thickening reduced both stresses (blue arrow points down and to the right in Fig. 5I), driving ED fiber stress below its setpoint before reaching the ES setpoint (arrow does not pass through the homeostatic point). Interestingly, we note that none of the multi-input growth laws reviewed here returned to homeostasis in prescribed-force simulations. Of course, neither a prescribed-stretch nor a prescribed-force simulation fully replicates the in vivo setting, where hemodynamic adaptations to overload change diastolic and systolic pressures and volumes even as growth proceeds. Furthermore, biologic examples of uncontrolled growth such as progressive dilation leading to heart failure after severe volume overloading suggest that growth may not always return the system to homeostasis.

One final mechanism that might limit growth in response to a mechanical perturbation is if the setpoint gradually resets. In the constrained mixture framework proposed by Humphrey and Rajagopal [76], this occurs as older material is degraded and replaced by new material with a different unstressed reference configuration. In the KUR growth law examined here, the authors achieved something similar by resetting the reference configuration, but this greatly reduced growth. A possible compromise might be a "fading-memory" approach to gradually update the reference configuration with some time lag relative to the current growth time step. One of the models reviewed here (KOM) limited the maximum rate of growth; biologically this might correspond to a maximum rate of new protein synthesis by the myocytes. Other laws reviewed here (GEG and GCG) simply prescribed a maximum level of growth. There has been some speculation that the pericardium could constrain the total amount of LV dilation, but Freeman and LeWinter [77] measured increased pericardial area with heart weight, suggesting that during cardiac hypertrophy the pericardium enlarges as well. At the cellular level, it seems plausible that transport distances could place an upper limit on cell size, but it is currently unclear how or whether transport limits the length, width, or volume of hypertrophying myocytes *in vivo*.

4.3 Limitations and Sources of Error

Many of the limitations of this study arise from the decision to simulate the mechanics of pressure overload and volume overload by applying prescribed stretches to a thin slab of myocardium. All of the growth laws were originally combined with a model of the entire left ventricle (ranging in complexity from compartmental to full finite-element models, Table 1) with a range of boundary conditions that were comparable to our simulations in some cases but not others. The most important difference is that many of the original papers prescribed LV diastolic and/or systolic pressures typical of control and overload states and assumed those pressures did not evolve with growth [9, 11, 12]. The closest analogy in our simple biaxial model would be to cycle between prescribed end-systolic and end-diastolic forces rather than between prescribed diastolic and systolic stretches (Fig. 1C). When we performed these simulations (Fig. 5), all simulations except KUR and GCG showed runaway growth; the reasons for this are discussed in more detail under Stopping or Limiting Growth. In some cases, this result may be due to the fact that our simplified biaxial simulation distorted the feedback of growth on the stimuli that incite it. For example, in a simple cylinder model with circumferentially aligned fibers, circumferential (fiber) growth will increase fiber stresses while radial growth (wall thickening) will decrease them. A prescribed-force biaxial simulation captures the second effect, but not the first. On the other hand, most of the original papers that prescribed pressure boundary conditions in geometrically appropriate LV models also predicted runaway growth unless they stopped their simulations at preset growth levels [11, 12] or employed changing reference configurations [10, 11] to prevent it. In reality, the LV does not operate independently but is connected to the circulation, which reacts and remodels when subjected to overload. Thus, even whole-ventricle models that prescribe LV pressure (stress or force) or stroke volume (strain) do not fully represent the physiology of *in vivo* overload. An alternative approach, utilized by Arts et al. [13], is to couple the LV to an adaptive model of the circulation.

When prescribing stretch profiles to simulate baseline and overload mechanics, we also chose to apply equal stretches in the circumferential and longitudinal directions, based on literature showing that midwall circumferential and longitudinal strains measured during diastolic filling or systolic ejection are approximately equal [16–21]. However, the exact ratio of circumferential to longitudinal stretch obviously varies somewhat with transmural depth, location within the LV, and with loading. In heart models, this ratio also depends on the overall choice of LV geometry. Thus, we repeated our simulations with a 2:1 ratio of prescribed circumferential to longitudinal stretch and observed no differences in the predictions of most growth laws and only small differences ($\leq 3\%$ change in fiber-radial growth ratios) for KOM and L&T. In no case did the qualitative trend predicted by any law change. These results suggest that the behavior of the growth laws reported here is relatively insensitive to the exact ratio of circumferential to longitudinal stretch prescribed in our simulations. In addition, studies that measured both short-axis and long-axis dimensions in the canine LV immediately following volume [31] or pressure [35] overload found no change in eccentricity (an index of shape), which suggests that overload does not induce major changes in the ratio of circumferential to longitudinal strains *in vivo*. Another limitation of our simulation is that it cannot capture the regional heterogeneity that occurs *in vivo* and can be significant in conditions such as myocardial infarction and dilated heart failure; however, in this respect our study differs less from the original papers proposing these growth laws, most of which assumed axisymmetric geometries and a single growth law and setpoint for the entire LV.

Another potential source of error in our study was estimating the magnitude of end-diastolic stretch relative to the unloaded configuration. We estimated this stretch through extrapolation of segment length data from Fomovsky et al. [25] to 1 mmHg, including in our analysis only data from the six animals where linear fits provided estimates with low variability. For comparison, we also estimated this value using an independent method based on published studies of sarcomere lengths. Rodriguez et al. [78] developed a method to reconstruct sarcomere lengths *in vivo*. They reported the midwall end-diastolic sarcomere length for a single dog as 2.28 μm; comparing this measurement to reported sarcomere length values in unloaded arrested dog hearts of 1.80–1.83 μm by Guccione et al. [79] and 1.9 μm by Spotnitz et al. [80] yields an estimated end-diastolic fiber stretch between 1.20–1.27, a bit lower than our mean value but within the range of the individual values we computed in 6 animals (Figure S2). While we sought to impose physiologically realistic stretches in these experiments, we recognize that they may differ substantially from the stretches computed in the various computational models employed by authors of the original growth laws. We partially accounted for these differences by identifying new setpoints for each law given the control stretches employed here, but could likely obtain a better quantitative match between the predictions of many of these laws and the corresponding myocyte size data if we re-parameterized each law for the specific simulations we performed.

Finally, when considering experimental measurements of growth, we focused on comparisons to myocyte dimensions since growth is a local phenomenon occurring on the myocyte level, even though some of the growth laws were originally validated against changes in LV dimensions. There appears to be a good basis for expecting growth in the fiber direction at the midwall and changes in LV radius or dimension to be congruent, since myocardial fibers at the midwall have been shown to run circumferentially [15]. However, it is less clear how to relate reported changes in myocyte width measured in isolated cells (where the relationship to *in vivo* orientation is lost) or cross-sectional area in histologic sections (includes both radial and cross-fiber components) to changes in LV wall thickness.

Acknowledgements This study was supported by the Hartwell Foundation (CMW) and the National Institutes of Health (U01 HL127654, JWH).

References

1. Anversa, P., Ricci, R., Olivetti, G.: Quantitative structural analysis of the myocardium during physiologic growth and induced cardiac hypertrophy: a review. J. Am. Coll. Cardiol. **7**, 1140–1149 (1986)

2. Cantor, E.J.F., Babick, A.P., Vasanji, Z., Dhalla, N.S., Netticadan, T.: A comparative serial echocardiographic analysis of cardiac structure and function in rats subjected to pressure or volume overload. J. Mol. Cell. Cardiol. **38**, 777–786 (2005)
3. Frey, N., Olson, E.N.: Cardiac hypertrophy: the good, the bad, and the ugly. Annu. Rev. Physiol. **65**, 45–79 (2003)
4. Heineke, J., Molkentin, J.D.: Regulation of cardiac hypertrophy by intracellular signalling pathways. Nat. Rev. Mol. Cell Biol. **7**, 589–600 (2006)
5. Rodriguez, E.K., Hoger, A., McCulloch, A.D.: Stress-dependent finite growth in soft elastic tissues. J. Biomech. **27**, 455–467 (1994)
6. Grossman, W., Jones, D., McLaurin, L.P.: Wall stress and patterns of hypertrophy in the human left ventricle. J. Clin. Invest. **56**, 56–64 (1975)
7. Emery, J.L., Omens, J.H.: Mechanical regulation of myocardial growth during volume-overload hypertrophy in the rat. Am. J. Physiol. **273**, 1198–1204 (1997)
8. Holmes, J.W.: Candidate mechanical stimuli for hypertrophy during volume overload. J. Appl. Physiol. **97**, 1453 (2004)
9. Lin, I.E., Taber, L.A.: A model for stress-induced growth in the developing heart. J. Biomech. Eng. **117**, 343–349 (1995)
10. Taber, L.A.: Biomechanical growth laws for muscle tissue. J. Theor. Biol. **193**, 201–213 (1998)
11. Kroon, W., Delhaas, T., Arts, T., Bovendeerd, P.: Computational modeling of volumetric soft tissue growth: application to the cardiac left ventricle. Biomech. Model. Mechanobiol. **8**, 301–309 (2009)
12. Göktepe, S., Abilez, O.J., Parker, K.K., Kuhl, E.: A multiscale model for eccentric and concentric cardiac growth through sarcomerogenesis. J. Theor. Biol. **265**, 433–442 (2010)
13. Arts, T., Delhaas, T., Bovendeerd, P., Verbeek, X., Prinzen, F.W.: Adaptation to mechanical load determines shape and properties of heart and circulation: the CircAdapt model. Am. J. Physiol., Heart Circ. Physiol. **288**, 1943–1954 (2005)
14. Kerckhoffs, R.C.P., Omens, J.H., McCulloch, A.D.: A single strain-based growth law predicts concentric and eccentric cardiac growth during pressure and volume overload. Mech. Res. Commun. **42**, 40–50 (2012)
15. Streeter, D.D., Spotnitz, H.M., Patel, D.P., Ross, J., Sonnenblick, E.H.: Fiber orientation in the canine left ventricle during diastole and systole. Circ. Res. **24**, 339–347 (1969)
16. Fomovsky, G.M., Holmes, J.W.: Evolution of scar structure, mechanics, and ventricular function after myocardial infarction in the rat. Am. J. Physiol., Heart Circ. Physiol. **298**, 221–228 (2010)
17. Villarreal, F.J., Waldman, L.K., Lew, W.Y.: Technique for measuring regional two-dimensional finite strains in canine left ventricle. Circ. Res. **62**, 711–721 (1988)
18. Lew, W.Y., LeWinter, M.M.: Regional comparison of midwall segment and area shortening in the canine left ventricle. Circ. Res. **58**, 678–691 (1986)
19. Omens, J.H., Farr, D.D., McCulloch, A.D., Waldman, L.K.: Comparison of two techniques for measuring two-dimensional strain in rat left ventricles. Am. J. Physiol. **271**, 1256–1261 (1996)
20. Omens, J.H., MacKenna, D.A., McCulloch, A.D.: Measurement of strain and analysis of stress in resting rat left ventricular myocardium. J. Biomech. **26**, 665–676 (1993)
21. Tobita, K., Schroder, E.A., Tinney, J.P., Garrison, J.B., Keller, B.B.: Regional passive ventricular stress-strain relations during development of altered loads in chick embryo. Am. J. Physiol., Heart Circ. Physiol. **282**, 2386–2396 (2002)
22. Glower, D.D., Spratt, J.A., Snow, N.D., Kabas, J.S., Davis, J.W., Olsen, C.O., Tyson, G.S., Sabiston, D.C., Rankin, J.S.: Linearity of the Frank-Starling relationship in the intact heart: the concept of preload recruitable stroke work. Circulation **71**, 994–1009 (1985)
23. Sodums, M.T., Badke, F.R., Starling, M.R., Little, W.C., O'Rourke, R.A.: Evaluation of left ventricular contractile performance utilizing end-systolic pressure-volume relationships in conscious dogs. Circ. Res. **54**, 731–739 (1984)
24. Rankin, J.S., McHale, P.A., Arentzen, C.E., Ling, D., Greenfield, J.C., Anderson, R.W.: The three-dimensional dynamic geometry of the left ventricle in the conscious dog. Circ. Res. **39**, 304–313 (1976)
25. Fomovsky, G.M., Clark, S.A., Parker, K.M., Ailawadi, G., Holmes, J.W.: Anisotropic reinforcement of acute anteroapical infarcts improves pump function. Circ. Heart Fail. **5**, 515–522 (2012)
26. Omens, J.H., Fung, Y.C.: Residual strain in rat left ventricle. Circ. Res. **66**, 37–45 (1990)
27. Rodriguez, E.K., Omens, J.H., Waldman, L.K., McCulloch, A.D.: Effect of residual stress on transmural sarcomere length distributions in rat left ventricle. Am. J. Physiol. **264**, 1048–1056 (1993)
28. Alyono, D., Ring, W.S., Anderson, M.R., Anderson, R.W.: Left ventricular adaptation to volume overload from large aortocaval fistula. Surgery **96**, 360–367 (1984)
29. Fujisawa, A., Sasayama, S., Takahashi, M., Nakamura, M., Ohyagi, A., Lee, J.D., Yui, Y., Kawai, C.: Enhancement of left ventricular contractility after opening of an arteriovenous fistula in dogs. Cardiovasc. Res. **18**, 51–59 (1984)

30. Nakano, K., Swindle, M.M., Spinale, F., Ishihara, K., Kanazawa, S., Smith, A., Biederman, R.W., Clamp, L., Hamada, Y., Zile, M.R.: Depressed contractile function due to canine mitral regurgitation improves after correction of the volume overload. J. Clin. Invest. **87**, 2077–2086 (1991)
31. Katayama, K., Tajimi, T., Guth, B.D., Matsuzaki, M., Lee, J.-D., Seitelberger, R., Peterson, K.L.: Early diastolic filling dynamics during experimental mitral regurgitation in the conscious dog. Circulation **78**, 390–400 (1988)
32. Kleaveland, J.P., Kussmaul, W.G., Vinciguerra, T., Diters, R., Carabello, B.A.: Volume overload hypertrophy in a closed-chest model of mitral regurgitation. Am. J. Physiol. **254**, 1034–1041 (1988)
33. Carabello, B.A., Nakano, K., Corin, W., Biederman, R., Spann, J.F. Jr.: Left ventricular function in experimental volume overload hypertrophy. Am. J. Physiol. **256**, 974–981 (1989)
34. Sasayama, S., Ross, J., Franklin, D., Bloor, C.M., Bishop, S., Dilley, R.B.: Adaptations of the left ventricle to chronic pressure overload. Circ. Res. **38**, 172–178 (1976)
35. Crozatier, B., Caillet, D., Bical, O.: Left ventricular adaptation to sustained pressure overload in the conscious dog. Circ. Res. **54**, 21–29 (1984)
36. Su, J.B., Crozatier, B.: Preload-induced curvilinearity of left ventricular end-systolic pressure-volume relations. Effects on derived indexes in closed-chest dogs. Circulation **79**, 431–440 (1989)
37. Karunanithi, M.K., Feneley, M.P.: Single-beat determination of preload recruitable stroke work relationship: derivation and evaluation in conscious dogs. J. Am. Coll. Cardiol. **35**, 502–513 (2000)
38. Zile, M.R., Tomita, M., Ishihara, K., Nakano, K., Lindroth, J., Spinale, F., Swindle, M., Carabello, B.A.: Changes in diastolic function during development and correction of chronic LV volume overload produced by mitral regurgitation. Circulation **87**, 1378–1388 (1993)
39. Sabri, A., Rafiq, K., Seqqat, R., Kolpakov, M.A., Dillon, R., Dell'italia, L.J.: Sympathetic activation causes focal adhesion signaling alteration in early compensated volume overload attributable to isolated mitral regurgitation in the dog. Circ. Res. **102**, 1127–1136 (2008)
40. Lee, J.D., Sasayama, S., Kihara, Y., Ohyagi, A., Fujisawa, A., Yui, Y., Kawai, C.: Adaptations of the left ventricle to chronic volume overload induced by mitral regurgitation in conscious dogs. Heart Vessels **1**, 9–15 (1985)
41. Newman, W.H., Webb, J.G.: Adaptation of left ventricle to chronic pressure overload: response to inotropic drugs. Am. J. Physiol. **238**, 134–143 (1980)
42. Wojciechowski, P., Juric, D., Louis, X.L., Thandapilly, S.J., Yu, L., Taylor, C., Netticadan, T.: Resveratrol arrests and regresses the development of pressure overload- but not volume overload-induced cardiac hypertrophy in rats. J. Nutr. **140**, 962–968 (2010)
43. Pawlush, D.G., Moore, R.L., Musch, T.I., Davidson, W.R.: Echocardiographic evaluation of size, function, and mass of normal and hypertrophied rat ventricles. J. Appl. Physiol. **74**, 2598–2605 (1993)
44. Guido, M.C., De Carvalho Frimm, C., Koike, M.K., Cordeiro, F.F., Moretti, A.I.S., Godoy, L.C.: Low coronary driving pressure is associated with subendocardial remodelling and left ventricular dysfunction in aortocaval fistula. Clin. Exp. Pharmacol. Physiol. **34**, 1165–1172 (2007)
45. Melenovsky, V., Benes, J., Skaroupkova, P., Sedmera, D., Strnad, H., Kolar, M., Vlcek, C., Petrak, J., Benes, J., Papousek, F., Oliyarnyk, O., Kazdova, L., Cervenka, L.: Metabolic characterization of volume overload heart failure due to aorto-caval fistula in rats. Mol. Cell. Biochem. **354**, 83–96 (2011)
46. Ryan, T.D., Rothstein, E.C., Aban, I., Tallaj, J.A., Husain, A., Lucchesi, P.A., Dell'Italia, L.J.: Left ventricular eccentric remodeling and matrix loss are mediated by bradykinin and precede cardiomyocyte elongation in rats with volume overload. J. Am. Coll. Cardiol. **49**, 811–821 (2007)
47. Chen, Y.W., Pat, B., Gladden, J.D., Zheng, J., Powell, P., Wei, C.C., Cui, X., Husain, A., Dell'Italia, L.J.: Dynamic molecular and histopathological changes in the extracellular matrix and inflammation in the transition to heart failure in isolated volume overload. Am. J. Physiol., Heart Circ. Physiol. **300**, 2251–2260 (2011)
48. Wilson, K., Guggilam, A., West, T.A., Zhang, X., Trask, A.J., Cismowski, M.J., de Tombe, P., Sadayappan, S., Lucchesi, P.A.: Effects of a myofilament calcium sensitizer on left ventricular systolic and diastolic function in rats with volume overload heart failure. Am. J. Physiol., Heart Circ. Physiol. **307**, 1605–1617 (2014)
49. Gladden, J.D., Zelickson, B.R., Guichard, J.L., Ahmed, M.I., Yancey, D.M., Ballinger, S., Shanmugam, M., Babu, G.J., Johnson, M.S., Darley-Usmar, V., Dell'Italia, L.J.: Xanthine oxidase inhibition preserves left ventricular systolic but not diastolic function in cardiac volume overload. Am. J. Physiol., Heart Circ. Physiol. **305**, 1440–1450 (2013)
50. Wei, C.C., Chen, Y., Powell, L.C., Zheng, J., Shi, K., Bradley, W.E., Powell, P.C., Ahmad, S., Ferrario, C.M., Dell'Italia, L.J.: Cardiac kallikrein-kinin system is upregulated in chronic volume overload and mediates an inflammatory induced collagen loss. PLoS ONE **7**, 1–14 (2012)
51. Miyamoto, M.I., del Monte, F., Schmidt, U., DiSalvo, T.S., Kang, Z.B., Matsui, T., Guerrero, J.L., Gwathmey, J.K., Rosenzweig, A., Hajjar, R.J.: Adenoviral gene transfer of SERCA2a improves left-ventricular function in aortic-banded rats in transition to heart failure. Proc. Natl. Acad. Sci. USA **97**, 793–798 (2000)

52. Litwin, S.E., Katz, S.E., Weinberg, E.O., Lorell, B.H., Aurigemma, G.P., Douglas, P.S.: Serial echocardiographic-Doppler assessment of left ventricular geometry and function in rats with pressure-overload hypertrophy: chronic angiotensin-converting enzyme inhibition attenuates the transition to heart failure. Circulation **91**, 2642–2654 (1995)
53. Kobayashi, S., Yano, M., Kohno, M., Obayashi, M., Hisamatsu, Y., Ryoke, T., Ohkusa, T., Yamakawa, K., Matsuzaki, M.: Influence of aortic impedance on the development of pressure-overload left ventricular hypertrophy in rats. Circulation **94**, 3362–3368 (1996)
54. Hao, J., Kim, C.H., Ha, T.S., Ahn, H.Y.: Epigallocatechin-3 gallate prevents cardiac hypertrophy induced by pressure overload in rats. J. Vet. Sci. **8**, 121 (2007)
55. Doenst, T., Pytel, G., Schrepper, A., Amorim, P., Färber, G., Shingu, Y., Mohr, F.W., Schwarzer, M.: Decreased rates of substrate oxidation ex vivo predict the onset of heart failure and contractile dysfunction in rats with pressure overload. Cardiovasc. Res. **86**, 461–470 (2010)
56. Shyu, K.G., Liou, J.Y., Wang, B.W., Fang, W.J., Chang, H.: Carvedilol prevents cardiac hypertrophy and overexpression of hypoxia-inducible factor-1alpha and vascular endothelial growth factor in pressure-overloaded rat heart. J. Biomed. Sci. **12**, 409–420 (2005)
57. Derumeaux, G., Mulder, P., Richard, V., Chagraoui, A., Nafeh, C., Bauer, F., Henry, J.P., Thuillez, C.: Tissue Doppler imaging differentiates physiological from pathological pressure-overload left ventricular hypertrophy in rats. Circulation **105**, 1602–1608 (2002)
58. Hwang, H.S., Bleske, B.E., Ghannam, M.M.J., Converso, K., Russell, M.W., Hunter, J.C., Boluyt, M.O.: Effects of hawthorn on cardiac remodeling and left ventricular dysfunction after 1 month of pressure overload-induced cardiac hypertrophy in rats. Cardiovasc. Drugs Ther. **22**, 19–28 (2008)
59. Ma, Y., Chen, B., Liu, D., Yang, Y., Xiong, Z., Zeng, J., Dong, Y.: MG132 treatment attenuates cardiac remodeling and dysfunction following aortic banding in rats via the NF-κB/TGFβ1 pathway. Biochem. Pharmacol. **81**, 1228–1236 (2011)
60. Xu, R., Lin, F., Zhang, S., Chen, X., Hu, S., Zheng, Z.: Signal pathways involved in reverse remodeling of the hypertrophic rat heart after pressure unloading. Int. J. Cardiol. **143**, 414–423 (2010)
61. Li, J., Li, P., Feng, X., Li, Z., Hou, R., Han, C., Zhang, Y.: Effects of Losartan on pressure overload-induced cardiac gene expression profiling in rats. Clin. Exp. Pharmacol. Physiol. **30**(11), 827–832 (2003)
62. Kume, O., Takahashi, N., Wakisaka, O., Nagano-Torigoe, Y., Teshima, Y., Nakagawa, M., Yufu, K., Hara, M., Saikawa, T., Yoshimatsu, H.: Pioglitazone attenuates inflammatory atrial fibrosis and vulnerability to atrial fibrillation induced by pressure overload in rats. Heart Rhythm **8**, 278–285 (2011)
63. Phrommintikul, A., Tran, L., Kompa, A., Wang, B., Adrahtas, A., Cantwell, D., Kelly, D.J., Krum, H.: Effects of a Rho kinase inhibitor on pressure overload induced cardiac hypertrophy and associated diastolic dysfunction. Am. J. Physiol., Heart Circ. Physiol. **294**, 1804–1814 (2008)
64. Spinale, F.G., Ishihra, K., Zile, M., DeFryte, G., Crawford, F.A., Carabello, B.A.: Structural basis for changes in left ventricular function and geometry because of chronic mitral regurgitation and after correction of volume overload. J. Thorac. Cardiovasc. Surg. **106**, 1147–1157 (1993)
65. Tsutsui, H., Urabe, Y., Mann, D.L., Tagawa, H., Carabello, B.A., Cooper, G., Zile, M.R.: Effects of chronic mitral regurgitation on diastolic function in isolated cardiocytes. Circ. Res. **72**, 1110–1123 (1993)
66. Gerdes, A.M., Campbell, S.E., Hilbelink, D.R.: Structural remodeling of cardiac myocytes in rats with arteriovenous fistulas. Lab. Invest. **59**, 857–861 (1988)
67. Liu, Z., Hilbelink, D.R., Crockett, W.B., Gerdes, A.M.: Regional changes in hemodynamics and cardiac myocyte size in rats with aortocaval fistulas. 1. Developing and established hypertrophy. Circ. Res. **69**, 52–58 (1991)
68. Gerdes, A.M., Clark, L.C., Capasso, J.M.: Regression of cardiac hypertrophy after closing an aortocaval fistula in rats. Am. J. Physiol. **268**, 2345–2351 (1995)
69. Zierhut, W., Zimmer, H.G., Gerdes, A.M.: Effect of angiotensin converting enzyme inhibition on pressure-induced left ventricular hypertrophy in rats. Circ. Res. **69**, 609–617 (1991)
70. Wang, X., Li, F., Gerdes, A.M.: Chronic pressure overload cardiac hypertrophy and failure in guinea pigs: I. Regional hemodynamics and myocyte remodeling. J. Mol. Cell. Cardiol. **31**, 307–317 (1999)
71. Campbell, S.E., Rakusan, K., Gerdes, A.M.: Change in cardiac myocyte size distribution in aortic-constricted neonatal rats. Basic Res. Cardiol. **84**, 247–258 (1989)
72. Korecky, B., Rakusan, K.: Normal and hypertrophic growth of the rat heart: changes in cell dimensions and number. Am. J. Physiol. **234**, 123–128 (1978)
73. Campbell, S.E., Korecky, B., Rakusan, K.: Remodeling of myocyte dimensions in hypertrophic and atrophic rat hearts. Circ. Res. **68**, 984–996 (1991)
74. Degens, H., Brouwer, K.F.J., Gilde, A.J., Lindhout, M., Willemsen, P.H.M., Janssen, B.J., van der Vusse, G.J., van Bilsen, M.: Cardiac fatty acid metabolism is preserved in the compensated hypertrophic rat heart. Basic Res. Cardiol. **101**, 17–26 (2006)

75. Mohammed, S.F., Storlie, J.R., Oehler, E.A., Bowen, L.A., Korinek, J., Lam, C.S.P., Simari, R.D., Burnett, J.C., Redfield, M.M.: Variable phenotype in murine transverse aortic constriction. Cardiovasc. Pathol. **21**, 188–198 (2012)
76. Humphrey, J.D., Rajagopal, K.R.: A constrained mixture model for growth and remodeling of soft tissues. Math. Models Methods Appl. Sci. **12**, 407–430 (2002)
77. Freeman, G.L., LeWinter, M.M.: Pericardial adaptations during chronic cardiac dilation in dogs. Circ. Res. **54**, 294–300 (1984)
78. Rodriguez, E.K., Hunter, W.C., Royce, M.J., Leppo, M.K., Douglas a, S., Weisman, H.F.: A method to reconstruct myocardial sarcomere lengths and orientations at transmural sites in beating canine hearts. Am. J. Physiol. **263**, 293–306 (1992)
79. Guccione, J.M., O'Dell, W.G., McCulloch, A.D., Hunter, W.C.: Anterior and posterior left ventricular sarcomere lengths behave similarly during ejection. Am. J. Physiol. **272**, 469–477 (1997)
80. Spotnitz, H.M., Sonnenblick, E.H., Spiro, D.: Relation of ultrastructure to function in the intact heart: sarcomere structure relative to pressure volume curves of intact left ventricles of dog and cat. Circ. Res. **18**, 49–66 (1966)

J Elast (2017) 129:283–305
DOI 10.1007/s10659-017-9640-7

Modelling Cardiac Tissue Growth and Remodelling

Vicky Y. Wang[1] **· Jagir R. Hussan**[1] **· Hashem Yousefi**[1] **·**
Chris P. Bradley[1] **· Peter J. Hunter**[1] **· Martyn P. Nash**[1,2]

Received: 10 September 2016 / Published online: 2 June 2017

Abstract Cardiac growth and remodelling are key phenomena influencing the physiological performance of the heart and progression of various cardiac diseases. Our basic knowledge of the characteristics of cardiac growth and remodelling have benefitted from the study of the corresponding changes in tissue structure. Computational modelling is capable of integrating multi-scale biological and physiological information of the heart to better delineate the mechanisms underpinning normal physiological and abnormal pathological cardiac processes. In this article, we summarise recent state-of-the-art modelling studies of cardiac growth and remodelling in health and disease, demonstrate the use of finite element modelling approach in simulating growth of an embryonic linear heart tube and towards understanding the roles of structural remodelling in heart failure.

Keywords Myocardial constitutive properties · Cardiac development · Cardiac growth · Structural remodelling

Mathematics Subject Classification (2010) 78M10

1 Introduction

Cardiac tissue, like all biological tissues, continually adapts to its changing mechanical and biochemical environment. At the cellular level, this adaptation can occur via changes in the gene transcription and translation that controls the proteins underpinning cellular and tissue function, and spatially, it can occur at both a subcellular level (leading to changes in the size or shape of cells, along with changes in their contractile function) and tissue level (e.g.,

✉ V.Y. Wang
vicky.wang@auckland.ac.nz

[1] Auckland Bioengineering Institute, University of Auckland, Level 6, UniServices House, 70 Symonds Street, Auckland 1142, New Zealand

[2] Department of Engineering Science, University of Auckland, Level 3, UniServices House, 70 Symonds Street, Auckland 1142, New Zealand

within the extracellular matrix). For a growing embryo, these changes are largely driven by a genetic programme, but with some dependence on environmental factors.

The goal of this paper is to review the physiological background and previous modelling work on cardiac growth, both in relation to growth of the embryonic heart and to hypertrophy in the remodelling adult heart (typically associated with heart failure), and also to discuss some of the mathematical and computational challenges in this scientifically and clinically important field. We begin by reviewing the background physiology of the heart and its pathologies, and the kinematics of embryological heart development (Sect. 2), and the background continuum mechanics based mathematical framework needed to analyse cardiac function (Sect. 3). We then examine first embryological growth (Sect. 4) and then the impact of tissue remodelling (Sect. 5) on myocardial mechanical behaviour during heart failure.

An appropriate mathematical framework for understanding and modelling cardiac tissue growth should include large deformation elasticity theory. Since cardiac tissue, like all biological tissues, is highly anisotropic, the description of material properties by constitutive relations (both stress-strain relations and growth 'laws') should be based on material axes that are defined by the underlying tissue microstructure. We discuss a number of these constitutive relations and how, in some cases, the parameters can be linked to observations of tissue structure. Note that we do not deal with dissipative mechanisms—just hyperelasticity and growth (for a discussion on including dissipative mechanisms into the continuum framework, see [1]). The tissue is also heterogeneous, in terms of both the orientation of material axes and the material properties—reflecting spatial variation in the underlying tissue microstructure. Finite element methods, capable of describing the complex three-dimensional anatomy of the heart, provide the necessary framework for describing the spatially and temporally varying continuum fields relevant to cardiac mechanics.

2 Physiological Background

The heart is the first working organ during the development of the embryo (embryogenesis). It is a four-chamber muscular organ that acts as a mechanical pump to actively distribute sufficient oxygenated blood throughout the body, to deliver deoxygenated blood from the body to the lungs for oxygenation, and to passively receive oxygenated blood returned from the lungs. It is perfused by the coronary blood vessels embedded within the myocardium. This highly coordinated mechanical pump originates from a single tube. Embryonic ventricular myocardium initially grows primarily by hyperplasia (cell replication with relatively constant cardiomyocyte volume). Then, at least in sheep, mice and rats, cardiomyocytes exit the cell cycle and form a second nucleus (binucleation). Mononucleated cells proliferate but do not enlarge (hypertrophy) [2, 3]. Binucleated cardiomyocytes, on the other hand, cannot divide but can hypertrophy [4]. The physiological environment in which embryogenesis occurs, which can be influenced by genetic mutations, placental size and other factors, affects the proportion of binucleated cardiomyocytes and hence the maturation process.

Extensive reviews of growth processes are available in [5] and [6]. One way of characterising these is to divide them into volume growth (e.g., nutrient driven tumour growth [7]), area growth (e.g., the growth of cortical folds in the brain [8]) or the fibre growth typical of organogenesis, which is the only type of growth discussed further here.

Cardiac development involves a multitude of events that occur at the molecular, cellular and tissue scales. Embryonic heart development begins with the formation of a linear heart tube (HT), which undergoes a series of morphological transformations to form the

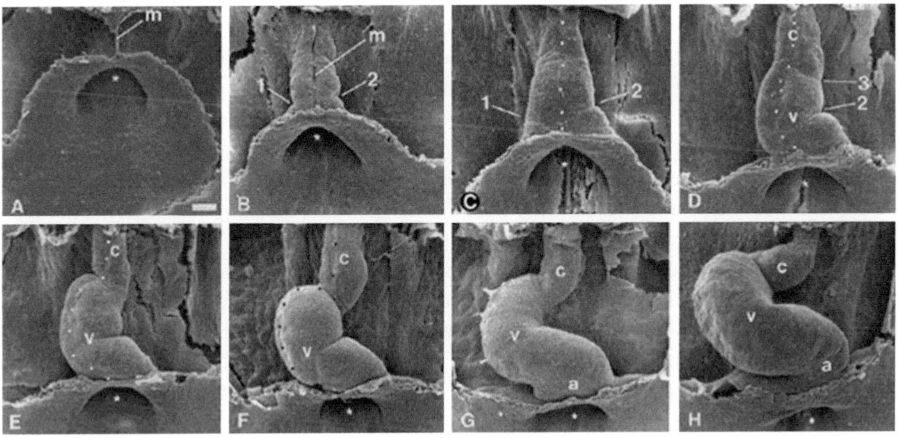

Fig. 1 C-looping in chicken embryo. **A**: the cardiac crescent. **B–G**: The C-looping process where the linear heart tube bends dextrally. **H**: The formation of the S-bend. Reproduced with permission from Fig. 1 of Manner [10]

fully developed heart. Morphologically, the cardiac development process can be categorised into the following main stages [9–11]: pre-looping [12–15]; rightward/dextral looping (C-shaped looping) [10, 16]; immature S-looping [13, 17]; mature S-looping [18]; and finally cardiac septation and valve formation [13, 17]. The biophysical and biological mechanisms proposed for growth during C-looping include growth induced buckling [19], cell proliferation [20], cardiac jelly swelling [21–23], active cell shape change via actin polymerisation [16, 24, 25], and cellular hypertrophy [13]. In this paper, we focus on the dextral looping phase in which the straight HT undergoes a combination of ventral bending and dextral torsion (see Fig. 1).

Determining the causal relationships between cardiac development, maturation and remodelling processes is clinically relevant. Heart failure is a leading cause of death throughout the world. It is a remarkably heterogeneous clinical syndrome and can appear in different forms depending on: (i) whether it is congenital or acquired; (ii) the cardiac phase during which failure occurs (e.g., diastolic heart failure versus systolic heart failure); and (iii) cardiac performance (e.g., heart failure with preserved ejection fraction versus heart failure with reduced ejection fraction). For congenital heart disease, impaired development of the heart during embryogenesis is the sole causal factor [26, 27]. Determining factors that impact cardiac development in ways that predispose to disease is an active area of research. Pathological cardiac development directly influences adult cardiac function and predisposes the heart to specific pathologies. For instance, as the number of binucleated cells within the myocardium increases, the generative potential of the myocardium decreases. On the other hand, if the population density of cardiac myocytes is reduced, each remaining myocyte increases in size to maintain adequate muscle volume needed to carry out the pumping function. This affects, for example, the final architecture of the coronary tree since in the mature myocardium there is a one-to-one capillary to myocyte ratio [28]. Therefore, fewer large cells mean fewer capillaries and thus the myocardium could be potentially more vulnerable to ischaemic damage [29]. Better understanding of the development of the normal heart can help to delineate the mechanisms that dysregulate this process.

For many forms of acquired cardiovascular disease seen in adult hearts (e.g., hypertrophic cardiomyopathy, dilated cardiomyopathy, dysynchronous heart failure, and myocardial infarction), abnormal structural remodelling of the left ventricle (LV), which involves altered myocyte dimensions (i.e., maladaptive cardiac hypertrophy) and/or microstructural constituents (i.e., myocardial fibrosis), has been identified as one of the key biomarkers [30, 31]. For example, ventricular hypertrophy in the form of concentric remodelling of the myocardium is often seen in patients with diastolic heart failure (e.g., hypertrophic cardiomyopathy). Cardiac hypertrophy generally results from compensatory remodelling [31] required to normalise elevated myocardial wall stress caused by pressure overload (i.e., hypertension), but such hearts eventually become incapable of working efficiently. This type of remodelling is typically characterised by a normal or increased LV wall mass, normal LV wall volume (with some exceptions), increased LV wall thickness, larger wall-thickness-to-radius ratio, elevated LV end-diastolic pressure [32–34], and normal contractile function. At the microscopic level, the gross increase of wall thickness is caused by increased myocyte diameter, parallel addition of myocytes, and elevated content of collagen [35, 36], which is an important constituent of the myocardial extracellular matrix [31, 37, 38]. In terms of functional remodelling, stiffening (an upward- and leftward-shift) of the diastolic pressure-volume relation (or force-length relation) has been observed at multiple scales, including at the chamber, trabecula and single myocyte scales, in both human and animal heart failure models [33, 35, 39].

On the other hand, patients with systolic heart failure due to dilated cardiomyopathy or myocardial infarction present with so-called "eccentric" remodelling [40] and global or regional LV dilatation with relatively unchanged or reduced wall thickness, and hence a smaller wall-thickness-to-radius ratio. The eccentric remodelling process implicates serial addition of myocytes [31], elongation of myocytes [41], replacement of myocytes with collagen [42], degradation of interstitial collagen [43] leading to realignment of myocytes [44], as well as reduced cross-linking of the myosin and actin proteins. Reducing the overlap between the thick (myosin) and thin (actin) filaments leads to compromised myofibril shortening [45], hence impaired contractile function. The diastolic properties also appear to be altered with a compliant chamber as indicated by a downward shift of the diastolic pressure-volume relation [35].

The precise trigger for the pathological hypertrophy and remodelling is not yet fully understood, but it can be initiated by changes in haemodynamic loading conditions [46], gene mutations [47], myocardial injury [48], and many other factors, which all result in global and/or regional LV shape changes, and ultimately functional changes. The main consequences of pathological hypertrophy and remodelling include changes in not only the structure (macro- and microscopic) of the LV, but also the constitutive properties of the myocardium and hence myocardial wall stress. This often worsens heart failure by elevating the mechanical energy expenditure of the heart [44]. *In vitro* experimental studies have shown that elevated myocardial stress and strain play essential roles by inducing gene mutation, growth-promoting factors and protein synthesis [49, 50]. Investigating these relationships under *in vivo* condition has proven to be difficult because direct *in vivo* measurement of myocardial stress is not feasible. Mathematical modelling tools offer an alternative means to examine the link between myocardial stress and hypertrophic growth and remodelling at the macroscopic level.

3 Mathematical Modelling of the Heart

We start by briefly describing myocardial tissue microstructure and how this is used to establish material coordinates relative to which the stress-strain constitutive relations and the material growth laws can be defined. We then describe the strain tensors, stress tensors and

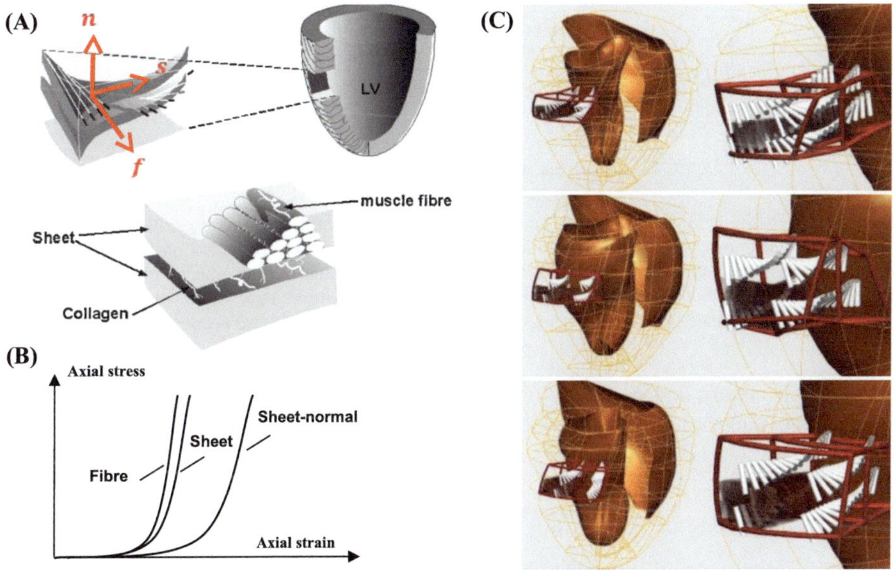

Fig. 2 The definition of a material coordinate system based on tissue microstructure. (**A**) The structure of myocardial tissue illustrating the fibre direction (f), the sheet direction (s) and the sheet-normal direction (n). (**B**) The highly nonlinear stress-strain characteristics of the passive muscle in each of the three directions defined by the tissue microstructure. (**C**) The orientation of the fibres (shown by the *grey rods*) in a myocardial sheetlet within a finite element model of the intact left and right ventricular myocardium. The finite element model is used to predict mechanical deformation throughout the cardiac cycle, and the orthogonality of the fibres in the reference state (*bottom panel at the right*) is lost when the heart fills (*middle panel*) and when it is at its fully contracted state (*top panel*)

the governing equations for large deformation mechanics based on the underlying conservation laws of physics. Finally we describe some very simple growth laws that are used in Sect. 4 for studying cardiac embryonic development.

3.1 Tissue Microstructure and Material Coordinate Systems

Cardiac muscle is 'orthotropic', meaning that the properties are different in three orthogonal material directions ('fibre', 'sheet' and 'sheet-normal') in the reference state [51]. The fibre direction coincides with the long axis of the cardiomyocytes, the cells responsible for contraction. These cells are coupled into branching sheetlets that are 3–5 cells thick and facilitate shearing between the sheet planes [52]. The 'sheet' axis lies orthogonal to the fibre axis within this sheetlet, and the 'sheet-normal' axis is orthogonal to the sheetlet. Figure 2(A) illustrates these material axes in relation to the tissue microstructure. Figure 2(B) shows typical axial stress-strain relations associated with the microstructural material axes, and Fig. 2(C) illustrates how the orientations of these material axes vary throughout the myocardium in a way that can be represented using finite element modelling.

3.2 Deformation Gradient and Strain Tensors

We define a system of material coordinates $\{f, s, n\}$ in relation to the tissue microstructure. A point in the undeformed (reference) or deformed body has Eulerian coordinates $\{X_i\}$

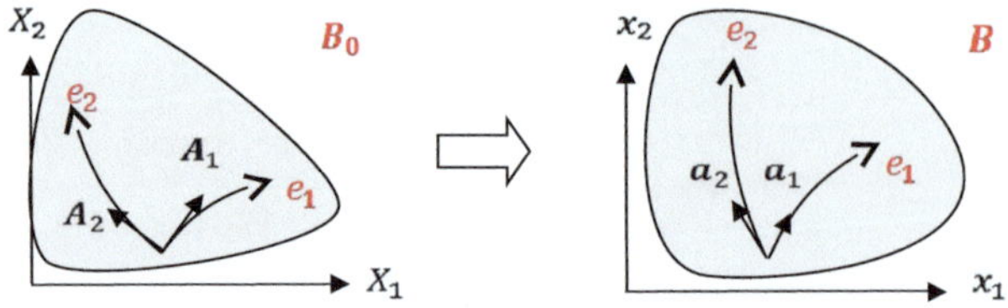

Fig. 3 Large deformation mechanics. A body $\mathcal{B}_0$ in the reference state deforms under the action of internal and external forces, to state $\mathcal{B}$. A material point in the undeformed body, whose position is defined by Eulerian coordinates $\{X_i\}$, deforms to position $\{x_i\}$. Material coordinates $\{e_\alpha\}$ reflect the underlying tissue microstructure (e.g., fibre-sheet coordinates in the heart). Orthogonal unit base vectors A_α tangent to these material axes in the undeformed body become (non-orthogonal, non-unit) base vectors a_α in the deformed body

or $\{x_i\}$, respectively, and g_i are the orthonormal base vectors for the Eulerian coordinates. Base vectors tangent to these material coordinates (see Fig. 3) are defined for the reference and deformed states of the body, respectively, as

$$A_\alpha = \frac{\partial X_i}{\partial \upsilon_\alpha} g_i, \quad \text{and} \quad a_\alpha = \frac{\partial x_i}{\partial \upsilon_\alpha} g_i \quad \text{(summation applies over repeated indices).} \tag{1}$$

Deformation is described by the mapping $\varphi : X \to x$, and $F^i_\alpha = \frac{\partial x_i}{\partial \upsilon_\alpha}$ are components of the deformation gradient tensor F [51]. The inner products of the base vectors in (1) give the metric tensors with respect to the material coordinates

$$A_{\alpha\beta} = A_\alpha \cdot A_\beta = \frac{\partial X_k}{\partial \upsilon_\alpha}\frac{\partial X_k}{\partial \upsilon_\beta}, \quad \text{and} \quad a_{\alpha\beta} = a_\alpha \cdot a_\beta = \frac{\partial x_k}{\partial \upsilon_\alpha}\frac{\partial x_k}{\partial \upsilon_\beta} \tag{2}$$

for the reference and deformed bodies, respectively. $a_{\alpha\beta}$ denote the covariant components of the right Cauchy deformation tensor $C = F^T F$.

The Green-Lagrange strain tensor links the two states, and being expressed in material coordinates, is independent of rigid body rotations and translations. Note that for the case where the material coordinate axes (υ_α) are chosen to be orthonormal at a point in the reference state (but still not coinciding with the reference Eulerian axes). $A_{\alpha\beta} = \delta_{\alpha\beta}$ (i.e., the Kronecker delta) and $E_{\alpha\beta} = \frac{1}{2}(a_{\alpha\beta} - \delta_{\alpha\beta})$ are the components of the Green-Lagrange strain tensor $E = \frac{1}{2}(C - I)$.

3.3 Stress Tensors

Cauchy's formula relates the (force/unit area) stress vector, τ, acting on a small surface area with normal $n = n_i g^i$ to a Cauchy stress tensor, σ, with components σ^{ij} referred to the base vectors g_i:

$$\tau = \sigma \cdot n, \quad \text{or} \quad \tau = \sigma^{ij} n_i g_j. \tag{3}$$

$T^{\alpha\beta}$ are the components of the 2nd Piola-Kirchhoff stress tensor, T, (the energy conjugate of the Green-Lagrange strain tensor $(E_{\alpha\beta})$. Note that $T^{\alpha\beta}$ is symmetric, and $S^{i\beta} = F^i_\alpha T^{\alpha\beta}$ $(S = FT)$ are the components of the 1st Piola-Kirchhoff stress tensor, S. The relationships between the two Piola-Kirchhoff stress tensors and the Cauchy stress tensor are

$$\sigma^{ij} = J^{-1} S^{i\alpha} \frac{\partial x_j}{\partial \upsilon_\alpha}, \quad \text{or} \quad \sigma = J^{-1} S F^T, \tag{4}$$

and

$$\sigma^{ij} = J^{-1} \frac{\partial x_i}{\partial \upsilon_\alpha} T^{\alpha\beta} \frac{\partial x_j}{\partial \upsilon_\beta}, \quad \text{or} \quad \sigma = J^{-1} F T F^T, \tag{5}$$

where $J = \det \boldsymbol{F} = \sqrt{I_3}$ (where I_3 is the determinant of the right Cauchy-Green deformation gradient tensor) is the ratio of tissue volumes in the deformed and undeformed states ($J = 1$ for incompressible materials).

3.4 Conservation Laws

There are three conservation principles that are fundamental to solving kinematics and mechanics of the heart:

(i) conservation of mass, only when there is no growth,

$$\frac{D(\rho J)}{Dt} = 0, \tag{6}$$

where $J = \det(\boldsymbol{F})$ is the Jacobian (see above), and ρ is the mass per unit volume of the material;

(ii) conservation of linear momentum,

$$\frac{1}{J}\frac{D(\rho \boldsymbol{v} J)}{Dt} = \nabla_x \cdot \boldsymbol{\sigma} + \mathbf{f}, \quad \text{or} \quad \frac{1}{J}\frac{D(\rho v_i J)}{Dt} = \frac{\partial}{\partial x_j} \cdot \sigma^{ij} + f^i, \tag{7}$$

where $\boldsymbol{v} = \{v_i\}$ is velocity, $\boldsymbol{\sigma} = \{\sigma^{ij}\}$ is the Cauchy stress tensor, and $\mathbf{f} = \{f^j\}$ is the acceleration; and under static equilibrium,

$$\nabla_x \cdot \boldsymbol{\sigma} + \mathbf{f} = \mathbf{0}, \quad \text{or} \quad \frac{\partial}{\partial x_j}\sigma^{ij} + f^i = 0; \tag{8}$$

(iii) conservation of angular momentum implies the symmetry of the Cauchy stress tensor, i.e.,

$$\boldsymbol{\sigma} = \boldsymbol{\sigma}^T, \quad \text{or} \quad \sigma^{ij} = \sigma^{ji}. \tag{9}$$

3.5 Growth Laws

For an extensive review of soft tissue growth, see [53]. Briefly, for growing tissue, the deformation gradient tensor can be expressed as the product of a growth tensor and an elastic deformation gradient tensor: $\boldsymbol{F} = \boldsymbol{F}^e \cdot \boldsymbol{F}^g$. We assume that growth occurs in the fibre, sheet and sheet-normal directions with growth factors $\lambda_1, \lambda_2, \lambda_3$, respectively:

$$\boldsymbol{F}^g = \begin{bmatrix} \lambda_1 & 0 & 0 \\ 0 & \lambda_2 & 0 \\ 0 & 0 & \lambda_3 \end{bmatrix} \tag{10}$$

and define a growth law:

$$\frac{d}{dt}\begin{bmatrix} \lambda_1 \\ \lambda_2 \\ \lambda_3 \end{bmatrix} = \begin{bmatrix} \alpha_1 \\ \alpha_2 \\ \alpha_3 \end{bmatrix}, \tag{11}$$

with fixed growth rates $\alpha_1, \alpha_2, \alpha_3$. These three growth rates can be set independently to control the relative amounts of growth in the fibre, sheet and sheet-normal directions. They can also vary spatially, and we show below that differential growth rates across the myocardium drive the observed C-looping and S-looping in the growing embryonic heart.

4 Models of Embryonic Growth

The most comprehensive modelling on cardiac looping has been the work of Taber [16, 24, 54–56], based primarily on experimental data from chick hearts. In the following analysis, we start with the heart tube and consider the subsequent growth that bends and twists the tube into the more familiar grown heart shape via initially C-looping and then S-looping. The mechanisms underlying this growth involve actin polymerisation, myocyte cell division, differential myocyte growth, and possibly myocyte contraction and external loads.

In this paper, we present two illustrative examples of mechanics with growth of an embryonic linear heart tube. These examples were solved with the OpenCMISS [57] computational package. Both examples considered the initial bending phases of heart development. The first example considered heterogeneous growth in a tube as shown in Fig. 4. The initial length of the tube was 6.0 units, with an inner radius of 1.25 units and an outer radius of 2.0 units. The tube was orientated so that the longitudinal axis of the tube was aligned with the z axis. Note that the finite element mesh is relatively coarse as the elements use tri-cubic Hermite basis functions, which have more degrees of freedom than lower order linear or quadratic Lagrange elements and can thus model higher order spatial gradients. The finite element model of this tube consisted of 4 elements in the circumferential direction, 4 elements in the longitudinal direction and 1 element through the wall of the tube. A fibre field was used to describe directions of anisotropy. In this case, the fibres ran circumferentially so that the sheet direction was in the longitudinal direction of the tube and the sheet-normal direction was the radial direction of the tube.

As described in Sect. 3, the total deformation of the tube is composed of a growth component defined by a growth tensor $\boldsymbol{F}^g$ and an elastic component defined by an elastic tensor $\boldsymbol{F}^e$. The growth component was modelled using a CellML[1] based model to describe the growth tensor rate defined by (11). The parameters of the growth equation were the growth rates in the fibre, sheet and sheet-normal directions. By using CellML, other growth models can be easily incorporated by importing the appropriate CellML description into the framework. The elastic component was modelled using finite deformation mechanics as described in Sect. 3. The constitutive equation for the material was specified using a CellML based constitutive model. In this example, an incompressible elastic Mooney-Rivilin material was used with a c_1 constant of 2.0 kPa and a c_2 constant of 6.0 kP. Other constitutive models can be used simply by specifying an appropriate CellML based model. Tri-cubic Hermite basis functions were used for interpolating the reference and deformed geometries and tri-linear Lagrange basis functions were used for interpolating the fibre directions and hydrostatic pressure used to ensure incompressibility. The boundary conditions for the simulation were such that the bottom ring of the tube was fixed in the z (longitudinal) direction. The nodes in the bottom ring were also constrained such that they could only move in a radial direction in order to fix rotations of the tube.

The OpenCMISS framework allows us to have arbitrary spatial variations in both the growth model and the growth rate parameters. In order to simulate the C- and S-bending phases of the embryonic heart tube, we specified two phases of heterogeneous growth rates in the tube. Ideally, the rates and models of growth for the two phases would be determined experimentally. Such data are presently unavailable, whoever, work is currently being undertaken to determine these rates however, at the time of writing, this data is unavailable. In order to provide a proof of concept generic rates are used with time units undetermined. It is

[1]CellML is an XML markup language used to encode ordinary differential and algebraic equation models in a curated, annotated and reproducible form. See www.cellml.org.

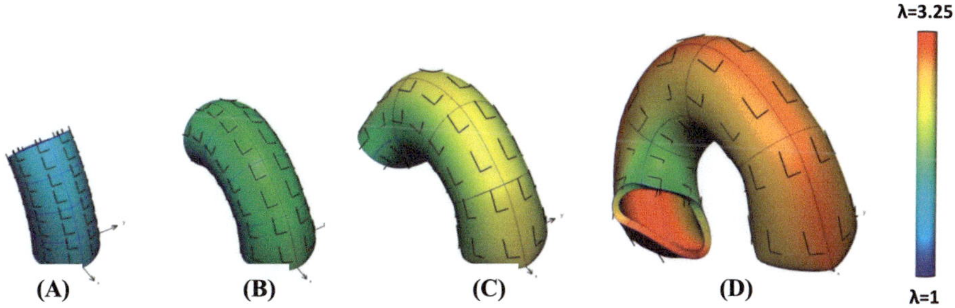

Fig. 4 Growth initiated bending in a linear heart tube with a free end. (**A**) The tube after 1 step at the beginning of the first phase (C-bending) of growth; (**B**) The tube after 30 steps at the end of the first phase (C-bending) of growth; (**C**) The tube after 60 steps during the second (S-bending) phase of growth; (**D**) The tube after 90 steps at the end of the second (S-bending) phase of growth. The pseudo-colour spectrum on the right indicates the total growth factor, λ, i.e., $\lambda = \sqrt{\lambda_1^2 + \lambda_2^2}$ where λ_1 is the fibre (or circumferential) growth factor and λ_2 is the sheet (or longitudinal) growth factor. The L shaped glyphs indicate the relative magnitudes of growth. The circumferentially oriented line segment in the L-glyph indicates the fibre growth factor (λ_1) and the longitudinally orientated line segment indicates the sheet growth factor (λ_2) values. The initial length of each line segment in the L-glyph at the beginning of the simulation is 1.0

thus important to consider the relative magnitudes of the rates in these simulations. For the initial C-bending phase, we set the base fibre (or circumferential) growth rate to 0.005, the base sheet (or longitudinal) rate to 0.015 and the base sheet-normal (or radial) growth rate to 0.0. The sheet rate was multiplied by a heterogeneous sheet factor of $1 + \cos\theta$ where θ was the angle in the xy plane measured in an anti-clockwise manner from the x axis. The beginning of this phase of growth is shown in Fig. 4(A) and the end of this phase (after 30 steps) is shown in Fig. 4(B). To model the next, S-bending phase, the heterogeneous pattern of growth rates were altered. The fibre (circumferential) and sheet-normal (radial) rates were left unchanged from the previous phase. The heterogeneous sheet factor was, however, changed to $(1 + \sin\theta) \cdot L^2$ where L was the normalised length coordinate of the tube (i.e., $L = 0$ at the bottom of the tube and $L = 1$ at the top of the tube). The result of this growth phase is shown after a further 30 steps in Fig. 4(C) and after a further 30 steps in Fig. 4(D).

The second example considered heterogeneous growth in a tube that was clamped at both ends as shown in Fig. 5(A). The parameters were similar to the first example, except as noted below. Here the tube was initially 20.0 units long, and the finite element model had 6 elements in the longitudinal direction. The boundary conditions were defined such that both the top and bottom rings were fixed in the z direction and were constrained to only move in the radial direction. The base growth rates were 0.03 in the fibre (circumferential) direction and 0.015 in the sheet (longitudinal) direction. Similar to the previous example, the growth was split into two phases. During the first phase, the fibre (circumferential) growth rate was multiplied by a heterogeneous factor of $L \cdot (1 - L)$ where L is the normalised length coordinate. The sheet growth rate was also multiplied by a heterogeneous factor of $(1 + \frac{1}{2}\cos\theta) \cdot (L + \frac{2}{3}) \cdot (1 - L)$. The first phase was set to grow for 45 time steps. The second phase then followed with the sheet heterogeneous factor changed to $(1 + \frac{1}{2}\sin\theta) \cdot (L + \frac{2}{3}) \cdot (1 - L)$. The second phase was allowed to grow for a further 45 time steps and the results are shown in Fig. 5(D). The final growth state illustrates C-bending and is comparable to Fig. 1(E).

Note that the growth models shown in this section are illustrative examples only. Work is currently underway to optimise the heterogeneously distributed parameters of the growth

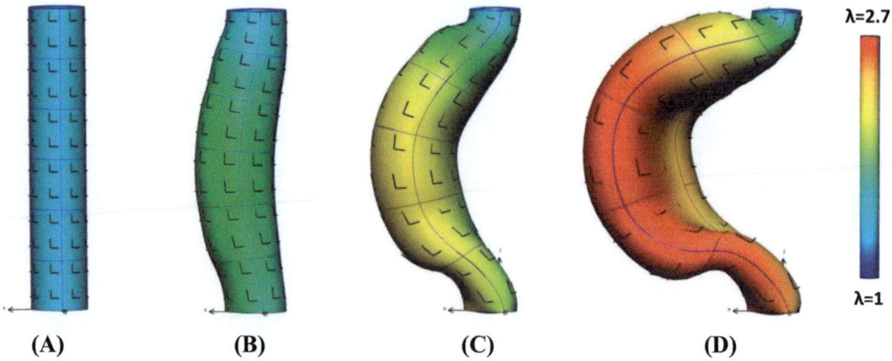

Fig. 5 Bending in a linear heart tube clamped at both ends. A tricubic Hermite mesh of the linear tube was specified with heterogeneous growth rates to achieve C-bending. (**A**) Initial mesh before growth. (**B**) The tube after 30 steps. (**C**) The tube after 60 steps. (**D**) The tube after 90 steps and the end of the growth phase illustrating C-bending. The pseudo-colour spectrum indicates the total growth factor, λ, i.e., $\lambda = \sqrt{\lambda_1^2 + \lambda_2^2}$ where λ_1 is the fibre (or circumferential) growth factor and λ_2 is the sheet (or longitudinal) growth factor. In addition to the colour field, the L shaped glyphs indicate growth. The circumferentially orientated line segment in the L-glyph indicates the fibre growth factor (λ_1) and the axially orientated line segment indicates the sheet growth factor (λ_2) values. The initial length of each line segment in the L-glyph in (**A**) is 1.0

laws to optimally reproduce the observed embryological tissue growth and to link these parameters to molecular pathway models using experimental data on cell replication and cell hypertrophy at the various stages of growth.

5 Modelling Cardiac Mechanics During Tissue Hypertrophy and Remodelling

We now consider the hypertrophy and remodelling of tissue in a fully grown heart and its physiological/pathological impact on ventricular mechanics. In this section, the term growth is used synonymously with hypertrophy and does not imply hyperplasia. In adult myocardium, the number of cells is approximately fixed and two types of remodelling can occur: cell growth (or shrinkage) and changes in the extracellular matrix. Studying the extent of cardiac remodelling and the underlying mechanisms in both human and animal heart failure models has been an area of intense research for decades [35, 58–60]. With the aid of imaging modalities, such as cardiac magnetic resonance imaging, echocardiography, and computed tomography, *in vivo* 3D images of the heart can be acquired routinely and provide accurate quantification of mal-adaptive growth on a personalised basis. Such subject-specific imaging data also enables important geometric information to be incorporated into personalised models of cardiac mechanics. On the other hand, detailed *in vivo* assessment of microstructural remodelling is not yet feasible on a whole heart basis. Instead, an estimate of the degree of microstructural remodelling can be gained through histological studies of excised myocardial tissue samples. Recently, T1 imaging, a variant of magnetic resonance imaging, has emerged as an innovative technique to evaluate fibrosis in heart failure patients [61]. T1 imaging makes use of the longer T1 relaxation time exhibited by non-viable myocardium (i.e., collagen) to demarcate fibrotic tissue from normal myocardium. This technique has been useful for providing qualitative evidence of diffuse and localised fibrosis [61, 62] in heart failure patients.

Over the past decade, our fundamental understanding of the different types of cardiac tissue remodelling has advanced significantly. What is missing, however, is quantitative insight into the mechanistic link between microstructural remodelling and mechanical dys-function during disease. This information has the potential to help in the design of mechanism-targeted treatment options for heart failure. At a basic level, while some aspects of the effect of geometric remodelling on passive ventricular mechanics can be reproduced using Laplace's law, the effects of remodelling of the myocardial tissue constituents are not well characterised. Collagen fibrils are the main load-bearing structures in the myocardium, and the extracellular matrix is the key determinant of the passive constitutive properties of myocardium. Therefore, remodelling of the collagen distribution can have significant detrimental effects on the mechanical behaviour of the heart [63, 64]. Despite extensive research on characterising myocardial fibrosis during heart failure, the fundamental biomechanical effects of myocardial tissue remodelling remain unknown. Modelling the constitutive properties of myocardium can help to explain the relationship between structural and functional remodelling.

5.1 Constitutive Relations for Myocardial Tissue Remodelling

At present, there are no established clinical protocols for measuring *in vivo* myocardial constitutive properties. Integrative inverse finite element modelling is a common research tool for non-invasive estimation of myocardial properties. In addition, this approach can be used to discern the relative contributions of the remodelling factors that lead to ventricular dysfunction. Central to this is the design of constitutive equations that can describe *in vivo* myocardial tissue behaviour. The anisotropic, hyperelastic mechanical response of normal myocardial tissue has been represented using several well-known phenomenological constitutive models based on the Fung-type exponential constitutive framework (e.g., models that reproduce transversely-isotropic [65] or orthotropic [66] mechanical behaviour). The formulations of these equations were motivated by knowledge of the gross microstructure and stress-strain relationships measured from excised myocardium. While capturing the hyperelastic behaviour has been the main focus of many experimental and modelling studies, it is becoming increasingly evident that myocardium exhibits viscous properties as supported by two key experimental studies [67, 68] which performed biaxial, simple shear and stress relaxation tests on excised porcine and human myocardial tissue blocks, respectively. This consequently motivated the design of orthotropic viscoelastic constitutive models [69, 70] in order to reproduce the stress relaxation behaviour, and the hysteresis observed in cyclic loading experiments. The research on viscoelastic model of myocardium is limited by the sparse experimental data at the whole heart level. Therefore, hyperelastic model remains as standard. Another important series of constitutive models has been developed by Ogden, Holzapfel and Gasser [71–74]. These include models of 2D and 3D fibre dispersion, fibre dispersion with rotational symmetry, and non-symmetric fibre dispersion. This family of constitutive equations is not discussed further here as they have not been applied to model healthy cardiac tissue, but such an application may be worthwhile to model diseased myocardium as demonstrated in [75]. Although some of the Fung-type constitutive models have been used extensively for modelling cardiac mechanics, it has become apparent that taking these models forward to characterise the constitutive properties of structurally remodelled myocardium remains a big challenge. Limiting factors include the phenomenological nature of the equations, and the lack of identifiability of the constitutive parameters [76]. While there is a great deal of literature on microstructural remodelling, this has not yet been integrated into constitutive modelling frameworks, and thus the links to functional remodelling remain poorly characterised. This has motivated the development of microstructure-based constitutive models.

Constitutive models that incorporate detailed descriptions of the mechanical properties and spatial distributions of collagen fibres, have been successfully applied to study mechanical behaviour of membranous tissues such as heart valves and skin [77]. However, the complex arrangement of collagen fibres in the myocardium makes it difficult to quantify microstructural indices that feature in the constitutive models. Therefore, translating this microstructural distribution approach to myocardium remains challenging, largely due to the lack of comprehensive data with which these types of models can be calibrated and validated, and due to the typically high computational demands, which can be prohibitive when applying these types of models to organ-scale simulations. In contrast, phenomenological models require less computational resource, and are already able to sufficiently reproduce *ex vivo* uniaxial and biaxial stretch data [78]. Nevertheless, progress on developing microstructure-driven constitutive models of myocardium has been made, as summarised in Table 1. For example, Nevo and Lanir [78] proposed that the passive stress response of myocardium could be formulated using an exponential function that included an undulation density distribution function, fibre extension ratio, and largest stress-free extension ratio. The total strain energy of the tissue was assumed to be the sum of contributions from individual collagen fibres with the key collagen types in the myocardium being endomysial and perimysial collagen. This formulation produced realistic pressure-volume loops, however some of the constitutive parameters were dependent on the collagen volume fraction, which was difficult to quantify on a whole heart basis. Sacks [79] demonstrated in pericardium that incorporating the angular distribution of the collagen fibres ($R(\theta)$), measured using a small angle light scattering technique, into the structural model proposed by Lanir [77] not only reproduced the response to both equibiaxial and non-equibiaxal loading, but also circumvented the multiple co-linearity difficulties inherent in biaxial testing protocols. Therefore, the addition of microstructural data into constitutive modelling frameworks can help to better interpret and represent experimental data in terms of the underlying biophysical mechanisms of myocardial mechanics.

In the past decade, the formulation of microstructure-driven constitutive models for healthy myocardium has gained considerable attention. Holzapfel and Ogden [80] proposed a variant of the separated Fung-type exponential constitutive model, based on the underlying morphology of the myocardium, to model the orthotropic passive mechanical behaviour of the heart. Recently, a constitutive model, that consists of mechanical contributions from myofibre, collagen fibre ensembles, and the interactions between them, was also put forward by Avazmohammadi et al. [81]. Using microstructural parameters quantified from histological studies, they characterised the passive mechanical response of right ventricular myocardium by also accounting for the degree of collagen fibre recruitment. After fitting the parameters of the constitutive models to bi-axial stress-strain data, they revealed the contributions of the different microstructural components on different parts of the stress-strain curve. The model also has the potential to be applied to examine changes in right ventricular myocardial mechanical behaviour under pulmonary hypertension. Constitutive models that represent the various aspects of microstructural remodelling are also being developed. Eriksson et al. [75] modelled the disarray in fibre and sheetlet orientations [82] commonly found in hypertrophic cardiomyopathy cases. By fitting distribution functions to myocyte and sheetlet orientations measured from the healthy heart, they observed a relatively small dispersion in the myocyte orientation ($\sim$12–15°), but a larger (nearly double) dispersion in sheetlet orientations. Conversely, the hypertrophic heart exhibited a fibre dispersion range of $\sim$65°. To model the effect of such microstructural disarray on global and regional LV mechanics, Eriksson et al.

Table 1 Summary of microstructure-based hyperelastic constitutive models of myocardium

Constitutive equations	References
$\overline{W} = \sum_{k=1}^{2} S_k \int_{\Omega} R_k(n) w_k^*(E_{ff}) d\Omega$, where S_k is the volumetric fraction of fibre type k in the tissue, w_k^* is the uniaxial stress-strain relationship of fibre type k, and Ω is the tissue domain.	Lanir [77]
$\boldsymbol{T}_c = \begin{cases} D(e^{C(\alpha_c-\alpha_r)} - 1) \cdot \alpha_c & \alpha_c > \alpha_r \\ 0, & \alpha_C \le \alpha_r, \end{cases}$ where $\boldsymbol{T}_c$ is the Cauchy stress, C and D are material parameters of the fibre, α_c is the fibre extension ratio, and α_r is the largest stress-free extension ratio.	Nevo and Lanir [78]
Gradual recruitment of fibres with fibre strain (ε_s) : $D(\varepsilon_s) = \frac{1}{\beta^{\alpha}\Gamma(\alpha)} \varepsilon_s^{\alpha-1} \exp(\frac{-\varepsilon_s}{\beta})$, Stress generated by one fibre: $\boldsymbol{T}_{11}^{f} = K \int_0^{\varepsilon} D(x) \frac{\varepsilon - x}{1+2x} dx$, where K is the fibre volume fraction, Total stress: $\boldsymbol{T} = \int_{-\pi/2}^{\pi/2} R(\theta) \boldsymbol{T}_{11}^{f}(\varepsilon)[N \otimes N] d\theta$.	Sacks [79]
$\overline{W} = \frac{a}{2b} \exp[b(I_1 - 3)] + \sum_{i=f,s} \frac{a_i}{2b_i} \{\exp[b_i(I_{4i}^* - 1)^2] - 1\}$ $+ \frac{a_{fs}}{2b_{fs}} [\exp(b_{fs} I_{8fs}^2) - 1]$, where $I_{4i}^* = \kappa_i I_1 + (1 - 3\kappa_i) I_{4i}$ and κ_i are the dispersion parameters for fibre and sheetlets.	Eriksson et al. [75]
$\overline{W} = \frac{t_{\text{endo}} a_f}{2b_f} [\exp(b_f(I_{4f} - 1)^2) - 1] + \frac{t_{\text{endo}} a_s}{2b_s} [\exp(b_s(I_{4s} - 1)^2) - 1]$ $+ \frac{t_{\text{peri}} a_n}{2b_n} [\exp(b_n(I_{4n} - 1)^2) - 1] + \frac{a_{fs}}{2b_{fs}} [\exp(b_{fs} I_{8fs}^2) - 1]$, where t_{endo} and t_{peri} represent the degree of thickening of endomysial collagen and perimysial collagen, respectively.	Wang et al. [83]

augmented their group's orthotropic constitutive model [80] with a term involving a dispersion parameter, which was directly fitted to the myocyte orientation distribution function (Table 1). This additional factor allowed examination of the mechanical consequences of fibre dispersion at the tissue and organ scales [80].

To investigate the link between myocardial tissue remodelling and impaired passive mechanical function in hypertensive heart failure hearts, Wang et al. [84] proposed an alternative augmentation of the Holzapfel and Ogden constitutive model [80]. As illustrated in Fig. 6, high resolution confocal imaging revealed extensive proliferation of both endomysial and perimysial collagen in hypertensive rat hearts [35]. This motivated the development of a new constitutive model (Table 1) that was expressed in terms of image-derived structural parameters, which represented the degree of remodelling and growth of endomysial (t_{endo}) and perimysial (t_{peri}) collagen, with values determined directly from high resolution confocal images [83]. The proposed constitutive model not only allowed microstructural remodelling information to be built directly into the mechanics analysis, but also removed the variability in material parameters due to non-uniqueness of the parameter identification procedure, which is a known issue in constitutive parameter estimation that can confound comparison of material properties between healthy and diseased subjects.

These new microstructural models have enabled progress towards understanding the impact of cardiac tissue remodelling at the continuum scale. However, this class of constitutive model is still largely phenomenological in nature. Consequently, some of the constitutive parameters are still not reflective of the underlying microstructural mechanisms of mechanics and remodelling. Moreover, some of the remodelling features that have been observed

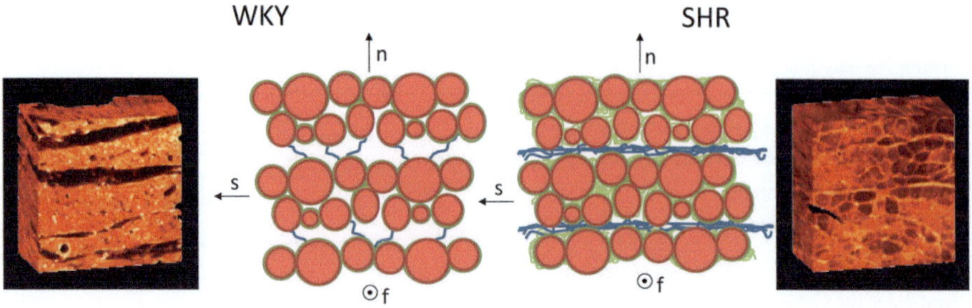

Fig. 6 3D confocal images of myocardial tissue blocks from the hearts of (*left*) a control Wistar-Kyoto rat (WKY), and (*right*) a diseased spontaneously hypertensive rat (SHR), both at 12 months of age. Centre: schematic representation of the WKY (*left*) and SHR (*right*) myocardium viewed along the myocyte axis. Endomysial collagen (*green*) surrounds individual myocytes (*red*). Perimysial collagen (*blue*) connects sheetlets of myocytes, with a much greater density in the SHR heart

in the images (such as the spatial re-organisation of the collagen network, and myocyte hypertrophy) have not yet been accounted for in any constitutive model. Therefore, to enhance our ability to represent the mechanical response of normal and diseased myocardium, there is further opportunity to develop new biophysical constitutive models that incorporate other key features of myocardial remodelling that affect myocardial mechanical function.

5.2 Modelling of Cardiac Mechanics During Tissue Growth and Remodelling

With the improved access to quantitative image-based observations on structural and functional remodelling in research and clinical environments, computational modelling is the most suitable tool to interpret this plethora of data in a consistent manner that adheres to the physical principles that govern tissue mechanics. In recent decades, mathematical modelling of cardiac mechanics has also evolved from generic to subject-specific analyses across multiple spatial and temporal scales. The ability of computer modelling to integrate a wealth of subject-specific structural and functional information measured at multiple scales from cell to organ provides a suitable platform for forward and inverse analysis of cardiac mechanics on a personalised basis. It also enables estimation of myocardial wall stress and constitutive properties—two key factors that can be affected by myocardial tissue remodelling, and that significantly affect myocardial growth [49].

Modelling the mechanics of myocardial growth at the continuum level was pioneered in the 1990s by Rodriguez et al. [85], who proposed a general continuum formulation to model stress-dependent finite volumetric growth (represented kinematically as a growth tensor), while allowing the development of residual stresses, which were thought to play an important role in ensuring spatial continuity of cell growth [86]. They demonstrated that residual stress in the LV wall was altered in the presence of transmurally uniform pure circumferential growth, a feature of eccentric ventricular hypertrophy. This formulation became the fundamental theoretic framework adopted by many studies to investigate maladaptive growth. For example, Göktepe et al. [87] developed a multi-scale modelling framework, using a regularly shaped bi-ventricular model, to simulate strain-driven growth (i.e., volume-overload) via serial alignment of myocytes and stress-driven growth (i.e., pressure-overload) via parallel addition of sarcomere units generated by sarcomerogenesis. Incorporating the growth tensor into the evolution equation for sarcomerogenesis enabled the prediction of morphological changes in cardiac myocytes and their impact on the ventricular growth. This framework was further extended by [88] to propose a constitutive model not only able to predict ventricular growth but also its reversal which presented opportunity to study the effect of several surgical procedures on the remodelling pattern of the left ventricle [89]. Klepach et al. [90] simulated growth in a patient-specific LV model derived from data before and after surgical removal of an infarcted apex, and they observed heterogeneous growth patterns in both cases. It was found that while the removal of the infarcted tissue allowed the passive constitutive properties (modelled using the Guccione transversely isotropic model [65]) to become homogenous as indicated by their inverse finite element analysis, the heterogeneous growth deformation predicted by their model also initiated further growth and remodelling. Given that the exact triggers for the different types of ventricular hypertrophy remain poorly characterised, Kerckhoffs et al. [91] proposed a strain-based growth law which hypothesised that increased fibre and cross-fibre strain were solely responsible for inducing cardiac myocytes to grow longer and thicker, respectively. In other words, the feedback of myocardial stress to growth was not a feature of this growth law. When incorporated into a canine ventricular mechanics analysis along with a closed-loop circulation model, this strain-based growth law was able to reproduce both concentric and eccentric hypertrophy.

In addition to the concept of a volumetric growth tensor, Kroon et al. [92] attempted to characterise growth using a scalar quantifying the relative volume change during deformation. The process of growth was modelled by simulating ventricular mechanics over several cardiac cycles, while assuming that the myocardial tissue was incompressible. They applied the resultant strain as a growth remodelling stimulus to simulate growth of the ventricle, while treating myocardial tissue as a compressible material. This approach enabled the analysis of mechanical strain, stress and growth within the same computational framework. Apart from volumetric growth, myofibre orientations within the ventricle were also allowed to adapt, which was a major point of difference in comparison to other studies. As predictions of growth at multiple scales have been attempted, calibrating constitutive models of growth and validating the resultant growth require further research attention. Recently, a new theoretical framework for modelling growth and remodelling based on the specific mechanical and load-dependent turnover kinematics of the key constituents of materials with uni-directional fibres was put forward by [93]. In contrast to previous framework which focused on modelling and reproducing the consequence of growth and remodelling, instead this new framework was entirely based on the mechanics of the tissue microstructure. Importantly, this framework does not rely on the assumption that the deformation of a growing material can be fully explained by the deformation gradient tensor. It was demonstrated that this type of mechanistic theoretical framework can also reproduce the observed growth phenomenon. At the core of such research is the requirement for quantitative experimental data, ideally acquired *in vivo* and non-invasively. Using biplane videofluoroscopy techniques, Tsamis et al. [94] were the first to obtain *in vivo* quantification of sheep LV wall growth, which was induced by myocardial infarction, over a course of seven weeks by tracking the displacement of an array of beads that were implanted in the wall of the LV. Such experimental data along with *in vivo* measurement of cavity pressure, albeit invasive, will be extremely useful to parameterise growth and myocardial constitutive models.

The number of studies that have modelled the gross effect of myocardial tissue remodelling on ventricular mechanical behaviour has expanded rapidly over recent years. In particular, constitutive modelling of myocardial mechanical behaviour after LV aneurysm (i.e., myocardial infarction) has been extensively investigated. Using finite element models of the LV fitted to geometric data captured by cardiac magnetic resonance imaging or echocardiography, helix angle data provided by diffusion tensor magnetic resonance imaging, and *in vivo* LV pressure data recorded via catheterisation, a number of studies have performed inverse mechanics analyses using a transversely-isotropic constitutive model [65] to estimate passive constitutive properties of myocardium in healthy remote, border and infarcted regions in sheep hearts [95–97]. Compared to the remote tissues, the infarcted regions were consistently characterised by stiffer constitutive properties, suggesting it is correlated with replacement fibrosis (i.e., collagen replaces viable myocytes). The stiffening of myocardium resulted in an increase in stresses along both the myocyte and cross-myocyte directions in the border region. The elevated myocardial stress was thought to play a role in LV remodelling, and has therefore been considered as a treatment target. Analysis of LV mechanics after linear repair of an aneurysm showed substantial reduction in end-systolic myofibre stress in the border zone throughout the LV, except for the septal region. Wall et al. [98] investigated the impact of bolus material injection on the mechanical behaviour of the LV by modifying the regional wall thickness of the LV geometric model to reproduce the bulging effect from the injection, and they found that the small wall volume change reduced border zone fibre stress. Using a similar mechanics modelling framework, Wenk et al. [97] studied the effects of injecting calcium hydroxyapatite-based tissue filler on the passive material response of infarcted LV using an finite element model of the sheep LV derived from echocardiography data. They showed that average myofibre stress in the treated LV was significantly reduced in both the border and remote regions. Therefore, the injection appeared to reduce stress in the LV, which was hypothesised to be preventative of further remodelling of the myocardium [97]. Finite element modelling was also applied to assist with designing the optimal patterns of polymeric injection for treating heart failure, as demonstrated in [99]. Analysing changes in myocardial mechanical behaviour in hypertrophic and dilated cardiomyopathy patients has also attracted intense research interest. Non-invasive quantification of myocardial stress and constitutive properties enabled by inverse mechanics modelling has provided quantitative insight into impaired mechanical behaviour due to myocardial infarction, and into the effects of ventricular surgery, which is not available from imaging alone. The interactions among the different remodelling factors during heart failure can also be interrogated using an image-driven modelling approach [84]. Wang et al. [83] first examined whether geometric remodelling alone could explain the change in the passive mechanical behaviour measured using pressure-compliance curves. This was achieved by estimating a single set of material constants for a group of hypertensive rat hearts (healthy and hypertensive animals each at 14 and 24 months of age) by matching to the subject-specific pressure-compliance data while neglecting microstructural remodelling. The results indicated that geometric remodelling alone (Fig. 7, dashed lines) could not explain differences in the passive mechanical function. However, by including subject-specific image-derived microstructural parameters, a single set of fitted material parameters was able to reproduce the four different pressure-compliance curves (Fig. 7, solid lines). Therefore, accounting for both geometric and microstructural remodelling was required in order to explain the impaired passive mechanical behaviour in hypertensive heart failure.

6 Challenges in Modelling Myocardial Growth and Tissue Remodelling

The development of mathematical models of cardiac growth is hampered by the paucity of experimental data. This paucity is due to the inaccessibility of the samples-mammalian embryos are well protected within the host and this complicates tracking of kinematics and the measurement of boundary conditions. Current approaches rely on the reconstruction of kinematics from imaging data. Multiscale models with biophysically meaningful parameters are then fitted to reproduce the observed kinematics. The model(s) for which the parameters fall within expected biophysical ranges are then chosen as candidates. Recent developments such as droplet-based sensors [100, 101] may help in quantitatively parameterising the

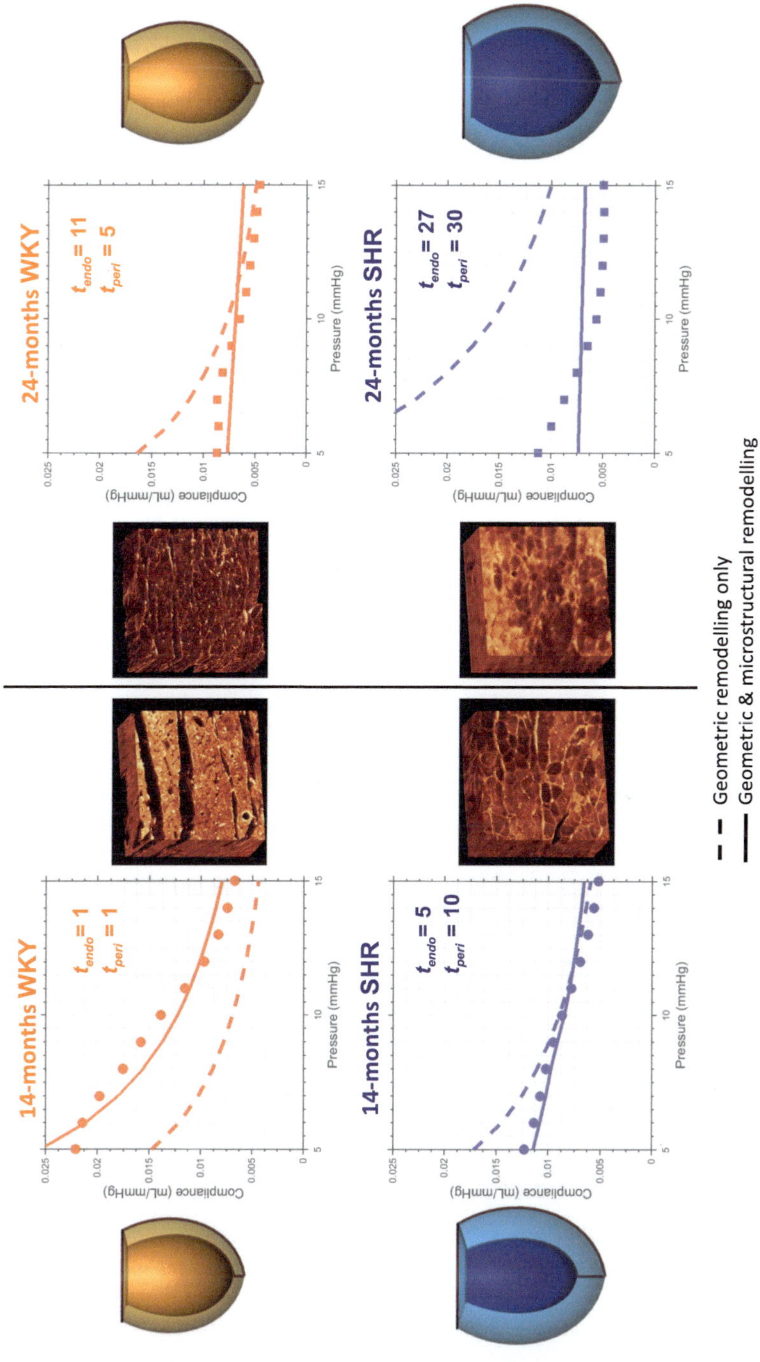

Fig. 7 LV finite element models and confocal images of a group of healthy and hypertensive rats (Wistar Kyoto (WKY) and spontaneously hypertensive rat (SHR) hearts each at 14 and 24 months of age), and their subject-specific pressure-compliance curves (*orange circles*: 14-months WKY; *orange squares*: 24-months WKY; *blue circles*: 14-months SHR; *blue squares*: 24-months SHR) along with the model predicted pressure-compliance curves when accounting only for geometric remodelling (*dashed lines*) and when accounting for both geometric and microstructural remodelling (*solid lines*)

models. These sensors measure mechanical stresses within living embryonic tissues using imaging and inverse modelling. We are presently gathering data on neonatal mouse hearts imaged with 0.5 μm microCT and confocal microscopy at multiple stages of growth during the looping phase. These data are allowing us to quantify the spatially varying extent of actin polymerisation, cell differentiation and cell growth, which we are then using to identify the spatial varying activation of subcellular signalling pathways. This will, in a future publication, provide a more mechanistic underpinning of the continuum growth laws used in this paper. Growth modelling offers a rich set of problems across multiple scales, solutions for which would not only advance our understanding of the cardiac development program but will also contribute towards understanding and diagnosing adult disease.

Despite decades of research on constitutive and microstructural modelling of myocardial mechanics, there is still a lack of mechanistic understanding of structure-mechanical function of myocardial tissue. One major issue is that majority of the myocardial tissue constitutive models are over-parameterised, and rely on fitted parameters that phenomenologically reproduce the mechanical behaviour of the myocardium, rather than basing constitutive behaviour on the underlying structure. The lack of biophysical relevance makes it difficult to interpret differences in fitted parameters. To this end, there is a need to develop microstructural-based constitutive model that incorporates observable parameters that characterise myocardial structure to improve the predictive power of constitutive modelling in the era of personalised medicine. To accurately model myocardial tissue mechanics, constitutive models need to contain sufficient microstructural detail, however, this is difficult to characterise and cannot be individually acquired. In addition, there is a mis-match between the large number of constitutive parameters in microstructural models versus the paucity of experimental data required to uniquely identify these parameters. To date, only two experimental studies have characterised mechanical function and structure of myocardial tissue blocks, which can be used for estimating material parameters [67, 68]. Although such data are extremely useful for quantifying the force versus displacement relations under controlled deformations, the invasive nature of such tests and the viability of tissue may have an impact on the estimated material parameters, which raises questions regarding the applicability of the parameters to *in vivo* myocardium. This challenge can potentially be overcome using model reduction techniques. For example, a comprehensive multivariate sensitivity analysis could be carried out on the detailed microstructural model to identify the parameters that have observable effects on cardiac mechanical behaviour. The parameters with large sensitivity could then be analysed and personalised to individual hearts to assist with understanding of the relationship between myocardial microstructure and mechanical function in health and disease.

7 Conclusions

This review summarises current research on modelling cardiac development, growth (concentric and eccentric hypertrophy), and microstructural remodelling (fibrosis). We discussed the power of computational modelling in simulating events such as C-looping of the chick heart during embryogenesis. With appropriate microstructural constitutive models, inverse finite element analysis can be used to analyse the factors that lead to impaired mechanical behaviour in heart failure. Computational modelling of cardiac mechanics can not only provide valuable insights into the mechanisms underpinning physiological and patho-physiological cardiac processes, but may also have the potential to be used as a non-invasive method to assess myocardial tissue mechanical properties which provide a specific biomarker for cardiac performance.

Acknowledgements VYW and MPN gratefully acknowledge New Zealand Government funding from the Marsden Fund (grant number 12-UOA-222) administered by the Royal Society of New Zealand and from the Health Research Council of New Zealand (grant number 13-317). JRH, HY, and PJH would also like to acknowledge funding from the MedTech Centre of Research Excellence and a Marsden Grant 'How does the heart grow?' funded by the Royal Society of New Zealand.

Conflict of Interest The authors declare that they have no conflicts of interest.

References

1. Rubin, M., Safadi, M., Jabareen, M.: A unified theoretical structure for modeling interstitial growth and muscle activation in soft tissues. Int. J. Eng. Sci. **90**, 1–26 (2015)
2. Sundgren, N.C., Giraud, G.D., Schultz, J.M., Lasarev, M.R., Stork, P.J., Thornburg, K.L.: Extracellular signal-regulated kinase and phosphoinositol-3 kinase mediate IGF-1 induced proliferation of fetal sheep cardiomyocytes. Am. J. Physiol., Regul. Integr. Comp. Physiol. **285**(6), R1481–R1489 (2003)
3. Sundgren, N., Giraud, G., Stork, P., Maylie, J., Thornburg, K.: Angiotensin II stimulates hyperplasia but not hypertrophy in immature ovine cardiomyocytes. J. Physiol. **548**(3), 881–891 (2003)
4. Paradis, A.N., Gay, M.S., Zhang, L.: Binucleation of cardiomyocytes: the transition from a proliferative to a terminally differentiated state. Drug Discov. Today **19**(5), 602–609 (2014). doi:10.1016/j.drudis.2013.10.019
5. Taber, L.A.: Biomechanics of growth, remodeling, and morphogenesis. Evolution **490**, 6 (1995)
6. Kuhl, E.: Growing matter: a review of growth in living systems. J. Mech. Behav. Biomed. Mater. **29**, 529–543 (2014). doi:10.1016/j.jmbbm.2013.10.009
7. Ambrosi, D., Mollica, F.: The role of stress in the growth of a multicell spheroid. J. Math. Biol. **48**(5), 477–499 (2004)
8. Bayly, P., Taber, L., Kroenke, C.: Mechanical forces in cerebral cortical folding: a review of measurements and models. J. Mech. Behav. Biomed. Mater. **29**, 568–581 (2014)
9. De La Cruz, M., Markwald, R.R.: Living Morphogenesis of the Heart. Springer, Berlin (1998)
10. Männer, J.: Cardiac looping in the chick embryo: a morphological review with special reference to terminological and biomechanical aspects of the looping process. Anat. Rec. **259**(3), 248–262 (2000)
11. Männer, J.: The anatomy of cardiac looping: a step towards the understanding of the morphogenesis of several forms of congenital cardiac malformations. Clin. Anat. **22**(1), 21–35 (2009)
12. Miquerol, L., Kelly, R.G.: Organogenesis of the vertebrate heart. Wiley Interdiscip. Rev.: Dev. Biol. **2**(1), 17–29 (2013)
13. Soufan, A.T., van den Berg, G., Ruijter, J.M., de Boer, P.A., van den Hoff, M.J., Moorman, A.F.: Regionalized sequence of myocardial cell growth and proliferation characterizes early chamber formation. Circ. Res. **99**(5), 545–552 (2006)
14. Srivastava, D., Olson, E.N.: A genetic blueprint for cardiac development. Nature **407**(6801), 221–226 (2000)
15. van den Berg, G., Abu-Issa, R., de Boer, B.A., Hutson, M.R., de Boer, P.A., Soufan, A.T., Ruijter, J.M., Kirby, M.L., van den Hoff, M.J., Moorman, A.F.: A caudal proliferating growth center contributes to both poles of the forming heart tube. Circ. Res. **104**(2), 179–188 (2009)
16. Ramasubramanian, A., Latacha, K.S., Benjamin, J.M., Voronov, D.A., Ravi, A., Taber, L.A.: Computational model for early cardiac looping. Ann. Biomed. Eng. **34**(8), 1355–1369 (2006)
17. Gruber, P.J., Wessels, A., Kubalak, S.W.: Development of the heart and great vessels. In: Pediatric Cardiac Surgery, pp. 1–26. Blackwell Sci. Oxford (2012)
18. Männer, J.: On the form problem of embryonic heart loops, its geometrical solutions, and a new biophysical concept of cardiac looping. Ann. Anat., Anat. Anz. **195**(4), 312–323 (2013)
19. Männer, J., Bayraktar, M.: Cardiac looping may be driven by compressive loads resulting from unequal growth of the heart and pericardial cavity. Observations on a physical simulation model. Front. Physiol. **5**, 112 (2014)
20. Sissman, N.J.: Cell multiplication rates during development of the primitive cardiac tube in the chick embryo. Nature **210**(5035), 504–507 (1966)
21. Manasek, F.J., Isobe, Y., Shimada, Y., Hopkins, W.: The embryonic myocardial cytoskeleton, interstitial pressure, and the control of morphogenesis. In: Nora, J., Takao, A. (eds.) Congenital Heart Diseases Causes Process, pp. 359–376. Futura Publishing, Mount Kisco (1984)
22. Baldwin, H.S., Solursh, M.: Degradation of hyaluronic acid does not prevent looping of the mammalian heart in situ. Dev. Biol. **136**(2), 555–559 (1989)

23. Linask, K.K., Han, M.D., Linask, K.L., Schlange, T., Brand, T.: Effects of antisense misexpression of CFC on downstream flectin protein expression during heart looping. Dev. Dyn. **228**(2), 217–230 (2003)
24. Latacha, K.S., Rémond, M.C., Ramasubramanian, A., Chen, A.Y., Elson, E.L., Taber, L.A.: Role of actin polymerization in bending of the early heart tube. Dev. Dyn. **233**(4), 1272–1286 (2005)
25. Manasek, F.J., Burnside, M.B., Waterman, R.E.: Myocardial cell shape change as a mechanism of embryonic heart looping. Dev. Biol. **29**(4), 349–371 (1972)
26. Nolan, G.P.: Cardiac development: transcription and the broken heart. Nature **392**(6672), 129–130 (1998)
27. Bruneau, B.G.: The developmental genetics of congenital heart disease. Nature **451**(7181), 943–948 (2008)
28. Zak, R.: Growth of the Heart in Health and Disease. Raven Press, New York (1984)
29. Li, G., Xiao, Y., Estrella, J.L., Ducsay, C.A., Gilbert, R.D., Zhang, L.: Effect of fetal hypoxia on heart susceptibility to ischemia and reperfusion injury in the adult rat. J. Soc. Gynecol. Investig. **10**(5), 265–274 (2003)
30. Braunwald, E.: Biomarkers in heart failure. N. Engl. J. Med. **358**(20), 2148–2159 (2008). doi:10.1056/NEJMra0800239
31. Hill, J.A., Olson, E.N.: Cardiac plasticity. N. Engl. J. Med. **358**(13), 1370–1380 (2008). doi:10.1056/NEJMra072139
32. Kitzman, D.W., Little, W.C., Brubaker, P.H., Anderson, R.T., Hundley, W.G., Marburger, C.T., Brosnihan, B., Morgan, T.M., Stewart, K.P.: Pathophysiological characterization of isolated diastolic heart failure in comparison to systolic heart failure. JAMA J. Am. Med. Assoc. **288**(17), 2144–2150 (2002). doi:10.1001/jama.288.17.2144
33. van Heerebeek, L., Borbély, A., Niessen, H.W.M., Bronzwaer, J.G.F., van der Velden, J., Stienen, G.J.M., Linke, W.A., Laarman, G.J., Paulus, W.J.: Myocardial structure and function differ in systolic and diastolic heart failure. Circulation **113**(16), 1966–1973 (2006). doi:10.1161/circulationaha.105.587519
34. Paulus, W.J., Tschöpe, C., Sanderson, J.E., Rusconi, C., Flachskampf, F.A., Rademakers, F.E., Marino, P., Smiseth, O.A., De Keulenaer, G., Leite-Moreira, A.F., Borbély, A., Édes, I., Handoko, M.L., Heymans, S., Pezzali, N., Pieske, B., Dickstein, K., Fraser, A.G., Brutsaert, D.L.: How to diagnose diastolic heart failure: a consensus statement on the diagnosis of heart failure with normal left ventricular ejection fraction by the Heart Failure and Echocardiography Associations of the European Society of Cardiology. Eur. Heart J. **28**(20), 2539–2550 (2007). doi:10.1093/eurheartj/ehm037
35. LeGrice, I.J., Pope, A.J., Sands, G.B., Whalley, G., Doughty, R.N., Smaill, B.H.: Progression of myocardial remodeling and mechanical dysfunction in the spontaneously hypertensive rat. Am. J. Physiol., Heart Circ. Physiol. **303**(11), 65 (2012). doi:10.1152/ajpheart.00748.2011
36. Tanaka, M., Fujiwara, H., Onodera, T., Wu, D.J.: Quantitative analysis of myocardial fibrosis in normals, hypertensive hearts, and hypertrophic cardiomyopathy. Br. Heart. J. **55**(6), 575–581 (1986)
37. Shirwany, A., Weber, K.T.: Extracellular matrix remodeling in hypertensive heart disease. J. Am. Coll. Cardiol. **48**(1), 97–98 (2006). doi:10.1016/j.jacc.2006.04.004
38. Leonard, B.L., Smaill, B.H., LeGrice, I.J.: Structural remodeling and mechanical function in heart failure. Microsc. Microanal. **18**(01), 50–67 (2012). doi:10.1017/S1431927611012438
39. Zile, M.R., Baicu, C.F., Gaasch, W.H.: Diastolic heart failure—abnormalities in active relaxation and passive stiffness of the left ventricle. N. Engl. J. Med. **350**(19), 1953–1959 (2004). doi:10.1056/NEJMoa032566
40. Opie, L.H., Commerford, P.J., Gersh, B.J., Pfeffer, M.A.: Controversies in ventricular remodelling. Lancet **367**, 9507 (2016), 356–367. doi:10.1016/S0140-6736(06)68074-4
41. Grossman, W., Jones, D., McLaurin, L.: Wall stress and patterns of hypertrophy in the human left ventricle. J. Clin. Invest. **56**(1), 56 (1975)
42. Marijianowski, M.M.H., Teeling, P., Mann, J., Becker, A.E.: Dilated cardiomyopathy is associated with an increase in the type I/type III collagen ratio: a quantitative assessment. J. Am. Coll. Cardiol. **25**(6), 1263–1272 (1995). doi:10.1016/0735-1097(94)00557-7
43. Ryan, T.D., Rothstein, E.C., Aban, I., Tallaj, J.A., Husain, A., Lucchesi, P.A., Dell'Italia, L.J.: Left ventricular eccentric remodeling and matrix loss are mediated by bradykinin and precede cardiomyocyte elongation in rats with volume overload. J. Am. Coll. Cardiol. **49**(7), 811–821 (2007). doi:10.1016/j.jacc.2006.06.083
44. Mann, D.L., Bogaev, R., Buckberg, G.D.: Cardiac remodelling and myocardial recovery: lost in translation? Eur. J. Heart Fail. **12**(8), 789–796 (2010). doi:10.1093/eurjhf/hfq113
45. Teeters, J.C., Alexis, J.D.: Systolic heart failure. In: Manual of Heart Failure Management, pp. 1–9. Springer, London (2009)
46. Diamond, J.A., Phillips, R.A.: Hypertensive heart disease. Hypertens. Res. **28**(3), 191–202 (2005)

47. Hershberger, R.E., Hedges, D.J., Morales, A.: Dilated cardiomyopathy: the complexity of a diverse genetic architecture. Nat. Rev. Cardiol. **10**(9), 531–547 (2013). doi:10.1038/nrcardio.2013.105
48. Savoye, C., Equine, O., Tricot, O., Nugue, O., Segrestin, B., Sautière, K., Elkohen, M., Pretorian, E.M., Taghipour, K., Philias, A., Aumégeat, V., Decoulx, E., Ennezat, P.V., Bauters, C.: Left ventricular remodeling after anterior wall acute myocardial infarction in modern clinical practice (from the REmodelage VEntriculaire [REVE] study group). Am. J. Cardiol. **98**(9), 1144–1149 (2006). doi:10.1016/j.amjcard.2006.06.011
49. Russell, B., Curtis, M.W., Koshman, Y.E.: Mechanical stress-induced sarcomere assembly for cardiac muscle growth in length and width. J. Mol. Cell. Cardiol. **48**(5), 817–823 (2010). doi:10.1016/j.yjmcc.2010.02.016
50. Ruwhof, C., van der Laarse, A.: Mechanical stress-induced cardiac hypertrophy: mechanisms and signal transduction pathways. Cardiovasc. Res. **47**(1), 23–37 (2000). doi:10.1016/s0008-6363(00)00076-6
51. Nash, M.P., Hunter, P.J.: Computational mechanics of the heart—from tissue structure to ventricular function. J. Elast. **61**(1), 113–141 (2000). doi:10.1023/a:1011084330767
52. LeGrice, I.J., Smaill, B.H., Chai, L.Z., Edgar, S.G., Gavin, J.B., Hunter, P.J.: Laminar structure of the heart: ventricular myocyte arrangement and connective tissue architecture in the dog. Am. J. Physiol., Heart Circ. Physiol. **269**(2), H571–H582 (1995)
53. Menzel, A., Kuhl, E.: Frontiers in growth and remodeling. Mech. Res. Commun. **42**, 1–14 (2012). doi:10.1016/j.mechrescom.2012.02.007
54. Wyczalkowski, M.A., Varner, V.D., Taber, L.A.: Computational and experimental study of the mechanics of embryonic wound healing. J. Mech. Behav. Biomed. Mater. **28**, 125–146 (2013)
55. Shi, Y., Yao, J., Young, J.M., Fee, J.A., Perucchio, R., Taber, L.A.: Bending and twisting the embryonic heart: a computational model for C-looping based on realistic geometry. Front. Physiol. **5**, 297 (2014)
56. Lin, I.E., Taber, L.A.: A model for stress-induced growth in the developing heart. J. Biomech. Eng. **117**(3), 343–349 (1995). doi:10.1115/1.2794190
57. Bradley, C., Bowery, A., Britten, R., Budelmann, V., Camara, O., Christie, R., Cookson, A., Frangi, A.F., Gamage, T.B., Heidlauf, T., Krittian, S., Ladd, D., Little, C., Mithraratne, K., Nash, M., Nickerson, D., Nielsen, P., Nordbø, Ø., Omholt, S., Pashaei, A., Paterson, D., Rajagopal, V., Reeve, A., Röhrle, O., Safaei, S., Sebastián, R., Steghöfer, M., Wu, T., Yu, T., Zhang, H., Hunter, P.: OpenCMISS: a multiphysics & multi-scale computational infrastructure for the VPH/Physiome project. Prog. Biophys. Mol. Biol. **107**(1), 32–47 (2011). doi:10.1016/j.pbiomolbio.2011.06.015
58. Conrad, C.H., Brooks, W.W., Hayes, J.A., Sen, S., Robinson, K.G., Bing, O.H.: Myocardial fibrosis and stiffness with hypertrophy and heart failure in the spontaneously hypertensive rat. Circulation **91**(1), 161–170 (1995)
59. Reinhardt, D., Sigusch, H.H., Henße, J., Tyagi, S.C., Körfer, R., Figulla, H.R.: Cardiac remodelling in end stage heart failure: upregulation of matrix metalloproteinase (MMP) irrespective of the underlying disease, and evidence for a direct inhibitory effect of ACE inhibitors on MMP. Heart **88**(5), 525–530 (2002). doi:10.1136/heart.88.5.525
60. Cohn, J.N., Ferrari, R., Sharpe, N.: Cardiac remodeling—concepts and clinical implications: a consensus paper from an international forum on cardiac remodeling. J. Am. Coll. Cardiol. **35**(3), 569–582 (2000). doi:10.1016/S0735-1097(99)00630-0
61. Iles, L., Pfluger, H., Phrommintikul, A., Cherayath, J., Aksit, P., Gupta, S.N., Kaye, D.M., Taylor, A.J.: Evaluation of diffuse myocardial fibrosis in heart failure with cardiac magnetic resonance contrast-enhanced T1 mapping. J. Am. Coll. Cardiol. **52**(19), 1574–1580 (2008). doi:10.1016/j.jacc.2008.06.049
62. Liu, C.-Y., Liu, Y.-C., Wu, C., Armstrong, A., Volpe, G.J., van der Geest, R.J., Liu, Y., Hundley, W.G., Gomes, A.S., Liu, S., Nacif, M., Bluemke, D.A., Lima, J.A.C.: Evaluation of age-related interstitial myocardial fibrosis with cardiac magnetic resonance contrast-enhanced T1 mapping MESA (Multi-Ethnic Study of Atherosclerosis). J. Am. Coll. Cardiol. **62**(14), 1280–1287 (2013). doi:10.1016/j.jacc.2013.05.078
63. Fomovsky, G.M., Thomopoulos, S., Holmes, J.W.: Contribution of extracellular matrix to the mechanical properties of the heart. J. Mol. Cell. Cardiol. **48**(3), 490–496 (2010). doi:10.1016/j.yjmcc.2009.08.003
64. Donekal, S., Venkatesh, B.A., Liu, Y.C., Liu, C.-Y., Yoneyama, K., Wu, C.O., Nacif, M., Gomes, A.S., Hundley, W.G., Bluemke, D.A.: Interstitial fibrosis, left ventricular remodeling and myocardial mechanical behavior in a population-based multi-ethnic cohort: MESA Study. CI, Cardiovasc. Imaging CIRCIMAGING. 113.001073 (2014)
65. Guccione, J.M., McCulloch, A.D., Waldman, L.: Passive material properties of intact ventricular myocardium determined from a cylindrical model. J. Biomech. Eng. **113**(1), 42–55 (1991)
66. Costa, K.D., Takayama, Y., McCulloch, A.D., Covell, J.W.: Laminar fiber architecture and three-dimensional systolic mechanics in canine ventricular myocardium. Am. J. Physiol., Heart Circ. Physiol. **276**(2), H595–H607 (1999)

67. Dokos, S., Smaill, B.H., Young, A.A., LeGrice, I.J.: Shear properties of passive ventricular myocardium. Am. J. Physiol., Heart Circ. Physiol. **283**(6), H2650–H2659 (2002)
68. Sommer, G., Schriefl, A.J., Andrä, M., Sacherer, M., Viertler, C., Wolinski, H., Holzapfel, G.A.: Biomechanical properties and microstructure of human ventricular myocardium. Acta Biomater. **24**, 172–192 (2015)
69. Cansız, B.F., Dal, H., Kaliske, M.: An orthotropic viscoelastic material model for passive myocardium: theory and algorithmic treatment. Comput. Methods Biomech. Biomed. Eng. **18**(11), 1160–1172 (2014). doi:10.1080/10255842.2014.881475
70. Gültekin, O., Sommer, G., Holzapfel, G.A.: An orthotropic viscoelastic model for the passive myocardium: continuum basis and numerical treatment. Comput. Methods Biomech. Biomed. Eng. **19**(15), 1647–1664 (2016). doi:10.1080/10255842.2016.1176155
71. Holzapfel, G.A., Gasser, T.C., Ogden, R.W.: A new constitutive framework for arterial wall mechanics and a comparative study of material models. J. Elast. **61**(1), 1–48 (2000). doi:10.1023/a:1010835316564
72. Holzapfel, G.A., Niestrawska, J.A., Ogden, R.W., Reinisch, A.J., Schriefl, A.J.: Modelling non-symmetric collagen fibre dispersion in arterial walls. J. Roy. Soc. Interface **12** 20150188 (2015). doi:10.1098/rsif.2015.0188
73. Holzapfel, G.A., Ogden, R.W.: Constitutive modelling of arteries. Proc. R. Soc. A, Math. Phys. Eng. Sci. **466**(2118), 1551–1597 (2010). doi:10.1098/rspa.2010.0058
74. Ogden, R.W.: Anisotropy and nonlinear elasticity in arterial wall mechanics. In: Holzapfel, G.A., Ogden, R.W. (eds.) Biomechanical Modelling at the Molecular, Cellular and Tissue Levels, pp. 179–258. Springer, Vienna (2009)
75. Eriksson, T.S., Prassl, A.J., Plank, G., Holzapfel, G.A.: Modeling the dispersion in electromechanically coupled myocardium. Int. J. Numer. Methods Biomed. Eng. **29**(11), 1267–1284 (2013)
76. Wang, V.Y., Nielsen, P.M.F., Nash, M.P.: Image-based predictive modeling of heart mechanics. Annu. Rev. Biomed. Eng. **17**(1), 351–383 (2015). doi:10.1146/annurev-bioeng-071114-040609
77. Lanir, Y.: Constitutive equations for fibrous connective tissues. J. Biomech. **16**(1), 1–12 (1983). doi:10.1016/0021-9290(83)90041-6
78. Nevo, E., Lanir, Y.: Structural finite deformation model of the left ventricle during diastole and systole. J. Biomech. Eng. **111**(4), 342–349 (1989). doi:10.1115/1.3168389
79. Sacks, M.S.: Incorporation of experimentally-derived fiber orientation into a structural constitutive model for planar collagenous tissues. J. Biomech. Eng. **125**(2), 280–287 (2003). doi:10.1115/1.1544508
80. Holzapfel, G.A., Ogden, R.W.: Constitutive modelling of passive myocardium: a structurally based framework for material characterization. Philos. Trans. R. Soc. Lond. A, Math. Phys. Eng. Sci. **367**(1902), 3445–3475 (2009)
81. Avazmohammadi, R., Hill, M.R., Simon, M.A., Zhang, W., Sacks, M.S.: A novel constitutive model for passive right ventricular myocardium: evidence for myofiber–collagen fiber mechanical coupling. Biomech. Model. Mechanobiol. **16**(2), 561–581 (2016). doi:10.1007/s10237-016-0837-7
82. Karlon, W.J., McCulloch, A.D., Covell, J.W.: Regional dysfunction correlates with myofiber disarray in transgenic mice with ventricular expression ofras. regional dysfunction correlates with myofiber disarray in transgenic mice with ventricular expression of ras. Am. J. Physiol., Heart Circ. Physiol. **278**(3), H898–H906 (2000)
83. Wang, V.Y., Wilson, A.J., Sands, G.B., Young, A.A., LeGrice, I.J., Nash, M.P.: Microstructural remodelling and mechanics of hypertensive heart disease. In: van Assen, H., Bovendeerd, P., Delhaas, T. (eds.) Proceedings of the 8th International Conference on Functional Imaging and Modeling of the Heart, FIMH 2015, Maastricht, The Netherlands. June 25–27, 2015 pp. 382–389. Springer, Cham (2015)
84. Wang, V.Y., Niestrawska, J.A., Wilson, A.J., Sands, G.B., Young, A.A., LeGrice, I.J., Nash, M.P.: Image-driven constitutive modelling of myocardial fibrosis. Int. J. Comput. Methods Eng. Sci. Mech. (2016). doi:10.1080/15502287.2015.1082675
85. Rodriguez, E.K., Hoger, A., McCulloch, A.D.: Stress-dependent finite growth in soft elastic tissues. J. Biomech. **27**(4), 455–467 (1994)
86. Skalak, R.: Growth as a finite displacement field. In: Proceedings of the IUTAM Symposium on Finite Elasticity, pp. 347–355. Springer, Berlin (1981)
87. Göktepe, S., Abilez, O., Parker, K., Kuhl, E.: A multiscale model for eccentric and concentric cardiac growth through sarcomerogenesis. J. Theor. Biol. **265**(3), 433–442 (2010). doi:10.1016/j.jtbi.2010.04.023
88. Lee, L., Genet, M., Acevedo-Bolton, G., Ordovas, K., Guccione, J.M., Kuhl, E.: A computational model that predicts reverse growth in response to mechanical unloading. Biomech. Model. Mechanobiol. **14**(2), 217–229 (2015)

89. Lee, L.C., Sundnes, J., Genet, M., Wenk, J.F.: An integrated electromechanical-growth heart model for simulating cardiac therapies. Biomech. Model. Mechanobiol. **15**(4), 791–803 (2016)
90. Klepach, D., Lee, L.C., Wenk, J.F., Ratcliffe, M.B., Zohdi, T.I., Navia, J.L., Kassab, G.S., Kuhl, E., Guccione, J.M.: Growth and remodeling of the left ventricle: a case study of myocardial infarction and surgical ventricular restoration. Mech. Res. Commun. **42**, 134–141 (2012). doi:10.1016/j.mechrescom.2012.03.005
91. Kerckhoffs, R.C.P., Omens, J.H., McCulloch, A.D.: A single strain-based growth law predicts concentric and eccentric cardiac growth during pressure and volume overload. Mech. Res. (2012). doi:10.1016/j.mechrescom.2011.11.004
92. Kroon, W., Delhaas, T., Arts, T., Bovendeerd, P.: Constitutive Modeling of Cardiac Tissue Growth. Constitutive Modeling of Cardiac Tissue Growth. Springer, Berlin (2007)
93. Lanir, Y.: Mechanistic micro-structural theory of soft tissues growth and remodeling: tissues with unidirectional fibers. Biomech. Model. Mechanobiol. (2015). doi:10.1007/s10237-014-0600-x
94. Tsamis, A., Cheng, A., Nguyen, T.C., Langer, F., Miller, C.D., Kuhl, E.: Kinematics of cardiac growth: in vivo characterization of growth tensors and strains. J. Mech. Behav. Biomed. Mater. **8**, 165–177 (2012). doi:10.1016/j.jmbbm.2011.12.006
95. Walker, J.C., Ratcliffe, M.B., Zhang, P., Wallace, A.W., Fata, B., Hsu, E.W., Saloner, D., Guccione, J.M.: MRI-based finite-element analysis of left ventricular aneurysm. Am. J. Physiol., Heart Circ. Physiol. **289**(2), H692–H700 (2005). doi:10.1152/ajpheart.01226.2004
96. Walker, J.C., Ratcliffe, M.B., Zhang, P., Wallace, A.W., Hsu, E.W., Saloner, D.A., Guccione, J.M.: Magnetic resonance imaging-based finite element stress analysis after linear repair of left ventricular aneurysm. J. Thorac. Cardiovasc. Surg. **135**(5), 1102.e1-2 (2008). doi:10.1016/j.jtcvs.2007.11.038
97. Wenk, J.F., Eslami, P., Zhang, Z., Xu, C., Kuhl, E.: A novel method for quantifying the in-vivo mechanical effect of material injected into a myocardial infarction. Ann. Thorac. Surg. (2011). doi:10.1016/j.athoracsur.2011.04.089
98. Wall, S.T., Walker, J.C., Healy, K.E., Ratcliffe, M.B.: Theoretical impact of the injection of material into the myocardium a finite element model simulation. Circulation (2006). doi:10.1161/CIRCULATIONAHA.106.657270
99. Wenk, J.F., Wall, S.T., Peterson, R.C., Helgerson, S.L., Sabbah, H.N., Burger, M., Stander, N., Ratcliffe, M.B., Guccione, J.M.: A method for automatically optimizing medical devices for treating heart failure: designing polymeric injection patterns. J. Biomech. Eng. **131**(12), 121011 (2009). doi:10.1115/1.4000165
100. Boukellal, H., Campás, O., Joanny, J.-F., Prost, J., Sykes, C.: Soft listeria: actin-based propulsion of liquid drops. Phys. Rev. E **69**(6), 061906 (2004)
101. Lucio, A.A., Ingber, D.E., Campàs, O.: Generation of biocompatible droplets for in vivo and in vitro measurement of cell-generated mechanical stresses. Methods Cell Biol. **125**, 373–390 (2015)

MIX
Papier aus verantwortungsvollen Quellen
Paper from responsible sources
FSC® C105338

FSC
www.fsc.org

If you have any concerns about our products,
you can contact us on
ProductSafety@springernature.com

In case Publisher is established outside the EU,
the EU authorized representative is:
Springer Nature Customer Service Center GmbH
Europaplatz 3, 69115 Heidelberg, Germany

Printed by Libri Plureos GmbH
in Hamburg, Germany